Lecture Notes in Physics

307

H. Richard Miller Paul J. Wiita (Eds.)

Active Galactic Nuclei

Proceedings of a Conference Held at the
Georgia State University, Atlanta, Georgia
October 28–30, 1987

Springer-Verlag Berlin Heidelberg GmbH

Editors
H. Richard Miller
Paul J. Wiita
Department of Physics and Astronomy, Georgia State University
Atlanta, GA 30303, USA

ISBN 978-3-662-13685-0 ISBN 978-3-540-39304-7 (eBook)
DOI 10.1007/978-3-540-39304-7

Originally published by Springer-Verlag Berlin Heidelberg New York in 1988
Softcover reprint of the hardcover 1st edition 1988

SCIENTIFIC ORGANIZING COMMITTEE

Mitchell C. Begelman

Alan H. Bridle

H. Richard Miller

Richard F. Mushotzky

Donald E. Osterbrock

Maarten Schmidt

Wayne A. Stein

Paul J. Wiita

PREFACE

The subject of active galactic nuclei, is, without a doubt, one of the most exciting and rapidly evolving in modern astrophysics. Thus it is clearly desirable to have frequent meetings devoted to this topic in order to bring researchers in the field together, thereby providing them with an opportunity for detailed discussions and a lively interchange of ideas. Because the previous meeting on this general area in the United States, at George Mason University in October 1986, was dominated by theoretical concerns, we felt that this conference, taking place a year later, ought to stress recent observational results. Nevertheless, some exciting new theoretical interpretations were presented and debated.

The over one hundred participants at the GSU conference interacted very well; several new collaborations were developed and some exciting new ideas were expressed. Thirteen invited speakers summarized observations in essentially all wave bands and also expounded varied theoretical opinions. A significant number of moderately long oral contributions were also given, with most other participants presenting poster papers. All of the oral papers presented at the meeting have been included in these proceedings, as have almost all of the poster papers, along with the bulk of the discussion after each oral paper.

Some of the key questions raised or addressed at this meeting included: How many fundamentally different types of active galactic nuclei (AGN) are there, and how are apparently different types related? What can (or cannot) variability of emission in different parts of the spectrum tell us about the size and nature of the central engine? How common are the wiggly, apparently filamentary, structures in radio jets, and do they imply the dominance of magnetic fields in the dynamics? What are the relations between the gas falling into AGN and the gas shooting out of them, especially in light of growing evidence for large scale effects of emerging jets on the appearance of galaxies? What is the nature of the AGN continuum in the millimeter and extreme ultraviolet regions, which have been barely touched by observations? What physical process (or processes) is (are) responsible for the AGN continuum observed over the entire electromagnetic spectrum? While discord remains concerning the answers to all of the above (and many more) questions, the data and interpretations presented at this conference have undoubtedly improved our level of understanding.

The other members of the scientific organizing committee were extremely helpful in suggesting speakers, determining the distribution of travel support, and chairing sessions; we also thank Diana Worrall for chairing a session. Organizational support was provided through the Division of Continuing Education at GSU, particularly by Lisa Coley and Liz Robinowich. Our colleagues at GSU, Ingemar Furenlid, Joseph Hadley Jr., Hal McAlister, and David Wingert, as well as our colleague at Agnes Scott College, Alberto Sadun, helped with logistical support. Student volunteers from GSU, Michael Carini, Alexander Rosen, Sethanne Howard, Ali al-Shukri, Edward Dombrowski, Tom Meylan and Wean Tsay drove the shuttle vans, showed the slides, and helped in many other ways. Agnes Scott students Jennifer Burger, Christy Cechman, Sarah Crane and Amy Lovell assisted with the recording of the discussions. We are most grateful to all of the above, who helped keep the conference running smoothly.

Finally, we express our gratitude to both the National Science Foundation (grant AST-8715488) and to the Division of Continuing Education of Georgia State University, without whose financial support this conference would not have been possible.

Atlanta, February 1988

H. Richard Miller
Paul J. Wiita

TABLE OF CONTENTS

I. SPECTRAL STUDIES IN THE OPTICAL, ULTRAVIOLET AND INFRARED

II. POLARIZATION AND CONTINUUM OBSERVATIONS

III. OPTICAL VARIABILITY STUDIES

IV. THEORIES OF THE CONTINUUM

V. X-RAY OBSERVATIONS AND MODELING

VII. COSMOLOGICAL AND EVOLUTIONARY CONSIDERATIONS

EMISSION-LINE SPECTRA AND THE NATURE OF ACTIVE GALACTIC NUCLEI

Donald E. Osterbrock
Lick Observatory
Board of Studies in Astronomy and Astrophysics
University of California
Santa Cruz, California 95064

ABSTRACT

Seyfert galaxy emission-line spectra provide clues to the structure of active galactic nuclei and the nature of their central sources. Classification of their spectra organizes a wide range of properties; they do not all fit into a single one-parameter sequence. The luminosity functions of the individual types provide constraints on their evolution and on possible geometries, but are difficult to determine because objective-prism and color surveys tend to be most efficient in finding blue objects. The IRAS survey has identified many more Seyfert 2 galaxies, which tend on the average to be more heavily reddened than the previously known objects. The discovery of very faint broad Hα emission components in many emission-line galaxies by Filippenko and Sargent indicates the continuity of the general AGN process to very low luminosity levels. The narrow-line profiles of all types of Seyfert galaxies have the same general form, indicating a single type of velocity field. It is evidently mostly radial, but not necessarily spherical. There are differences in detail, which on the average tend to be correlated with the spectral type, but these are by no means invariably followed at the level of individual galaxies. The broad-line profiles show a great range in properties. Combinations of radial and rotational velocity fields give the best fits with the observed profiles, but there are theoretical difficulties with this picture, particularly with rotational flow.

I. INTRODUCTION

Emission lines provide clues to the nature and properties of the objects which emit them. Much of what we know about H II regions and planetary nebulae depends, in one way or another, on such data. Concerning active galactic nuclei, I have been quoted in print as saying "If you can't stick your head under the hood and see the engine, you better look at what's coming out of the tail pipe," and I suppose that this is as good a way as any of expressing the thought that clues from their emission-line spectra are also important for these objects.

My own research in recent years has been devoted almost entirely to the Seyfert galaxies, and in this review I will deal chiefly with subjects I know and understand myself. Continuity suggests and the information we have on luminosity functions seems to confirm quantitatively that Seyfert galaxies and QSOs form a continuous sequence. This is indicated schematically in Figure 1, reprinted with permission from Morgan and Dreiser (1983). In recent years various authors have made many apparent detections of galaxies around objects formerly described as "quasistellar," from images and spectra obtained with CCDs and other current detectors which have high quantum efficiency, digital readout, and subtraction capabilities. Two recent excellent reviews of this subject, containing many references to the original papers, are by Hutchings (1983) and Miller (1985). Schmidt and Green (1985) in their Bright Quasar Survey simply adopted absolute magnitude $M_B = -23$ as the division between Seyfert 1 galaxies and QSOs as a practical working definition.

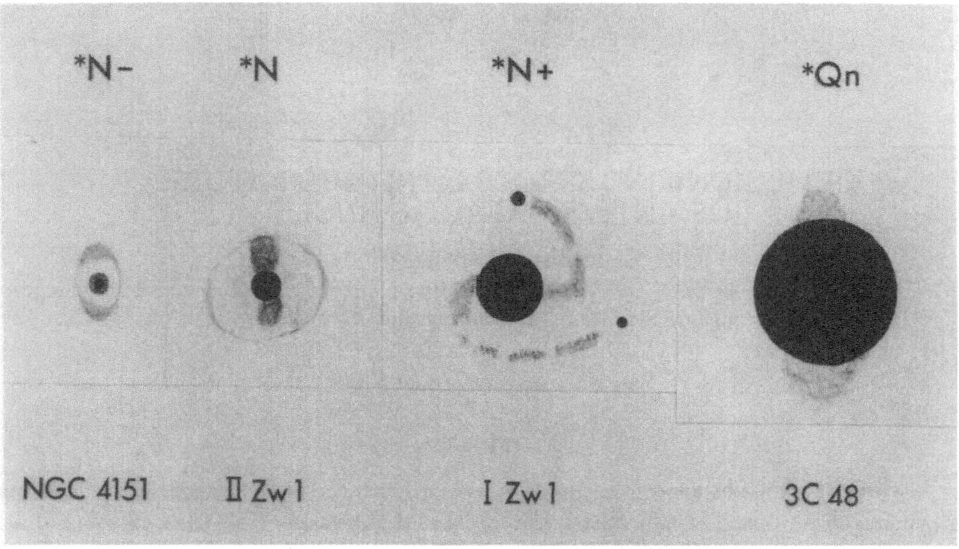

Figure 1. One-dimensional sequence of four AGNs, illustrating the range from Seyfert galaxies to QSOs (W. W. Morgan and R. D. Dreiser, *Ap. J.*, **269**, 438, 1983).

Continuity is important, but it must always be remembered that it does not necessarily imply that all AGNs are alike, any more than continuity of stellar properties implies that all stars are alike. For instance an O star and a K star are both in gravitational equilibrium, and both produce energy by nuclear reactions, but beyond these common properties of stars they are quite different in their structures and in their specific nuclear-energy-production reactions. The same is true of a K giant and a K dwarf. We must not assume that anything we learn from one AGN is true for all AGNs, any more than we assume that all stars have mass 1 $M_{\odot}$, or produce energy by the proton-proton cycle.

II. SPECTRAL CLASSIFICATION

The spectral classification of Seyfert galaxies into two types, Seyfert 1, with broader H I, He I and He II emission lines, and narrower forbidden lines, and Seyfert 2, with H I, He I and He II emission with essentially the same widths as the forbidden lines, both comparable with the widths of the forbidden lines in Seyfert 1 galaxies, is well known from the papers of Khachikian and Weedman (1974). We later introduced the additional intermediate type Seyfert 1.5, meaning an object whose H I emission lines have composite profiles, with both broad and narrow components easily recognizable (Osterbrock and Koski 1976; Osterbrock 1977). Still later I defined two further types, Seyfert 1.8 and 1.9, in which in addition to the narrow components, very weak but recognizable broad components of Hβ and Hα, or of Hα alone, respectively, are present (Osterbrock 1981). These additional types have been used by some but not all researchers.

Classification ideally is a strictly observational process, independent of theory, that different people can do (with good data), and get the same answers. The types may be defined by standards, as in the MK system of stellar spectral classification, or by definitions, as in the Khachikian and Weedman system of Seyfert galaxy spectral classification. These definitions however, have been modified in the course of time; for instance "narrow" line widths of all the emission lines (FWHM < 300 km s^{-1}) is no longer considered a criterion to eliminate a candidate galaxy from the Seyfert 2 class. The reason is that many galaxies have been discovered with this property whose emission-line spectra are otherwise indistinguishable from the classically known Seyfert 2's (Veron 1981; Phillips, Charles and Baldwin 1983).

III. PHYSICAL PICTURES

Though classification is an empirical process, we always hope that it will isolate physically significant types of objects, and will lead to understanding them, as in the case of stars. From many studies we understand the main features of Seyfert galaxy spectra as resulting from photoionization by a spectrum that extends to high-energy photons. The gas is in clouds with a small filling factor, the dense, broad-line gas closest to the source, and the less-dense, narrow-line gas on the average at larger distances. The diagnostic line ratios of the narrow-line spectrum, and the equivalent widths of the broad-plus narrow-line spectra agree with this picture. (For a recent review of this material see *e.g.* Osterbrock 1984.)

An observational model that fits many of these data and others connected with the differences in optical spectra between Seyfert and radio galaxies is shown schematically in Figure 2 (Osterbrock 1978). The dense (BLG) and low-density (NLG) gas components are shown in cylindrical and spherical distributions, respectively, about the central photoionizing source, presumably connected with an accretion disk and black hole. Ionized regions are indicated by cross hatching; most highly ionized regions closest to the source, where the ionization parameter is largest, for a fixed density. In this model, both the ionizing photons and the radio plasma can escape through the BLG most easily along the axis of the cylinder, where the amount of material encountered is the smallest, and correspondingly the ionized NLG has a roughly conical distribution about this axis. In this picture pure Seyfert 1 galaxies are those in which the amount of BLG is large and few ionizing photons reach the NLG; pure Seyfert 2 galaxies are objects in which the amount of BLG is small or nonexistent and nearly all the ionizing photons reach the NLG. Intermediate objects in the classification scheme are those which have intermediate amounts of BLG, according to this model.

Furthermore, there are many indications that the active galactic nucleus may be tipped with respect to the main body of the galaxy in which it occurs. This is suggested by the lack of correlation between the FWHMs of the broad-emission lines and inclination of the parent galaxy, that would be expected if the velocity field were rotational, or indeed had any other type of cylindrical symmetry (Tohline and Osterbrock 1982). Also, there are many observational indications that the structures of the NLG and the radio plasma are closely connected, and high-resolution radio-frequency maps show that the jets or axes of the nuclear structure are often misaligned with the parent galaxy (Wilson and Ulvestad 1982; Ulvestad and Wilson 1984; Wilson, Baldwin and Ulvestad 1985).

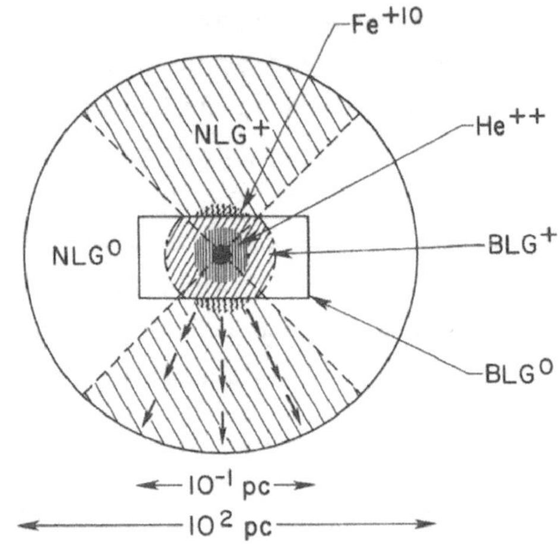

Figure 2. Schematic observational model of an AGN as described in text. Ionized gas is shown in various cross-hatchings as indicated, neutral gas is shown as plain white.

A schematic drawing of such a model is indicated in Figure 3 (Osterbrock 1983). The small cylindrically symmetric BLR, at the center, is tilted with respect to the larger cylindrically symmetric NLR. (As with Figure 2, these models cannot be drawn to scale correctly, for the ratio of sizes of the two regions is $\sim 10^3$.) The core about the axis of the BLR, in which most of the ionizing photons that get to the NLR escape, is shown by the dashed lines, and the units of the structure implied by the small filling factor are schematically indicated as spherical, homogeneous, uniform-density clouds. The various shadings and hatchings indicate that the ionization is highest on the faces of the clouds closest to the ionizing source (under the simplifying assumptions stated), and drops to nearly zero at the back (or side away from the ionizing source) of the optically thick clouds. The picture is simplified; the BLR must have an inhomogeneous structure also, that can be mimicked by similar but denser clouds on a much smaller scale. They cannot be drawn in the figure, but coincidental "tunnels" between them may allow ionizing photons to escape in some few directions outside the primary cone, as shown by the ionized clouds in the NLR, near the projection of the equator of the BLR.

In addition, there are many observational indications from line-profile studies that the division into two regions, BLR and NLR, is a gross oversimplification. In reality there is a wide range of density, extending more or less continuously from what we call the "BLR" through the "NLR" (De Robertis and Osterbrock 1984). Correspondingly, the ionized gas in the AGN probably has a smoothly warped distribution, with the greatest misalignment at the center, in the densest, smallest part of what is conventionally called the BLR, and merging into the plane of the galaxy at the outer part of the "NLR". The density distribution everywhere must be highly inhomogeneous, or "cloud"-like, so the picture is a complicated one indeed.

In testing whether or not there is a correlation between emission-line width and inclination of the parent galaxy, Keel (1980) discovered that there is a paucity of nearby edge-on Seyfert 1s. This indicates that there exist a fairly large number of galaxies, with the physical properties of Seyfert 1s, not detected or known because of extinction by the dust outside their nuclei, but in their principal planes. The most nearly edge-on nearby Seyfert 1 galaxy, NGC 4235, indeed has an unusual spectrum, in which both the broad and narrow emission lines show the effects of strong reddening, as does the continuum (Abell, Eastmond and Jenner 1978; Osterbrock 1984).

In addition to the conventional NLR, direct images and spectra to very faint limits show the presence of gas photoionized by hard photons from the central source and at distances of several kpc from it, in a few very well studied active galaxies, as will be reported, for instance, by Pogge at this Symposium. The distribution of this ionization appears to be more or less conical about an axis that can be associated with the central source (Unger *et al.* 1987; Pogge 1988). This is a strong argument in favor of models of the general type represented in Figures 2 and 3.

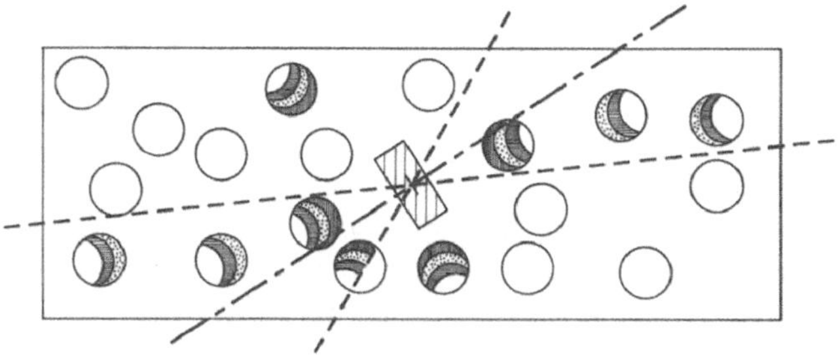

Figure 3. Schematic, slightly more sophisticated observational model of an AGN as described in text. Tilt of BLR with respect to NLR, and ionization structure of clouds in NLR are indicated.

In Seyfert 1.8 and 1.9 galaxies, the ratio of the intensities of the broad emission-line components $H\alpha/H\beta$ is often large, typically perhaps $8 - 10$. Though this ratio can have a range of values under BLR conditions, depending on density (collisional excitation and deexcitation), size (line optical depths) and velocity fields (ditto), the fact that it is so large in Seyfert 1.8 and 1.9s suggests that dust extinction may be especially significant in them (Osterbrock 1981). Likewise, there are X-ray galaxies with very weak broad $H\alpha$ components, so weak that they would not be classified Seyfert 1.9 on spectra of normal signal-to-noise ratios; they may be objects in which the dust extinction is even stronger (Veron *et al.* 1980; Shuder 1980). This idea has been carried still further by the suggestion that all Seyfert galaxies have approximately the same structure, and the same relative amounts of BLG and NLG, but have different amounts of extinction as seen from different directions. According to this hypothesis, the Seyfert 2s are objects in which the BLR is completely hidden by extinction, from the direction in which we see it (Lawrence and Elvis 1982).

A very great step was the recognition of the "hidden Seyfert 1" spectrum seen in the polarized flux from NGC 1068 (a Seyfert 2 in the total flux) by Antonucci and Miller (1985). From this observation they developed a picture of the BLR within a tipped, cylindrically symmetric torus, containing dust, that is optically very thick in its equatorial plane, but much thinner or transparent along its axis. It has many similarities to the earlier models described above, but goes far beyond them. As these observations and ideas will be described by Miller himself at this Symposium, I will go into no more detail here. One most interesting aspect of this model is the question of whether it applies to all, or most Seyfert 2 galaxies, that is, if they all have "hidden Seyfert 1" broad-line regions within them. Miller and his collaborators have made spectropolarimetric observations of several more Seyfert 2s, and has found similar broad emission-line features in the polarized light of some, but not all of these, as he explains in his paper. Krolik and Begelman (1986) have proposed a picture according to which all Seyfert 2s are of this type, and Krolik will describe further progress along these lines in his paper at this Symposium.

Certainly there is a wide range of properties among observed Seyfert 2s, and among Seyfert 1s also. One interesting group is the narrow-line Seyfert 1 galaxies, as described by Osterbrock and Pogge (1985). We have found more of these objects among the AGN candidates isolated by the IRAS survey (Osterbrock and De Robertis 1985) and the PG survey (Osterbrock and Pogge 1987). The latter provides fairly good statistics in the relative number of galaxies of this type; if we adopt the definition that a Seyfert 1 or 1.5 galaxy with broad line FWHM $<$ 2000 km s^{-1} is to be called a narrow line Seyfert 1, approximately 15% of the Seyfert 1 and 1.5 galaxies belong to this group. I Zw 1 is one member, Mrk 507, recently described by Halpern and Oke (1987), is another. One immediate possibility is that the narrow-line Seyfert 1s may be the most nearly "pole-on" members of the Seyfert 1 class, objects in which the component of the velocity field (if it is rotational, or has some other origin but is largely in a plane) along the line of sight is small. Observational aspects of this possible working hypothesis will be discussed by Goodrich at this Symposium.

These examples, and others, show observationally that all Seyfert galaxies do not fit into a single, one-parameter sequence, or even into a two-parameter grid. Observationally, the parameters include luminosity, relative strengths of the BLR and NLR, line widths of the broad lines, line widths of the narrow lines, level of ionization, and relative strength of the Fe II emission lines. (Their relative strength seems to be partly but not completely anticorrelated with high ionization.)

On the other hand the physical parameters, in addition to the overall physical structure, include mean ionization parameter Γ, mean electron density N_e, the variation or both of these with distance (in the simplest spherical models), or with ρ, z (for the simplest cylindrically symmetric models), or with r, θ, z in general, plus the fluctuations or distributions around these mean values (the "filling factor" is one of the simplest possible descriptions of these), and the internal velocity field. Our aim is to understand AGNs by recognizing the connections between the observational and physical parameters.

IV. LUMINOSITY FUNCTIONS

Luminosity functions provide constraints on the physical models of AGNs, or on the evolution of AGNs under an evolutionary hypothesis. Probably the best available published luminosity function is that of Meurs and Wilson (1984), which is based on a sample of Seyfert galaxies drawn from the Markarian surveys. According to it the maximum of the Seyfert 1 luminosity function is at $M_B = -21.0$, of the Seyfert 2 luminosity function at $M_B = -20.2$, and there are approximately equal numbers per unit volume of space. Because of the difference in absolute magnitudes, this means the relative number observed down to a given apparent magnitude is expected to be about Seyfert 1/Seyfert 2 = 3/1, which agrees with the statistics of observed Markarian galaxies.

But in this survey and hence in the luminosity functions the numbers of Seyfert 2 galaxies is too low. They are only very incompletely represented on the Markarian lists, because this survey, based on blue color or ultraviolet excess, misses many Seyfert 2 galaxies, which have relatively faint featureless continua. This was pointed out by Rieke (1978), and discovered or confirmed observationally by Huchra, Wyatt and Davis (1982), who found many "additional," previously unknown Seyfert galaxies, especially Seyfert 2s, in the Center for Astrophysics redshift survey. They commented that this meant the space density of Seyfert 2s is significantly higher than that implied by the Meurs and Wilson luminosity function, but did not give a specific numerical estimate. Also, many examples of relatively bright, previously unknown Seyfert 2 galaxies were found by Phillips, Charles and Baldwin (1983), in a sample of emission-line galaxies known from previous slit spectra to have relatively strong [O III] $\lambda\lambda$4959, 5007. They estimated that the relative numbers of Seyfert 2 and Seyfert 1 galaxies (down to a given apparent magnitude limit) might be 2/1 or larger.

We are attempting to study in depth the Seyfert galaxies and Seyfert galaxy candidates in the Wasilewski field, comprising 825 square degrees centered on the North Galactic Pole (Osterbrock and Shaw 1988). Wasilewski (1983), in a moderate-dispersion objective-prism survey based on emission *lines* rather than ultraviolet color excess. In this field he found 36 previously known emission-line galaxies and 96 "new" ones. From the objective-prism spectra he classified $\sim$ 15% of the emission-line galaxies as Seyferts, and commented on the relatively large number of Seyfert 2s among them. Our slit spectra confirm that many (but not all) of Wasilewski's candidates are Seyfert 2s. Our preliminary analysis gives for the relative numbers per unit volume of space of Seyfert (1 + 1.5): (1.8 + 1.9): 2 = 0.1: 0.l: 0.8. The main errors come from the statistical uncertainty due to the small number of active galaxies in this small field; the actual numbers of Seyferts down to the limit of Wasilewskis's survey are 4 Seyfert 1 and 1.5s, 2 Seyfert 1.8 and 1.9s and 9 Seyfert 2s.

On the simplified hypothesis that all Seyfert galaxies evolve through all these types, the times spent in each stage would be in the ratios 0.1: 0.l: 0.8, whether they evolved from Seyfert 1 and 1.5 through 1.8 and 1.9 to 2, or vice versa, or back and forth between them. On the simplified hypothesis that they all had the same torodial structure but were seen in different orientations, the Seyfert 1s and 1.5s seen nearly pole-on, so the BL spectrum is essentially unobscured; the Seyfert 2s are seen nearly equator-on, so the BL spectrum is nearly completely obscured; and the Seyfert 1.8 and 1.9s at intermediate orientations so that the BL spectrum is partly but not completely obscured, the angle θ between the line of sight and the axis would be $0 < \theta < 29°$ for Seyfert 1 and 1.5; $29° < \theta < 39°$ for Seyfert 1.8 and 1.9; and $39° < \theta < 90°$ for Seyfert 2 (or their supplements). This corresponds to a rather thick (or high) torus. Further details are given by Osterbrock and Shaw (1988).

Work is continuing on using the data obtained in the CfA survey to determine the Seyfert 1 and Seyfert 2 luminosity functions. It is cited as "Huchra and Berg (1987), in preparation" by Edelson (1987), but I have not seen a preprint of it. However, a list of the 42 galaxies in this sample was published by Edelson (1987). We have not analyzed it in detail, but the results that will be obtained from this sample over a much larger area of the sky, but to a brighter magnitude limit, totalling nearly three times as many Seyfert galaxies as ours, appear to be qualitatively similar to our results quoted above.

V. FINDING AGNs

Slit spectra show that of the order of 1% of field galaxies are Seyfert galaxies, with emission lines roughly as strong as in the spectra of the classical objects of this type. Thus to find appreciable numbers of them to a faint magnitude limit by taking individual slit spectra would require prohibitive amounts of large-telescope time. The Markarian or First Byurakan Survey was an objective-prism survey based largely on ultraviolet excess, and about 10% of the galaxies isolated by it have turned out to be Seyfert galaxies (Lipovetsky, Markarian and Stepanian 1987). The other 90% are largely galaxies with H II-region type spectra. The Second Byurakan Survey, now underway, covers part of the same area with the same telescope, but with fine-grain plates in three different spectral regions (blue, green and red) to a fainter magnitude limit, and with various orientations of the prisms. Follow-up slit spectra of Seyfert candidates are then obtained with the 6-m telescope (Markarian, Stepanian and Erastova 1987). At Lick Observatory we have obtained spectra of some of the fainter galaxies identified as Seyferts in the SBS. In all these cases we agree that they are Seyfert galaxies, and in most cases of nearly the same types as classified by the Soviet astronomers.

Other low-dispersion objective-prism surveys have been carried out with the Michigan Schmidt telescope at Cerro Tololo by MacAlpine and various collaborators (see MacAlpine and Williams 1981), and with the Case Schmidt at Kitt Peak by Sanduleak and Pesch (e.g. 1987).

These surveys use relatively low dispersion, reach a relatively faint limit, and are based *both* on ultraviolet excess and the presence of emission lines, but the fraction of them that are Seyfert galaxies is significantly smaller than in the FBS.

Defining the "limit" of an objective-prism survey aimed at finding emission-line galaxies is difficult. It is *not* simply a single magnitude, but rather a combination of magnitude and the relative strength of the strongest emission line(s) with respect to the continuum. This point has been discussed most recently by Gratton and Osmer (1987) and by Salzer (1987).

A three-color photographic survey is being carried out with the Kiso Schmidt telescope of the Tokyo Astronomical Observatory by Takase and Miyauchi-Isobe (e.g. 1987). This survey also has led to the discovery of many blue galaxies, a large fraction of which have emission-line spectra. Again, most of these are H II-region type spectra, and only a small fraction are Seyferts (Maehara *et al.* 1986; Takase *et al.* 1987).

The IRAS color indices of known Seyfert galaxies indicate that they can be used to segregate partially these objects from other, "normal" galaxies (Miley, Neugebauer and Soifer 1985). This was expected from previous infrared studies of Seyfert galaxies, which showed that they contain large amounts of "warm" dust, near the active nucleus and heated by it. Many additional candidate Seyfert galaxies have been isolated using the IRAS color-index criteria developed in this way (de Grijp, Miley and Lub 1987). Slit spectra of these candidates have shown that a very large fraction of them are Seyfert 2s, many more heavily reddened than typical previously known objects of this type. Only a few previously unknown Seyfert 1s are among them, confirming that the optical objective-prism surveys have provided more nearly complete samples of these. All the candidates isolated by the IRAS color-index criteria are not Seyfert galaxies, but to date of those for which adequate slit spectra have been obtained and published, about 60% have proved to be Seyferts, about 10% to 15% LINERs, and about 25% to 30% reddened H II region galaxies (Carter 1984; de Grijp *et al.* 1985; Osterbrock and De Robertis 1985). The IRAS color-index criteria do *not* recover all previously known Seyfert galaxies.

In the Wasilewski field Shaw has to date found from the IRAS colors only one previously unknown Seyfert, IRAS 14434+2714, a Seyfert 2. I understand that Halpern, Bothun and Lonsdale have found two Seyfert 2s in this field from their IRAS color indices, but I have not seen a preprint of their paper. One of them may be the object named above.

VI. LOW-IONIZATION AND LOW-LUMINOSITY AGNS

Heckman (1980) recognized the class of emission-line galaxies he called LINERs as low-ionization AGNs. They seem to be objects photoionized by the same type of hard spectrum, extending to high energies, as in Seyfert galaxies, but with a lower mean ionization parameter Γ (Ferland and Netzer 1983; Keel 1983*a*; Halpern and Steiner 1983). This can be due either to a lower-luminosity central source, or greater mean distance of the ionized gas from the source; probably both situations appear in nature. A careful study of a well defined sample of spiral galaxies by Keel (1983*b*) shows that approximately half of these objects have observed Seyfert (5%) or LINER (50%) spectra.

Among the LINERs he studied, Keel (1983*a*) found several cases in which very weak, barely detectable broad Hα emission components are present in addition to the narrow emission lines. Another example of this type of object is Mrk 883, whose spectrum was published by Osterbrock and Dahari (1983).

In another very interesting study, Filippenko and Sargent (1985) made a special search for such weak broad Hα emission components in a large number of LINERs. This necessitated accurate subtraction of the galaxy integrated-stellar absorption line spectrum, to detect the weakest possible emission components. They found that about a significant fraction of the LINERs analyzed in this way have a measurable broad Hα emission component. Of course these observational results strongly suggest that *all* AGNs have some BL gas within them, whether it is bright enough to be detected in their spectra or not.

Among the LINERs in which Filippenko and Sargent (1985) detected weak, broad Hα emission are the best known elliptical galaxies with emission lines in their spectra, NGC 1052 and NGC 4278. This is the most convincing evidence known to me to date that the elliptical-galaxy LINERs belong to the same general type of AGNs photoionized by a hard spectrum extending to high energies, as in spiral-galaxy nuclei.

VII. CLASSIFICATION DIAGRAMS

Many galaxies that are not AGNs have strong emission lines in their spectra; only about 10% of the Markarian galaxies are Seyferts and the remaining 90% are nearly all H II region galaxies. A significant fraction of the latter have quite strong [O III] $\lambda\lambda$4959, 5007, as in Seyfert 2 galaxies. To segregate the AGNs from the H II region galaxies is one of the main goals of emission-line galaxy spectral classification. A straightforward way to do this is by using numerical criteria based on ratios of emission line strength, as first proposed by Baldwin, Phillips and Terlevich (1981). They explored many different pairs of ratios; Veron (1981), at about the same time, used [O III] λ5007/Hβ vs [N II] λ6583/Hα, which has the advantage that both ratios are nearly independent of the exact reddening, and of a highly accurate calibration of the sensitivity as a function of wavelength. Keel (1983*b*) used the three ratios [N II] λ6583/Hα, [S II] (λ6716 + λ6731)/Hα and [O I] λ6300/Hα for the same reason. Veilleux and Osterbrock (1987) obtained, collected and analyzed a large body of data involving all three of these ratios as functions of [O III] λ5007/Hβ.

The results are shown in graphical form in Figures 4, 5, and 6. In all three diagrams the solid and open circles are the previously classified AGNs and H II regions respectively. It can be seen that the two types of objects are well segregated in these diagrams. (The few objects marked by half-filled circles are cases in which there is some question about the data or the prior classification.) The [O I]/Hα and [S II]/[O I] vs [O III]/Hβ segregations shown in Figures 4 and 5, respectively, can be understood directly in terms of the differences between the photoionizing spectra of O stars and of AGNs, while the [N II]/Hα vs [O III]/Hβ segregation is less straightforward but is well reproduced by the models. The solid line in all three diagrams shows the best adopted division between H II region galaxies and AGNs. The galaxies not previously classified, indicated by stars, were put into one class or the other on the basis of these diagrams. For nearly all of these galaxies, all the ratios measured give the same classification. The [O III]/Hβ vs [O I]/Hα diagram (Figure 4) is the best single criterion for separating AGNs from H II region galaxies, but because [O I] λ6300 is weak even in AGNs, it requires the best signal-to-noise ratio on the spectra. If possible it is best to use all three ratios; if they do not all agree the average classification should not be adopted, but instead the object should be regarded as peculiar and deserving further study.

The dashed lines in all three diagrams are from a series of AGN photoionization models calculated by Binette (1985) for a power-law input spectrum, of the form $F_\nu \propto \nu^{-n}$, with $n =$ 2. The ionization parameter decreases along the series, from $\Gamma = 10^{-2.2}$ at the top end of each of the dashed lines to $\Gamma = 10^{-8}$ at the bottom end. The abundances of the elements are solar, and the mean electron density is $N_e \approx 10^{2.5}$ cm^{-3}.

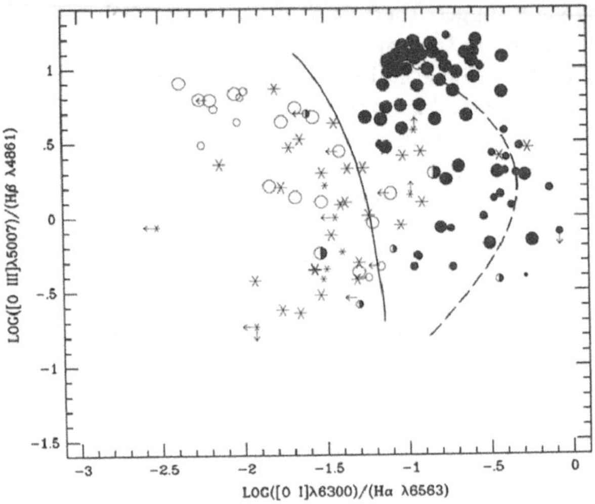

Figure 4. [O III]/Hβ vs [O I]/Hα line ratios for H II region galaxies (open circles), AGNs (filled circles), and unclassified objects (asterisks). The half filled circles are objects for which the data or published classification is in doubt. Dashed line is the sequence of AGN photoionization models described in the text; solid line is adopted division between H II region galaxies and AGNs.

VIII. VELOCITY FIELD

The nature(s) of the velocity field(s) in AGNs is one of the most important subjects of present research. Veron (1981), Pelat, Alloin and Fosbury (1981), Atwood, Baldwin and Carswell (1982), Filippenko and Halpern (1984), De Robertis and Osterbrock (1984; 1986), Vrtilek and Carleton (1985), Whittle (1984*a, b, c*), and others have investigated the line profiles in Seyfert 2s and the narrow-line profiles in Seyfert 1s. They are generally broader than the emission lines in H II region galaxies, but some AGNs have lines as narrow as in some H II region galaxies (FWHM ≈ 200 km s^{-1}). The forms of the narrow-line profiles are almost invariably asymmetric, with a longer tail to the blue than to the red. The profiles of all the different narrow lines in any given AGN are quite similar; the profiles in different AGNs are also similar to one another. They can all be described by one general form. This strongly suggests that the same physical mechanism is important in all these objects.

Radial flow plus extinction will clearly produce the observed asymmetry, and has been suggested as the mechanism in many of the papers listed above. Simple arguments, however, cannot be used to settle the question as to whether the sense of the flow is radially inward or outward, though most of the observers have assumed it to be an outwardly flowing wind. Vrtilek (1985) has calculated several series of kinematic models on this basis, and shown that they can fit the observed [O III] profiles well. Many of the specific features of the profiles can be reproduced by a simplified dynamical "catapult" model, in which the clouds move under the combined forces of gravity and the ram pressure from the central source (Mardaljevic, Raine and Smith 1986; Mobasher and Raine 1987). These dynamical, photoionization models appear promising; a crucial test is that one model reproduce the relative strengths and profiles of all the lines in an observed AGN spectrum.

All the models published to date assume spherical symmetry, largely to simplify the calculations. Likewise, discussions of the direction of the velocity field (radially inward or outward) as determined from the observed profiles, such as that to be given by De Robertis and Shaw

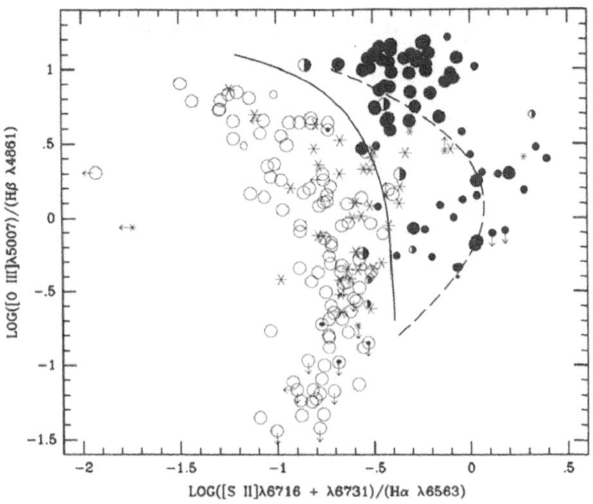

Figure 5. [O III]/Hβ vs [S II]/Hα line ratios for H II regions and AGNs. Symbols and curves as in Figure 4.

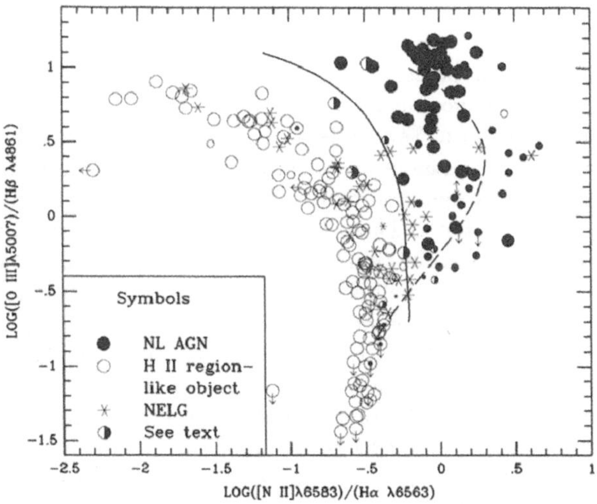

Figure 6. [O III]/Hβ vs [N II]/Hα line ratios for H II regions and AGNs. Symbols and curves as in Figure 4.

at this symposium, have all been based on the assumption of spherical symmetry. Actually, as described above, there are many observational indications of cylindrical symmetry, with much of the dust responsible for the extinction concentrated to the plane of symmetry. Thus models, such as that schematically indicated in Figure 7, in which the velocity of the ionized gas is radially outward from the central nucleus, but there is strong extinction by dust in dense, mostly neutral gas in a plane, should also be explored.

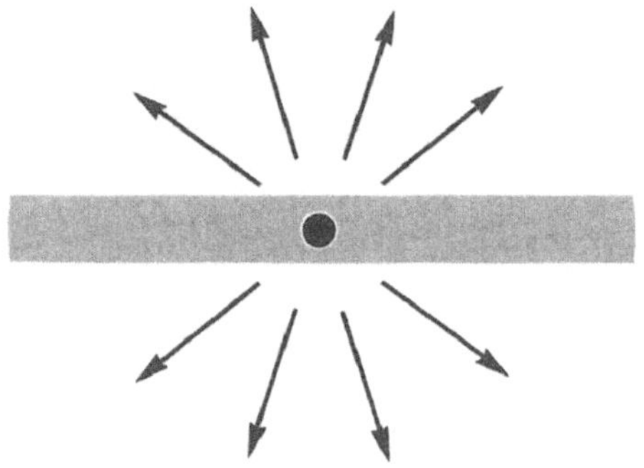

Figure 7. Schematic model of AGN with ionized gas flowing radially outward in spherical or conical distribution, and heavy obscuration in a plane, leading to line profiles with asymmetric wings extended to blue.

In many Seyfert galaxies the widths of the emission lines are correlated either with ionization potential of the ions concerned or with critical electron density for collisional deexcitation of the ionic energy levels concerned. Correlations in which the FWHMs increase with increasing potential are more prevalent in Seyfert 1s, while correlations in which the FWHMs increase with increasing critical density are more prevalent in Seyfert 2s. However, both types of correlations also occur in some Seyfert galaxies of the other type, and in still others there are not significant correlations of either type. These results strongly suggest that the mean electron density decreases outward, away from the central source, which agrees with (and is no doubt related to, or an alternate description of) the density decrease from the BLR to the NLR. These different types of correlations indicate different types of mean density distributions; on the average more nearly constant with distance from the nucleus in Seyfert 1s, and decreasing more steeply with distance in the Seyfert 2s (De Robertis and Osterbrock 1986).

However, very high-resolution line profiles recently obtained by Veilleux, with the Hamilton coudé spectrograph at Lick Observatory show that many of these narrow line profiles have irregularities or multiple components at this velocity scale, as he will report at this Symposium. Such features were previously described, particularly by Pelat and Alloin (1982).

Good signal-to-noise ratio measurements of broad-line profiles of Seyfert 1 galaxies have been obtained by Osterbrock and Shuder (1982) and by Crenshaw (1986), and of QSOs by Wilkes (1984), among others. In contrast with narrow-line profiles, the broad-line profiles have no preferred direction of asymmetry. Roughly equal numbers of Seyfert 1s have broad-line profiles that are asymmetric to the blue and to the red, and a sizable fraction have approximately symmetric profiles. Radial flow alone will not produce such profiles, unless there are large deviations from symmetry in the density distribution in the BLR, such as might be described as a relatively few, separate large clumps. Symmetric kinematical models combining outward flow along an axis or in a cone, combined with rotation in a disk can fit the observed profiles (De Robertis 1985).

However, there are severe theoretical difficulties with the concept of a Keplerian rotational velocity field under the gravitational force of the central black hole, because of the calculated shearing and destruction of clouds (Mathews 1986). Yet, if this problem of stability could be solved, a typical average velocity $v \approx 2500$ km s^{-1}, a typical black-hole mass $M \approx 10^8 M_\odot$, and a typical average radius $R \approx 0.035$ pc agree with the gravitational-force equation $v^2 = GM/R$. Furthermore, analysis of the broad-line profiles shows that the line ratios Hβ/Hα, He I λ5876/Hα and He II λ4686/Hα all increase with increasing velocity, all as predicted if the velocity and density both increase inward toward the central source, as in a rotational velocity field (Shuder 1982).

Furthermore, the radial flow has generally been assumed to be due to an outward wind of a hot, confining medium, which carries the clouds with it. However, a recent theoretical analysis shows that the temperature in any such wind $T \lesssim 10^7$ °K, but that at such temperatures, the wind cannot confine the clouds (Mathews and Ferland 1987). Evidently severe discrepancies still exist between observational data on the velocity field in the BLR and our theoretical understanding of this region. One recent suggestion is that the BLR may be pervaded by a magnetic field of sufficient strength to confine the observed BL clouds, and that it (with a low-density plasma) may thus replace the hot wind in the models (Rees 1987).

IX. FUELING AND BLACK HOLES

The energy source of AGNs is most probably a rotating accretion disk about a massive black hole. If the galaxy has a black hole at or near its nucleus, a continual source of mass delivered to it, with nearly zero angular momentum on the scale of the galaxy, is required. Several recent studies agree in confirming earlier ideas that gravitational interaction with a nearby companion or slowly orbiting galaxy is often the mechanism by which this fueling process occurs (Adams 1977; Simkin, Su and Schwarz 1980; Kennicutt and Keel 1984; Dahari 1984). However, very strongly interacting and apparently disrupted galaxies tend not to be Seyferts (Dahari 1985). Evidently in these cases the perturbations or interactions are too strong to bring appreciable amounts of mass to the center with near-zero velocity.

Some small, loose groups of galaxies also contain several AGNs; evidently in such groups the interactions can have the effect of delivering gas close enough to the nucleus to refuel the accretion disk (Kollatschny and Fricke 1985). On the other hand, rich, dense clusters contain practically no AGNs, again indicating that strong or frequent interactions inhibit, rather than promote, the formation of AGNs (Dressler, Thompson and Schechtman 1985).

There is strong evidence for the presence of a black hole of mass $M \approx 5 \times 10^6 M_\odot$ in the nucleus of our own Galaxy (Genzel *et al.* 1985). This is much smaller than the black-hole masses necessary to provide, through their accretion disks, the energies radiated in typical observed AGNs. Though we cannot observe the nucleus of our Galaxy directly in the optical wavelength region, because of the very great extinction in the galactic plane, far infrared and radio measurements confirm that it is only very mildy active. In the nucleus of M 31 there is evidence for a black hole of mass $M \approx 10^7 M_\odot$ (Dressler and Richstone 1988, Kormendy 1988). There is little if any evidence of an AGN in it (Münch 1960; Rubin and Ford 1986). Yet the fact that both these galaxies, whose nuclei we can observe in such great detail, contain small black holes, suggests strongly that there is a continuous range of black hole masses, from QSOs and Seyfert galaxies, probably down through LINERs, through lower and lower luminosity AGN-like nuclei in which similar physical processes are going on, but at a level too faint for us to detect. Probably the simplest working hypothesis is that every galactic nucleus contains a black hole, and that those which have unusually massive black holes, that are unusually well supplied with fuel, are observed as AGNs.

I am very grateful to W. G. Mathews, J. S. Miller, R. A. Shaw, R. W. Goodrich, R. W. Pogge and S. Veilleux for many stimulating discussions of the subjects treated in this review paper, and to W. W. Morgan and R. D. Dreiser for permission to reproduce Figure 1 from their paper. I am also grateful to the National Science Foundation for partial support of this research under Grant AST 86-11457.

REFERENCES

Abell, G. O., Eastmond, J. S., and Jenner, D. E. 1978, *Ap. J. (Letters)*, **221,** L1.

Adams, T. F. 1977, *Ap. J. Suppl.,* **33,** 19.

Antonucci, R. R. J., and Miller, J. S. 1985, *Ap. J.,* **297,** 621.

Atwood, B., Baldwin, J. A. Carswell, R. F. 1982, *Ap. J.,* **257,** 559.

Baldwin, J., Phillips, M. M., and Terlevich, R. 1981, *Pub. A.S.P.*, **93,** 5.

Binette, L. 1985, *Astr. Ap.,* **143,** 334.

Carter, D. 1984, *Astr. Express,* **1,** 61.

Crenshaw, D. M. 1986, *Ap. J. Suppl.,* **62,** 821.

Dahari, O. 1984, *A. J.,* **89,** 966.

________. 1985. *Ap. J. Suppl.,* **57,** 643.

de Grijp, M. H. K., Miley, G. K., Lub, J. 1987, *Astr. Ap. Suppl.,* **70,** 95.

de Grijp, M. H. K., Miley, G. K., Lub, J., and de Jong, T. 1985, *Nature,* **314,** 240.

De Robertis, M. M. 1985, *Ap. J.,* **289,** 67.

De Robertis, M. M., and Osterbrock, D. E. 1984, *Ap. J.,* **286,** 171.

________. 1986, *Ap. J.,* **301,** 727.

Dressler, A., and Richstone, D.O. 1988, *Ap. J.* , in press.

Dressler, A., Thompson, I. A., and Schechtman, S. A. 1985, *Ap. J.*, **288,** 481.

Edelson, R. A. 1987, *Ap. J.,* **313,** 651.

Ferland, G. J., and Netzer, H. 1983, *Ap. J. (Letters),* **264,** 105.

Filippenko, A. V., and Halpern, J. P. 1984, *Ap. J.,* **285,** 458.

Filippenko, A. B., and Sargent, W. L. W. 1985, *Ap. J. Suppl.,* **57,** 503.

Genzel, R., Watson, D. M., Crawford, M. K., and Townes, C. H. 1985, *Ap. J.,* **297,** 766.

Gratton, R. G., and Osmer, P. S. 1987, *Pub. A.S.P.,* **99,** 899.

Halpern, J. P., and Oke, J. B. 1987, *Ap. J.,* **312,** 91,

Halpern, J. P., and Steiner, J. E. 1983, *Ap. J. (Letters),* **269,** L37.

Heckman, T. M. 1980, *Astr. Ap.,* **87,** 152.

Huchra, J. P., Wyatt, N. F., and Davis, M. 1982, *A. J.,* **87,** 1628.

Hutchings, J. B. 1983, *Pub. A.S.P.,* **95,** 799.

Keel, W. C. 1980, *A. J.,* **85,** 198.

________. 1983*a*, *Ap. J.,* **269,** 466.

________. 1983*b*, *Ap. J. Suppl.,* **52,** 229.

Kennicutt, R. C., and Keel, W. C. 1984. *Ap. J. (Letters),* **279,** L5.

Khachikian, E. Ye., and Weedman, D. W. 1974, *Ap. J.,* **192,** 581.

Kollatschny, W., and Fricke, K. J. 1985, *Astr. Ap.,* **143,** 393

Kormendy, J. 1988, *Ap. J.*, in press.

Krolik, J. H., and Begelman, M. C. 1986, *Ap. J. (Letters)*, **308,** L55.

Lawrence, A., and Elvis, M. 1982, *Ap. J.,* **256,** 410.

Lipovetsky, V. L., Markarian, B. E., and Stepanian, J. A. 1987, in *IAU Symposium 121, Observational Evidence of Activity in Galaxies,* eds. E. Ye. Khachikian *et al.,* (Dordrecht: Reidel), p. 25.

Mac Alpine, G. M., and Williams, G. A. 1981, *Ap. J. Suppl.,* **45,** 113.

Maehara, H., Noguchi, T., Kondo, M., Miyauchi-Isoba, N., and Takase, B. 1986, in *Structure and Evolution of Active Galactic Nuclei,* eds. G. Giuricin *et al.* (Dordrecht: D. Reidel), p. 619.

Mardaljevic, J., Raine, D. J., and Smith, M. D. 1986, *M.N.R.A.S.,* **218,** 651.

Markarian, B. E., Stepanian, J. A., and Erastova, L. K. 1987, in *IAU Symposium 121, Observational Evidence of Activity in Galaxies,* eds. E. Ye. Khachikian *et al.,* (Dordrecht: Reidel), p. 25.

Mathews, W. G. 1986 , *Ap. J.,* **305,** 187.

Mathews, W. G., and Ferland, G. J. 1987, *Ap. J.,* **323,** in press.

Meurs, E. J. A., and Wilson, A. S. 1984, *Astr. Ap.,* **136,** 206.

Miley, G. K., Neugebauer, G., and Soifer, B. T. 1985, *Ap. J. (Letters),* **239,** L11.

Miller, J. S. 1985, in *Astrophysics of Active Galaxies and Quasi-Stellar Objects,* ed. J. S. Miller (Mill Valley: University Science Books), p. 367.

Mobasher, B., and Raine, D. J. 1987, *M.N.R.A.S.,* **228,** 159.

Morgan, W. W., and Dreiser, R. D. 1983, *Ap. J.,* **269,** 438.

Münch, G. 1960, *Ap. J.,* **131,** 250.

Osterbrock, D. E. 1977, *Ap. J.,* **215,** 733.

________. 1978. *Proc. Natl. Acad. Sci.,* **75,** 540.

________. 1981. *Ap. J.,* **249,** 462.

________. 1983. in *IAU Symposium 103, Planetary Nebulae,* ed. D. R. Flower (Dordrecht: Reidel), p. 473.

________. 1984. *Q.J.R.A.S.,* **25,** 1.

Osterbrock, D. E., and Dahari, O. 1983, *Ap. J.,* **273,** 478.

Osterbrock, D. E., and De Robertis, M. M. 1985, *Pub. A.S.P.,* **97,** 1129.

Osterbrock, D. E., and Koski, A. T. 1976, *M.N.R.A.S.,* **176,** 61 P.

Osterbrock, D. E., and Pogge, R. W. 1985, *Ap. J.,* **297,** 166.

________. 1987, *Ap. J.,* **323,** in press.

Osterbrock, D. E., and Shaw, R. A. 1988, *Ap. J.,* **327,** in press.

Osterbrock, D. E., and Shuder, J. M. 1982, *Ap. J. Suppl.,* **49,** 149.

Pelat, D., and Alloin, D. 1982, *Astr. Ap.,* **105,** 335.

Pelat, D., Alloin, D., and Foxbury, R. A. S. 1981, *M.N.R.A.S.,* **195,** 787.

Phillips, M. M., Charles, P. A., and Baldwin, J. A. 1983, *Ap. J.,* **266,** 485.

Pogge, R. W. 1988, *Ap. J.,* **328,** in press.

Rees, M. J. 1987, *M.N.R.A.S.,* **228,** 47P.

Rieke, G. H. 1978, *Ap. J.,* **226,** 550.

Rubin, V. C., and Ford, W. K. 1986, *Ap. J.,* **305,** L35.

Salzer, J. J. 1987, University of Michigan, Ph.D. Thesis.

Sanduleak, N., and Pesch, P. 1987, *Ap. J. Suppl.,* **63,** 809.

Schmidt, M., and Green, R. F. 1983, *Ap. J.,* **269,** 352.

Shuder, J. M. 1980, *Ap. J.,* **240,** 32.

________. 1982, *Ap. J.,* **259,** 48.

Simkin, S. M., Su, H. J., and Schwarz, M. P. 1980, *Ap. J.,* **237,** 404.

Takase, B., Maehara, H., Noguchi, T., and Miyauchi-Isoba, N. 1987, in *IAU Symposium 121, Observational Evidence of Activity in Galaxies,* eds. E. Ye. Khachikian *et al.* (Dordrecht: D. Reidel), p. 37.

Takase, B., and Miyauchi-Isobe, N. 1987, *Ann. Tokyo Astr. Obs.,* **21,** 363.

Tohline, J. E., and Osterbrock, D. E. 1982, *Ap. J. (Letters),* **252,** L49.

Ulvestad, J. S., and Wilson, A. S. 1984, *Ap. J.,* **285,** 439.

Unger, S. W., Pedlar, A., Axon, D. J., Whittle, M., Meurs, E. J. A., and Ward, M. J. 1987, *M.N.R.A.S.,* **228,** 671.

Veilleux, S., and Osterbrock, D. E. 1987, *Ap. J. Suppl.,* **63,** 295.

Veron, M.-P. 1981, *Astr. Ap.,* **100,** 12.

Veron, P., Lindblad, P. O., Zuiderwijk, E. J., Veron, M.-P., and Adam, G. 1980, *Astr. Ap.,* **87,** 245.

Vrtilek, J. M. 1985, *Ap. J.,* **294,** 121.

Vrtilek, J. M., and Carleton, N. P. 1985, *Ap. J.,* **294,** 106.

Wasilewski, A. J. 1983, *Ap. J.,* **272,** 68.

Whittle, M. 1985*a*, *M.N.R.A.S.,* **213,** 1.

________. 1985*b*, *M.N.R.A.S.,* **213,** 33.

________. 1985*c*, *M.N.R.A.S.,* **216,** 817.

Wilkes, B. J. 1984, *M.N.R.A.S.,* **207,** 73.

Wilson, A. S., Baldwin, J. A., and Ulvestad, J. S. 1985, *Ap. J.,* **291,** 627.

Wilson, A. S., and Ulvestad, J. S. *Ap. J.,* **263,** 439.

QUESTIONS

J. Krolik:

Were enough of the galaxies used in the Meurs and Wilson sample observed by IRAS so that a revision of the luminosity function could be made using a better estimate of L_{bol}?

D. E. Osterbrock:

I believe that many of these galaxies were observed by IRAS and that what you suggest – finding the bolometric luminosity function using the optical plus infrared data – could be done.

W. A. Stein:

How have relative numbers of AGN/normal galaxies changed through results of recent surveys?

D. E. Osterbrock:

The number of known Seyfert 2s per unit volume of space is considerably larger than in the Meurs and Wilson luminosty function based on the Markarian surveys while the number of Seyfert 2s is not greatly changed. Hence the number of Seyferts is considerably greater than previously calculated. Also, LINERs are AGNs not included in the Meurs and Wilson luminosity function, and hence add still more AGNs. Keel's survey shows that about half of all spiral-galaxy nuclei are AGNs.

J. W. Sulentic:

You showed an interesting plot (due to Dahari) of the incompleteness in the number of Seyferts above redshift $z \sim 0.04$. Do you know what a similar n *versus* z plot would look like for the LINERs?

D. E. Osterbrock:

No I do not, but I believe that the statistics of LINERS are even more incomplete than for Seyferts.

J. S. Miller:

Is there evidence that when Seyfert 1.9s change to, say, Sy ls or 1.5s, there is a change in reddening or extinction? Or are the changes likely to result from ionization changes?

R. Antonucci:

Certainly in the case of NGC 4151, the change between Type 1.8 and 1.5 was accompanied by an increase in observed flux (but not a shape change) of the ionizing continuum and therefore probably ionizing photon density. In at least one other case Cohen has concluded based on the spectral changes that a 1.9 to 1.2 transition cannot be attributed to a change in reddening. So the steep decrements in at least some 1.8s are probably attributable to low ionizing photon density rather than high reddening.

C. M. Gaskell:

There is a general result that Balmer decrements are flatter when objects are brighter and steeper when they fade. This is just what is predicted by photoionization models (e.g. those of G. Ferland).

O. Dahari:

What is the distribution of the broad Balmer decrement among Seyfert 1s?

D. E. Osterbrock:

$H\alpha/H\beta$ ranges from ~ 3 to ~ 8; in the cases in which $H\gamma$ can also be measured the $H\alpha/H\beta/H\gamma$ ratios follow very roughly a reddening line, with large scatter about it, no doubt resulting from collisional excitation from levels $n \geq 2$ and radiative transfer effects (Osterbrock 1977).

D. Kazanas:

Are there any other distinguishing characteristics between radio-loud and radio-quiet objects besides the Fe II emission you have mentioned?

D. E. Osterbrock:

Yes. In radio-loud objects, the broad H I profiles are broader, and more irregular than in the typical radio-quiet Seyfert 1s. The broad-line radio galaxies have relatively stronger narrow H I emission-line components, so they are more similar to Seyfert 1.5 than Seyfert 1 spectra. In BLRGs the $H\alpha/H\beta$ broad emission-line ratios tend to be, on the average, larger than in Seyfert 1s and 1.5s.

C. M. Urry:

There is also a distinction between radio loud and radio quiet AGN found in the X-ray spectra, at least for the higher luminosity objects. As found by Wilkes and Elvis, radio-loud objects have a mean spectral index of $\alpha = 0.5$, while radio-quiet objects have $< \alpha > = 1.0$. In this connection, what is the effect of running photoionization models with these different spectra?

D. E. Osterbrock:

The models calculated and published to date show the flatter spectral index of the radio-quiet objects predicts a larger partly ionized region and hence stronger Fe II emission lines, exactly the opposite of what is observed.

EMISSION LINE REGIONS IN ACTIVE GALACTIC NUCLEI: A UNIFIED VIEW

Julian H. Krolik
Department of Physics and Astronomy, Johns Hopkins University
Baltimore, Maryland 21218 USA

I. INTRODUCTION

Since the discovery of quasars nearly twenty-five years ago, a significant fraction of the effort expended by astronomers on all topics in the study of active galactic nuclei has been devoted to the analysis of their emission lines. It will come as no surprise to anyone, therefore, to announce that genuine advances have been made. Nor will it surprise anyone familiar with the frustrations of scientific work in any field that much remains to be done. Looking back over this effort, what *is* surprising, however, is the degree to which it has been dominated by two conceptual schemes, one of them now twenty years old, and the other dating back fifteen years. This realization appears especially striking when set against the typical style of discourse in the field, which, taking its cue from the almost botanical diversity of phenomena observed, divides up the subject matter into a great many subcategories, each considered independently of the others. It is the object of this review to highlight the unities of this subject, and, using that simple conceptual framework as a guide, to both set out some recent results of particular interest and point the way to problems I see as ripe for solving in the near future.

Why is it that emission lines have received so much attention? If the consensus view is correct, the truly exciting phenomena take place a few gravitational radii from the central black hole whose accretion flow is responsible for all the power release, not out at 10^5 or 10^8 gravitational radii where we believe the emission lines come from (I will justify these estimates shortly). Initially the answer to this question was a paraphrase of Edmund Hilary's remark about Everest: "Because they are there". Emission lines often visually dominate plots of the optical/UV spectra of AGN's. The extremely large widths which can be seen (often $\sim 10,000$km s^{-1}) bespeak violent, dramatic motions. Moreover, unlike most everything else in AGN's, they are generated by recognizable, calculable, well-understood physical processes. Black hole accretion is still a mysterious physical regime; atomic physics is something we think we know all about.

As we began to learn something about these emission line regions, a second answer to the question became apparent. The emission line regions could be used as diagnostics of nearby, unseen matter. It is easy to show (as I will in §II B) that the gas responsible for these lines occupies only a very small fraction of the volume of the overall region which it inhabits. Something must fill the rest of the volume surrounding the line-emitting material; using well-known principles of dynamics we should be able to infer some of that unseen matter's properties. Perhaps this volume-filling medium is blowing outward in a wind, or, even more interestingly, perhaps it is a portion of the accretion flow. In that case, we would have our hands on information of more fundamental interest.

One of the major goals of this review is to set forth a third answer. The unified framework I've mentioned leads us to eliminate the Balkanization of the subject, in which misleading boundaries are drawn according to line width or photon energy. Once that is done, I believe we will be able to use emission lines to learn about the accretion flow as it passes through a number of different locales, connect all of them dynamically, and very likely also learn about the mutual interaction of the active nucleus and its host galaxy. In the long run we may be able to use such studies to uncover what makes certain galaxies possess active nuclei in the first place.

I. THE TWO PARADIGMS

A. Onion-Skin Geometry

At this point it is appropriate to set out the two organizing principles which have historically dominated this field. The first is geometric: AGN's have been visualized as crudely spherical objects whose different regimes are characterized by the logarithm of their distance from the center. In this "onion skin" model, only certain special radial shells have drawn attention because they are the ones from which we receive photons. Working inwards (in roughly declining order of understanding), we first encounter a domain at tens to hundreds of parsecs from the center in which emission lines are made in gas of comparatively small velocity dispersion (only hundreds of km s^{-1}). This is the so-called "Narrow Line Region" (NLR). We then follow conventional practise to skip over a dark region two or three decades thick and land in the zone where lines of are made in gas having truly large velocities (thousands of km s^{-1}), the "Broad Line Region" (BLR). This region is generally thought to lie $0.1 - 1$ parsec from the center (both emission line region radii scale as $L^{1/2}$, where L is the continuum luminosity). Next, there are another three decades about which we are totally ignorant because they seem to emit nothing reaching us. Somewhere around 100 AU is probably the far infrared-millimeter band synchrotron photosphere (*cf.* Begelman's paper in this volume), and somewhere inside that there should be an accretion disk, terminating near the marginally stable orbit around a central black hole possibly ~ 1 AU in size. It is possible that the accretion disk also extends much farther out, but we have at present no way of telling where its outer boundary is.

One of the most striking results in the whole field of AGN's in the past several years is the discovery by Antonucci and Miller [1] of a new element in this geometrical picture. By means of careful spectropolarimetry (see Miller's article in this volume), they were able to show that, at least in one Seyfert galaxy, there is a geometrically and optically thick torus filling much of the region between the broad and narrow emission line zones. Subsequent work (which Miller will discuss) has enabled us to generalize the picture they invented. They also demonstrated that there is a warm, ionized plasma filling the central hole of this torus and extending some ways above its top. As I will discuss in more detail later in this review, this discovery is the cornerstone of the new, unified view of AGN emission line regions.

B. Photoionization

The second fundamental idea of the subject is that of photoionization. There is now a strong consensus that photoionization by the observed continuum is responsible for the bulk of the line emission seen in AGN's; in fact, this idea has had hardly any rivals since its first suggestion in 1964 [2]. It explains in a very natural way why we see high ionization species such as C^{+3} and N^{+4}, while the temperature is no greater than $\simeq 2 \times 10^4$K. It also provides a mechanism for the dominant energy channel in these systems, the continuum radiation, to supply the power required by a weaker channel, the emission lines. Although it doesn't require that the line luminosity be roughly proportional to the continuum luminosity over the sample, it is easily consistent with that observed fact. Finally, detailed calculations [3,4,5,6,7,8,9,10] are remarkably successful. If the lines are divided crudely according to their profiles (*i.e.*, broad separated from narrow), nearly all the relative line strengths are reproduced to within a factor of two with a minimum number of free parameters, namely the three required to define a single set of physical conditions for each profile class.

A long-term program for working out its consequences comes as an immediate corollary to the conception of the physical idea of photoionization. In order to make clear how successful that program has been, and also where holes still remain, I will set out the principal logical elements of that program. These are: the ability to compute predictions of line emissivities; the assembly of line emissivity predictions from models with some range of free parameters into a model prediction for what observations should see; a crude match against observations which determines the mean values of the free parameters; and detailed fits to observations.

i.) The physical calculation

The first step is construction of a computer code. This involves considerable library work gathering atomic data, and then a long spell of tedious programming. Five physical problems must be solved by this code: ionization balance, temperature balance, excited state population balance for a number of species, line and continuum radiation transfer, and internal dynamical equilibrium. Only the last is essentially trivial. Into the ionization balance equations go rates for photoionization, spontaneous and stimulated radiative recombination, collisional ionization, dielectronic and three-body recombination, and charge exchange. Temperature balance must include energy transfers associated with ionization as well as collisional excitation and deexcitation of bound-bound transitions, and recombination cascade. The current standard for broad line region codes requires excited state population balance (including at least a few states) for H, He, O, and Fe^{+1}. Radiation transfer is generally handled by assuming pure absorptive opacity for the continuum (with a modified on-the-spot approximation for photons resulting from radiative recombination to the ground state) and escape probabilities for the lines. Finally, the normal dynamical assumption is that the clouds have zero internal pressure gradient, counting the pressure of trapped line photons as an addition to the isotropic gas pressure. Although this assumption is somewhat problematic—most dynamical models require significant internal pressure gradients—not enough is known about the true situation to choose any particular configuration over the others.

ii.) Assembly of emissivities into testable predictions

Such a calculation requires fixing three free parameters: the ionization parameter Ξ ($\equiv 2.3F_{ion}/(cp_g)$) where F_{ion} is the continuum flux between 1 and 1000 Rydbergs; the gas pressure p_g; and the column density N. Once these are set, the calculation, which is customarily done in plane-parallel symmetry, produces a list of emitted line surface brightnesses. To make model predictions, one must define a set of distribution functions for these three physical quantities. Clearly, the simplest such set is a unique value for each of Ξ, p_g, and N. These would then be interpreted as a sort of flux-weighted mean over the real distributions.

iii.) Crude fits to the data

To the great surprise of everyone in the field, it was discovered that this simplest possible model gave "reasonable" agreement with the relative line strengths typically observed. In practise, most workers have defined "reasonable" to mean "the strongest N lines within a factor of 2", where N is anywhere from half a dozen to a dozen. The informality of this procedure has meant that all the inferred numbers are necessarily left uncertain to within factors of three in either direction. In the case of the broad lines, the best match is for $\Xi \simeq 0.2$, $p_g \simeq 0.03$ dyne cm^{-2}, and $N \gtrsim 3 \times 10^{22}$ H cm^{-2}[5] . The narrow lines are not as well described by a single set of physical conditions, but to the degree that they can be modelled by a single zone, they have $\Xi \simeq 0.1$, $p_g \simeq 10^{-7}$ dyne cm^{-2}, and $N \sim$ few $\times 10^{21}$ H cm^{-2}[10].

Three interesting consequences may be derived from this exercise. First, we are able to estimate the typical distance between the source of the ionizing continuum and the line-emitting material: $\langle r \rangle = [2.3 L_{ion}/(4\pi c p_g \Xi)]^{1/2}$. The BLR radius is then $\simeq 1 L_{ion,46}^{1/2}$pc, the "typical" NLR radius is $\simeq 500 L_{ion,46}^{1/2}$pc. Second, we can find the areal covering fraction C for each line-emitting region. For the BLR it is $\simeq 0.1$ (though there may well be a global scaling with luminosity), while for the NLR it is $\simeq 1$. Thirdly, we can estimate the volume filling factor f of line-emitting material. It is $\sim 10^{-5}$ in the BLR and 10^{-4} in the NLR. These small filling factors are the essential part of the argument for organizing the line-emitting material into clouds. They might equally well justify placing the line-emitting regions on the surface of a disk, but there are other arguments restricting the plausibility of that model.

Once we know the characteristic size of these regions, we quickly recognize that the corresponding light travel timescale (for the broad line region) is definitely in the realm of human observations. Since the response time of this material is very short (the recombination timescale is ~ 15 *minutes*), the time variability properties of the line emission are completely controlled by light-travel time effects. If the lines are truly driven by photoionization by the continuum, then line variations should directly reflect continuum variations convolved with a delay due to geometry. Thus, the joint study of line and continuum variations should permit a direct observational study of the geometry of the broad line region (and §III of this review is devoted to it).

iv.) Detailed comparison with observations: line flux ratios

Despite these successes, there are still a number of unsolved problems even at the level of these one-zone models. Perhaps the one which has attracted the greatest attention in the last few years has to do with the strength of FeII emission relative to the sum of all other broad lines [11,12,9,13]. Netzer has argued [11] that either there is an additional heat source beyond photoionization, or else that the ionizing continuum the line-emitting material sees is not the same as the one we see, on the ground that the ratio of the *energy* conveyed in the total line flux (of which he claims the largest single component is FeII) to the *number* of Lα photons is a fixed property of the shape of the ionizing continuum, and does not match up to the observed shape. However, Malkan [12] and Kallman and myself [kk88] have shown that the ratio of total energy in the ionizing spectrum to the number of Lα photons produced can vary by as much as a factor of 2 depending on physical conditions; such a swing could bring the ratio back into balance. One way to look at this controversy is in terms of the efficiency of Lα photon production relative to the pure recombinational "Case B" number of 1 per ionizing photon. Because both collisional excitation and deexcitation are important, this efficiency is no longer locked to be identically unity. Collin-Souffrin and Joly [13] have presented the issue in a different, somewhat more model-dependent way: they argue that realistic calculations never produce enough FeII relative to Hβ without either supplementary heating or an extraordinary Fe abundance. The difficulty of doing reliable calculations at high density and large optical depth, as well as the considerable uncertainty in the FeII collision strengths, are likely to keep the pot boiling on this question for a while to come.

A second unsolved problem is one that we thought for a while we had licked, namely the notorious "Lα/Hβ" problem [14,15,16]. By the early 1980's there was a consensus that this problem was solved by X-ray heating of very optically thick portions of the line-emitting material, much of whose reradiation went into collisionally excited Balmer lines, while Lα radiation was quenched by thermalization in those zones. Unfortunately, those calculations were based on what was then the only measured AGN X-ray spectrum, that of 3C 273. Since that time, we have learned that 3C 273 has a particularly strong X-ray component in its spectrum, and that in the average case not enough heat is available in those very optically thick regions to bring Hβ all the way up to its observed strength relative to Lα. Just how far short it still falls depends a bit on detailed choices about spectral shapes, calculational techniques, and collections of line data.

v.) More detailed fits to the observations: distributions of physical conditions

The quality of one-zone models is both *a priori* and *a posteriori* surprising. We had no reason to expect such a narrow rnage of physical conditions ahead of time, and, having found better agreement than expected, we still have no explanation. Moreoever, there are some observational reasons why the quality of the fit may be illusory. Now that data of higher and higher quality and resolution are becoming available, we are learning that the profiles of different lines from a single object, within both the broad and narrow categories, are not always so similar [17,18]. If that is the case, the line ratios are functions of position within the profile, and different isovelocity surfaces must have different mean physical conditions. Such a situation demands more attention be given to finding the distributions of the physical conditions, and their correlations with velocity. Indeed, we already know that among the narrow lines there are strong correlations of physical properties like critical deexcitation density with velocity width [18].

Before proceeding with this effort as it applies to the BLR, however, it is quite possible that it will be necessary to solve another complicated technical problem, that of computing photoionization models at very high density (*i.e.*, $\gg 10^{10}\text{cm}^{-3}$) and to very high column density ($\gtrsim 10^{24}\text{cm}^{-2}$). Obviously, if the spread of physical conditions present extends very far at all towards higher densities and column densities, we will needs results in those regions of parameter space.

Extending the domain of calculation to higher densities requires treating the excited state population balance for a large number of new ionic species, for at such high densities more and more ground-state transitions are thermalized. Consideration of excited state populations in C^{+2}, C^{+3}, N^{+4}, Si^{+3}, Mg^{+1} will be required, and possibly in other species as well. This means much more work for the (human) computer because the volume of atomic data to be gathered must grow substantially, and, of course, the additional bookkeeping in the computer code requires many new lines to be written and debugged.

Extending the calculation to higher column densities requires improving the quality of the radiation transfer solutions both for the lines and the continua. The simple escape probability approximation fails for many lines because such large column densities mean that their scattering takes place in the damping wings. When that is the case, the photons diffuse in space as well as in frequency before either leaving the region altogether or being absorbed. If there is little reabsorption, a revised form of the escape probability is satisfactory. However, when reabsorption is significant, the escape probability, which is an essentially *local* quantity, is no longer useful because energy can be transferred between two widely-separated zones. Real radiation transfer solutions are required. The new difficulty encountered in continuum transfer has to do with the excited state ionization continua of hydrogen. When they become optically thick, they also provide a way of transferring energy from place to place; the "on-the-spot" approximation, valid for the extremely optically thick Lyman continuum is entirely inappropriate for them. What is worse is that the different excited state continua overlap in energy; rather than treat each excited state continuum as a single photon class, instead numerous finely-divided photon energy classes must be each considered on its own.

The situation at the moment is that neither problem has been solved. Several groups have begun work: Ferland and Netzer, and separately Zheng, have begun to accumulate the mountain of atomic data necessary to treat higher densities; Collin-Souffrin and Dumont [35] have done calculations in which they assume an ionization and temperature structure in order to concentrate on doing an accurate radiation transfer solution. More work is clearly required on both parts of the problem.

When these technical problems have been solved, it will be possible to approach the analysis via either of two routes: the study of detailed line profiles, and the study of line flux variations. The two approaches are conceptually very similar, but overlap only partially in their capabilities. In both cases, we use the facts of geometry to project out a family of special surfaces within the unresolvable emission line region, and then solve for the line flux produced on each surface within the family. Likewise, in both cases, once given the flux emitted over each member of a set of surfaces, it is necessary to insert symmetry assumptions to achieve a more detailed map of the emitting region. Given a particular choice of symmetry, one then solves an integral equation to arrive at the final answer. The differences between the two methods arise because each has its own set of special surfaces, and because the extent of our *a priori* knowledge about these special surfaces differs between the two cases.

When we do profile analysis, we suppose that the line flux measured at each velocity offset comes from the emitting material lying on a surface defined by the requirement that the local velocity projected on our line of sight be the same as the velocity offset. The thickness of these surfaces is proportional to the width of the local velocity dispersion. Profile analysis is relatively easy in the sense that we can go directly from the measured profile to the line emission produced on a single isovelocity surface; it is comparatively difficult because local velocity dispersions can make the isovelocity surfaces overlap, and because we have no *apriori* knowledge of the actual shape of the isovelocity surfaces.

When analyzing time series of line flux variations, the special surfaces are the paraboloids of equal delay. We assume that all the variability is driven by the variability of the ionizing continuum, and that we know how to relate fluctuations in the continuum directed towards us to continuum fluctuations seen from other directions. Because we understand light kinematics very well, we know exactly where these isodelay surfaces are. The difficulty in doing time series analysis is that the flux observed at any given time is an integral over the fluxes produced on all the isodelay surfaces at different times in the past. That is, we must solve a deconvolution problem. In principle, the Fourier method suggested by Blandford and McKee [39] can be used to perform this deconvolution; in practice, it does not yield useful results unless the sampling can be made quite uniform and noise levels kept under strict controls. Because considerable effort has been given to developing alternate methods of time series analysis recently, I will expand on this topic in its own section, §III.

III. THE USE OF TIME SERIES

A particularly popular technique in the study of line flux variability is that of cross-correlation analysis. In this method, one monitors, over a stretch of time T, a particular AGN simultaneously in a line (or lines) and the continuum. Given that stream of data, one computes the cross-correlation between line and continuum fluxes as a function of the lag τ:

$$C_{lc}(\tau) = \frac{1}{\sigma_l \sigma_c} \int dt \, \left[F_l(t) - \bar{F}_l\right] \left[F_c(t-\tau) - \bar{F}_c\right] \tag{1}$$

where subscript l, c labels a quantity as either line or continuum, and $F(t)$ is the measured flux as a function of time, σ is the standard deviation of the sample of measured fluxes and $\bar{F}$ is the sample mean. The hope is that the result will be a smooth curve with a well-defined peak, and that the position of that peak can be identified with the light-travel time across a characteristic scale of the line-emitting region.

There are many good reasons to use this method, but the conclusions drawn by its practitioners to date are, I believe, too uncritically arrived at, and more sweeping than the data truly justify. The difficulties come in two parts: reliably calculating the cross-correlation function, and then physically interpreting it.

Two problems complicate the observational determination of a cross-correlation function: noise, especially correlated noise; and sampling. Absolute spectrophotometry to better than 10 – 20% is very difficult, but it is a rare object whose characteristic fluctuation magnitude is as much as 50%. Therefore, the effective "signal-to-noise" of most monitoring programs so far has been at best 2:1 or 3:1. This problem is exacerbated by the fact that generally speaking both the line flux and the continuum flux are measured relative to the *same* calibration standard. If an error is made in the measurement of the standard flux, it is conveyed to both the line and the continuum flux with zero lag. These correlated errors have the effect, of course, of producing a spurious positive correlation at zero lag. As I will explain in the next two paragraphs, the usual techniques for dealing with irregular sampling then spread this spurious correlation to non-zero lags of magnitude comparable to the mean sampling interval.

If the observations are taken at regular intervals, then it is easy to compute the cross-correlation at any lag which is a multiple of the spacing interval. One chooses the Riemann sum discretization of the integral in equation (1) so that the integrand is evaluated at times which are multiples of the spacing, and then real measurements exist both for $F_l(t)$ and for $F_c(t-\tau)$. However, when the spacing is irregular, there is no way to discretize the integral in such a way that actual measurements are available for more than a few of the points of integrand evaluation. The standard choice hitherto [19] has been to solve this problem by linear interpolation between the nearest measured points. This method is tolerable if the sampling is fine enough that there are few gaps large compared to the natural periods of the system. When there are many gaps large compared to the characteristic variation timescales, any interpolation scheme amounts to unguided invention of data.

Our misfortune is that ground-based data on non-circumpolar objects *always* has gaps of order half a year, and this is exactly the characteristic timescale predicted for the Seyfert galaxy (Akn 120) with the most extensive monitoring. Moreover, since interpolation has the consequence of spreading the "influence" of measured points over times comparable to the sample spacing, correlated errors are also spread over this characteristic width in the cross-correlation function. In the case of Akn 120, it has been claimed [19] that the peak occurs at a lag of 7 days; the mean sampling interval for the data on which that claim is based is 38 days, while the rms fluctuations in both the continuum and the line fluxes are only twice as large as the stated measurement errors, and half the measurement errors are correlated.

Both the problem of correlated error and the problem of highly irregular sampling can be alleviated by a recently-invented method called the "Discrete Correlation Function" [20]. This technique also has the advantage of producing well-defined error bars for the cross-correlation function. Its key idea is to accumulate all the actual pairs of point-by-point samples of the cross-correlation $[F_l(t_j)-\bar{F}_l]\times[F_c(t_k)-\bar{F}_c]$ (where $t_{j,k}$ are both actual measurement times) and associate each with the lag $t_j - t_k$. Because the cross-correlation function can be expected to be smooth even when the individual time series are not, it is fair to bin sample cross-correlations at nearby lags, and average them. The result is a cross-correlation function whose computation has required *no* interpolation, and hence *no* fabrication of data. If each point-by-point sample were independent of all the others, the standard deviation in the mean for each bin would be the square root of the variance within the bin divided by the number of points (minus 1) accumulated in the bin. In this case, we can expect some correlations; by studying the groupings of pairs it is possible to correct for these in evaluating the standard deviation in the mean. Thus this method provides a measure of the cross-correlation free of spurious features due to correlated errors, free of arbitrarily invented data, and with a well-defined uncertainty.

The second class of problems in using the cross-correlation method stem from interpreting the result once it is computed. As many have noted, *e.g.*, [21], whenever the fluctuation spectrum of the exciting continuum has power over a broad range of timescales, the cross-correlation, even with perfect data, is intrinsically biassed in favor of small scales because they are able to respond to more of the continuum fluctuations' frequency components. Simulations [20] show that this effect is so strong that in order for material at larger radius to be noticeable in the cross-correlation relative to material at smaller radius, its relative contribution to the total flux must be larger by at least the ratio of radii.

What has not been previously recognized, however, is that the peak of the cross-correlation function, even when bias is not an issue, may not be the best indicator of the physical scale, and that the feature which is the clearest indicator depends on the power spectrum of continuum fluctuations. A particularly simple example of this is given by the cross-correlation functions associated with line emission produced by single spherical shells. When the power spectrum of the continuum is $\propto f^{-1}$, *i.e.*, "flicker" noise, the cross-correlation function peaks at $\tau = r/c$, but it is a low, broad peak; when the continuum power spectrum is white noise, *i.e.*, independent of frequency, the cross-correlation is a square wave—with *no* peak. Indeed, in both these cases, the more reliable indicator of the underlying scale may be the position of the large lag shoulder, which is roughly at $2r/c$ for flicker noise and exactly at that location for white noise.

The upshot of all this is that although cross-correlations have a definite place in the repertory of emission line study techniques, a number of issues regarding both their quantitative reliability and their interpretation need greater care than has hitherto been applied.

IV. EMISSION LINE GAS LIFE-CYCLE

One of the most durable elements in our picture of line-emitting material over the years has been that it is organized in filaments or clouds. As remarked earlier, the essential argument behind this picture is that the volume filling factor is so small. The problem which has dogged the cloud model throughout its life, and prevented its complete acceptance, is that although it is easy to invent mechanisms which destroy such clouds, it is much harder to invent mechanisms which create them [22].

A partial advance was made a number of years ago when it was recognized that, in the context of a static model for the emission line region, clouds could at least be preserved, if not created, by the presence of a surrounding hot, rarefied medium which confined them by its pressure [23,24,27,18]. In the steady state, the temperature of this intercloud medium would be fixed by the shape of the continuum spectrum through Comptonization. However, the heating and cooling timescales are so long relative to likely dynamical timescales (in both the BLR and the NLR), that there is no particular reason for the intercloud medium to be anywhere near radiative equilibrium. In fact, recent work on the distribution of AGN spectral shapes has suggested that the Compton equilibrium temperature may in many cases be so low that a number of its more desirable features disappear [25].

It is possible that a solution to the problem of creating the clouds (in the BLR, anyway) may grow out of this very problem of disequilibrium [26]. If we estimate the physical conditions in the BLR intercloud medium by taking observational limits on the ionization parameter and temperature, then it appears that the intercloud medium is most likely to be in a state of net radiative cooling. Its high temperature must then be attributed either to non-radiative processes (compression during infall? shock waves?) or an earlier passage through a zone of strong radiative heating. In this state of net radiative cooling, it is thermally unstable, but the growth of perturbations is only algebraic in time, not exponential. Consequently, perturbations must begin at small, but finite amplitude in order to grow to substantial size. Now in order for the instability to get anywhere at all, the cooling time for the intercloud medium must be comparable to the dynamical time. This requirement can be translated into a constraint on the pressure of the medium. Remarkably, the pressure at which thermal instabilities grow best is just about what has been inferred by photoionization modelling for the broad line region. It is also encouraging that the nonlinear development of these instabilities gives a natural scale of cool condensations which is about right: $\sim 10^{23}\mathrm{cm}^{-2}$.

If this model is correct, it has the immediate consequence that the cloud velocities can never stray very far from that of the local intercloud medium. Once the velocity difference becomes more than a few hundred km s^{-1}, the clouds are fairly rapidly destroyed hydrodynamically. Thus, emission line profile analysis would, in this case, also provide a diagnostic for intercloud medium dynamics.

It is natural to ask at this point about the origin of the clouds in the narrow line region. It turns out that if one substitutes the observationally estimated values of the ionization parameter and temperature for the intercloud medium in that zone, both the expected cloud thickness and cloud pressure come out in the right range. Unfortunately, this mechanism is unlikely to apply to that region. Whereas only a minority of the total mass in the broad line region is in clouds, the cloud and intercloud masses in the narrow line region are probably comparable [27]. Applied to the broad line region, this mass ratio constraint merely requires a modest stirring amplitude; applied to the narrow line region, almost no imaginable stirring suffices.

V. THE INTERMEDIATE ZONE

At the beginning of this talk I promised a unified view of all AGN emission line regions. In terms of local physics, they are united by the common mechanism of photoionization. Similarly, they all also probably share the common cloud structure. In this section I will discuss the geometrical and dynamical tie between the broad and narrow line regions, namely the recently named "intermediate zone".

For many years there was almost nothing we could intelligently say about the region between the broad and narrow emission line regions because no photons could be identified as coming from there. This sort of stumbling block is, after all, one of the fundamental ground rules of astronomy (consider even cosmology: it only became a real subject after the microwave background was discovered). The big break in this area took place two years ago with the work of Antonucci and Miller [1]. I'm sure Miller will go into detail on this and subsequent observational work, so here I will just remind you of what I remarked on earlier in this talk: spectropolarimetry of the archetypical type 2 Seyfert galaxy, NGC 1068 demonstrated that a type 1 nucleus was hidden inside a toroidal obscuring region of great optical depth whose characteristic scale is intermediate between the broad and narrow emission line regions. The inner nucleus and broad line region are visible to us only through scattering off warm electrons which fill the inner hole of this torus and extend some ways above its top surface. A comparison of the reflected flux to the bolometric flux from the galactic nucleus suggests that the Thomson optical depth of this reflection region is $\lesssim 0.1$; the breadth of the reflected Balmer emission puts an upper limit on its temperature of 10^6K.

Even these purely geometrical facts begin to tie together the different emission line regions. The obscuring torus collimates ionizing photons, so that the only clouds in the NLR which are lit up by photoionization are those lying along the torus's axis. In addition, the obscuring torus obstructs our view of the far side of the NLR, an effect which may explain the well-known characteristic asymmetry of the narrow line profiles [36,37,38].

A. The Reflection Region

With such strong observational hints, theoretical progress has been rapid. Begelman and I showed last year [28] that the characteristics of the reflection region could be readily understood as due to a thermal wind driven by the continuum radiation of the type 1 nucleus. The ionizing radiation first evaporates cold material off the inner surface of the torus by photoionization, and then heats it to escape by Compton scattering. A semi-quantitative analysis based on this model predicts an optical depth and temperature in the wind which match very nicely the observationally-derived numbers given in the previous paragraph. The corresponding mass loss rate is about $0.5 M_{\odot}$ yr^{-1}.

This mass loss rate also provides a suggestive link to the NLR. On entirely independent grounds, it has been estimated [27] that the rate of mass flow through the NLR is $\sim 1 M_{\odot}$ yr^{-1}. Perhaps this wind is the source of that flow. Its velocity, a few hundred km s^{-1}, is also of the right magnitude.

More recently, Kallman and I [29] have shown that this reflection region is itself an emission line source, provided one broadens the definition of the emission line band to include X-rays. Because the Fe K-edge opacity is always within a factor of 2 of the Thomson opacity independent of ionization state (for solar abundance, and ionization low enough that most Fe has at least one electron remaining), the fraction of Fe-ionizing photons absorbed is almost the same as (and is certainly proportional to) the fraction of all photons scattered. The fluorescent yield of Kα photons after photoionization is a weak function of ionization state, but even that only introduces a play of another factor of 2. The result is that this reflection region should be a moderately strong source of Kα line emission. Compared to the direct X-ray flux, the equivalent width expected is measurable but not huge: $\lesssim 0.1$keV. However, compared to the *reflected* X-ray flux, the equivalent width is very large indeed: $\simeq 1$ keV. In type 2 Seyfert galaxies, the only X-ray flux visible is the reflected part, so this model makes a very strong prediction about the equivalent width of Kα emission. In fact, the proportionality between Fe K-edge and Thomson opacity means that the only dependence of the predicted Kα equivalent width in type 2 Seyferts is on the Fe abundance; all model-dependent physical quantities drop out.

B. The Obscuring Torus

Turning our attention to the obscuring torus leads to yet another new kind of emission line region. The chances are [30] that the obscuring matter in this torus is clumped into dusty, molecular clouds. These clouds are extremely opaque to visible and UV light—$A_V \sim 1000$mag, but only moderately optically thick to hard X-rays. Consequently, their interiors are warmed ($T \simeq 1000$K) and ionized (electron fraction $\sim 10^{-3}$) by heat deposited in Compton scatterings of hard X-rays from the nucleus [31].

Most of the energy absorbed in these clouds is ultimately reradiated by dust as a quasi-thermal continuum, but a significant amount goes into near-IR vibrational lines of H_2 and mm-band high-J rotational transitions of CO. These emission lines provide another potentially powerful diagnostic of physical conditions in the vicinity of AGN's [31].

Yet another kind of emission line probably originates in these dusty clouds, but our ignorance about its production under more mundane circumstances prevents us from saying very much about it. Extremely powerful H_2O maser emission has already been detected from a number of Seyfert galaxies [32]. VLBI observations show that it comes from a region no more than a few parsecs in size, within a few tens of parsecs of the nucleus (the main uncertainty in the distance to the center comes from the difficulty of identifying the position of the nucleus on that scale [33]). Thus, it is very likely generated in the torus. Conditions in the torus would seem very propitious for pumping masers because there are quite a number of non-equilibrium processes already known to be present, but our ignorance of H_2O maser pumping in our own Galaxy forestalls a calculation predicting its strength in Seyfert galaxies.

Torus dynamics is intrinsically a difficult and complicated subject because a number of dynamical effects are comparable in magnitude: the orbital timescale, the cloud-cloud collision timescale, an individual cloud's self-gravitational timescale, and a typical cloud's sound-crossing and Alfven wave-crossing timescale are all the same to within a factor of ten. So far, work in the field can best be described as exploratory.

Consider first the radial mass flow through the torus. We already know that the rate at which mass is evaporated off its inner edge is $\sim 1M_\odot$ yr^{-1}. Because angular momentum is transported as a result of cloud-cloud collisions, mass flows inward through the torus at a rate of $\sim 1M_\odot$ yr^{-1} when its equatorial covering fraction is ~ 1, its mean Compton optical depth is ~ 1, the orbital speeds are ~ 100km s^{-1}, the velocity dispersion is a half or so of the orbital speed, and the radial scale of the torus is measured in parsecs [30]. These are, of course, the parameters which are required observationally for different reasons.

Indeed, this match can be turned around: the position of the inner edge is fixed by the balance between evaporation of material and the rate at which mass can be fed in from the outer parts of the torus. It is extremely striking when one realizes that this balance is unstable unless the luminosity is in part determined by the position of the torus's inner edge, as would happen, for example, if it is due to accretion of a small part of the evaporated wind [30]. Stabilization merely requires that the capture fraction decrease with increasing radius of the inner edge, a very plausible circumstance. The wide spread in the distribution function of cloud velocities in the vertical direction strongly suggests a similarly wide spread in the azimuthal and radial directions; if only the few percent of the evaporated gas with the least specific angular momentum is accreted, it would suffice to fuel the central engine. On its way in, it might also become the intercloud medium of the broad line region.

The toughest nut to crack in the area of torus dynamics is the source of the stirring necessary to maintain that large vertical velocity dispersion. The power derivable from stellar processes may be scaled in terms of the rate of mass inflow because nearly all stellar stirring mechanisms are associated with short-lived stars; this rate fails by several orders of magnitude [30]. It is possible that the stirring instead comes from turning the shear energy stored in orbital motions into random motions by partially elastic collisions (see, *e.g.*, [34]). One rather speculative way in which the elasticity might be great enough to permit this to happen is if the clouds are moderately strongly magnetized, *i.e.*, if the ratio of gas pressure to magnetic field pressure is $\lesssim 0.1$ [30], but I emphasize the speculative character of this proposed explanation.

OUR WORK ASSIGNMENTS

I hope by now it is clear how close the physical connections are between all the different emission line regions, and also how widely-separated bands in which line emission occurs can all contribute to our understanding of active galactic nuclei. In closing, I think it most useful to outline the areas I would like to see singled out for hard work by theorists in the next few years.

In the area of basic theory, three projects demand our attention. The first is to extend our ability to make predictions about line emission to yet higher pressures. In the BLR, this means adding many new metastable states to the long list whose populations are already calculated, and inventing new techniques for radiation transfer (§II B v). It is possible that we may need to break our VAX addiction and move to supercomputers in order to accomplish this task. In the obscuring torus, this means being able to properly approach the thermalization limit, while in the NLR the requirement is for inclusion of more forbidden and semi-forbidden transitions with higher critical densities. The second project is further study of cloud dynamics and life-cycle. We need to know more about the hydrodynamical properties of clouds when they are subjected to a confining pressure made non-uniform on their scale because of motion relative to the intercloud medium. More careful work is also required to better understand the nonlinear development of thermal instabilities. Finally, there is the problem of the dynamics and thermodynamics of the intercloud medium, a subject which time constraints have prevented me from more than touching on.

Formally, both profile analysis and time series analysis are exercises in data analysis. However, they present so many methodological problems that much care must be given to study of the techniques before they can be applied to actual data. For both, stabler methods of integral equation inversion are a must, as is also a clarification of the degree to which results are model-dependent. Time series analysis also requires further study of how best to choose the sampling rate, and how to recover from imperfect sampling.

While doing all this, it will be well to keep in mind our ultimate goals: to quantitatively understand the physical mechanisms responsible for all the emission lines; to be able to tie the emission line regions to the global mechanics of active nuclei; and to use knowledge gleaned from emission lines to learn what place active nuclei have in the long-term evolution of galaxies.

REFERENCES

1. Antonucci, R.R.J. and Miller, J.S., 1985, *Ap. J.* **297**, 621.
2. Osterbrock, D.E. and Parker, R., 1966, *Ap. J.* **143**, 268.
3. Davidson, K., 1972, *Ap. J.* **171**, 213.
4. MacAlpine, G.M., 1972, *Ap. J.* **175**, 11.
5. Kwan, J.Y. and Krolik, J.H., 1981, *Ap. J.* **250**, 478.
6. Weisheit, J.C., Shields, G.A., and Tarter, C.B., 1981, *Ap. J.* **245**, 406.
7. Mushotzky, R.F. and Ferland, 1984, *Ap. J.* **278**, 558.
8. Kwan, J.Y., 1984, *Ap. J.* **283**, 70.
9. Krolik, J.H. and Kallman, T.R., 1988, *Ap. J.* in press.
10. Ferland, G., 1981, *Ap. J.* **249**, 17.
11. Netzer, H., 1985, *Ap. J.* **289**, 451.
12. Malkan, M., 1987, in preparation
13. Collin-Souffrin, S., Joly, M., Pequignot, D., and Dumont, S., 1986, *Astr. Ap.* **166**, 27.
14. Baldwin, J., 1977, *M.N.R.A.S.* , **178**, 67P.
15. Davidsen, A., Hartig, G., and Fastie, W., 1977, *Nature* **269**, 203.
16. Kwan, J. and Krolik, J., 1979, *Ap. J. Lett.* **233**, L91.
17. Wilkes, B.J., 1986, *M.N.R.A.S.* **218**, 331.
18. Filippenko, A. and Halpern, J., 1984, *Ap. J.* **285**, 458.
19. Gaskell, C.M. and Peterson, B., 1987, *Ap. J. Suppl.* **65**, 1.
20. Edelson, R.A. and Krolik, J.H., 1988, *Ap. J.* submitted
21. Gaskell, C.M. and Sparke, L.S., 1986, *Ap. J.* **305**, 175.

22. Osterbrock, D.E. and Mathews, W.G., 1986, *Ann. Rev. Astron. Astrop.* **24**, 171.
23. McCray, R.M., 1979, in *Active Galactic Nuclei*, ed. by C.R. Hazard and S.A. Mitton (Cambridge: Cambridge University Press)
24. Krolik, J.H., McKee, C.F., and Tarter, C.B., 1981, *Ap. J.* **249**, 422.
25. Mathews, W.G. and Ferland, G.J., 1987, *Ap. J.* in press.
26. Krolik, J.H., 1988, *Ap. J.* in press.
27. Krolik, J.H. and Vrtilek, J.M., 1984, *Ap. J.* **279**, 521.
28. Krolik, J.H. and Begelman, M.C., 1986, *Ap. J. Lett.* **308**, L55.
29. Krolik, J.H. and Kallman, T.R., 1987, *Ap. J. Lett.* **320**, L5.
30. Krolik, J.H. and Begelman, M.C., 1988, *Ap. J.* in press
31. Krolik, J.H. and Lepp, S., 1988, in preparation.
32. Claussen, M.J., Heiligman, G.M., and Lo, K.-Y., 1984, *Nature* **310**, 298.
33. Claussen, M.J. and Lo, K.-Y., 1986, *Ap. J.* **308**, 592.
34. Goldreich, P. and Tremaine, S., 1978, *Icarus* **34**, 227.
35. Collin-Souffrin, S. and Dumont, S., 1986, *Astr. Ap.* **166**, 13.
36. Heckman, T.M., Miley, G.K., van Bruegel, W.J.M., Butcher, H.R., 1981, *Ap. J.* **247**, 403.
37. Vrtilek, J.M. and Carleton, N.P., 1981, *Ap. J.* **294**, 106.
38. Whittle, M.,1985, *M.N.R.A.S.* **213**, 1.
39. Blandford, R. D. and McKee, C.F., 1982, *Ap. J.* **255**, 419.

QUESTIONS

ISAAC SHLOSMAN: It seems to me that cloud-cloud collisions are strongly dissipative and cannot, therefore, support the vertical dispersion velocities in the obscuring torus. Would an additional energy source beyond viscous dissipation then be necessary? Some objects have constant $L_{IR} > L_{non-thermal}$. Could you comment on the possible nature of this additional source?

KROLIK: I agree that one's first guess would be that ordinary, purely hydrodynamic cloud-cloud collisions are not sufficiently elastic to account for the large vertical velocity dispersion in the torus. Begelman and I suggested that the clouds might be so strongly magnetized precisely as a way to possibly achieve greater elasticity, but I regard that as a suggestion which is far from being proven. If the orbital energy cannot be tapped for the vertical velocity dispersion, I don't know where to turn.

MARTIN GASKELL: I want to disagree with your conclusions about our cross-correlation method. Let me first say what I agree with. Firstly, I agree that the cross-correlation method is dominated by gas close in. In fact, in our original paper proposing the cross-correlation approach Linda Sparke and I said: "If the BLR does not consist simply of a shell but is extended in radius, then the line emission from material closest to the central source will respond most quickly and with the largest amplitude to continuum fluctuations. Thus the peak in the line-continuum cross-correlation function will indicate light-travel time typical of the inner parts of the line-emitting region." We have done further work on this problem and believe that there is something like an r^{-2} weighting.

The second thing I agree with is the result that the inversion process to recover the "transfer function" is unstable (a couple of years ago Linda Sparke did almost exactly the same numerical experiments you have shown us this morning with similar results). This is why we have stuck with the cross-correlation function. Now although there can be gas at very large radii which this method is not telling us about, many broad lines vary substantially and on short time-scales. This is telling us that there is gas close in.

Now let me say the main thing I disagree with. This is your implication that our cross-correlation method is giving the wrong results (quite apart from the distribution of the gas question just mentioned). Brad Peterson and I have looked into this both analytically and via Monte Carlo simulations in our most recent published paper. Taking care to simulate closely the actual continuum variability (we matched continuum auto-correlations) and using exactly the spacing of the real observations, we find from the simulations that the cross-correlation peak is giving the right answer (*i.e.*, the one we initially put in) to within the expected errors 90 – 95% of the time. This is the simulation that matters and it answers the question "What fraction of the time do we get the right answer?" In this paper you have shown us things that can go wrong (and I agree that they can go wrong), but I think our simulations show that they usually do not go wrong.

My final comment is that the methods of time series analysis and profile analysis can be combined. I have shown how the cross-correlation technique can be extended to studying profile variability and investigating the kinematics of the BLR gas. Preliminary results for the quasar in NGC 4151 indicate that the highest velocity gas is closest to the center and inflowing (to be published in *Ap. J.* next year).

KROLIK: I'll respond to your comments in the order in which you made them. I am glad you agree about the intrinsic bias of the cross-correlation method. I discussed the point first to quantitatively clarify the strength of the bias, and second to emphasize its presence. Too often in discussion of this topic, people either give lip service to the bias toward small scales, or else ignore it altogether. Since the photoionization predictions refer to the mean conditions in the gas, cross-correlations suggesting smaller scales than would be indicated by photoionization arguments are not necessarily in direct conflict with the photoionization arguments.

There are several reasons why, despite your reassuring Monte Carlo simulations, I am not so confident about the reliability of your results on Akn 120. The first is the matter of correlated error at zero lag. I believe I discussed that adequately in my talk, explaining how its sinister effects propagate out, in your case, to lags as large as 40 days or so. The second is that when I recompute the cross-correlation by the discrete method, and so eliminate the correlated errors, I find that the magnitude of the cross-correlation is disturbingly small, and quite consistent with what one would expect from correlating independent random signals. The third is the very large error bars associated with the cross-correlation in the neighborhood of 150 – 200 days, which is just where the photoionization model predicts a peak. The reason for those very large error bars is simply that you had very few real pairs of points separated by lags in that range because you were faced with one of the fundamental limitations of ground-based astronomy: objects are typically only observable for half the year.

Edelson and I have also done extensive simulations to test the sensitivity of various methods. Our experience, also using the actual sampling pattern of the Akn 120 data, is less sanguine than yours. We might wish to discuss that privately. I doubt, by the way, that your measured continuum autocorrelation function is all that reliable, for it suffers from almost all the same problems as the linearly interpolated cross-correlation. Correlated error is, of course, no longer a worry, but the coarse spacing is almost certainly inadequate to resolve the true variability.

Finally, I certainly agree that the techniques of profile analysis and time series can be combined. Time series of profiles are perhaps the most complete set of structural data we can ever hope to have. I concentrated here on flux time series because I thought there were plenty enough subtleties to sort out before moving on to a more complicated problem.

DEMOS KAZANAS: Is the bottom line of these statistical arguments that the observed variability in Akn 120 is not real?

KROLIK: That's a delicate question. The nominal signal/noise ratio that Peterson claims for his data is around 2:1, *i.e.*, ratio of rms fluctuation level to measurement error. Simulations Edelson and I have done cooked up to mimic these conditions suggest that the resulting cross-correlation, although it is noisy and not of very large magnitude, still is a bit clearer to interpret than the Akn 120 cross-correlation.

EMISSION-LINE REGION STRUCTURE FROM VARIABILITY STUDIES

Bradley M. Peterson
Department of Astronomy
The Ohio State University
174 West 18th Avenue
Columbus, OH 43210

In principle, the physical and kinematic structure of the broad-line region (BLR) in AGNs can be determined from emission-line variability. As continuum pulses propagate outwards from the central source of ionizing radiation, they produce excitation inhomogeneities in the BLR and these should be detectable as changes in the line profile, provided that the BLR velocity field is ordered. Indeed broad emission-line profiles have been found to vary with time, though it has not yet been unambiguously demonstrated that line profile variations are attributable to light-travel time effects within the BLR.

Over the last several years, considerable effort has been devoted to trying to determine the BLR size and structure from variability studies (see the recent review by Peterson 1988). In this contribution, we give examples of the impact that variability studies have had on our perception of the BLR structure, but show that no clear picture has yet emerged.

The Size of the BLR

In terms of understanding the physics of the BLR, the most surprising result has been that the integrated emission-line fluxes respond to continuum variations on time scales much shorter than previously imagined. For brighter Seyfert 1 galaxies ($\sim 10^{44}$ erg s^{-1}), the size of the BLR can be estimated as about 1 light year on the basis of ionization equilibrium arguments, assuming parameters (electron density: $\log n_e \approx 9.5$; ionization parameter: $\log U \approx -2$) from photoionization models (cf. Kwan and Krolik 1981). However, the rapid response of the emission-line fluxes to continuum variations suggests that the size of the BLR is considerably smaller than this. Gaskell and Sparke (1986) showed that cross-correlation analysis of line and continuum light curves could be used to determine the BLR radius since the emission-line light curve is just the continuum light curve shifted and broadened by a transfer function. Gaskell and Sparke showed that the "lag", or phase shift, between the light curves could be identified as the light-travel time across the BLR τ_{LT} if the BLR can be modeled as a thin spherical shell (for which the transfer function is rectangular and non-zero in the range $0 \le t \le 2\tau_{LT}$). Cross-correlation analysis for a few galaxies has led to estimates of the BLR radius ranging from several light weeks (e.g., $\tau_{LT} \approx 30$ light days for NGC 5548 [Peterson and Gaskell 1986]; $\tau_{LT} \approx 25$ - 45 light days for 3C 390.3 [Clavel and Wamsteker 1987]) down to a few light days (e.g., $\tau_{LT} \lesssim 7$ light days for Akn 120 [Gaskell and Sparke 1986; Peterson and Gaskell 1986]). However, Gaskell and Peterson (1987) examined the errors and limitations of the cross-correlation method as applied to real data and concluded that the small size inferred for the BLR in Akn 120 by the cross-correlation method was not reliable because of limitations imposed by the quality of the data, the most important effect arising from the small correlated component in the errors in the continuum and broad-line fluxes. The other major limitation affecting *all* existing data sets is undersampling, since even the best data sets typically have median temporal resolution of around one month. Gaskell and Peterson nevertheless conclude that the BLR radius in Akn 120 is not likely to be larger than ~30 light days. Another concern which has not yet been addressed in the literature is that the cross-correlation method is biased towards the inner part of the BLR, which responds more coherently to continuum changes. In other words, if the BLR is modeled as a series of concentric shells, the transfer function is narrowest for the innermost shells, which therefore dominate the response of the emission lines. Cross-correlation estimates of the BLR radius are thus weighted toward detection of the BLR clouds closest to the continuum source. Sizes inferred from cross-correlation estimates might then represent something closer to a lower limit rather than a representative BLR radius. However, it must still be pointed out that the amplitude of variation in the lines is generally nearly as large as the amplitude of continuum variation, which suggests that we are observing *most* of the line-emitting material responding to continuum changes on a short time scale, unless the amplitude of variation is much larger in the ionizing continuum than in the observed continuum or the line strength is a very sensitive function of the continuum flux. It is probably also significant that the line *profiles* do not change nearly as fast as the integrated line flux, which again could suggest that the entire BLR is responding rapidly.

The Structure of the BLR

Any model of the structure of the BLR must explain (1) the nearly logarithmic line profiles, (2) the line asymmetries (which can go in either direction), (3) why some galaxies have double-peaked profiles, and (4) the variability of the line profiles. The logarithmic profiles (Blumenthal and Mathews 1975; Capriotti, Foltz, and Byard 1980; Hubbard and Puetter 1983) and the asymmetries can be explained by a variety of models. Differences in line profiles within a given object (say, between C III] $\lambda1909$ and C IV $\lambda1549$) and in the variability of different lines has led to suggestions of multiple component BLRs which show ionization stratification through the line-emitting region (Ulrich and Boisson 1983; Ulrich et al 1984; Gaskell and Sparke 1986). Multiple components, such as a double-peaked emission from a rotating disk (van Groningen 1984; Alloin, Boisson, and Pelat 1987; Stirpe, de Bruyn, and van Groningen 1987), or physically distinct BLRs associated with the components of a supermassive binary system (Gaskell 1983; Peterson, Korista, and Cota 1987), have been invoked to explain the double peaked emission lines and the variable line profiles.

One of the best studied cases is NGC 5548. Peterson (1987) noted that during a continuum outburst in 1984, the total line flux changed significantly without any accompanying change in profile. However, by the 1985 observing season, a second broad component appeared in Hβ and Hγ, blueshifted by ~2000 km s^{-1} relative to the systemic velocity of the galaxy. There was also weaker, somewhat dubious evidence for a weak third component, redshifted by ~2000 km s^{-1} in Hβ. By 1986, the blue component had become very strong, making the broad lines highly asymmetric. A simple attempt to deconvolve the Hβ line profile into a brighter main component (close to the systemic redshift) and a weaker blueward component (Peterson, Korista, and Cota 1987) revealed that the flux in the main component is correlated with the continuum brightness with a lag of ~16 $\pm$ 8 days, but the flux in the blueward component is formally uncorrelated with the continuum variability. Peterson, Korista, and Cota (1987) suggest that these two broad components might represent physically distinct BLRs, each surrounding a member of a supermassive binary. The mass estimates based on this assumption are not at variance with mass estimates based on virial arguments.

Stirpe, de Bruyn, and van Groningen (1987), however, find evidence for a *third*, redshifted (~2000 km s^{-1} relative to systemic) broad component in the difference between Hα scans taken in 1985 June and 1986 July. This may be related to the weak feature detected by Peterson (1987) in Hβ. In light of this result, we have recently reexamined all of our spectra of NGC 5548 to search for a redshifted component. Our highest quality data from 1984 - 1986 show only the very weak feature in Hβ reported by Peterson (1987). Moreover, the difference between two Hα scans obtained in 1986 March and 1985 March shows only the blueward component (Fig. 1). Evidently, a marked decrease in the flux in the red component of Hα took place between 1986 March and 1986 June, but the red component apparently must be constant throughout our observations between early 1984 and early 1986.

We also examined carefully some earlier lower quality Hβ scans of NGC 5548. A redshifted component may be present in at least some spectra obtained between 1981 and 1983 (Fig. 2 shows the clearest example), but in most cases it is difficult to isolate any possible Hβ component from the blended emission feature just longward of Hβ.

The Hα difference spectrum shown by Stirpe, de Bruyn, and van Groningen (1987) clearly shows components at ±2000 km s^{-1}, and this led them to suggest that the emission arises in a rotating disk. However, our data indicate that the red component does not vary in the same way as the blue component, which seems to argue against a single structure. On the other hand, a supermassive binary scenario also seems unlikely, since it appears that the profiles are best modeled with *three* broad components, not just two.

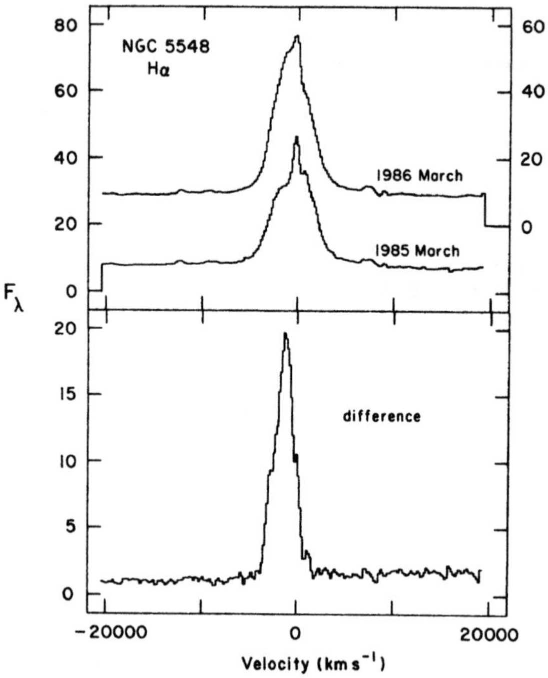

Figure 1

Summary

Recent results have shown that emission-line variability studies are potentially valuable for determining the size and structure of the BLR. However, this potential has not been fully realized, primarily on account of insufficient data. It is clear that size determinations will require better temporal resolution since the BLR is apparently smaller than estimated earlier. On the other hand, experience with NGC 5548 suggests that a very long baseline of observations will be required to investigate the kinematic structure of the region. Considerable emphasis needs to be placed on long-term spectroscopic monitoring programs that provide good temporal resolution.

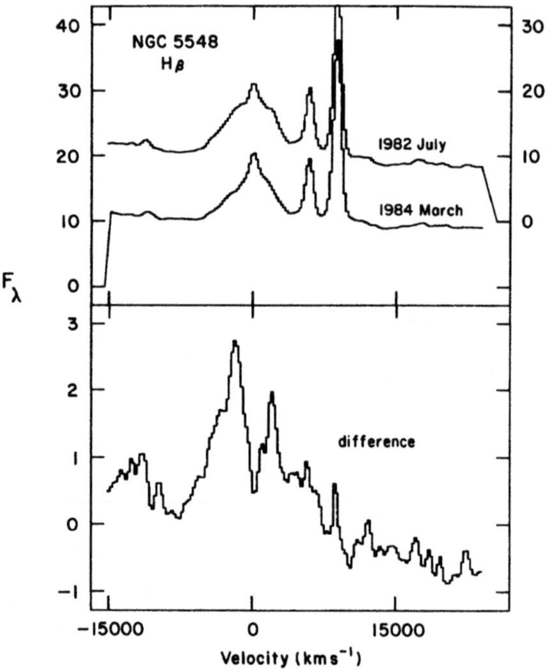

Figure 2

Studies of emission-line variability are supported at The Ohio State University by the NSF under grant AST 87-02691 and by NASA under grant NAS 5-826. The author is grateful for this support.

References

Alloin, D., Boisson, C., and Pelat, D. 1987, *Astron. Astrophys.*, in press.
Blumenthal, G.R., and Mathews, W.G. 1975, *Astrophys. J.*, **198**, 517.
Capriotti, E., Foltz, C., and Byard, P. 1980, *Astrophys. J.*, **241**, 903.
Clavel, J., and Wamsteker, W. 1987, *Astrophys. J. (Letters)*, **320**, L9.
Gaskell, C.M., and Peterson, B.M. 1987, *Astrophys. J. Suppl.*, **65**, 1.
Gaskell, C.M., and Sparke, L.S. 1986, *Astrophys. J.*, **305**, 175.
Hubbard, E.N., and Puetter, R.C. 1983, *Astrophys. J.*, **265**, 35.
Kwan, J., and Krolik, J.H. 1981, *Astrophys. J.*, **250**, 478.
Peterson, B.M. 1987, *Astrophys. J.*, **312**, 79.
____________. 1988, *Publ. Astron. Soc. Pac.*, **100**, in press.
Peterson, B.M., and Gaskell, C.M. 1986, *Astron. J.*, **91**, 552.
Peterson, B.M., Korista, K.T., and Cota, S.A. 1987, *Astrophys. J. (Letters)*, **312**, L1.
Stirpe, G.M., de Bruyn, A.G., and van Groningen, E. 1987, *Astron. Astrophys.*, in press.
Ulrich, M.H., and Boisson, C. 1983, *Astrophys. J.*, **267**, 515.
Ulrich, M.H. et al. 1984, *M.N.R.A.S.*, **206**, 221.
van Groningen, E. 1984, Ph.D. dissertation, Leiden University.

QUESTIONS:

A. Wandel:

Is there any evidence for correlation of the line response time to continuum variability with ionization potential or with critical density? Such a correlation could indicate stratification of the BLR.

B.M. Peterson:

There is, of course, no correlation seen with critical density since C III] λ1909 is generally the only forbidden line unambiguously detected from the BLR (see however, Crenshaw and Peterson 1986, *Publ. Astron. Soc. Pac.*, **98**, 185). On the basis of cross-correlation analysis, Gaskell and Sparke (1986) found evidence that the higher ionization lines (C IV λ1549 and He II λ4686) in NGC 4151 respond to continuum variations faster than lower ionization lines (Balmer lines and Mg II λ2798). However, estimates of the errors in the cross-correlation method reveal that the differences in the response times for various lines are not statistically significant (Peterson and Cota, preprint). While the BLR may indeed have a stratified structure, the data do not yet prove this.

G. Benford:

Is it too heretical to inquire whether shock or particle ionization might be present as well as photoionization?

B.M. Peterson:

I don't think so. A number of investigators are now incorporating shocks into their models, but these provide more or less fine tuning of the spectrum. The bulk of the evidence suggests, however, that photoionization provides the principal source of energy in the BLR.

D.E. Osterbrock:

If you attribute variations in the strengths of the emission lines to shock-wave heating, this would make the observed correlation between the continuum and emission-line variations a coincidence.

J.H. Krolik:

If something other than the continuum were to power the lines, there would be no reason to expect any correlation between continuum and line fluctuations. Also, it is quite possible that because of the biassing effect Brad described, material at a small fraction of the total line luminosity could produce the entire cross-correlation peak at a small lag.

B.M. Peterson:

It is still significant, I believe, that the line variations are often large, usually close to the amplitude of the continuum variations, and sometimes involve a significant fraction of the total line flux. While Julian and I agree that there is probably a distribution of BLR gas between a few light weeks and, say, a light year of the ionizing source, I believe that most of the gas is closer rather than farther away whereas Julian believes the opposite.

D. Kazanas:

The question is whether C III] λ1909 also varies on the same time scale as the permitted lines, not simply that it is there. Do you see such a variation?

B.M. Peterson:

In standard model clouds, which are supposed to form all of the lines, the question of the existence of C III] is relevant. This line is present in all of the well-studied galaxies in which small BLRs are inferred from cross-correlation results. In NGC 5548, C III] λ1909 varies in basically the same fashion as C IV λ1549 and the Balmer lines (see the paper by Reichert and Peterson in this volume).

UV AND OPTICAL SPECTROSCOPY OF NGC 5548

G. A. Reichert
Astronomy Programs, Computer Sciences Corporation, Beltsville, MD

B. M. Peterson
Dept. of Astronomy, Ohio State University, Columbus, OH

We present studies of *IUE* spectra of NGC 5548 from nine epochs of observation from mid–1982 to present. Eight of the epochs were within two weeks of optical observations. Our goal is to study variability and structure in the ultraviolet emission lines, and to relate these to variability and structure in the optical emission lines, in particular 1) the 8–fold increase in very broad HeII λ4686 observed during 1984 (JD ~2445800), and 2) the two–component structure seen in broad Hβ λ4861 (Peterson and Ferland 1986, hereafter PF; Peterson 1987; Peterson, Korista, and Cota 1987; hereafter PKC). Our approach is to use the optical emission lines as a guide to deconvolving the ultraviolet emission lines.

For each epoch, we subtracted power law fits to the continuua, and fitted the resulting emission lines to combinations of a narrow line profile, to remove the contribution from the narrow line region, plus up to three broad component profiles. Figure 1 shows the deconvolution into the various components for CIV, MgII, and CIII]. The narrow line templates were derived from the [OIII] λ5007 profile, convolved to the *IUE* resolution at the

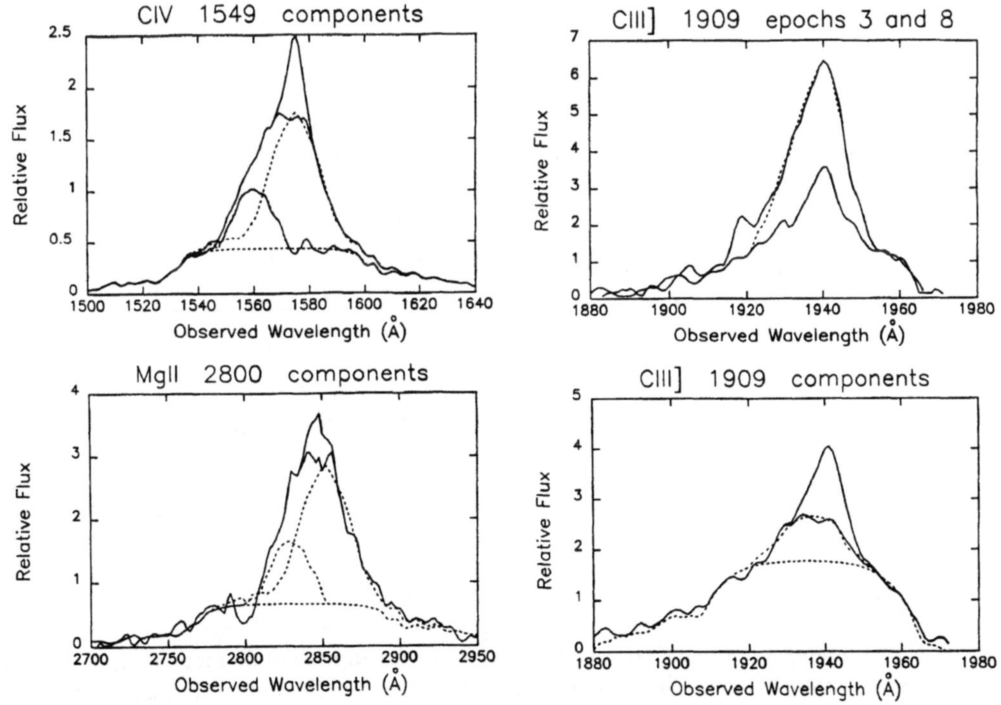

Figure 1: Decomposition of a) CIV 1549Å , b) MgII 2798Å , and d) CIII] 1909Å into narrow and broad components. Flux (in units of 10^{-14} ergs cm^{-2} s^{-1} Å^{-1}) plotted against observed wavelength. Solid lines show entire emission line (from total spectrum summed over all epochs), emission line after subtraction of narrow line component, and (for CIV) emission line after subtraction of narrow line and red broad components. Dashed lines show the individual broad components. 1c) shows the two spectra (epochs 3 and 8) used to construct the CIII] broad templates. Components used for MgII are the same as those derived for CIV, appropriately scaled in wavelength.

appropriate wavelengths. For CIV λ1549, we used three broad components – a red component derived from PKC's red Hβ profile, a very broad component, and a blue component. The very broad and blue components were derived by subtracting the red template from the total spectrum summed over all epochs, after shifting and scaling the template to match the red side of CIV. The same three components were fitted to MgII λ2798; this feature also shows MgII absorption from within our Galaxy (W_λ ~5Å). The CIV components were too broad to describe the CIII] λ1909 line profiles. For CIII], we used two broad components – a broad "main" component, and a very broad component. The "main" component was constructed from the difference between the spectra from epochs 8 and 3, when the "main" component was strong and weak, respectively, but the very broad component was about the same strength (as of course was the narrow line component). For HeII λ1640, we used a single very broad component, derived from PF's very broad HeII λ4686 profile.

The templates were fitted to each individual spectrum until minimum χ^2 were found. The scale factors and central wavelengths were allowed to vary, but it was not necessary to adjust the widths. Uncertainties in the data were estimated from differences in the contiguous images when possible, and from the scatter about the continuum fits. Deviations from the models were always comparable to or less than the uncertainties. Flux limits were

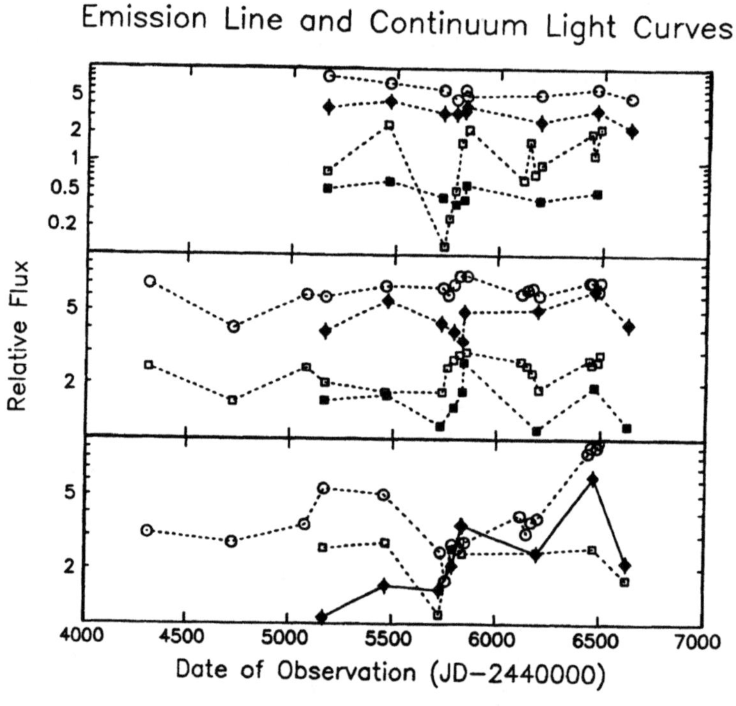

Figure 2: Light curves for continuum and broad line component fluxes. All line fluxes are given in units of 10^{-12} ergs cm^{-2} s^{-1}, continuum fluxes are given in units of 10^{-14} ergs cm^{-2} s^{-1} Å^{-1}.
a) Relative fluxes in very broad CIV (open circles), 1640Å feature (solid diamonds), and HeII 4686Å (open squares), and ratio of nonstellar 5500Å to 1775Å continuum fluxes (solid squares). CIV, 1640Å , and 4686Å fluxes, and continuum ratio were scaled by factors of 1.5, 40, 100, and 0.65, respectively. Average uncertainties are ~8% for very broad CIV, ~10% for the 1640Å feature, and ~30% for HeII 4686Å .
b) Relative fluxes in red Hβ (open circles) and red CIV (solid diamonds), and nonstellar continuum fluxes at 5500Å (open squares) and 1775Å (solid squares). Hβ, CIV, and continuum fluxes were scaled by factors of 10, 1.5, and 0.4, respectively. Average uncertainties for red CIV are ~8%.
c) Relative fluxes in blue Hβ (open circles), blue CIV (open squares), and "main" CIII] (solid diamonds). Hβ, CIV, and CIII] fluxes were scaled by factors of 25, 2.5, and 11, respectively. Average uncertainties are ~15% for blue CIV and ~40% for "main" CIII].

derived from the values of scale factor which increased χ^2 by 4.7, 3.5, or 2.3, corresponding to 1σ confidence for 4, 3, or 2 parameters of interest (*i.e.*, the scale factors), respectively. As expected, the narrow line component fluxes were consistent with constant values (5.1, 0.8, and 1.4 $\times 10^{-13}$ ergs cm^{-2} s^{-1} for CIV, MgII, and CIII]).

Figure 2 shows resulting light curves for some of the broad component fluxes. From these data, we derive the following conclusions:

1) The feature at (rest) 1640Å is almost certainly not (primarily) due to HeII, because its strength is approximately constant, even during early 1984 when the flux in HeII λ4686 changed by a factor of 8. At the time of minimum HeII λ4686 flux, the flux in the 1640Å feature was more than 40 times higher; from atomic physics we would expect a ratio of 10 or less (if there is reddening). We believe this feature to be primarily OIII] λ1663 blended with a well-known fixed pattern artifact at ~1660Å (Hackney, Hackney, and Kondo 1982). Interpreting the ~1640Å fluxes as upper limits to HeII, we derive $E_{B-V} \gtrsim 0.06$, in agreement with other reddening estimates (Wampler 1968; Barr, Willis and Wilson 1983; Ulrich and Boisson 1983; De Zotti and Gaskell 1985).

2) The HeII λ4686 fluxes are strongly correlated with the continuum flux at 1775Å, especially so during early 1984. The ratio of nonstellar 5500Å to 1775Å continuum fluxes also varied during this time. Together with the lack of large increases in very broad Hβ and HeI λ5876, this may imply an even larger change in the far ultraviolet continuum.

3) The light curves for the very broad components are similar, and follow the ultraviolet continuum fairly well. We do not see large changes corresponding to the 1984 change in HeII λ4686. This is difficult to understand, particularly so for CIV.

4) Red CIV (and red MgII) follow red Hβ, but blue CIV is not simply related to blue Hβ. The situation for blue MgII is unclear. The "main" component of CIII] seems to follow blue Hβ, at least after mid-1984, when blue Hβ became stronger. Broad CIII] may therefore be composed of two (or more) unrelated components, showing a structure similar to that of CIV, MgII, and Hβ, but narrower.

Clearly further studies and additional observations are required to unravel the structure, kinematics, and physical conditions in the broad line regions of this object.

Acknowledgments: Support for this work was provided by NASA Contract NAS5–28749 to the Computer Sciences Corporation, and by NASA Grant NAG5–826 and NSF Grant AST87–02691 to the Ohio State University.

References:

Barr, P., Willis, A. J., and Wilson, R. 1983, M.N.R.A.S., **203**, 201.

De Zotti, G., and Gaskell, C. M. 1985, Astron. Ap., **147**, 1.

Hackney, R. L., Hackney, K. R. H., and Kondo, Y. 1982, *Advances in Ultraviolet Astronomy: Four Years of IUE Research*, NASA Conference Publication 2238, 335.

Peterson, B. M. 1987, Ap. J., **312**, 79.

Peterson, B. M., and Ferland, G. J. 1986, Nature, **324**, 345.

Peterson, B. M., Korista, K. T., and Cota, S. A. 1987, Ap. J. (Letters), **312**, L1.

Ulrich, M. H., and Boisson, C. 1983, Ap. J., **267**, 515.

Wampler, E. J. 1968, Ap. J. (Letters), **154**, L53.

The Circumnuclear Environment of the Nearby Non–Interacting Seyfert Galaxies NGC 5273 and NGC 3516

Richard W. Pogge
Lick Observatory, Santa Cruz, California, U.S.A.

Introduction

The conditions that prevail in the regions immediately surrounding Seyfert nuclei are a matter of considerable speculation. I have made observations of a sample of 20 nearby non–interacting Seyfert galaxies, combining CCD interference filter images isolating the bright emission lines of Hα+[N II]$\lambda\lambda$6548,6583 and [O III]λ5007, with long–slit low–resolution spectrophotometry to determine the spatial distribution and physical conditions present in ionized gas regions (if any) surrounding the Seyfert nuclei. All data were obtained with the 1 meter Nickel telescope and the 3 meter Shane telescope at Mt. Hamilton. To provide a basis for comparison with non–Seyfert spirals, a CCD interference filter imaging survey of the central regions of a statistically significant sample of bright Shapley–Ames spirals meeting the same criteria of non–interaction has been done along side the Seyfert observations.

This paper presents preliminary results for two galaxies which are representative of the range of phenomena encountered among the Seyferts in my non–interacting sample. The full analysis is in progress.

NGC 5273

NGC 5273 is an intermediate type Seyfert 1.5 galaxy (Huchra, Wyatt, and Davis 1982). The continuum subtracted Hα+[N II] and [O III]λ5007 CCD images obtained at Lick show no evidence of extended structure. The profiles of the nuclear light are indistinguishable from the profiles of field stars on the images. A lack of extended circumnuclear structure is a common feature of the 9 Seyfert 1 and 1.5 galaxies observed in this study. Rich structures like those seen in NGC 3516 (below) and NGC 4151 (Heckman and Balick 1983) seem to be exceptional cases.

NGC 3516

NGC 3516 is a Seyfert 1 (Seyfert 1943) in an SB0 host galaxy (Adams 1977). In Adams' ultraviolet and redshifted Hα filter photos there is a hint of an extended emission region lying nearly perpendicular to the stellar bar, described as a "curved filament" reaching about 10$''$ to the southwest. Ulrich and Pequinot (1980) found evidence of non–rotational motions in the extended emission region NE of the nucleus and that the extended gas is in a rather high ionization state, which they have modeled as photoionization by the nuclear power–law continuum. Goad and Gallagher (1987) have found similar evidence of non–rotational motions.

Both the continuum subtracted Hα+[N II] and [O III]λ5007 CCD images obtained at Lick (Fig. 1) have the same basic morphology: a fat, stubby spiral of emission extending to 10″ from the nucleus, with only a slight indication of clumpiness. Assuming that the Seyfert 1 nucleus is an unresolved point source, 75% of the [O III]λ5007 line emission flux arises from the extended regions.

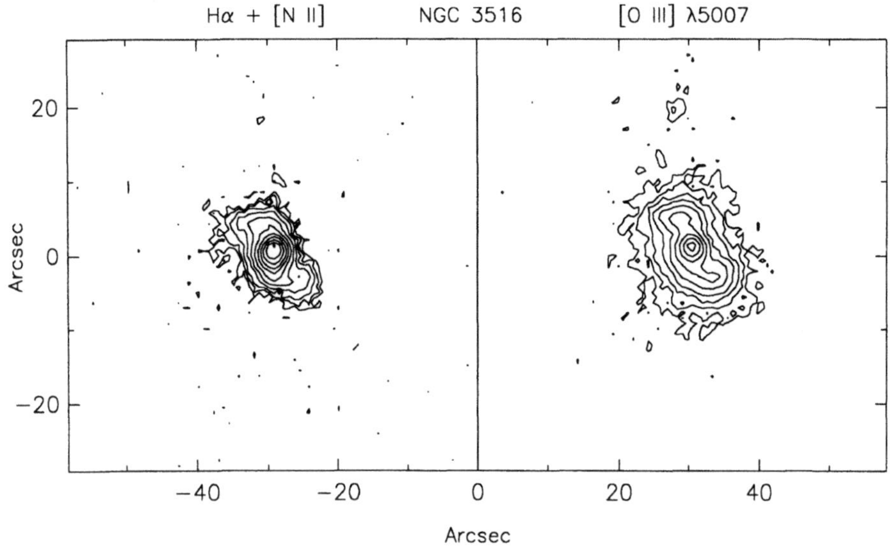

Figure 1

The nuclear spectrum (Fig. 2a) is typical of a Seyfert 1. The spectra of the extended regions shows many high ionization lines, including He II λ4686 and weak [Ne V]λ3426 (Fig. 2b). With the exception of the rather weak [O I]λ6300 line, the line ratios are typical of a Seyfert 2 nucleus. A synthetic spectrum approximating what would be seen if a 16″ round aperture were used is in Fig. 2c. This is equivalent to a spectrum obtained with a conventional 4″ aperture if the galaxy were 4 times further away. This spectrum is typical of an intermediate–type Seyfert 1.5 nucleus.

Some mysteries remain. The spiral shape of the extended nebulosity is similar to what is seen in density maps obtained from numerical simulations of gas flow in a stellar bar potential (*e.g.*, Sanders and Tubbs 1980). Alternatively, the shape is suggestive of a bent radio jet, possibly due to shearing of the radio plasma as it rams into the ISM of the rotating bar. Wilson and Ulvestad (1983) list NGC 3516 as an unresolved radio source without extended structure. A deeper, higher dynamic range VLA map is clearly needed. Unfortunately, no detailed kinematical data are available. A velocity field measured using either Fabry–Perot or long slit, high resolution spectral data would be most interesting. Interpretation will be tricky, as non–rotational motions are not indicative *a priori* of outflow, as in a radio jet, as shown by work on gas motions in bar potentials.

Figure 2 - Spectra of NGC 3516.

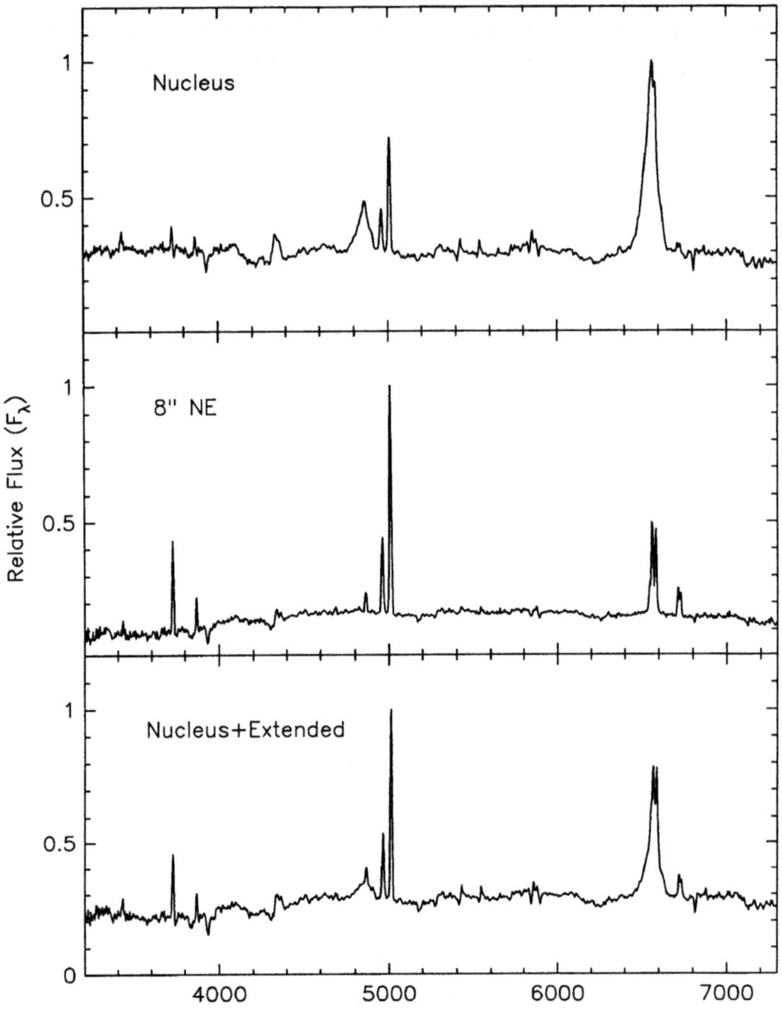

References

Adams, T.F. 1977, *Ap.J.Suppl.*, **33**, 19.

Goad, J.W. and Gallagher, J.S. 1987, *A.J.*, **94**, 640.

Heckman, T. and Balick, B. 1983, *Ap.J.*, **268**, 102.

Huchra, J.P., Wyatt, W.F., and Davis, M. 1982, *A.J.*, **87**, 1628.

Sanders, R.H., and Tubbs, A.D. 1980, *Ap.J.*, **235**, 803.

Seyfert, C.K. 1943, *Ap. J.*, **97**, 28.

Ulrich, M.–H. and Pequignot, D. 1980, *Ap.J.*, **238**, 45.

Wilson, A.S., and Ulvestad, J.S. 1983, *Ap. J.*, **275**, 8.

LONG-SLIT SPECTROSCOPY OF STAR-BURST GALAXIES

Richard A. Shaw
Lick Observatory, University of California
Santa Cruz, CA 95064
and
Michael M. De Robertis
Dept. of Physics, York University
Downsview, Ont. M3J-1P3

ABSTRACT

We present high-resolution, long-slit spectra of nine optically bright nuclear star-burst galaxies (SBGs) selected from the sample of Balzano from which we measure line strengths, line widths, and velocity profiles along the slit in the Hα, [N II], and [S II] emission lines. We show that these galaxies have strong nebular emission (and presumably star formation) extending over several hundred to more than a thousand parsecs from their centers. We find that the peak of the continuum and line emission profiles are displaced from the center (as defined by the rotation curve) in many of the galaxies. Gas densities and ionization levels in the SBGs fall in the range expected for giant H II region complexes, although the line widths are often much larger. The velocity profile in each galaxy is consistent with a rotation curve in which emission-line material is confined to a disk coaligned with that of the parent galaxy. The star-burst therefore likely consists of from tens to thousands of massive H II regions rather than one giant, spherical complex. The line widths would then result from an intrinsic velocity dispersion and rotational broadening.

Though SBGs and active galaxies (AGNs) occur in similar environments and have similar space densities, we find little similarity (spectroscopically or morphologically), nor any direct evidence for an evolutionary link between them. Timescale arguments suggest that if such a link does exist, then the AGN phase must succeed the SBG phase, though probably not in the manner suggested by Weedman.

INTRODUCTION

Since they were first examined as a class, SBGs have attracted much attention as galaxies with intense nuclear activity and as possible precursors of AGNs (see *e.g.* Weedman *et al.* 1981; Feldman *et al.* 1982; Balzano 1983; Weedman 1983). Indeed, many similarities have been reported. Weedman (1983) pointed out that the space density for optically bright SBGs is roughly the same as for Seyfert galaxies. Balzano (1983) noted that about 30% of SBGs apparently have companion galaxies, which is similar to that found for AGNs but significantly higher than that for normal galaxies. In addition, several authors have found prominent star formation activity

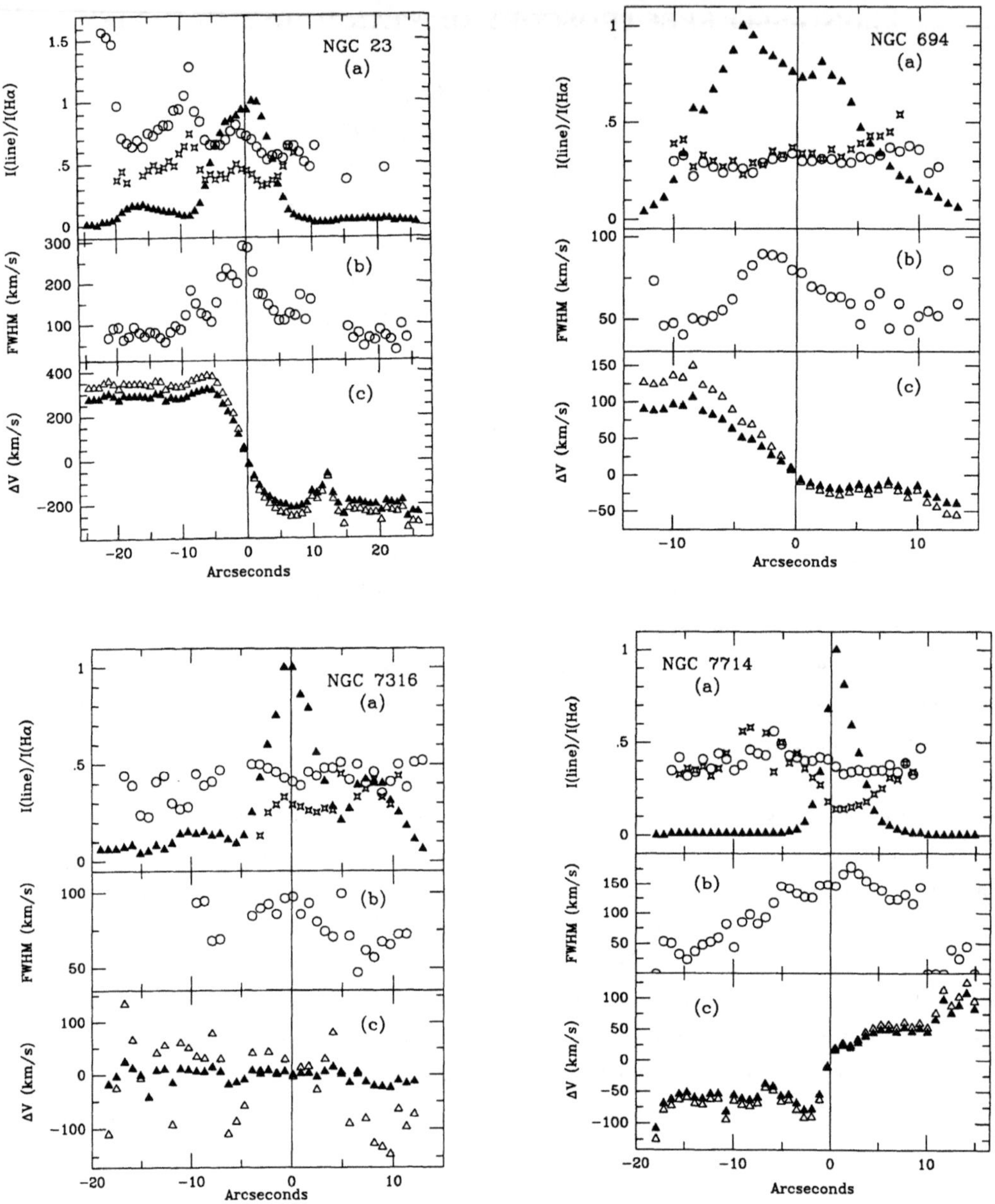

Part *a*. Emission-line intensities of [N II] λ6583 (open circles) and [S II] $\lambda\lambda$6716+6731 (open stars), relative to Hα (shown as filled triangles, and normalized to Hα_{max}) *vs.* radius in arcsec for 4 SBGs. North is to the left, and south is to the right. Vertical line indicates the maximum in the continuum emission profile.

Part *b*. Corrected FWHM for Hα *vs.* radius in arcsec. Vertical line as in Part *a*.

Part *c*. Hα velocity as measured (filled symbols), and corrected for projection (open symbols), *vs.* radius in arcsec. Vertical line as in Part *a*.

surrounding the nuclei of Seyferts. While it is well known that the activity in AGNs is probably confined to a volume less than 10 parsecs in diameter, the spatial extent of star formation in SBGs has not been well studied. According to Balzano (1983) the most dominant morphological feature of SBGs, as for AGNs, is a bright, star-like nucleus.

In an effort to examine possible connections between SBGs and AGNs, we have obtained high-resolution, long-slit CCD spectra in the (redshifted) Hα region of 9 SBGs (of which 4 are presented here) selected from Balzano's (1983) survey, including the prototype, NGC 7714 (Weedman *et al.* 1981). We analyzed several properties, including the kinematics of the emission-line material through the measurement of line widths, intensity and line profiles, and radial velocity profiles.

MAJOR CONCLUSIONS

- Star formation is not confined to the centers of SBGs, but extends over many hundreds or even thousands of parsecs.
- The gas density in the S^+ emitting volume, and the level of ionization are both consistent with those found in giant H II region complexes.
- Evolution schemes which predict the AGN and SBG phases to occur contemporaneously also imply the existence of composite systems. And yet large differences exist in emission-line widths, levels of ionization, and infrared colors between AGNs and SBGs, although the median L(Hα) of SBGs and Seyfert 2s are nearly the same. While bursts of star formation sometimes accompany an active nucleus, the features seem to be dominated by the active nucleus.
- Weedman's (1983) model for the evolution of a SBG into an AGN does not appear to be viable, since he assumed a typical dimension for the nuclear star-burst which is a factor of 5 to 10 too small. The gas and massive stellar remnants must therefore remain in equilibrium with the galaxy potential, and no collapse to a small (1 pc^{-3}) volume, followed by the onset of nuclear activity, will take place.
- The velocity profiles (especially that for the nearly face-on NGC 7316) suggest that the star formation occurs in a disk-shaped volume which is nearly coaligned with the galactic disk and is supported by rotation. The star-forming disk would contain up to 10^3 separate, massive H II region complexes.

REFERENCES

Balzano, V. 1983, *Ap. J.*, **268**, 602.

Feldman, F. R., Weedman, D. W., Balzano, V. A., and Ramsey, L. W. 1982, *Ap. J.*, **256**, 427.

Weedman, D. W. 1983, *Ap. J.*, **266**, 479.

Weedman, D. W., Feldman, F. R., Balzano, V. A., Ramsey, L. W., Sramek, R. A., and Wu, C.-C. 1981, *Ap. J.*, **248**, 105.

VOYAGER FAR UV OBSERVATIONS OF MARKARIAN 509

T. E. Carone
Department of Physics and
Lunar and Planetary Laboratory-West
University of Arizona

and

M. A. Malkan
Department of Astronomy, UCLA

ABSTRACT

Voyager far ultraviolet (500-1700 Å) observations of the Seyfert 1 galaxy Markarian 509 are presented. Markarian 509 was observed for 4.2 hours on day 87/014, for 8.8 hours on day 87/016, for 2.6 hours on day 87/019 and for 4.2 hours on day 87/023 with the ultraviolet spectrometer on board the Voyager 2 spacecraft. The average flux at 1050 Å during these observations of Markarian 509 was 3.67 x 10^{-12} ergs cm^{-2} sec^{-1} $Å^{-1}$. This flux value is approximately ten times higher than the brightest IUE SWP image of Markarian 509. The continuum of both observations is similar and is flat throughout the far UV spectral region and no flux shortward of 912 Å is seen. This result for Markarian 509 implies that either this galaxy was unexpectedly bright during the Voyager observations or that the spectrum measured is instead due to a nearby luminous accretion disk system, such as a cataclysmic variable. However, an optical survey of the region has not produced any potential candidates.

INTRODUCTION

Since launch, the ultraviolet spectrometers (UVS) on board both Voyager spacecraft have been used to conduct astronomical observations from 500-1700 Å of hot stars, subluminous stars, cataclysmic variables, low-mass and massive x-ray binaries and, recently, active galaxies. Voyager 2 successfully observed 3C 273 (Reichert et al. 1987) and found that the quasar showed a significant drop in flux below 1060 Å; 1056 Å corresponds to 912 Å in the rest frame of 3C 273. This drop is consistent with an absorption edge in an accretion disk within the nucleus of 3C 273 but it is also *not* inconsistent with a turnover in the underlying continuum.

The Voyager active galaxy program has been expanded to other well-known Seyferts and quasars. Observations of NGC 4151, Fairall 9, Markarian 279, ESO 141-G55 and PG 1211+143 have been obtained in addition to Markarian 509. There is a possible detection of Fairall 9 and for now we have just upper limits of 2.0 x 10^{-13} ergs cm^{-2} sec^{-1} $Å^{-1}$ for PG 1211+143, Markarian 279 and NGC 4151. The observations of ESO 141-G55 were just done and the data is not yet in hand. In this paper we present the results of the Voyager far UV observations of Markarian 509.

OBSERVATIONS

The UVS and their in-flight performance are described elsewhere (Broadfoot et al. 1977; Broadfoot et al. 1981). The UVS on each spacecraft are nearly identical objective grating spectrometers which cover the 500-1700 Å spectral range at a dispersion of 9.26 Å per detector channel with an effective spectral resolution of 15 Å for point sources and 20-25 Å for diffuse sources. A mechanical collimator defines the field of view. The field of view has a full-width at half-maximum of 0°.097 in the dispersion direction and 0°.87 in the cross-dispersion direction. The instrument sensitivity is highest shortward of 1200 Å. The absolute calibration used here is described elswhere (Holberg et al. 1982).

Observations of Markarian 509 were obtained in the high-rate data mode in which a spectrum was taken every 3.84 seconds. During these observations the spacecraft scan platform was fixed in the direction of the galaxy while the spacecraft limit cycle motions allowed the UVS field of view to wander on and off the target. Using information from the spacecraft attitude control system, it is possible to sort spectra obtained at various positions in the limit cycle into spatial bins along the dispersion direction. Spectra acquired while each galaxy was on axis can be accurately calibrated. In determining absolute fluxes for the high data rate only spectra obtained within ±0°.025 of the center of the field of view were used.

For faint sources such as active galaxies, an accurate subtraction of the background is vital. The UVS background consists of two components, a continuous one from the radioisotope thermoelectric generator and a line spectrum (Lyman α, Lyman β, Lyman γ and He I λ504 Å) from the backscattered solar flux. A half millon second observation of the background was obtained from observations done before and after the galaxy observations and was used in the processing of the Markarian 509 observations. A more detailed description of these reduction techniques are discussed elsewhere (Drilling et al. 1984; Polidan and Holberg 1987).

RESULTS

The far UV spectrum of Markarian 509 is shown in Figure 1. The continuum is flat through-out the 912-1150 Å spectral region and the average flux at 1050 Å is 3.67 x 10^{-12} ergs cm^{-2} sec^{-1} $Å^{-1}$ (the excess flux just shortward of 912 Å is due to background subtraction inaccuracies). The spectrum has been dereddened using E(B-V) = 0.06. The flux level found is ~ 10 times that of 3C 273 (Reichert et al. 1987).

The flux level detected for Markarian 509 was unexpected - it is about 10 times higher than its brightest IUE SWP image so that this flux level is unlikely to be a normal one for Markarian 509. There are two possibilities. The first is that Markarian 509 was in a high state for at least 10 days and Voyager was fortunate enough to be on it. This high a flux level has never been seen by IUE or by ground-based observations. On the other hand, there may be a nearby luminous accretion disk system, such as a cataclysmic variable, which was the true object detected by Voyager. The Voyager spectrum of Markarian 509 does not resemble a hot star but it does resemble the luminous accretion disk systems which have been observed with Voyager in the past. In July of this year, an optical survey of the Markarian 509 region was done using the TI 3PCCD and the Boller and Chivens spectrograph on the University of Arizona 2.3 meter. This survey turned up no viable candidates whose far UV spectrum would resemble that observed. For the future, two approaches will be done. During the month of January, 1988, we plan to monitor Markarian far ultraviolet spectrometer. IEH is an objective grating spectrometer containing two gratings. The first grating covers the extreme UV (400-860 Å) and is blazed at 600 Å. The second grating covers the far UV (840-1300 Å) and is blazed at 1075 Å. IEH uses two intensified CCDs (one for each grating), has a spectral resolution of 1 Å and a spatial resolution of 4.5 arcseconds.

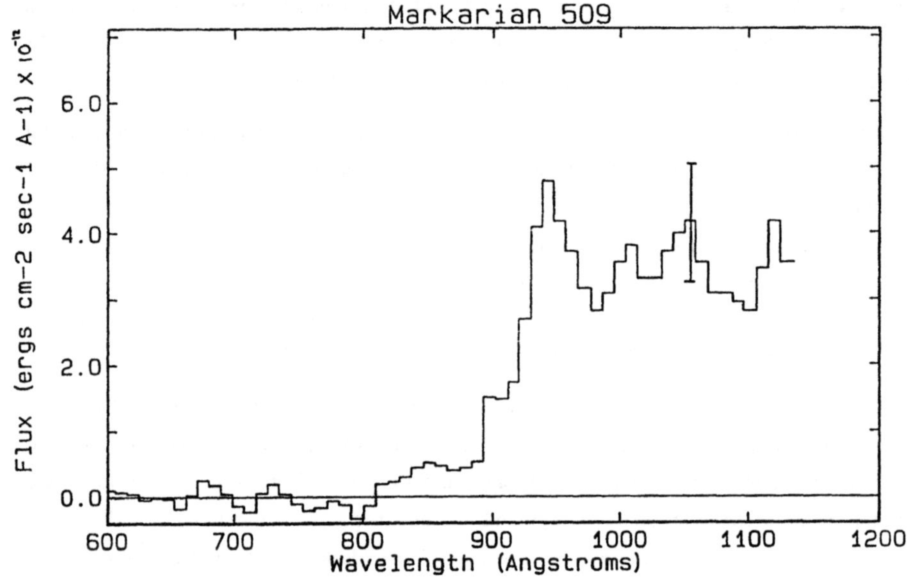

Figure 1

REFERENCES

Broadfoot, A.L. *et al.*, 1977, *Space Sci. Rev.*, **21**, 183.
Broadfoot, A.L. *et al.*, 1981, *J. Geophys. Res.*, **86**, 8259.
Drilling, J.S. *et al.*, 1984, *Astrophys. J.*, **283**, L67.
Holberg, J.B. *et al.*, 1982, *Astrophys. J.*, **257**, 650.
Polidan, R.S. and J.B. Holberg, 1987, *MNRAS*, **225**, 131.
Reichert, G.A., R.S. Polidan, C.-C. Wu and T.E. Carone, 1987, *Astrophys. J.*, **325**, in press.

HIGH RESOLUTION STUDY OF NGC 4151

Sylvain Veilleux
Lick Observatory, Santa Cruz, California, 95064

A) Observations

The classic Seyfert galaxy NGC 4151 was observed with the new Hamilton echelle spectrograph (Vogt 1987) installed at the coudé focus of the Shane 3-meter telescope. The spectral region covered by the data ranges from 3700 Å to 7600 Å at an average resolution of about 10 km/sec. Four setups were required to cover this spectral range and the total exposure times were typically of two hours. Although the galaxy was observed through a 2.5" x 2.5" aperture the effective size of the aperture was set by the poor seeing and guiding errors to about 4". The data were reduced using a reduction package developed at Lick where each image is flat-fielded to remove the pixel-to-pixel variations, the spectrum of each order is extracted and the background (= dark count + scattered light) removed. A short-decker flat was used to remove the fringing effects longward of about 6000 Å. These data were then wavelength calibrated using a thorium-argon lamp spectrum. The resulting accuracy of this calibration is approximately 0.02 Å. The spectrum of a standard star was obtained with the same setup as the galaxy so it will be possible in the near future to flux the spectra and merge them together. This has not been done for the present data. Consequently, some residual low-frequency color terms are probably still present in the data and may affect the relative intensity of lines within the same order.

B) Results

Using the line list of Boksenberg *et al.* (1977) nearly fifty emission lines were identified in the present data. Some of these emission lines are presented in figures 1 through 6. We note that the blue and red "shoulders" already observed in [O III] λ5007 by Pelat and Alloin (1982) and in Hα + [N II] by Kielkopf *et al.* (1985) are present in essentially all the lines. These features are attributed to Ulrich's (1973) clouds II and III (Pelat and Alloin 1982). The relative intensity of these components may vary from one line to the other but it is not possible to make any definite statement before the data are fluxed. Aperture effects may also affect the relative intensity of these components although data taken a few weeks apart and under different atmospheric conditions do not show any apparent differences in the line profiles.

A close inspection of the profile of [O III] λ5007 (Fig. 6) shows that most of the features identified by Pelat and Alloin are also present in our data. [Fe VII] λ4989.4 is detected near λ5004 and so is the feature around λ5010 which they attribute to Fe II λ4993.35. The redshifted component at λ5028 in the [O III] λ5007 profile and λ4980 in the [O III] λ4959 profile is produced by Ulrich's cloud IV. It is also observed in the red wing of Hα, [O I] λ6300 (partially blended with [S III] λ6312), Hβ, [N II] λ6583, and [S II] λ6731. [Fe III] λ5032.7 is not present in our data and Fe II λ5018.43 is very weak if at all present. Figure 4 shows the [O I] λ6364 + [Fe X] λ6375 blend. Using [O I] λ6300 as a template for [O I] λ6364 (after removing of course the [S III] λ6312 contribution), valuable profile information on the high-ionization line [Fe X] λ6375 will become available.

No absorption features were detected in the blue wing of Hβ (Fig. 5). This agrees with the data of Kielkopf *et al.* (1985) and confirms the well-known variability of these absorption features (Cromwell and Weymann 1970). Any statement about Hα absorption will have to wait the fluxing of the Hα + [N II] blend and the measurement of [N II] λ6583/λ6548 (Schulz 1987). Hγ absorption may be present near λ4340 but is very uncertain. The low signal-to-noise of our data near λ4000 prevents us to make any comments about the presence of He I λ3889 absorption lines.

This study is part of a long-term project involving approximately twenty bright Seyfert galaxies. The kind of substructures observed in the profile of the lines of NGC 4151 is also observed in many other galaxies such as NGC 1068 (extreme case), NGC 4051, NGC 4388, NGC 5506, Mrk 3, Mrk 533, Mrk 1066, and Mrk 1073 to name just a few.

Acknowledgements. I am very grateful to Dr. D. E. Osterbrock for numerous discussions and to Dr. R. A. Shaw for assistance in obtaining these data. I thank the members of the Time Allocation Committee for their generosity, as well as the Natural Sciences and Engineering Research Council of Canada for financial support. This work has also been partially supported by the NSF grants AST 83-11585 and AST 86-14510.

References

Boksenberg , A., Shortridge, K., Allen, D. A., Fosbury, R. A. E., Penston, M. V. and Savage A. 1975, MNRAS, **173**, 381.

Cromwell, R. and Weymann, R. 1970, Ap. J. Letters, **159**, L147.

Kielkopf, J., Brashear, R. and Lattis, J. 1985, Ap. J., **299**, 865.

Pelat, D. and Alloin, D. 1982, Astr. Ap., **105**, 335.

Schulz, H. 1987, Astr. Ap., **178**, 7.

Ulrich, M.-H. 1973, Ap. J., **181**, 51.

Vogt, S. S. 1987, Pub. A. S. P., **621**, in print.

FIGURE 1

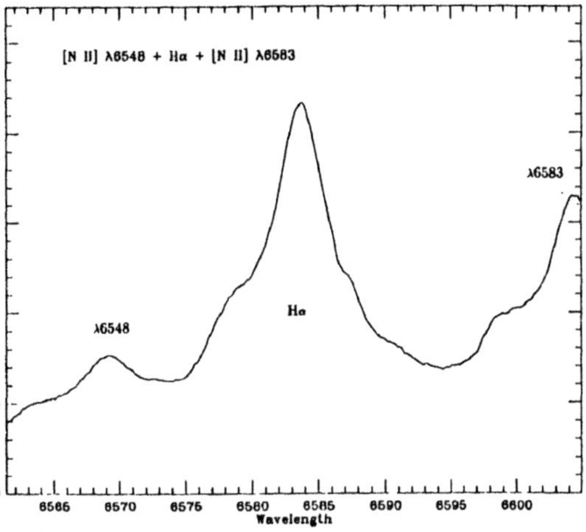

FIGURE 2

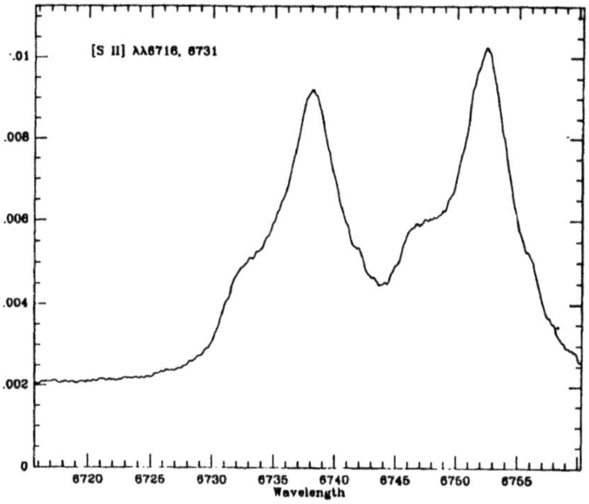

FIGURE 3

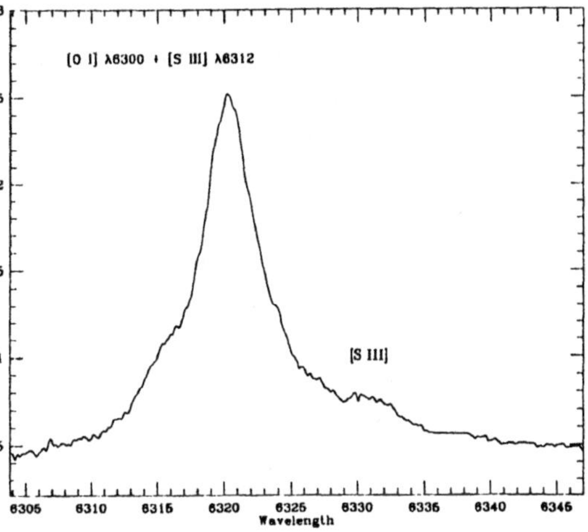

FIGURE 4

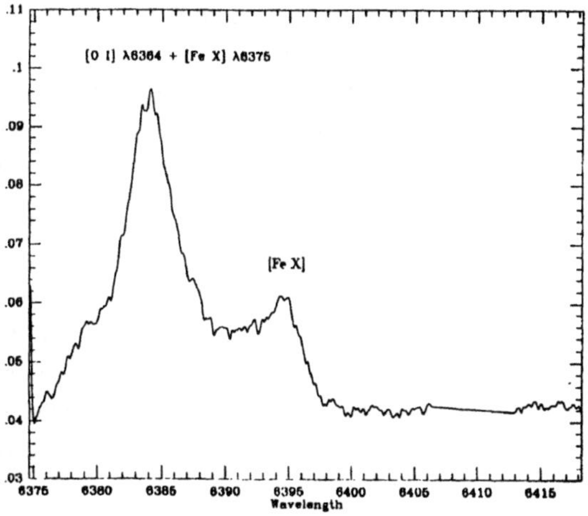

FIGURE 5

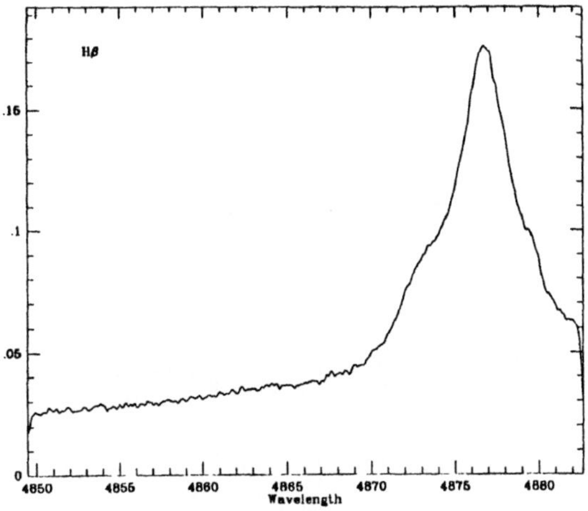

FIGURE 6

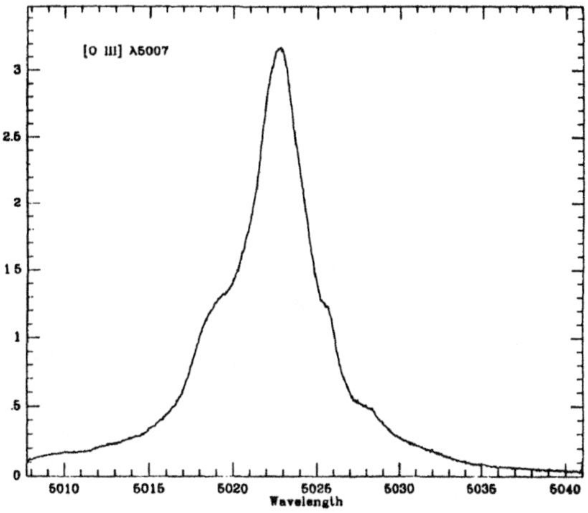

Figures: 1. Hα + [N II]λλ6548, 6583; 2. [S II] λλ6716, 6731; 3. [O I] λ6300 + [S III] λ6312; 4. [O I] λ6364 + [Fe X] λ6375; 5. Hβ; 6. [O III] λ5007.

LONG-SLIT ECHELLE SPECTROGRAMS OF SEYFERT GALAXY NUCLEI

Jean W. Goad
NOAO/ADP, P. O. Box 26732, Tucson, AZ 85726 USA

John S. Gallagher, III
Lowell Observatory, Mars Hill Road, 1400 W., Flagstaff, AZ 86001 USA

Observational Material

Long-slit spectrograms have been obtained for ten Seyfert nuclei with the Kitt Peak National Observatory echelle spectrograph on the 4-meter Mayall Telescope. We included seven Seyfert 1s (NGC 3227, NGC 3516, NGC 7469, Mk 6, MCG 8-11-11, NGC 7603, and 3C 120) and three Seyfert 2s (Mk 3, Mk 348, and Mk 78). All observations were made in December 1984, with an RCA C33063 image intensifier and IIIa-J plates. With the 79 l/mm echelle and 226-1 cross-disperser, the spectra were covered from approximately 4700 Å to 7000 Å at a resolution of ~20 km s^{-1}.

We show here sections of two of the echellograms, a 90-minute exposure of Mk 3 (PA 195°) and a 50-minute exposure of NGC 7469 (PA 176°). To the left of each is a photograph of the galaxy reproduced from the *Palomar Observatory Sky Survey* plates. The upper section of each echellogram shows five orders including the [S II]λλ6717, 6731 doublet, Hα, [N II]λλ6548, 6583, and (in the case of Mk 3) the [O I]λλ6300, 6364 lines. The lower section includes [O III]λλ4959, 5007 and Hβ, but note that for NGC 7469 the λ4959 line is vignetted at the righthand side and a very broad Hβ is vignetted at the left. Within each order south is at the top, and the small thorium-argon comparison dots are separated by 21.25 arc seconds.

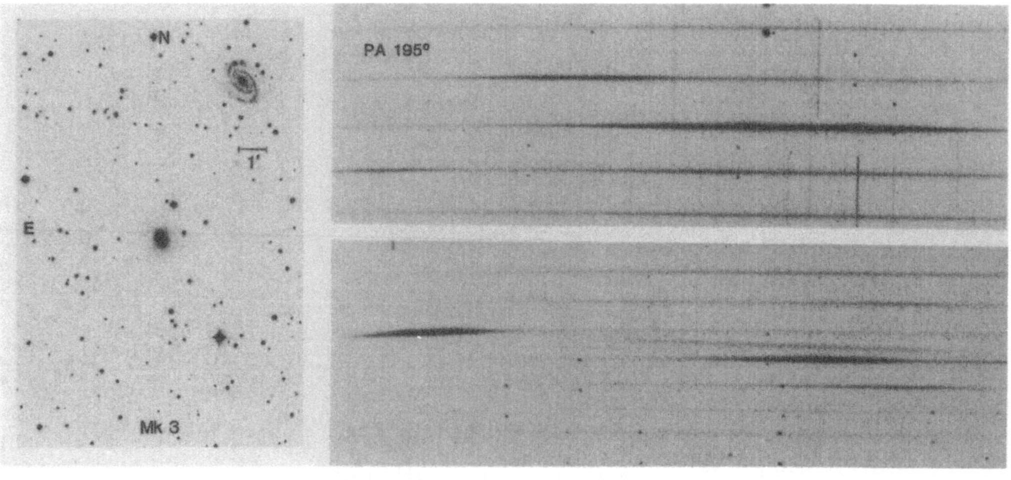

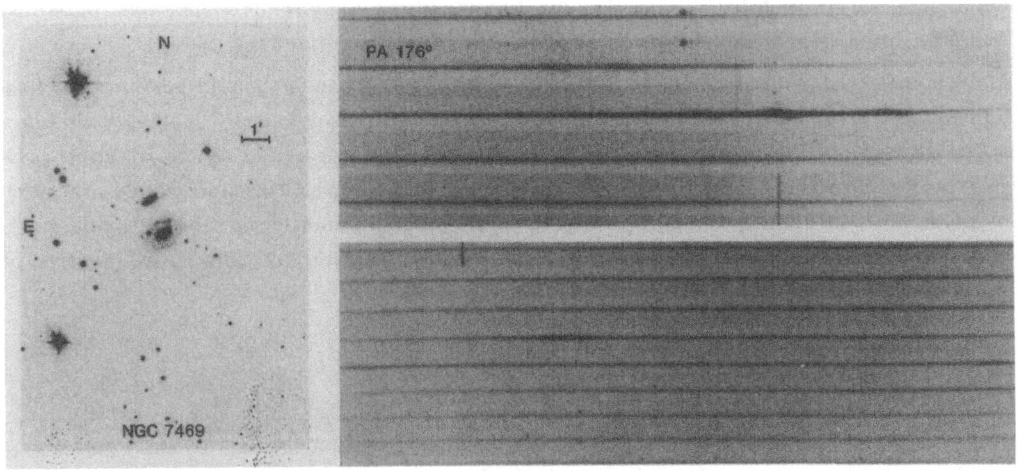

The Virtues of Echelle Spectroscopy

Compared to single-grating spectrographs, the echelle offers two obvious advantages. First, the high spectral dispersion sets the narrow emission lines apart from the very strong nuclear continuum and broad emission lines. A clear comparison is available for NGC 3516, for which an echellogram is published by Goad and Gallagher (1987, Fig. 1 [Plate 59]) and single-grating spectrograms are published by Ulrich and Péquignot (1980, Figs. 2 and 3 [Plates 3 and 4]). [*Note*: The orientation of the slit given by Goad and Gallagher is in error by 180°. Every "NE" should be replaced by "SW," and vice versa.]

The second advantage of echelle spectroscopy is its broad spectral coverage: a single exposure displays the differences between the blue-green and the red lines, as in NGC 7469. For this we require the 2×2-inch format of a photographic plate, which of course has the standard nonlinearities. For spectrophotometry, one would do well to trade off the spectral coverage for the linearity and greater dynamic range of a CCD.

Kinematical Information

Echelle spectroscopy excels in showing the kinematics of the circumnuclear gas. Two of the Seyferts we observed have no evidence of circumnuclear emission lines (Mk 6, NGC 7603) but the other eight show some extranuclear extension of the narrow-line region (NLR) or extended narrow-line region (ENLR; see De Robertis and Pogge 1986). The two most obvious interpretations of the ENLR kinematics are disk-like rotation or radial flows. NGC 3516 evidently shows both (Goad and Gallagher 1987). Anderson (1973) has made a convincing case that the ENLR in NGC 7469 is primarily in rotation rather than expansion or contraction.

The spectra of Mk 3 and NGC 7469 are obviously quite different: NGC 7469 is a Seyfert 1 while Mk 3 is a Seyfert 2 (but may contain a hidden type 1 region; see Miller 1988); and Mk 3's nuclear emission lines are doubled and asymmetric (Weedman 1970, Whittle *et al.* 1988). Yet the ENLR in Mk 3 appears kinematically similar to the ENLR in NGC 7469. The tilt of the circumnuclear lines is most easily seen by viewing the spectrogram edge-on, along the dispersion. Unlike NGC 7469, Mk 3 has visible extensions from the [O III] lines as well as from Hα and [N II].

The character of the motions in Mk 3 (rotation or outflow/inflow) could best be established by mapping the velocities in several position angles as Anderson did for NGC 7469. Whittle *et al.* suggest that in Mk 3 the "outer narrow emission defines an apparently normal rotation curve," while De Robertis and Shaw's (1988) models suggest that the NLR emission clouds are moving predominantly radially inward. Both processes could be going on at once. The accumulating evidence favors the presence of at least one type of organized motion in the inner few hundred parsecs of most Seyfert nuclei.

Acknowledgments

We thank Glynn Pickens, Joyce DuHamel, and Mark Hanna for their expert work in the NOAO photographic laboratory.

References

Anderson, K. S. *Astrophys. J.*, **182**, 369-380 (1973).

De Robertis, M. M., and Pogge, R. W. *Astron. J.*, **91**, 1026-1037 (1986).

De Robertis, M. M., and Shaw, R. A. In *Active Galactic Nuclei: Proceedings of the Georgia State University Conference* (this volume) (1988).

Goad, J. W., and Gallagher, J. S., III. *Astron. J.*, **94**, 640-643 (1987).

Miller, J. S. In *Active Galactic Nuclei: Proceedings of the Georgia State University Conference* (this volume) (1988).

Ulrich, M.-H., and Péquignot, D. *Astrophys. J.*, **238**, 45-53 (1980).

Weedman, D. W. *Astrophys. J.*, **159**, 405-413 (1970).

Whittle, M., Pedlar, A., Meurs, E. J. A., Unger, S. W., Axon, D. J., and Ward, M. J. Preprint (1988).

DOUBLE PEAKED BROAD LINE PROFILES -- EDGE ON ACCRETION DISKS OR DOUBLE QUASAR NUCLEI?

C. Martin Gaskell

Astronomy Department, Dennison Building, University of Michigan,
Ann Arbor, Michigan 48109-1090.

ABSTRACT

It is pointed out that a significant fraction (over 50%) of quasars do not show smooth symmetric "logarithmic" broad line profiles, but instead show double broad line peaks, usually unequal in brightness, always one blueshifted and one redshifted relative to the systemic velocity. The problems with the former explanations of obscuration effects or variability are summarized and the case is made for there being *two* active objects in these galactic nuclei. It is argued that line variability data (and other considerations) firmly rule out the recently suggested alternative hypothesis that these double peaked broad line objects are due to edge on accretion discs.

INTRODUCTION

At a cursory glance the majority of quasar broad line profiles seem to be rather simple. Over 12 years ago Blumenthal and Mathews (1975) showed that radiative acceleration naturally produces line profiles that are logarithmic (i.e. $I(\Delta\nu) \propto log(\Delta\nu)$) over most of their profiles and which closely match many observed profiles. Capriotti, Foltz and Byard (1980) showed that accretion of optically thick clouds, and a limited number of other scenarios, could also produce logarithmic profiles. Important requirements, however, were (approximate) spherical symmetry and radial motion of the cloudlets and Capriotti *et al* found in particular that "models with rotation and/or expansion confined to a disk seem to be excluded on the basis of observed profile shapes". The discovery of the blueshifting asymmetry of high ionization lines (Gaskell 1982) requires that at least the high ionization cloudlets be moving radially.

Although the majority of quasar broad lines *are* symmetric and "logarithmic", there are some glaring exceptions and if we are to really understand quasar broad line gas motion these too have to be understood. I will argue in this paper that these exceptions are readily understood as active nuclei in which there are *two* quasars.

OCCURRENCE OF DISPLACED BROAD LINE PEAK OBJECTS

Apart from a few cases of self-absorption (e.g. in broad absorption line quasars) I believe that these exceptions consist of either a single broad line region (BLR) peak displaced in velocity from the narrow line region (NLR) redshift, or (much less frequently), two BLR peaks (usually of unequal strength), one blueshifted and the other redshifted with respect to the NLR. In the most obvious cases the peaks are shifted by several thousand km s^{-1}. The most spectacular cases tend to be what are usually called "radio galaxies" (e.g. 3C390.3 - Sandage 1966, Lynds 1968), but it must be emphasized that they occur in *all* members of the quasar family. Examples include: the normal Seyfert galaxy NGC 5548 (Anderson 1971; Peterson, Korista and Cota 1987), the low redshift (z = 0.2109) quasar 2141+175 = OX 169 (Gaskell 1981) and the high redshift (z = 1.970) quasar PKS 0119-046 (Gaskell 1983a).

In a systematic study of the wavelengths of broad and narrow lines (Gaskell 1982) it was found that a large fraction of high luminosity flat radio spectrum quasars in complete samples had broad lines at very different redshift from the systemic velocity. More recent work on the same intermediate resolution data (Gaskell 1988) suggests that at least 73% of such quasars have the peaks of their broad lines differing in redshift from the narrow lines by more than 270 km s^{-1} and 19% by over 1000 km s^{-1}. These percentages might seem surprisingly large but it should be remembered that a typical quasar spectrum with a resolution of 9A at λ5000 has a velocity resolution of 540 km s^{-1} and that the Balmer lines have strong NLR components which mislead the eye at this resolution. In the UV (i.e. at high redshifts) there are no strong forbidden lines and the permitted NLR lines suffer from heavy extinction so it is difficult to determine the systemic velocity and so identify displaced BLR peak objects (the solution to this problem is to do spectroscopy in the near IR -- as has been done by Junkkarinen 1986 who confirms that the peaks of the UV lines of high redshift quasars frequently do not correspond to the systemic velocity).

THE BINARY MODEL

Initially these asymmetric line profiles were explained as either obscuration effects (Ferland, Netzer and Shields 1979; Capriotti, Foltz and Byard 1981) or the result of light echoes in the line profiles as a result of variability of the ionizing continuum (Capriotti, Foltz and Peterson 1982). Gaskell (1983b), however, showed that these explanations did not explain the general phenomenon of the displaced peaks. The problem with obscuration explanations was that blue and red displaced peaks are equally prevalent. In obscuration models this requires 50% of quasars to have BLR inflow and 50% to have BLR outflow -- an unlikely situation. The problem with the light echo explanation was that the displaced peaks should move systematically in wavelength and cross the whole line profile on a BLR light crossing time. Photoionization arguments showed that BLRs should be between a few light months to a few light years across and yet in a some objects the wavelengths of the displaced peaks have not changed (to ± 100-300 km s^{-1}) over almost two decades. Today the case for the light echo explanation is even worse since line variability studies (Gaskell and Sparke 1986) show that BLRs are even smaller than was thought (light weeks to light months). In no case is there any evidence of a displaced BLR peak changing its wavelength.

Gaskell (1983b) suggested instead that the displaced peaks were due to the whole BLRs moving by up to several thousand km s^{-1} with respect to the host galaxy. For the moving BLR to survive the massive black hole must be moving with it. It is possible that a single black hole can wander around in a galactic nucleus at low velocity (less than the stellar velocity dispersion) but where high bulk high velocities are observed they require that the black hole be orbiting another compact massive object (presumably another black hole). The existence of such binaries had previously been proposed (for other reasons) by Komberg (1968), Collins (1980) and Begelman, Blandford and Rees (1980).

The most plausible way to get binary black holes is through mergers of galaxies (as first proposed by Begelman, Blandford and Rees 1980). We know that galaxies can merge. We know that a high percentage of galaxies show at least some signs of quasar activity (see Keel 1985). If one believes that there is a black hole associated with each quasar then one is led inescapably to the result that at least some nuclei will have more than one black hole. One might argue that although a large fraction of active galaxies (especially high luminosity QSOs) are disturbed, many others show no obvious signs of interaction, but I have suggested (Gaskell 1985) that a collision with a dwarf elliptical companion (like M32 say) is all that is needed.

The early stages of a merger can give two spatially resolved quasars in the same galaxy. The well known multiple jet radio galaxy 3C 75 (= Abell 400 cD; Owen *et al* 1985) is a beautiful example of the early stages of the merger of two active nuclei. When the black holes are still far apart one can see two separate NLRs. The double peaked narrow line Seyfert Mrk 78 (de Robertis 1987) could well be an example of this and Mrk 266 certainly is. 6m spectra taken by Petrosian, Sahakian and Khachikian (1980) show that the two Seyfert 2 nuclei (Osterbrock and Dahari 1983), separated by about 10" are spinning in opposite senses and Kollatschny and Fricke (1984) have used the IUE to show that they are each have an ionizing UV continuum.

The evolution of a supermassive black hole binary has been sketched by Begelman, Blandford and Rees (1980) and investigated numerically by Roos (1981). They describe the evolution as a function of separation but if one translates this into velocity one can see that the binary black hole model very naturally predicts the observed velocities of the displaced broad line peaks. Slow orbits will get sped up fairly quickly by dynamical friction until the orbital velocities exceed the stellar velocity dispersion and dynamical friction becomes less efficient. The binary will now get "stuck" at this velocity for what can be of the order of a Hubble time. Higher velocities (of the order of a few percent of the speed of light) are excluded because gravitational radiation will make the orbit decay very rapidly. We therefore expect to observe such binaries when their velocities are of the order of a few thousand km s^{-1}.

Gaskell (1983b) estimated the orbital periods to be of the order of centuries. This estimate was made by assuming that the two BLRs are fairly distinct. Now that we know that BLRs are smaller than we thought shorter periods are possible (decades perhaps). Smaller BLRs mean that smaller orbits are *possible* (but not *required* -- the orbit size depends on the black hole masses if the velocity is fixed). The lack of change in wavelength of the displaced peaks can constrain the parameters of the binary model. Halpern and Filippenko (1988) feel that the restrictions on the model for Arp 102B are quite severe and, if the black hole masses are too low, might exclude the binary model.

EDGE ON ACCRETION DISKS?

It is possible that the accretion onto black holes which is believe to power quasars involves a disk structure. Disk geometries for the emission line regions have been considered for almost 30 yrs now, but detailed investigations always showed that they produced profiles incompatible with observed BLR profiles (e.g. Shields 1978, Capriotti *et al* 1980, Mathews 1982). In a number of *stellar* systems in our galaxy which feature accretion onto compact objects double-peaked lines *are* seen which arise from accretion disks. It has recently been proposed by many authors that perhaps the double BLR peaks in quasars are *also* edge on accretion discs (e.g. Oke 1987, Alloin,

Boisson and Pelat 1987, Halpern and Filippenko 1988, and Perez *et al* 1988,). I will conclude by showing why this idea is untenable.

First, quasars where both BLR peaks are equally prominent (as is sometimes the case in 3C 390.3 considered by Oke 1987 and Perez *et al* 1988) are *rare*. The more usual situation is that there is just *one* prominent displaced peak with the second one being hard to detect. This is *not* what disc profiles (such as those calculated by Mathews 1982) look like. They *always* have *two* peaks. Very unequal peaks are however exactly what are predicted by the binary model since the quasar luminosity function is a power-law with no preferred luminosity so one expects two quasars in the same nucleus to have very different luminosities.

Second, all the major problems with using disks to explain "normal" line profiles remain. One cannot invoke disks only to explain double peaked profiles.

The third, and I think most decisive, argument against BLR disks comes from variability. Probably all quasars vary and so do their broad emission lines (and the displaced BLR peak quasars are no exception). The emission line variations closely follow the continuum variations showing that the lines are caused by photoionization by a variable central source. In low luminosity quasars the BLRs are a light week to a light month across (Gaskell and Sparke 1986, Peterson 1988). It should be noted that some BLR disk models were actually proposed to offer an alternative energy input mechanism to photoionization (Collin-Souffrin *et al* 1979). The strong emission line/ continuum variability correlation does not support this idea, instead it requires photoionization of the lines seen to vary (the Balmer lines in particular). If the double peaked Balmer lines are arising in a disc then the double peaks *must always* vary together in phase. In no well observed double-peaked BLR object is this the case. 3C 390.3, the main example of a suggested accretion disc BLR, is a good illustration. The observations of Oke (1987, his Fig. 1) show that the ratio of red to blue components varies smoothly over a few years (which is *much* longer that the expected light crossing time) with first the red and then the blue component being stronger. The observed continuum is varying on the same timescale but there is no clear connection between it and the red/blue ratio. The two displaced peaks in Akn 120 are another similar example (Capriotti, Foltz and Peterson 1982). The best studied case to date is NGC 5548 where, again, at different epochs either the red or the blue component can dominate. Peterson, Korista and Cota (1987) have shown that actually the flux variation of *only one* of the peaks is correlated with the continuum variability. This completely contradicts the disk model but is exactly what one expects in the binary black hole model. For NGC 5548 the

size implied (from a cross-correlation analysis) for the component which *does* vary is about two light weeks, while the other BLR component must be at least a few light months away.

Alloin, D., Boisson, C. and Pelat, D. 1987, *Ast. and Ap.*, 185.
Anderson, K. S. 1971, *Ap.J.*, 169, 449.
Begelman, M. C., Blandford, R. D. and Rees, M. J. 1980, *Nature*, 287, 307.
Blumenthal, G. R. and Mathews, W. G. 1975, *Ap.J.*, 198, 517.
Capriotti, E. R., Foltz, C. B. and Byard, P. 1980, *Ap.J.*, 241, 903.
Capriotti, E. R., Foltz, C. B. and Byard, P. 1981, *Ap.J.*, 245, 396.
Capriotti, E. R., Foltz, C. B. and Peterson, B. M. 1982, *Ap.J.*, 261, 35.
Collins, G. W. Jnr. 1980, *Ap. and Space Sci.*, 73, 355.
Collin-Souffrin, S., Joly, M., Heidmann, N. and Dumont, S. 1979, *Ast. and Ap.*, 72, 293.
Crenshaw, D. M. and Peterson, B. M. 1985, *Ap.J.*, 291, 677.
De Robertis, M. M. 1987, *Ap. J.*, 316, 597.
Ferland, G. J., Netzer, H. and Shields, G. A. 1979, *Ap.J.*, 229, 274.
Gaskell, C. M. 1981, *Ap.J.*, 251, 8.
Gaskell, C. M. 1982, *Ap.J.*, 263, 79.
Gaskell, C. M. 1983a, *Ap.J. Letts.*, 267, L1.
Gaskell, C. M. 1983b, *Liege Astrophysical Colloquia*, 24, 473.
Gaskell, C. M. 1985, *Nature*, 315, 386.
Gaskell, C. M. 1988, in preparation.
Gaskell, C. M. and Sparke, L. S. 1986, *Ap.J.*, 305, 175.
Halpern, J. P. and Filippenko, A. V. 1988, *Nature*, 331.
Junkkarinen, V. T. 1986, *Bull. A.A.S.*, 18, 914.
Keel, W. C. 1985 in *Astrophysics of Active Galaxies and Quasi-Stellar Objects* ed. J. S. Miller (University Science Books: Mill Valley), p 1.
Kollatschny W. and Fricke, K. J. 1984, *Ast. and Ap.*, 135. 171.
Komberg, B. V. 1968, *Soviet Ast. J.*, 11, 4.
Lynds, C. R. 1968, *A.J.*, 73, 888.
Mathews, W. G. 1982, *Ap.J.*, 258, 425.
Oke, J. B. 1987, in *Superluminal Radio Sources*, ed. J. A. Zensus and T. J. Pearson, (Cambridge Univ. Press), p 267.267.
Osterbrock, D. E. and Dahari, O. A. 1983, *Ap.J.*, 273, 478.
Owen, F. N., O'Dea, P., Inoue, M. and Eilek, J. A. 1985, *Ap.J. Letts.* 294, L85.
Perez, E., Penston, M. V., Tadhunter, C., Mediavilla, E. and Moles, M. 1988, *M.N.R.A.S.* in press.
Peterson, B. M. 1988, *P.A.S.P.*, 99, in press.
Peterson, B. M., Korista, K. T. and Cota, S. A. 1987, *Ap.J. Letts.*, 312, L1.
Petrosian, A. R., Sahakian, K. A. and Khachikian, E. Ye. 1980, *Astrofiz.*, 16, 360.
Roos, N. 1981, *Ast. and Ap.*, 104, 218.
Sandage, A. R. 1966, *Ap.J.*, 145, 1.
Shields, G. A. 1978, in *Pittsburgh Conference on BL Lac Objects*, ed. A. M. Wolfe (Univ. of Pittsburgh: Pittsburgh), p 258.

M. J. Ward: Could the double peaks be due to jets?

C. M. Gaskell: I don't think so because the correlation with radio type is the wrong way round. The best cases of double peaks are in lobe-dominated objects where one doesn't expect the jet to be pointing at you, rather than in bright core-dominated quasars. This implies that the bulk motion is in a plane *perpendicular* to the radio jets. I've suggested elsewhere (Gaskell 1983b) that the presence of a supermassive binary might be essential for the formation of well collimated large scale radio jets (*alla* SS433).

R. J. Weymann: Are there any examples of bright quasars showing double peaks?

C. M. Gaskell: Quite a number. 2141+175 is an example.

R. J. Weymann: Are there any high redshift examples?

C. M. Gaskell: There are certainly high redshift examples of single displaced peaks (e.g. 0119-046) and probably some ones with evidence for a second peak but these are harder to see (especially since the high ionization UV lines are much broader).

D. E. Osterbrock: The profiles you have shown certainly showed well marked double peaks. Other well observed broad profiles with good signal to noise ratio data show structure that can be represented as the superposition of two or in some cases three components.

C. M. Gaskell: I haven't seen a convincing case yet of more than two peaks (I'm not counting the NLR contribution). Sure one often sees very lumpy looking line profiles, but it's always a signal to noise problem or the effect of not subtracting off the underlying starlight properly (see Crenshaw and Peterson 1985). When you look at the super high resolution, high signal to noise ratio data taken by quasar absorption line people the line profiles are *very* smooth. I'd be interested in seeing any convincing case of more than two displaced BLR peaks that anyone can produce. You certainly don't see them in, say, the high quality profiles of Mrk 509 by Atwood, Baldwin and Carswell or the survey by Filippenko and Halpern.

D. E. Osterbrock: In narrow line profiles, recent high-resolution profiles by Veilleux show other incipient double profiles in addition to Mrk 78 *(*that you mentioned*)* and in other galaxies narrow line profiles that can be regarded as superpositions of two or three unresolved components. Some of these were reported earlier by Alloin and Pelat.

C. M. Gaskell: I'm not suggesting that all off-nucleus NLR emission or structure in narrow line profiles is due to a second compact object, although this obviously is the case in Mrk 266.

OJ287 AS A BINARY SYSTEM

M. Valtonen[1], A. Sillanpää, S. Haarala, L. Valtaoja, E. Valtaoja
Turku University Observatory, 20520 Turku, Finland

B. Sundelius
Department of Astronomy, University of Gothenburg
41296 Gothenburg, Sweden

G. G. Byrd
Department of Physics and Astronomy, University of Alabama
Tuscaloosa, Alabama 35487, USA

H. Teräsranta and S. Urpo
Radio Laboratory, Helsinki University of Technology
02150 Espoo, Finland

Abstract. The historical optical light curve of OJ287 shows repeated outbursts at 11.65 yr intervals. Making use of computer modeling, we interpret this as resulting from a 9 yr orbital period in a binary system. Variability time scales are used to construct a mass model of the binary system, with derived masses of $5 \cdot 10^9$ $M_\odot$ and $2 \cdot 10^7$ $M_\odot$. We discuss how binary systems such as in OJ287 may result from mergers of galaxies. We infer that the system must contain also a third supermassive black hole.

1. Introduction. At high radio frequencies, we presumably have a fairly unobstructed view of the central engines of active galactic nuclei (AGN). Unfortunately, the resolution of VLBI is still several orders of magnitude too low in order to give a clear picture of the central 10^{-3} pc or so of the AGN. We are restricted to the indirect information which may be deduced from total and polarized flux measurements and in particular from the flux variations.

We have used the Metsähovi 14-m radio telescope of Helsinki University of Technology at 12, 22, 37 and 77 GHz to monitor the flux variations in about 40 quasars and related sources during the past ten years. At the same time, we have followed the evolution of optical brightness of a large fraction of these sources using the Tuorla 70 cm Schmidt telescope of University of Turku. The measurements have been supplemented by data from the Crimean astrophysical laboratory at the same spectral frequencies. For a general description of the results, we refer the reader to Valtaoja et al. (1987) and Valtaoja et al. (1988).

---

[1]Present address: Observatory, 00130 Helsinki, Finland

2. <u>The optical light curve</u>. **We have collected data on optical magnitudes of OJ287 from literature from about 1890 up to present (see Sillanpää et al. 1988 for references). The resulting light curve, in linear scale, is shown in Fig. 1.**

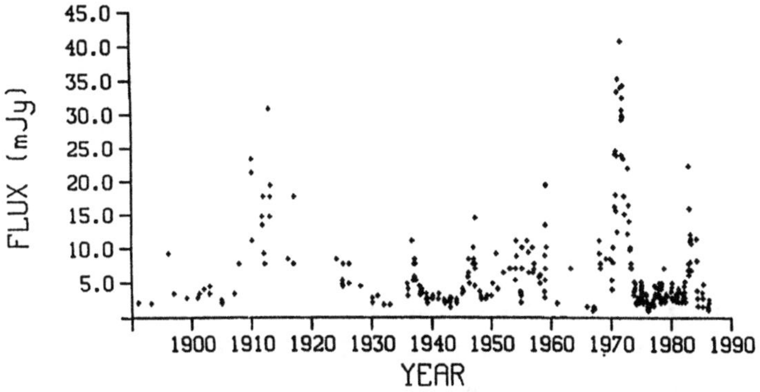

Fig. 1. The optical light curve.

The striking feature of the light curve is the repeated character of periods of strong activity. Seven active periods at average intervals of 11.65 yrs may be identified. In the initial compilation of the data (Sillanpää et al. 1986) the work of Kurochkin (1971 a,b) and Tsessevich (1972) was inadvertently overlooked, and only five of the seven peaks were seen. Even then, the periodic character of outbursts was striking enough that a binary model for OJ287 was proposed. The subsequent confirmation of the two missing outbursts strengthened this proposal considerably.

Similarities exist in the behavior of OJ287 during its 1972 and 1983 optical outbursts (Sillanpää et al. 1985). Roughly one month long brightness oscillations appear to be superposed on the otherwise steady decline after the main event. On the other hand, a very similar rapid rise and a slow decline with superposed oscillations is seen also in the <u>inflow rate</u> of mass to the centre of a tidally perturbed self-gravitating disk (Byrd et al. 1986b, 1987). This suggests that the main optical outbursts in OJ287 (and possibly also in other variable radio sources) are tidally driven by a companion, which at regular intervals excites spiral density waves in the accretion disk, surrounding the black hole, and that this wave feeds the "monster" at the centre. Further numerical simulations demonstrate that repeated inflow bursts (corresponding to outbursts in light output) may indeed be generated in this way (Sillanpää et al. 1988).

Figure 2 gives a schematic view of the OJ287 system in light of the information described above.

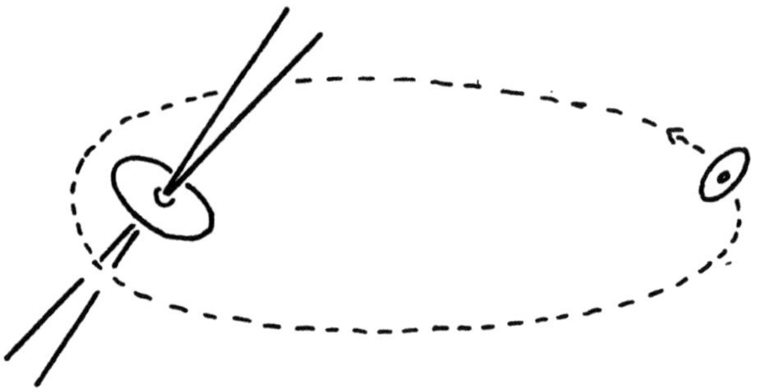

Fig. 2. The binary model of OJ287.

What would be the physical scale of the system? We only know the period (9 yrs in the frame of OJ287) of the orbit, but neither the major axis nor the masses are known. The only handles on the masses at present are the variability time scales. We may assume that the fastest flux variations are related to orbital periods P at few Schwarzschild radii from the black holes. If the mass of the black hole in millions of solar masses is M_6, then $P \approx 50\ M_6$ sec.

There are two much discussed variability time scales in OJ287. The short time scale is of the order of 15-30 min (Valtaoja et al. 1985, Dreher et al. 1986, Miller 1988), with probably a periodic component associated with it. Considering the large luminosity of OJ287, it is unlikely that this period could be related to the mass of the whole system. It is more likely to tell us about the mass of the smaller of the two black holes. A longer time scale variation of 8 days was discussed by Kinman et al. (1974). Our radio data indicate a possible 5 day period in one six week span of 77 GHz observations, and autocorrelation analysis of 22 and 37 GHz observations indicate significant correlations from about 3 days upwards. Uncertain as these estimates are, they anyhow establish a reasonable mass scale for the larger black hole system of about $5 \cdot 10^9\ M_\odot$. The shorter period would correspond to about $2 \cdot 10^7\ M_\odot$ black hole. Assuming these mass values, the semimajor axis of the binary is 0.1 pc.

3. Mergers of galaxies. How could a binary system originate in OJ287? We assume, as is common, that the quasar lies in the centre of a major galaxy. One way to bring two black holes together in the centre of a galaxy is through a merger of two galaxies, both of which contain originally a single black hole. According to

simulations, galaxies in groups and clusters suffer typically several mergers during the Hubble time (Aarseth and Fall 1980, Roos 1981a,b, Navarro et al. 1987). On the other hand, there is increasing evidence that radio sources prosper in an environment where they have half-a-dozen or dozen companions (Heckman et al. 1985). Many of these companions will presumably merge with the radio galaxy some time in the future. Consequently, it is reasonable to assume that a powerful radio source such as OJ287 was born as a result of merger where two galaxies and two or more black holes came together.

The way a merger of two identical elliptical galaxies takes place was studied by Miller and Smith (1980). Figure 3 illustrates the relative orbits in one of their simulations.

Fig. 3. The relative motion of the mass centers of two galaxies. Solid line: without dynamical friction; dashed line: with dynamical friction.

In Miller and Smith's mergers, the relative orbit is slowed down in the initial encounter. A very strong frictional force sets in at the next apogalacticon, forcing a head-on collision between the galaxies. The same behavior is seen also in the simulations of Byrd et al. (1986a).

The stars of the two galaxies are mixed in the head-on collision, but the black holes keep on going along the center-of-mass path of the former galaxies. If the impact parameter between the black holes is large enough ($\gtrsim$ 10 pc), the incoming black hole remains oscillating about the center of the primary galaxy well over the Hubble time for reasonable black hole masses. Typical oscillation amplitudes may be $\sim$ 10 kpc. Possible observational consequences of oscillating black holes have been discussed elsewhere (Valtonen and Byrd 1980).

However, in OJ287 we require that the impact distance of the incoming black hole has been small enough for effective two body scattering. In the scattering, the velocity of the oscillator is reduced by $\sim m_0/M_0$ per nuclear crossing (the exact value depends on the impact distance), if m_0 is the mass of the oscillator and M_0 the mass of the larger black hole. Thus, the smaller black hole will be slowly but surely, over the time scale of 10^9 yrs, brought to the center of the primary galaxy where a binary is formed.

4. Evolution of black hole binaries. The evolution of black hole binaries has been studied by Begelman et al. (1980) and Roos (1981b). The larger black hole collects a cusp of stars around it, and the smaller black hole goes in gradually shrinking orbit around it. The binary orbit shrinks until effective interaction with the surrounding stars ceases; according to Roos (1981b), the orbital radius of the binary becomes $a_0 \approx 0.4\ (M_0/10^9 M_\odot)$pc.

In order to decrease the orbital radius further, new stars should be placed to those orbits which have been depopulated by ejections of the black hole binary. As was pointed out by Begelman et al., the stellar relaxation time is too long to be a significant help in the process, and the normal rate of infall of stars through the cross-section of the binary is too small to provide enough energy carriers to absorb the binary energy further. Moreover, energy gains and energy losses balance each other in randon, nearly parabolic encounters of small bodies with a massive binary, unlike in encounters of three nearly equal bodies (Huang and Valtonen 1987). Thus, even if there would be ways to divert new stars to binary crossing orbits by e.g. collective effects, as proposed by Begelman et al., these stars would not carry away the binary energy.

Another mechanism for reducing the binary major axis is via gravitational radiation. The lifetime T_c to gravitational radiation becomes

$$T_c \approx 1.4 \cdot 10^{10}\ (M_0/m_0)(M_0/10^9\ M_\odot)\ \text{yr} \quad ,$$

i.e. it is much greater than the Hubble time in the OJ287 case.

What mechanism could bridge the gap between the orbital radius of $a_0 \sim 2$ pc which is expected in the $5 \cdot 10^9\ M_\odot$ binary, and the observationally deduced semi-major axis of 0.1 pc? A clue to the solution of the problem comes from the eccentricity of the binary orbit. In order to produce periodic outbursts of observed duration, we estimate that the orbital eccentricity of OJ287 binary must be about 0.7. But gravitational radiation tends to circularize the orbit in time scales which are less than the lifetime of the binary ($\sim 10^5$ yr), and therefore the orbit cannot have evolved simply by losing energy to gravitational radiation. The orbital eccentricity hints to the presence of a third massive body (black hole?) in the system. In the three-body interactions, the remnant binaries or temporary binaries tend to be eccentric, with the mean eccentricity around 2/3. Thus, we suspect that the reason for both the eccentricity and the small size of the orbit is a third body which has carried away a major part of the binary energy.

In the scenario of mergers of galaxies, there are necessarily cases where one or both of the galaxies contain binaries. The objects may be brought to dynamical interaction if the impact parameter of their relative orbit is small enough. Then a three- or four-body interaction follows, which leads to ejections: in the three-body process, a single object is ejected and a binary recoils in the opposite direction; in the four-body process, two binaries are usually ejected to opposite directions (Mikkola 1983). Depending on the ejection speeds, the ejected object or objects may quickly relax back to the center of the galaxy, or they may even escape from the galaxy.

The energy for ejections is extracted from binaries, and thus the binary semi-major axes decrease. At the same time, binaries become eccentric; this increases gravitational radiation losses and leads to a rapid decline of the binary semi-major axis.

In the case of OJ287, we suspect a dynamical three-body interaction which took place less than 10^5 yrs ago. With a typical ejection speed of a few thousand km s^{-1}, the third object should still be within 1 kpc from the binary, and possibly much closer. It would be an interesting project to look for evidence of another compact radio core or the trail of the ejection (a "jet", not related to the nuclear jet) in OJ287.

OJ287 provides important insight to the evolution of active nuclei in general. It suggests that many-body processes play a role in the radio sources. Perhaps they even determine the main structural characteristics of the sources (Saslaw et al. 1974).

References

Aarseth, S.J. and Fall, S.M. 1980, Ap.J., 236, 43.
Begelman, M.C., Blandford, R.D. and Rees, M.J. 1980, Nature 287, 307.
Byrd, G.G., Saarinen, S. and Valtonen, M.J. 1986a, Mon. Not. Roy. astr. Soc. 220, 619.
Byrd, G.G., Valtonen, M.J., Sundelius, B. and Valtaoja, L. 1986b, Astron. Astrophys. 166, 75.
Byrd, G.G., Sundelius, B. and Valtonen, M.J. 1987, Astron. Astrophys. 171, 16.
Dreher, J.W., Roberts, D.H. and Lehar, J. 1986, Nature 320, 239.
Heckman, T.M., Carty, T.J. and Bothun, G.D. 1985, Astrophys. J. 288, 122.
Huang, T.Y. and Valtonen, M.J. 1987, Mon. Not. Roy. astr. Soc. (in press).
Kinman, T.D., Wardle, J.F.C., Conklin, E.K., Andrew, B.H., Harvey, G.A., Macleod, J.M. and Medd, W.J. 1974, Astron. J. 79, 349.
Kurochkin, N.E. 1971a, Astron. Tsirk. 644, 1.
Kurochkin, N.E. 1971b, Astron. Tsirk. 654, 7.
Mikkola, S. 1983, Mon. Not. Roy. astr. Soc. 203, 1107.
Miller, H.R. 1988, this volume.
Miller, R.H. and Smith, B.F. 1980, Astrophys. J. 235, 421.
Navarro, J.F., Mosconi, M.B. and Garcia Lambas, D. 1987, Mon. Not. Roy. astr. Soc. 228, 501.
Roos, N. 1981a, Astron. Astrophys. 95, 349.
Roos, N. 1981b, Astron. Astrophys. 104, 218.
Saslaw, W.C., Valtonen, M.J. and Aarseth, S.J. 1974, Astrophys. J. 190, 253.

Sillanpää, A., Haarala, S. and Valtonen, M.J. 1986, Turku Univ. Obs. - Informo, No. 108.
Sillanpää, A., Haarala, S. Valtonen, M.J., Sundelius, B. and Byrd, G.G. 1988, Astrophys. J. (in press).
Sillanpää, A., Teerikorpi, P., Haarala, S., Korhonen, T., Efimov, Y. and Shakhovskoy, N.M. 1985, Astron. Astrophys. 147, 67.
Tsessevich, V.D. 1972, Astron. Tsirk. 688, 7.
Valtaoja, E., Haarala, S., Lehto, H., Valtaoja, L., Valtonen, M., Moiseev, I.G., Nesterov, N.S., Salonen, E., Teräsranta, H., Urpo, S. and Tiuri, M. 1988, submitted to Astron. Astrophys.
Valtaoja, E., Lehto, H., Teerikorpi, P., Korhonen, T., Valtonen, M., Teräsranta, H., Salonen, E., Urpo, S., Tiuri, M., Piirola, V. and Saslaw, W.C. 1985, Nature 314, 148.
Valtaoja, L., Sillanpää, A. and Valtaoja, E. 1987, Astron. Astrophys. 184, 57.
Valtonen, M.J. and Byrd, G.G. 1980, Astrophys. J. 240, 442.

Discussion

I. Shlosman: If you bring in the second black hole by means of dynamical friction, it is difficult to understand why this will result in a stable binary system, i.e. why the dissipation mechanism will stop to operate.

Valtonen: When the binary system is small, it is better to think about the binary interacting with individual stars rather than being influenced by friction in a continuous medium. The binary ejects initially stars from the immediate neighborhood of its orbit. For a sufficiently small binary orbit, the mass in this neighborhood is much less than the binary mass. At this binary radius, the dissipation mechanism ceases to operate. Even though stars cross the binary orbit also at later times, they do not change the binary energy on the average (see Section 4 above).

J.R. Webb: At Rosemary Hill Observatory, we have monitored OJ287 for 15 years. A time series analysis of our data reveal a 11.9 year period and a 5.3 year period. How can your model account for periods other than the orbital period, especially if they are not half period effects?

Byrd: Our computer simulation of the black hole interaction with the primary black hole and its disk shows a subpeak roughly halfway between the major peaks. Use of more particles in our simulations may reveal additional subpeaks. In general, the major peaks in the simulations occur at intervals of the pair orbital period. Then there are subpeaks at shorter intervals resulting from dynamical events within the disturbed disk.

M. Gaskell: The eccentric orbit in your model will decay rapidly because of gravitational radiation. What is the time scale for this decay?

Valtonen: It is about 10^5 years.

Line Profiles and the Kinematics of the Narrow-Line Region

M.M. De Robertis
Dept. of Physics, York University, North York, Ontario, Canada

R.A. Shaw
Lick Observatory, UCSC, Santa Cruz, CA

I. Observations

It is well known that narrow-line profiles in active galactic nuclei (AGN) are predominantly blueward asymmetric (*e.g.,* Heckman *et al.* 1981; Whittle 1985). This evidence suggests that the emission-line clouds are moving radially, *and* that there is differential extinction within the narrow-line region (NLR). By itself, this observation provides only limited information on the kinematics of the NLR.

Recently, however, observations of Liners (Filippenko 1985), and Seyfert 1s and 2s (De Robertis and Osterbrock 1984; 1986) have revealed that there is often a correlation between line width and critical density (N_{cr}) or ionization potential (IP) in AGN. These findings argue strongly that the NLR is in fact "stratified"– that is, to first order, emission for a given ion arises in a shell-like volume from optically thick clouds. Further, the dispersion velocity and number density of the clouds must be inversely proportional to the distance from the continuum source.

Combining these two important observations, it is possible to determine fundamental NLR properties like the direction of cloud motion and the distribution of dust by measuring relative asymmetries for all optical emission lines in individual AGNs. (The relative asymmetry, AI, at an intensity level i– where i is expressed as a percentage of the maximum intensity– is defined, $\mathrm{AI}i = (\Lambda_c - \lambda_c(i))/w(i)$, where $\lambda_c(i)$ and $w(i)$ are the line center and width at level i, and Λ_c is the centroid of the upper 20% of the profile.) Answers to these questions would provide useful information not only for the NLR, but also, for example, dictate how reddening corrections should be applied to the broad-line and continuum radiation.

There are a number of important observational constraints that must be imposed in a project of this sort. These include:

a) selecting only high-ionization AGNs. Our sample was taken from the Seyferts studied by De Robertis and Osterbrock (1984,6).

b) omitting galaxies with noticeable Fe II emission, or with very broad permitted lines so that we would not require (subjective) deblending techniques to isolate the NL profiles.

c) using high spectral resolution (~ 1 Å) in order to minimize instrumental effects.

d) using a long slit with a low read-noise CCD detector in order to obtain good sky subtraction and linear data, and to check whether the emission-lines are resolved.

e) obtaining absorption-line template galaxies in order to remove the effects of stellar features which can have a significant effect on the asymmetry measurements of many NL profiles (though a smaller effect on line width measurements).

In total, 10 Seyfert Galaxies (primarily 2s) were observed with the Cassegrain UV spectrograph on the Shane 3-m telescope at Lick Observatory in October 1986 and May 1987 using a TI 800×800 CCD. Four exposures were taken of each galaxy (using a slit width of 1.″5) in order to cover the (rest frame) wavelengths $\lambda\lambda 3400 - 6800$ (excluding $\lambda\lambda 5400 - 5900$) with a spectral resolution of ≈1.5 Å. The reduction techniques will be described elsewhere. Instead, we only give preliminary results for two Seyfert 2s, Mrk 3 and Mrk 533 which have relatively broad, narrow line profiles. In fact both galaxies show fairly good line width, critical density (and ionization potential) correlations. Moreover, in both galaxies,

(1) the relative asymmetry *increases* with *increasing* N_{cr} and IP.

(2) the low ionization (critical density) lines have non-zero asymmetries.

The implications of these observations will now be discussed.

II. Models

In order to demonstrate the potential for such observations to critically probe conditions in the NLR, we now present simple, spherically symmetric, two-component (numerical) models of the NLR. We consider two limiting cases; one in which the extinction is confined completely to the inter-cloud medium (and hence the clouds must be outflowing to produce blueward asymmetries), and the other in which the extinction arises entirely within the emission-line clouds (and hence the clouds must be inflowing). We note that even though the conditions hypothesized for the inter-cloud medium (*e.g.,* Krolik and Vrtilek 1984) are hostile to the formation and survival of dust which probably causes the reddening and profile asymmetries (see the chapter by Dahari and De Robertis in this volume), simple sputtering calculations show that dust may in fact survive for a significant fraction of the NLR crossing time by the inter-cloud medium.

Consider the case of outflowing clouds. We first modeled the outer NLR (*i.e.,* the low critical density lines) by uniformly filling a shell between inner and outer radii R_i and R_o with N_c emission-line clouds. For each cloud, we selected a radial velocity from a Gaussian distribution with appropriate mean (V_o) and dispersion (σ) velocities. (Note that our intention was not to describe exactly a "true" narrow-line profile, but only to provide a reasonable profile for illustrative purposes.) The entire NLR (from $0 - R_o$) was then filled with attenuating material according to either a constant, $1/r$, or $1/r^2$ density law, to a specified maximum extinction in magnitudes, Δm. (The parameter Δm is defined as the total extinction along the observer's line of sight through the symmetry axis.) The effective optical depth of the material depends upon the density distribution and path length through the NLR. The individual cloud δ-function profiles were then convolved with an instrumental profile and summed to produce the "observed" profile. The widths and asymmetries of this profile were then measured. The entire process was then repeated for a shell representing the inner NLR (*i.e.,* high critical density lines); that is, for R_i and R_o a factor of 10 smaller, all other parameters remaining the same.

The top left and right panels in Figure 1 show the profiles obtained from a constant, and $1/r^2$ dust distribution respectively for the outer NLR. The different curves show the effect of varying the total extinction in the NLR. The corresponding asymmetries are tabulated with

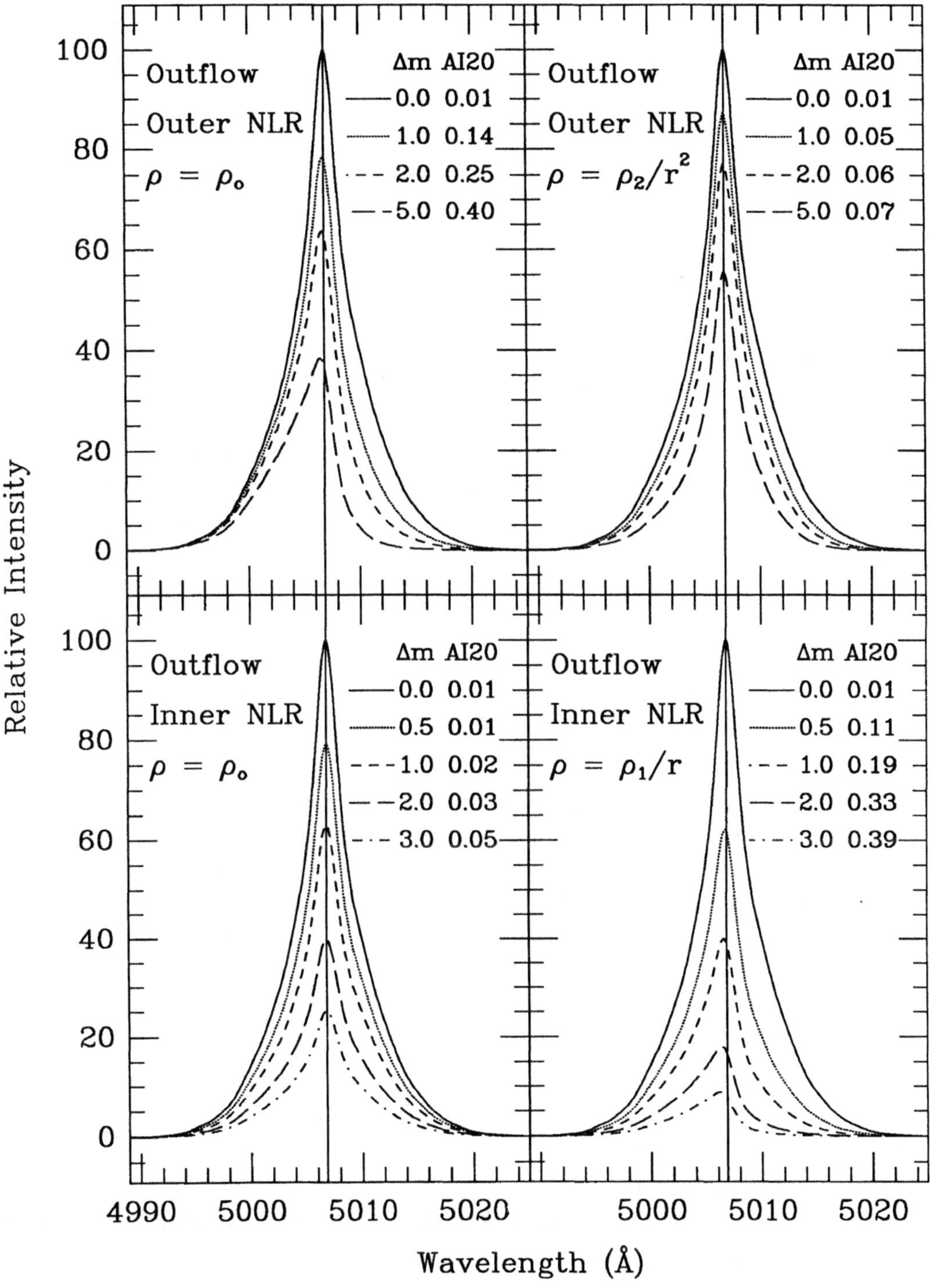

Figure 1. Line profiles for outer (top panels) and inner (bottom panels) NLR for models described in text. Maximum extinction (Δm) and asymmetry parameter (AI20) are tabulated for each model.

the maximum extinction values in magnitudes. (In every experiment, 25000 clouds were used along with a Gaussian velocity distribution having $V_o = 325$ km/s and $\sigma = 250$ km/s.) It can be seen that to generate a sizeable AI20 (the relative asymmetry parameter at 20% intensity) for the outer NLR, a dust distribution which is not centrally condensed seems to be required. The bottom left and right panels in Figure 1 show the same experiment (for a constant, and $1/r$ distribution respectively) for the inner NLR. (The rest wavelengths of all the lines in Figure 1 are the same to facilitate comparison.) In this case, to produce large asymmetries, the dust distribution must be centrally condensed. Thus, it may not be possible to satisfy the observations for both the low and high critical density profiles simultaneously with this model unless a third component ("accretion disk?") is introduced.

In the case of inflow where the dust is confined to the clouds, there is no apparent inconsistency. The only difference in the treatment between this case and the one discussed above is that amount of attenuation is no longer a function of path length through the NLR, but on the optical depth through the emitting cloud along the line of sight. For convenience we assumed that the optical depth is proportional to $(1 + \cos\theta)^n$; where θ is the angle made by the cloud to the line of sight along the symmetry axis and $n \approx 1$. One unfortunate characteristic of inflow models is that the exact "attenuation law" (*i.e.,* cloud shape, ionization structure, etc.) is required to treat them correctly. The law adopted is sufficiently general to be representative, however. In any event, one obtains rather reasonable profiles for inflowing clouds *and* larger asymmetries for higher critical density lines *if* the emission-line clouds are not substantially different between the inner and outer NLR. It is easy to show under this assumption that clouds closer to the continuum source will present a larger column density of dust to the observer and therefore produce larger asymmetries.

In summary, it appears that simple models in which emission-line clouds are radially infalling, *i.e.,* the dust is within the clouds, are more easily able to explain the observed run of asymmetry with critical density (*i.e.,* radius). Finally, with the direction of motion and the general distribution of dust identified, it should now be possible to construct realistic NLR models and so carry out a more careful investigation of the distribution and kinematics of the NLR.

References

De Robertis, M., and Osterbrock, D.E. 1984, *Ap. J.*, **286**, 171.

De Robertis, M., and Osterbrock, D.E. 1986, *Ap. J.*, **301**, 98.

Filippenko, A.V. 1985, *Ap. J.*, **289**, 475.

Heckman, T.M., Miley, G.K.., van Breugel, W.J.M., and Butcher, H.R. 1981, *Ap. J.*, **247**, 403.

Krolik, J.H., and Vrtilek, J.M. 1984, *Ap. J.*, **279**, 521.

Whittle, M. 1985, *M.N.R.A.S.*, **213**, 33.

REDDENING OF NARROW LINE REGIONS

Adam Wysota
Astronomy Program, ESS Building, State University of New York,
Stony Brook, NY 11794-2100.

C. Martin Gaskell
Astronomy Department, Dennison Building, University of Michigan,
Ann Arbor, MI 48109-1090

ABSTRACT

We make a critical comparison of a number of techniques for estimating narrow line region (NLR) reddenings and discuss their accuracy. We confirm that there is a preferred case B' Hα/Hβ ratio of 3.1 for the NLR. All the reported scatter can be explained by known observational errors and contrary to some earlier claims, there is no evidence for intrinsic spread in the Hα/Hβ ratio. This implies that the ionizing continua and NLR abundances in quasars are very similar.

INTRODUCTION

In order to use emission lines to investigate physical conditions in quasars it is first necessary to correct the line ratios for reddening. Even for the relatively low density gas in narrow line regions (NLRs) this has proved controversial. Classically, the Balmer decrement has been used to correct NLR ratios for reddening. However, Ferland and Mushotzky (1982) and Netzer (1982) questioned whether the standard Menzel-Baker case B was applicable under NLR conditions. They suggested that line transfer and collisional effects could significantly increase the Hα/Hβ ratio above its case B value (of about 2.86). Gaskell (1982) examined the reliability of the NLR Balmer decrement as an indicator of reddening by making a comparison with non-hydrogenic reddening indicators, and concluded that the intrinsic decrement was close to the case B value. Malkan (1983) independently made a similar analysis but concluded that the common practice of deriving NLR reddenings by assuming a case B Hα/Hβ ratio must be abandoned. An important method used by both Gaskell (1982) and Malkan (1983) was the "Allen" [O II]/[S II] method (Allen 1979). Mihalszki and Ferland (1983) challenged Allen's assertion that the [O II]/[S II] lines could be used as a reddening indicator.

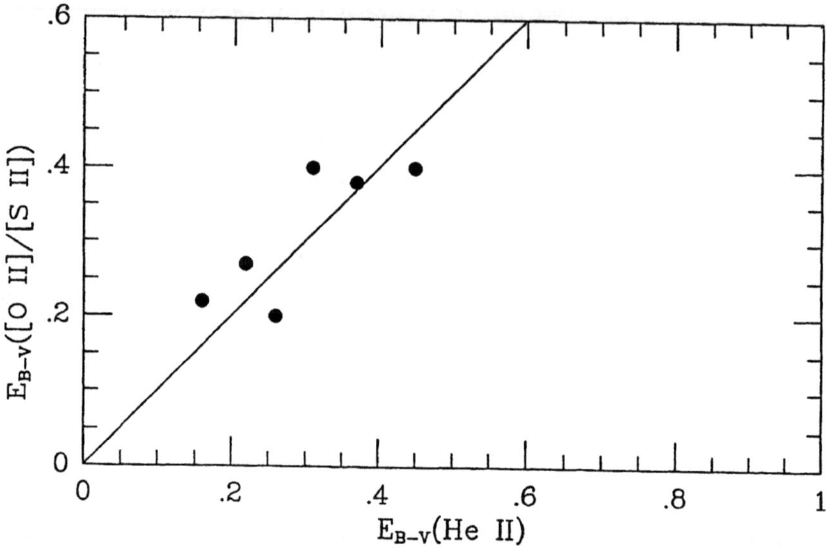

Fig. 1. Comparison of reddenings deduced by the Allen method and from the He II lines. The 45° line shows where all points would be if both methods agreed exactly.

THE BALMER DECREMENT

Fig. 2 shows the reddenings deduced using the case B' decrement (Hα/Hβ = 3.1) compared with the He II reddenings. As can be seen the agreement is excellent. The small scatter can be explained by the known observational errors (except for Mrk 176, which we believe needs re-observing). Comparisons with reddenings deduced from the other methods show comparable agreement (see Gaskell and Wysota 1988) and again the scatter can be explained by observational errors.

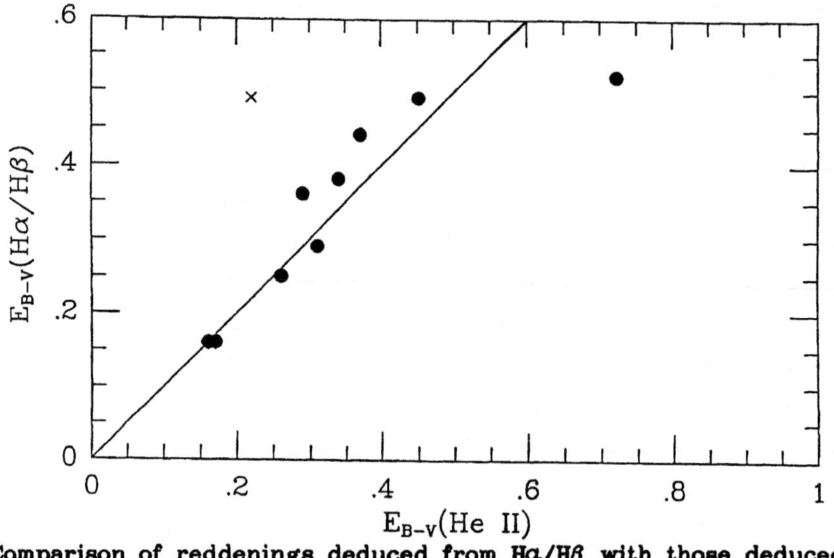

Fig. 2. Comparison of reddenings deduced from Hα/Hβ with those deduced from He II. The "×" is Mrk 176. The 45° line again represents perfect agreement between the two methods.

A re-examination of the observational data by Gaskell (1984) showed that, despite the problems anticipated by Mihalski and Ferland, the Allen method was indicating reddening and that the data favored a preferred Hα/Hβ ratio of about 3.1 with no evidence for intrinsic scatter.

Halpern (1982) and Halpern and Steiner (1983) had used photoionization models to calculate the NLR Hα/Hβ ratio and found that for a wide range of physical conditions the Hα/Hβ ratio was about 3.1. Halpern has coined the term case B' for this. A theoretical investigation by Gaskell and Ferland (1984) failed to find the self-absorption effects Netzer (1982) had thought would increase the Hα/Hβ ratio and instead confirmed the Halpern case B' value. They also showed how a Lyα/Hβ *vs.* Hα/Hβ diagram could be used to distinguish between reddening and collisional effects.

An important check on these results is provided by the He II lines since most of the problems other line ratios have are absent (MacAlpine 1981, Eastman and MacAlpine 1985). Gaskell (1982) used a couple of He II ratios to show that Hα/Hβ was not far from case B. Malkan (1986) has provided additional observations of the He II line ratios. We have used published He II line ratios as a reference to investigate the reliability of the various reddening methods and in this paper we report briefly on some of our main results. For a fuller account of this work see Gaskell and Wysota (1988). In all cases we will be using the reddening curve of Seaton (1979).

METHODS OF DETERMINING REDDENING

a) He II. We have used the recombination ratios of Seaton (1978).
b) S II doublet method. Dopita (1978) showed how the trans-auroral λ4071 lines and the red λ6731/λ6716 doublet ratio could be used as a reddening indicator. This method gives reddenings systematically too low by 0.12 (Gaskell 1984).
c) Allen Method. This method uses a combination of [S II] and [O II] lines to solve independently for the density and reddening. Fig. 1 shows that there is no systematic offset between reddenings derived via the He II and Allan methods. The scatter in the correlation is consistent with the known measuring errors.
d) Lyman α. Gaskell and Ferland (1984) suggested a method of deducing reddenings using a Lyα/Hα/Hβ line ratio diagram. We found no systematic offset between reddenings deduced from the Lyman α observations of Ferland and Osterbrock (1985) and those deduced using non-hydrogenic lines.

REFERENCES

Allen, D. A. 1979, *M.N.R.A.S.*, 186, 1P.

Dopita, M. A. 1978, *Ap.J. Suppl.*, 37, 117.

Eastman, R.G. and MacAlpine, G. M. 1985, *Ap.J.*, 299, 786.

Ferland, G. J. and Mushotzky, R. F. 1982, *Ap.J.*, 262, 564.

Ferland, G. J. and Osterbrock, D. E. 1985, *Ap.J.*, 289, 105.

Gaskell, C. M. 1982, *P.A.S.P.*, 94, 891.

Gaskell, C. M. 1984, *Ap. Letts.*, 22, 175.

Gaskell, C. M. and Ferland, G. J. 1984, *P.A.S.P.*, 96, 393.

Gaskell, C. M. and Wysota, A. 1988, in preparation.

Halpern, J. P. 1982, Ph.D. thesis, Harvard University.

Halpern, J. P. and Steiner, J. E. 1983, *Ap.J. Letts.*, 269, L37.

MacAlpine, G. M. 1981, *Ap.J.*, 251, 465.

Malkan, M. A. 1983, *Ap.J. Letts*, 264, L1.

Malkan, M. A. 1986, *Ap.J.* 310, 679.

Mihalszki, J. S. and Ferland, G. J. 1983, *M.N.R.A.S.*, 187, 871.

Netzer, H. 1982, *M.N.R.A.S.*, 198, 589.

Seaton, M. J. 1978, *M.N.R.A.S.*, 185, 5P.

Seaton, M. J. 1978, *M.N.R.A.S.*, 187, 738.

M. Ward: Can you say anything about form of the reddening curve?

A. Wysota: Not really. We just assumed the Seaton curve.

M. M. De Robertis: Where did you obtain the data for Mrk 176?

A. Wysota: Alan Koski's thesis.

M. M. De Robertis: There is a large uncertainty in the Hβ emission-line flux -- it is very weak and there is a large correction for Hβ absorption -- so it should be re-observed.

O. A. Dahari: The emission of the Balmer, He II and [S II] lines may not come from the same physical region, and this may explain the scatter you see in your diagrams.

C. M. Gaskell: It *could* cause some scatter, but we must stress that all the scatter we see is amply explained by the measuring errors. Also, the methods using lines from very different ionization regions agree remarkably well.

D. M. Whittle: Is it not possible to turn the argument around:- the fact that the Balmer decrement, He II and [S II] techniques *do* agree implies that the dust is *not* located within the emission-line region but is external to it -- so all emission lines suffer the same absorption.

A. Wysota: Yes, this could be the case.

ULTRAVIOLET AND OPTICAL SPECTRA OF BROAD-LINE RADIO GALAXIES

D. Michael Crenshaw (Computer Sciences Corp. and IUE Observatory)
Bradley M. Peterson and R. Mark Wagner (Ohio State University)

I. Introduction

The study of broad emission lines in the ultraviolet spectra of Seyfert 1 galaxies is hampered somewhat by the resolution of IUE at low dispersion, which is about 1250 km sec^{-1} (FWHM) at the position of Lα. At this resolution, the broad and narrow components are difficult to separate. In broad-line radio galaxies (BLRG), however, the broad-line profiles are wider and more flat-topped than those in Seyfert 1 galaxies, and therefore the two components can be deconvolved. The only clear identification of a composite broad/narrow Lα profile in BLRG has been for 3C 390.3 (Ferland et al. 1979). Thus we decided to obtain ultraviolet and optical spectra of three additional BLRG to determine separate line ratios for both the broad-line region (BLR) and narrow-line region (NLR).

In late 1986 and early 1987, SWP spectra with the International Ultraviolet Explorer and spectra of the Hβ region with the OSU image dissector scanner on the Perkins 1.8 meter telescope at Lowell Observatory were obtained. In addition, a spectrum of the Hα region was obtained for 3C 445. Large apertures were used (20"x10" for IUE, 7" in diameter for the IDS), and the nights were photometric.

II. Results

Each spectrum was corrected for reddening in our galaxy, and transformed to the rest frame of the BLRG; only 3C 382 has a substantial amount of galactic reddening (E_{B-V} = 0.15). The narrow components of Hβ and Hα were removed by using the [O III] λ5007 profile as a template. The intrinsic width of the [O III] λ5007 profile was estimated by assuming that the intrinsic and instrumental profiles are Gaussians, and the width of the narrow component of Lα was calculated by assuming that the intrinsic profile has the same width as the [O III] line. This calculation increased the FWHM of the line from an instrumental value of 5 Å to a value close to 6 Å in all cases. Since the IUE instrumental profile is a Gaussian to a good approximation (Cassatella et al. 1985), a Gaussian profile with a width of 6 Å was used as a template to remove narrow Lα. The template profiles were scaled in intensity and subtracted from the profiles by trial-and-error until a smooth broad-line profile was obtained.

The profiles before and after subtraction of the narrow component are shown in Figure 1. Although the profiles are somewhat noisy, the decomposition of the profiles into narrow and broad components is satisfactory for 3C 382 and PKS 2349-01. For 3C445, the subtraction leaves a residual component for Hβ and Hα, which may arise from a region kinematically distinct from both the narrow [O III] emitting region and the region associated with the very broad component.

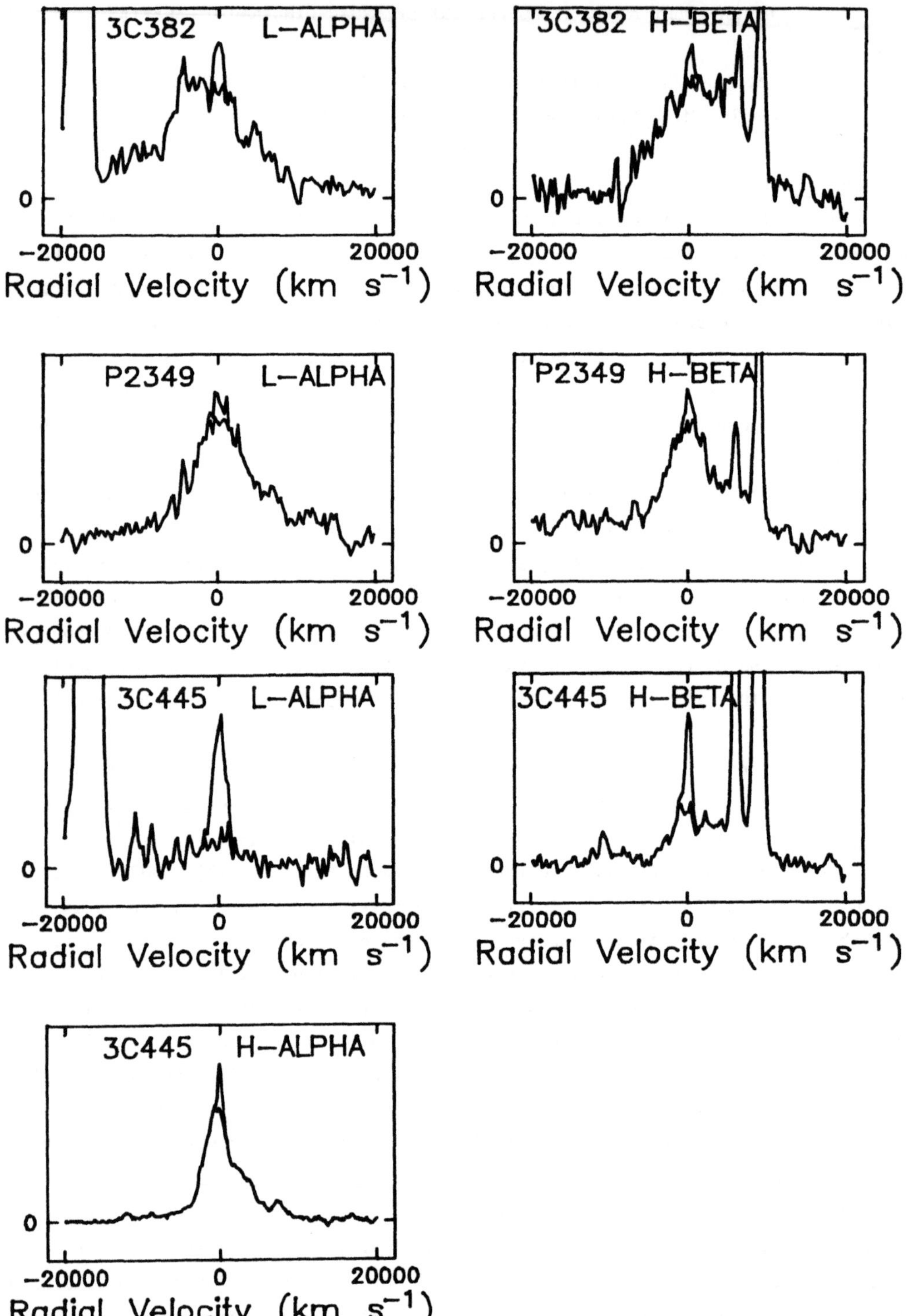

Figure 1. Ultraviolet and optical profiles in three broad-line radio galaxies.

Table 1 presents the reddening-corrected hydrogen line ratios for both the broad- and narrow-line regions. All of the values for 3C 390.3 are from Oke and Goodrich (1981), and the Hα/Hβ ratios for 3C 382 and PKS 2349-01, which are not simultaneous with our Lα/Hβ ratios, are from other sources.

III. Conclusions

With the exception of 3C445, the narrow line ratios are similar to those obtained by Ferland and Osterbrock (1986) for active galaxies with only narrow lines. These ratios lie in the region of their Lα/Hβ vs. Hα/Hβ diagram that they associate with low intrinsic reddening ($E_{B-V} < 0.25$).

The broad Lα/Hβ ratios cover a very wide range (1 - 22), and exceed the range covered by Seyfert 1 galaxies (2 - 15, see Wu et al. 1983). Thus the large Hβ/Hα ratio seen in BLRG (4 - 8) as compared to that in Seyferts (2.5 - 4.5) is in general not an effect of additional reddening in the BLR. If it is assumed that the intrinsic UV/optical continuum slope is similar in the BLRG, the observed UV continuum flux in 3C 445 is down by at least a factor of 10. If the broad-lines were corrected for this much reddening, the intrinsic Lα/Hβ and Hα/Hβ ratios would be 12 and 5.3 respectively, which are close to the average of the ratios presented. Thus, in the specific case of 3C 445, the continuum source and BLR may indeed be heavily reddened.

Table 1

Reddening-Corrected Line Ratios

BLRG	Galactic	Broad-Line Region		Narrow-Line Region	
	E(B-V)	Lα/Hβ	Hα/Hβ	Lα/Hβ	Hα/Hβ
3C 382	0.15	22.3[a]	4.5[d]	24.3[a]	3.6[d]
3C 390.3	0.03	6.9[c]	6.1[d]	27.0[c]	4.1[d]
3C 445	0.015	1.2[a]	7.9[a]	5.5[a]	1.9[a]
PKS 2349-01	0.015	18.2[a]	4.1[b]	12.4[a]	4.1[b]

[a]this paper, [b]Grandi and Osterbrock 1978, [c]Oke and Goodrich 1981, [d]Osterbrock, Koski, and Phillips 1975.

REFERENCES

Cassatella, A., Barbero, J., and Benvenuti, P. 1985, Astron. Astrophys., 144, 335.

Ferland, G. J. and Osterbrock, D. E. 1986, Ap. J., 300, 658.

Ferland, G. J., Rees, M. J., Longair, M. S., and Perryman, M. A. C. 1979, M.N.R.A.S., 187, 65P.

Grandi, S. A. and Osterbrock, D. E. 1978, Ap. J., 220, 783.

Oke, J. B., and Goodrich, R. W. 1981, Ap. J., 243, 445.

Osterbrock, D. E., Koski, A. T., and Phillips, M. M. 1975, Ap. J., 206, 898.

Wu, C. C., Boggess, A., and Gull, T. R. 1983, Ap. J., 266, 28.

LONG SLIT CCD OBSERVATIONS OF ACTIVE AND NORMAL GALAXIES

T. E. Carone
Department of Physics and
Lunar and Planetary Laboratory-West
University of Arizona

ABSTRACT

We present the first results of a low resolution, long slit study of 16 active and 15 normal galaxies. The galaxies of each sample chosen have similar Hubble types and similar distributions in redshift space and Holmberg magnitude. Using an 800 x 800 TI 3PCCD we were able to cover the 4500-7500 Å spectral region at a resolution of 10 Å; the 2.5 arcsecond by 4.0 arcminute long slit allowed us to cover both nuclear, near-nuclear and far off-nuclear emission line regions.

INTRODUCTION

Most studies of active galaxies have concentrated on their unresolved nuclear regions because this is where their prodigous energy output originates (Osterbrock and Mathews 1986). Theories of AGN have generally been based on these types of observations. Recent work has centered on determining how an AGN affects its host galaxy. A number of such studies have been carried out for individual objects (Balick and Heckman 1982; Whittle et al. 1986; Wilson, Baldwin and Ulvestad 1985). To isolate the effect of an AGN on its host galaxy, a comparison between a set of galaxies containing active nuclei and a set of normal galaxies would be useful. In principle, if the active and normal galaxy samples are chosen properly, then any systematic differences found between the two samples in the line emission and/or continuum properties of the host galaxy can be attributed to nuclear activity.

We have obtained long slit CCD observations of such a sample of active and normal galaxies. The use of the long slit allows us to take full advantage of the two dimensional format of the CCD and to acquire spectra from each point along the galaxy major axis. Our aim is to measure, in addition to the strong nuclear emission lines and continuum, the faint off-nuclear emission lines that arise in the galaxy's interstellar medium. By using relative line strengths, physical conditions in the galaxy interstellar medium can be determined as a function of distance from the nucleus (Veilleux and Osterbrock 1987).

OBSERVING PROGRAM

The instrumentation and data reduction techniques are discussed in greater detail elsewhere (Carone et al. 1987; Leach 1987). For this observing project we have selected two galaxy samples. The first sample consists of normal galaxies taken from the CfA Redshift Survey (Huchra et al. 1983) and the second consists of active galaxies (Veron-Cetty and Veron 1985). The samples are listed in Table I. The galaxies in both samples are chosen according to the following criteria:

1. 0.01 < z < 0.05
2. Hubble type Sa, Sb, Sab, SBa, SBb, SBab
3. $-19 < M_V < -21$
4. |b| > 20°

where z is the redshift, M_V is the absolute photographic magnitude (Holmberg 1975) and b is the galactic latitude. These criteria attempt to insure that the only difference between the two samples is nuclear activity.

Each galaxy is observed in the same manner. The long slit is placed along the galaxy major axis as defined in the UGC catalog (Nilsson 1974). For these observations we use a long slit which is 2.5 arcseconds wide and 4 arcminutes in length. The exposure times are chosen to obtain approximately the same signal-to-noise ratio for each galaxy in its strongest emission line. However, an upper limit of 40 minutes for a single exposure is imposed because, for exposure times greater than this, the number of cosmic rays detected becomes too high (>5 in the data region). We use a 300 line per millimeter grating blazed at 6690 Å. The resultant dispersion is 3.72 Å/pixel with an instrument resolution of 10 Å or 2-3 pixels. This allows a wavelength coverage of ≈ 3000 Å. For the observations presented here, the spectral region sampled is 4500-7500 Å. We can therefore typically measure HeII λ4686, Hβ, [OIII] λλ4959, 5007, HeI λ5876, [OI] λ6300, Hα, [NII] λλ6548,6584 and [SII] λλ6717,6731. Since the development of techniques to improve the blue and UV response (Leach 1987), we have changed the spectral range covered to include Hγ λ4341. Our spatial resolution is, of course, dependent on the seeing. Each pixel is 0.8 arcseconds and the typical seeing is between 1.5-3.5 arcseconds (2 to 4 pixels) FWHM.

RESULTS

The first results derived from this study are shown in Figures 1-4. We show here the Hα distributions for the active galaxies UGC 3223 and 3C 120 and the normal galaxies NGC 7771 and NGC 4500. The nuclear emission is off-scale because we wish to show the off-nuclear emission regions. The derivation of the emission line spatial distributions for the remaining objects is in progress as well as the determination of the radial dependence of the line ratios. To date, approximately 25% of the objects in each sample show no obvious extended emission.

REFERENCES

Balick , B., and T. M. Heckman, *Ann. Rev. Astron. Astrop.*, **20**, 431 (1982).

Carone, T. E., Simon L. Morris and R. W. Leach, *Optical Engineering*, **26**, No. 10, 1043 (October 1987).

Holmberg, E., in Stars and Stellar Systems, V. 9, ed. A. Sandage, M. Sandage and J. Kristian, The University of Chicago Press (1975).

Huchra, J., M. Davis, D. Latham and J. Tonry, *Ap. J. Supp.*, **52**, 89 (1983).

Leach, R. W., *Optical Engineering*, 26, No. 10, 1061 (October 1987).

Nilsson , P., "Uppsala General Catalogue of Galaxies," Uppsala Observatory Report No. 5 (1974).

Osterbrock, D. E., and W. G. Mathews, *Ann. Rev. Astron. Astrop.*, **24**, 171 (1986).

Veilleux, S., and D. E. Osterbrock, *Ap. J. Supp.*, **63**, 295 (1987).

Veron-Cetty, M.-P. and P. Veron, European Southern Observatory Scientific Report No.4 (2nd edition), April 1985.

Whittle, M., C. A. Haniff, M. J. Ward, E. J. A. Meurs, A. Pedlar, S. W. Unger, D. J. Axon and B. A. Harrison, *M. N. R. A. S.*, 222, 189 (1986).

Wilson, A. S., J. A. Baldwin and J. S. Ulvestad, *Ap. J.*, **291**, 627 (1985).

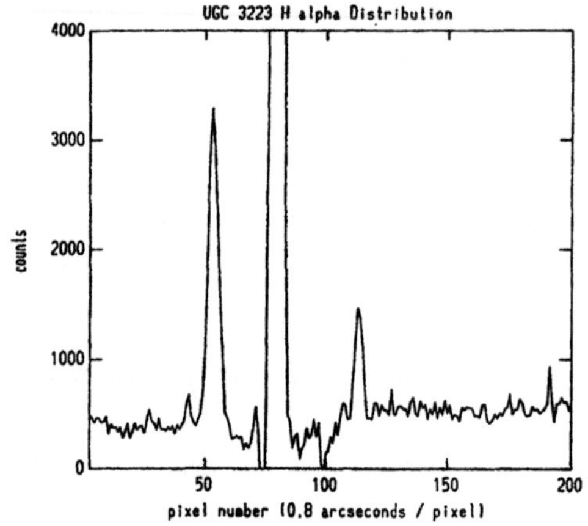

Fig. 1. Hα spatial distribution in the Seyfert galaxy UGC 3223.

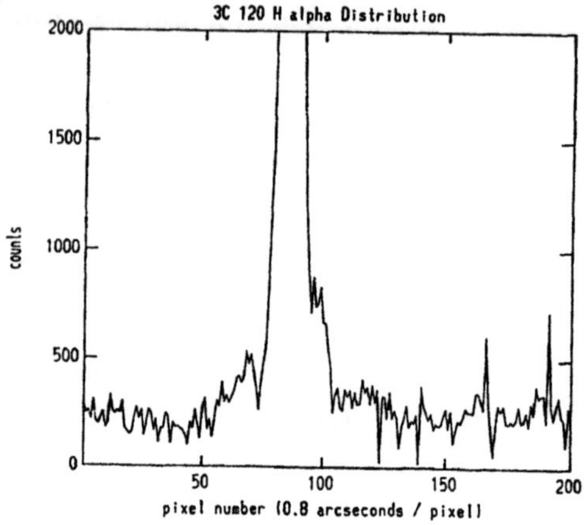

Fig. 2. Hα spatial distribution in the Seyfert galaxy 3C 120.

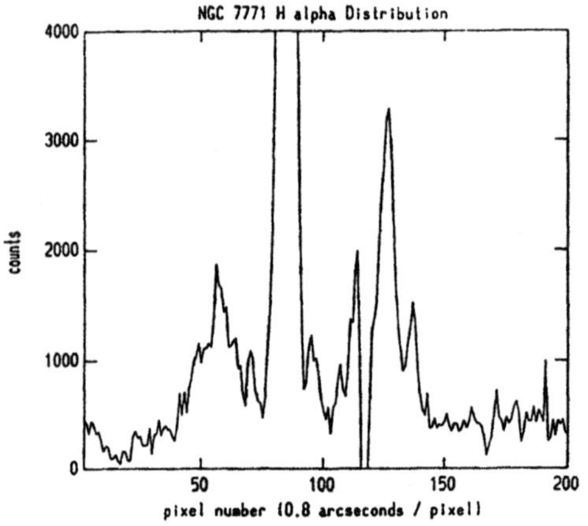

Fig. 3. Hα spatial distribution in the normal galaxy NGC 7771.

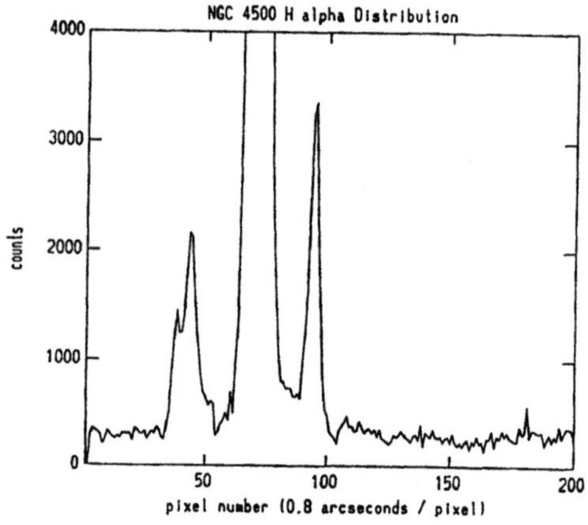

Fig. 4. Hα spatial distribution in the normal galaxy NGC 4500.

IMAGING SPECTROPHOTOMETRY OF IONIZED GAS IN ACTIVE GALAXIES

Gerald Cecil
Institute for Advanced Study, Princeton, NJ 08540
Joss Bland & Brent Tully
Institute for Astronomy, Honolulu, HI 96822

Our observational program aims to constrain the kinematic organization and dominant excitation mechanisms of ionized gas in active galaxies. We study both the large-scale patterns produced in starbursts (prototype M82), and those within the compact, but spatially extended, circumnuclear Narrow-Line regions (NLR). Particular questions we address are: how are *extranuclear* emission-line full-widths of >1500 km s^{-1} generated; what causes the observed gradients in excitation and ionization of the gas; how are these patterns interrelated to the extranuclear distribution of the radio and X-ray continua? To maximize spatial resolution, we select nearby active systems independent of luminosity but known to possess interesting morphologies and/or high-velocity extranuclear ionized gas. Imaging Fabry-Perot interferometers and low-noise (5 e^- readout) CCDs have been used in 1″ or better seeing at the U. Hawaii 2.2m and CFH 3.6m telescopes to obtain monochromatic images with velocity resolution of ~65 km s^{-1}. To avoid inter-order confusion, we use an etalon of 4000 km s^{-1} free spectral range. We map [NII]$\lambda\lambda$6548,6583 and Hα, with [OIII]λ5007 soon practical. The images are assembled into grids of line profiles at sub-arcsecond increments across a 3′ field. The profiles are of spectrophotometric quality, and the kinematic and spatial dependence of line ratios, hence gas excitation, can be mapped at the seeing limit.

Our datacubes $I(\alpha,\delta,\lambda,\pi)$ have 5-dimensions: 1 photometric (I), 2 spatial (α,δ), 1 spectral (λ) and 1 polarimetric (π). To handle the ~20,000 - 300,000 useful spectra that arise from a typical night's work, we have developed a complete analysis and reduction package for VAX and Sun image workstations. For starburst galaxies, reduction involves parametrization of ~10^8 raw data points to a few maps (e.g. velocities of each kinematic subsystem, continuum-free line fluxes) containing ~10^5 pixels. For NLRs we also examine the point to point variations of the synthesized line profiles. Much of the data in our massive spectral cubes become redundant once kinematic or structural symmetries have been recognized. The combination of monochromatic images and full spatial sampling of line profiles has allowed us to isolate such symmetries and has led to reliable kinematic deprojections. We find organized patterns even in emission distributions hitherto thought to be highly irregular. In the NLRs that we have studied, the morphology of the ionized gas distribution changes dramatically as the velocity resolution improves from the 1000 km s^{-1} (25 Å) passbands traditionally employed in interference imaging to our resolution of 65 km s^{-1}. Emission frequently concentrates into high-velocity, sharply defined, large-scale structures that attain great prominence. These resemble not static, photoionized nebulae, but rather the termini of hydrodynamic flows. We now summarize results for individual galaxies, selected to illustrate the application to AGN kinematics of a newly realized observational technique.

M 51

(For a detailed discussion see Cecil, 1988 May 15 *Ap. J.*) N^+ kinematics were mapped across the NLR (9″ radius). Two Hα velocity channels were also recorded. Complementary long-slit CCD spectroscopy provides S^+ densities and N^+ profiles with high signal-to-noise ratios at subarcsecond increments. Line intensity, line width, gas excitation and the radio continuum peak in a spatially resolved, extranuclear cloud (XNC) ~4″ south of the nucleus. Here, line profiles and channel maps show N^+ emission enhanced in a shell of unresolved thickness (<40 pc). The Hα distribution is more uniform, indicating that the N^+ enhancement is driven by temperature/ionization variations rather than by mass. The [NII]/Hα ratio has a distinct velocity-dependent component and peaks in a region occupied by relativistic charged particles, suggesting that cosmic-ray electrons are an important heat source. Emission is detected over a 1500 km s^{-1} range along a bright, arcuate surface near the southeastern boundary of the N^+ shell. The surface opens toward the interior of the XNC and incorporates ~4000 M_0 of high-velocity ionized gas. The nuclear isopleths are elliptical in line emission and the radio continuum, with the major axis aimed at this surface. The kinematics of this surface are deprojected to show that two superimposed cloud systems with velocity dispersion ~140 km s^{-1} are expanding at ~500 km s^{-1} in the disk of the galaxy. Thus in this NLR, line broadening and the asymmetries of the emission line profiles arise from *geometry, not selective obscuration.* The emission morphology, excitation distribution, and positional variation of N^+ kinematics in the XNC resemble those of galactic Herbig-Haro objects, and plausibly arise from thermalization of the kinetic energy of a directed nuclear outflow along well-defined oblique shocks in the ambient medium. Models of high-velocity, steady flow, planar shocks can fully account for the observed Hα luminosity only if the average preshock density is >50 cm^{-3}. Thus gas in the XNC must be exposed to an additional, centrally concentrated but nonstellar ionizing source if this density is confirmed to be ~1 cm^{-3}, as implied by the atomic hydrogen column.

M 82

(For further discussion, see Bland & Tully 1988, submitted to *Nature.*) For 25 years, the nature of the optical halo and filaments has been controversial. One view (Lynds and Sandage 1963) explains the filaments as fast-moving, line-emitting gas powered by an explosion at the nucleus (Axon & Taylor 1978). The other camp, originating with the discovery of well-ordered linear polarization by Elvius (1963), attributes the halo and filaments to scattered light from a bright, central source (Solinger & Markert 1975). These contradictory explanations have not hitherto been reconciled.

Our high spectral and spatial resolution [NII] and Hα observations now demonstrate that the line and continuum emission above and below the galactic disc arise from two distinct components: (1) filamentary, narrow-line emission (FWHM < 250 kms^{-1}) which is organized into a bipolar outflow (~600 kms^{-1}) over the surfaces of two elongated bubbles; (2) smooth, diffuse, broad-line (FWHM > 300 kms^{-1}) and continuum emission from a faint, exponential halo which rotates slowly (~100 kms^{-1}) with the disc. Our data provide strong evidence that M82 is a scaled-up bipolar reflection nebula: successive supernova ejections in the starburst nucleus have thermalized the surrounding medium which breaks out along the spin axis of the thick, gaseous disc as a high velocity, bipolar flow. The elongated bubbles extend to ~130″ (2 kpc) in radius along a position angle (~150°) almost parallel to the minor axis (~157°) of the disturbed disc. The narrow-line emission covers these surfaces in a complex filamentary network. Moreover, [NII]/Hα varies from 0.3 to 0.5 over individual filaments in contrast to the broad-line component where [NII]/Hα is ~0.8 everywhere. The low excitation, filamentary emission likely arises from cooling instabilities behind the moving shock surface set up by the hot wind as it sweeps up the interstellar medium. The broad-line emission and continuum appear to be light from the nuclear starburst that has preferentially escaped along the spin axis and is scattered to us by dust in an exponential halo. In this case, the scattered component is the probable source of the circularly symmetric, linear polarization observed about the galaxy.

NGC 1068

(See Cecil, Bland, & Tully, submitted to *Ap. J.*) Two datacubes have been acquired in N^+ and Hα: a high spatial resolution (0.8″) cube at *f*/8.6 on the U. Hawaii 2.2m telescope and a cube which is more sensitive to fainter emission at *f*/2 on the CFHT. Two striking trends are evident in both cubes:

(1) The brightest emission in the NLR "linearizes" along the axis of the radio jet as the channel velocity moves redward from systemic velocity (Fig. 1). The most straightforward kinematic deprojection suggests that ionized gas is expanding radially away from the jet axis in cylindrical symmetry. While linearization is most pronounced in the collisionally excited [NII] line, it is also apparent in Hα.

(2) The brightest line emission near systemic velocity is also associated with the radio jet: in particular, the NE radio lobe is enveloped by a sheath of line emission, visible as bright prongs in Fig. 1. Wilson and Ulvestad (1987) have modeled this lobe as a bow-shock in the ambient medium, and we suggest that the associated line emission arises in the post-shock recombination zone. The low velocity dispersion (~60 kms^{-1} FWHM) is explained if the radio ejecta

are colliding with the molecular ring mapped by Myers & Scoville (1987). Long-slit spectroscopy (Alloin *et al.* 1983) shows that [OIII]/Hβ ~17 for this component. This and other components are evident in Figure 2, a grid of our [NII]λ6583 profiles across the NLR. Variations on sub-arcsecond scales agree well with those more coarsely mapped with long-slit spectrographs (Baldwin, Wilson & Whittle 1987; Meaburn & Pedlar 1986).

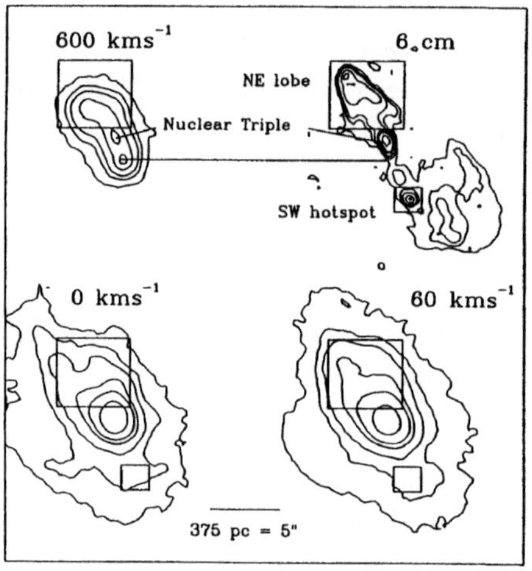

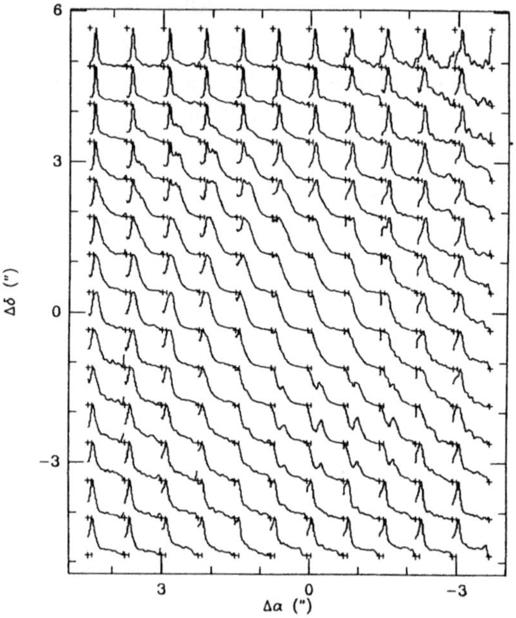

Fig. 1 (left) 5-minute exposure [NII]λ6583 channel maps of the NLR of NGC 1068. Velocities are relative to systemic.
Fig. 2 (right) Spectra of the red wing of [NII]λ6583 at 0.75 × 0.75" increments across the inner NLR of NGC 1068. A total velocity range of 2000 kms^{-1} is shown. Spectra are autoscaled at each point and smoothed to 100 kms^{-1} resolution.

NGC 4151

The NLR Hα emission concentrates in the SW quadrant where an extended component of soft X-ray emission has also been identified (Elvis *et al.* 1983). The Hα flux distribution, which closely resembles the [OIII]λ5007 image of Heckman and Balick (1983), is confined to a linear region with dimensions 13" × 30" roughly aligned with the axis of the radio "jet" (NE-SW). Within 8" of the nucleus, the NLR rotates rapidly in an inclined disk. We observe well-defined line splitting (< 300 kms^{-1}) fanning out NW and SE (offset by 5") from the nucleus: this is in direct contrast to the [OIII] line splitting which aligns with the jet axis (Schulz 1986). To the NE, the line profiles (FWHM < 350 kms^{-1}) are skewed and exhibit low velocity wings; to the SW, the profiles skew predominantly to higher velocities. Along the jet axis, the profiles become narrow (FWHM < 150 kms^{-1}) beyond 6" from the nucleus. Thus our observations separate the NLR into two distinct kinematic subsystems: a rapidly rotating disk, and ionized gas directly associated with the radio ejecta.

Our research is supported by NSF grant AST 86-15974 (Hawaii) and NASA grant NAS5-29225 (Princeton).

Alloin, D. *et al.* 1983, *Ap. J.*, **275**, 493.
Axon, D. & Taylor, K. 1978, *Nature*, **274**, 37.
Baldwin, J.A., Wilson, A.S., & Whittle, M. 1987, *Ap. J.*, **319**, 84.
Elvis, M. *et al.* 1983, *ApJ.* **268**, 105.
Elvius, A. 1963, *Low. Obs. Bull.*, **5**, 281.
Heckman, T.M. & Balick, B. 1983, *Ap. J.*, **268**, 102.
Lynds, C.R. & Sandage, A.R. 1963, *Ap.J.*, **137**, 1005.
Meaburn, J. & Pedlar, A. 1986, *Astr. Ap.*, **159**, 336.
Myers, S.T. & Scoville, N.Z. 1987, *Ap. J. (Lett.)*, **283**, 81.
Solinger, A.B. & Markert, T. 1975, *Ap. J.*, **197**, 309.
Wilson, A.S. & Ulvestad, J.S. 1987, *Ap. J.*, **319**, 105.

SOME SPECTROSCOPIC PROPERTIES OF MASS-EJECTING AND RADIO-LOUD QUASARS

Ray J. Weymann, Simon L. Morris, and Scott F. Anderson
Mount Wilson and Las Campanas Observatories, Carnegie Institution of Washington
813 Santa Barbara Street, Pasadena, CA. 91101

I. INTRODUCTION

The spectroscopic properties of quasars – continuum shape, emission line profiles and intensities, and absorption line strengths and profiles – have been under intense study in the two and a half decades since the discovery of quasars. The majority of the absorption features, especially those lines that are narrow and have redshifts significantly smaller than the emission line redshift, are thought to arise in clouds at cosmological distances from the QSO. Their study will lead, it is hoped, to an understanding of the distribution and properties of proto- and pre-galactic material, as well as material in galactic halos. We exclude from the present discussion this latter class of absorption lines. This class of absorption lines, along with all others, were the topic of a recent workshop (Blades, Norman, and Turnshek 1988) to which the reader is referred.

In the present discussion we concentrate on two classes of absorption systems in quasars which are thought to be physically associated with the quasar: The broad absorption line quasars and quasars with "associated absorption" systems.

II. BROAD ABSORPTION LINE QUASARS

The class of quasars that exhibit broad, blueward-displaced absorption lines in their spectra we refer to as broad absorption line quasars or "BALQSOs". Recent reviews of the properties of BALQSOs can be found in Weymann and Foltz (1983), Weymann, Turnshek, and Christiansen (1985), Turnshek (1986), and most recently in Turnshek (1988). A large sample of BALQSO spectra will be found in the first and the last of these references. Since the review in Turnshek (1988) is quite recent and comprehensive, we refer the reader to that review for much of the background to this topic, and in a sense the present review can be considered as commentary on several aspects of Turnshek's review.

1. What is the fraction of QSOs which show broad absorption line characteristics?

We, in collaboration with several colleagues, are engaged in a large survey (the "APM" survey) of optically-selected QSOs based upon UK Schmidt objective prism plates which are

scanned with the Cambridge Automatic Plate Measuring facility. The resulting one-dimensional spectra are then passed through a sieve of algorithms to select QSO candidates. Since a preliminary report on this survey has just been published (Foltz et al. 1987), and since some aspects of this survey were the subject of a poster paper at this conference (MacAlpine et al. 1988), we will not describe the survey again here. We only remark that one of its main virtues is that it allows quantitative estimates to be made of selection effects. In particular, while we have not yet completed a detailed analysis, preliminary results suggest that: (a) except possibly for very red "blazar"-type objects and about 10% of the objects for which the spectra on the plate overlap, this technique finds essentially all QSOs selected by all other techniques down to our magnitude limit of $m_J = 18.5$, provided the redshift is in the range $0.3 < z < 3.3$; and, (b) this technique does not miss BALQSOs which have properties similar to those of previously known BALQSOs. In this sense our survey is "complete" for both normal and BAL QSOs, although it may not be "complete" in the sense discussed recently by Netzer (1987), whose work we will discuss briefly below.

Including the objects published by Foltz et al. (1987) and QSOs identified during subsequent but still unpublished portions of the survey, we have examined about 320 QSOs. In order to classify a QSO as either a BALQSO or normal QSO, we require that the redshift be at least 1.5 so that the continuum is reasonably well-observed down to the SiIV-OIV] emission line.

It would be desirable to define some completely empirical and continuous index of "BALnicity", ranging from 0.0 in objects for which high signal-to-noise spectra show no trace whatsoever of absorption troughs, up to 1.0 for obvious cases such as those illustrated in the references cited above. This has not been done yet, and therefore, there are subjective judgments about whether to classify an object as a BALQSO or not, especially when the signal-to-noise is relatively poor. The two spectra in figure 9 of Turnshek (1988) provide a good illustration of this problem.

For the few objects in our survey which appeared to be borderline cases, for the present we subjectively assigned weights of 1/3, 1/2, or 2/3. All other objects with emission line redshifts greater than 1.5 were either assigned weights of 0.0 (non-BALQSOs) or 1.0 (definitely BALQSOs). We then find that the fraction of BALQSOs in the APM sample of optically-selected QSOs having emission line redshifts greater than 1.5 and weighted as described is 0.09, with a standard deviation of 0.03.

2. What is the fraction of all QSOs that are intrinsic BALQSOs?

By "intrinsic" BALQSOs we mean the following: Imagine that we could examine each QSO from every possible line of sight. If from any one of these lines of sight we could classify the QSO as a BALQSO then the QSO is an intrinsic BALQSO. The average fraction of 4π steradians filled by these lines of sight, or equivalently the average fraction of the sky covered by the BAL-producing clouds as seen from the continuum source, we refer to as the mean covering factor, which we denote here by $\langle\Omega/4\pi\rangle$. Evidently, the fraction of QSOs which are observed to be BALQSOs, f_{obs}, and the fraction of intrinsic BALQSOs, f_{intr} are related by

$$f_{intr} = f_{obs}/\langle\Omega/4\pi\rangle \quad (1)$$

Since the mean covering factor cannot exceed unity, f_{intr} must be at least as large as 0.09, and since f_{intr} must be $> f_{obs}$, it follows that $\langle\Omega/4\pi\rangle$ must be in the range

$$0.09 < \langle\Omega/4\pi\rangle < 1.0 \quad (2)$$

There has been considerable discussion about whether the actual mean covering factor is closer to the lower limit or the upper limit in equation (2). In the *former* case, the BAL-producing clouds cover only a small portion of the sky and would probably be confined to disks or cones. [In principle, one could have them uniformly distributed in solid angle, but patchily. In practice, the very large number of individual clouds inferred to be along the line of sight to an absorption trough, and the relatively common occurrence of multiple troughs make this possibility unlikely; cf. the discussion following Turnshek (1988).] In the *latter* case, the clouds completely cover the source in a more-or-less spherical manner and only a small fraction of the QSOs are intrinsic BALQSOs. If this latter view is approximately correct, BALQSOs are a different kind of object than normal QSOs, either because of some inherent but unknown property or, more likely, because of an evolutionary phase through which all luminous QSOs pass (cf. Hazard et al. 1984).

As discussed by many authors, one can in principle distinguish between these two extreme cases by considering the net deficit of photons as measured by the difference between the equivalent width of the emission line of the transition in question and the equivalent width of the corresponding absorption. One first subtracts the estimated number of photons created by collisional excitation within the BAL-producing clouds, and within a "normal" broad emission line region interior to the BAL-producing clouds. (This can be done, for example, by measuring the strength of the CIII] emission line for which scattered photons are not important,and adopting some assumed ratio between the collisionally-excited component of CIII] and the line in question, e.g., CIV.) There is some chance for systematic bias here, however, because the CIII]/AlIII/FeII emission complex may be anomalous in BALQSOs, as discussed by Hartig and Baldwin (1986).

When this is done, there can be a net deficit of scattered photons reaching the observer only if : (i) some of the scattered photons are collisionally destroyed within the BAL-producing clouds before escaping; (ii) some of the scattered photons are destroyed by being scattered into much denser or dust-containing regions before escaping; or, (iii) the covering factor of the BALQSO clouds is less than unity, so that more photons are scattered out of the line of sight than are scattered back into it by other BALQSO clouds not in the line of sight.

In some discussions of the problem (cf. Surdej and Hutsemekers 1987; Turnshek 1988), one sometimes sees considerations having to do with the velocity field of the BAL clouds (e.g., the degree of "turbulence", or decelerating vs. accelerating flows) cited as additional factors which influence the net equivalent width of the line. Such factors merely influence the number of scatterings that the photons must undergo before escaping but cannot, of course, actually destroy photons.

In many BALQSOs, the net equivalent width of the CIV emission line is sufficiently close to zero even prior to subtraction of the collisionally excited component that the uncertainties of this subtraction make it difficult to establish that there is a net deficit of photons in the resulting profile. RS23 is an example of such a BALQSO. However, in other objects – probably the majority of BALQSOs – there is an obvious deficit of photons even without subtracting off the collisionally excited component. The BALQSO shown in Figure 1 provides a good illustration of this.

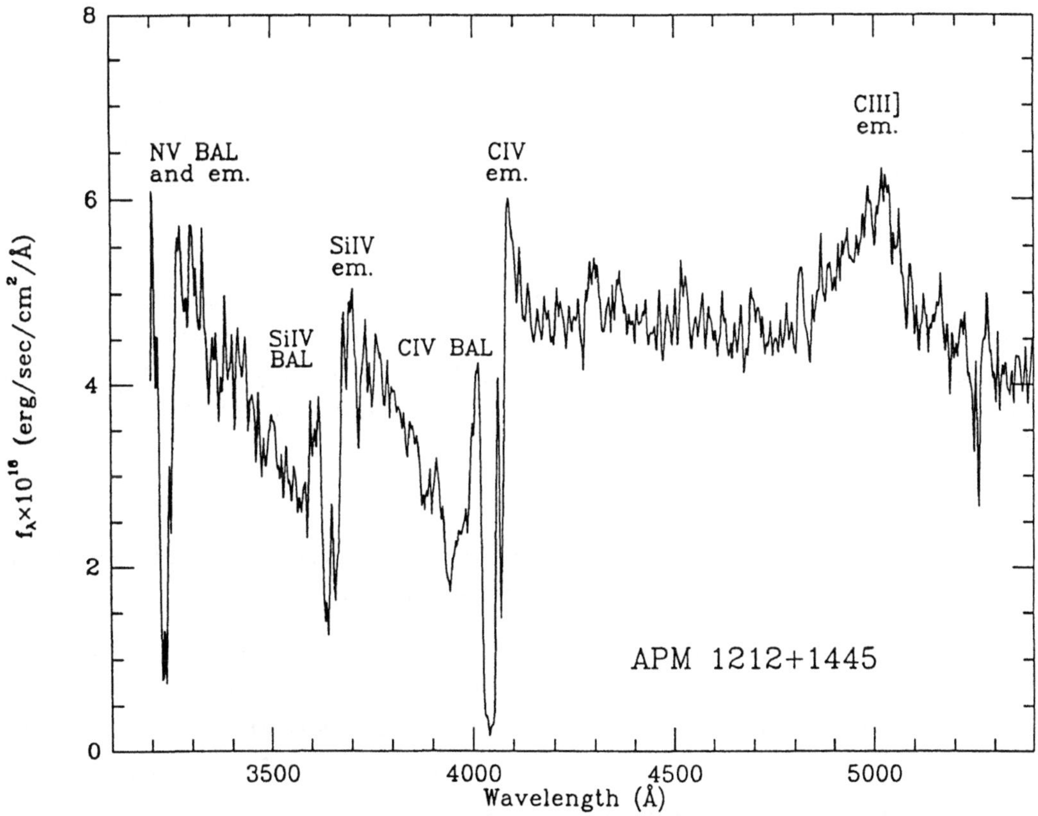

Figure 1. BALQSO (z=1.638) from the APM survey (Foltz et al. 1987).

Clearly, at least in BALQSOs like the one shown in Figure 1, one or more of the factors (i)-(iii) enumerated above must strongly apply.

Unless the current estimates of the densities and optical depths of the BAL clouds are off by several orders-of-magnitude, mechanism (i) cannot be of importance. Mechanism (ii) is hard to exclude. It may very well be that the back portions of the optically-thick clouds in the broad emission line region which are strongly shielded from ionizing radiation may contain dust. It is also possible, but we think unlikely, that the densities in the low-ionization portions of an accretion disk are sufficiently high that some collisional de-excitation takes place. In either event, the question then is: what is the covering factor presented to a scattered photon escaping from the broad absorption line clouds by such dust-containing regions? This is very hard to estimate. It seems likely that the broad emission line region, or at least that portion of it responsible for the Lyα line, is significantly smaller than the region producing the broad absorption in the NV ion since, whenever the velocity of the NV absorption is large enough to extend from a rest wavelength of 1240Å down to 1216Å, the Lyα emission seems almost completely obliterated. Because of this, we think it unlikely that destruction of photons scattered in the broad absorption line clouds can be accounted for by absorption somewhere in the broad emission line region, or, for similar reasons, in either the thermal and non-thermal continuum-producing regions. However, it is possible that some high-luminosity QSOs are surrounded by a large "dusty torus" similar to that discussed at this conference in connection with Seyfert II galaxies (e.g., Miller 1988). If such a dusty torus were to fill a large portion of the sky as seen by one of the broad emission line clouds, then a significant fraction of the scattered photons could be destroyed. We think it somewhat unlikely though, that this mechanism alone could account for the profiles of objects like the one shown in Figure 1.

We believe that the remaining possibility (iii above) is the most likely, and that therefore the covering factor associated with the broad-line absorption clouds is substantially less than unity. This in turn implies, through equation (1), that the intrinsic fraction of QSOs which are BALQSOs, f_{intr}, is substantially greater than the observed fraction, $f_{obs} = 0.09$, *and in fact, the majority of all QSOs may be ejecting matter and energy just as the BALQSOs are observed to do.*

3. How can the differences between the emission line properties of normal QSOs and BALQSOs be accounted for?

If we accept the fact that the fraction of QSOs which are intrinsic BALQSOs may be nearly unity, then it seems at first sight as if the distribution of total strengths of the emission lines of a given ion in BALQSOs should be about the same as in normal QSOs, whereas the distributions of the properties of the emission line profiles (e.g., widths) might differ. One might imagine, for example, that the observed BALQSOs are just the subclass of QSOs having a particular orientation in which the outflowing material in a disk is seen edge-on, so that the emission line widths might be broader than in normal QSOs. But if one assumes that the emission is emitted isotropically, then the strength of the emission should be immune to such orientation effects.

Several authors have therefore made such comparisons. Hartig and Baldwin (1986) conclude, for example, that the FeII emission tends to be stronger in BALQSOs than in normal QSOs, while Surdej and Hutsemekers confirmed a previous result of Turnshek's that the NV in BALQSOs tends to be stronger than in normal QSOs. (This was not confirmed by Hartig and

Baldwin, however. Our own impression is also that the NV is generally stronger in BALQSOs. However, with the exception of the recent work by Junkarrinen, Burbidge, and Smith (1987), the measurements of the integrated strengths of NV to date have not been very sophisticated and need to be improved.)

These results obviously contradict the simple expectations above. The assumption that the emission is isotropic may not be valid, however. For example, Turnshek (1988) has presented an ingenious model in which individual BALQSO clouds are in thin sheets oriented normal to the direction of outflow, with the whole ensemble of the BAL clouds confined to a disk. Such an arrangement might be expected, for example, if there are cooling shocks in a flow which is confined to an equatorial region. The optical depth for NV photons emitted normal to the sheets is observed to be of order unity, but could easily be an order of magnitude larger for photons traveling nearly in the plane of the sheets. As we have confirmed by simple Monte Carlo calculations, the angular distribution of the emitted photons in such a case is strongly anisotropic.

One can imagine an even simpler way of avoiding the apparent contradiction described above: It must certainly be true that the BAL covering factor varies from QSO to QSO. Suppose that there is a distribution of the broad absorption line region covering factor with an (observed) expectation value of 0.09. Since the probability of observing a QSO to be a BALQSO is proportional to the BAL covering factor, a sample of QSOs observed to be BALQSOs will have a larger mean covering factor than a sample of QSOs which are observed not to be BALQSOs. Since the amount of collisionally excited NV as well as the amount of continuum and Lyα photons which are scattered by the BAL clouds (and which therefore contribute to the observed NV emission line strength) is also proportional to the covering factor, it follows that the BALQSOs will have systematically stronger NV emission line strengths than the non-BALQSOs.

Whether either of these two (or any other) models can reproduce the distribution of the Lyα plus NV profile and strength characteristics among both BALQSOs and non-BALQSOs remains to be seen.

It also remains to be seen whether variants of these models can reproduce the different distributions of FeII strengths between the BALQSOs and non-BALQSOs. This will be difficult, because the explanation for the FeII strengths even among the normal QSOs themselves is not at all understood (Netzer 1987). Netzer also points out an additional possible selection effect which might bear upon the observed distribution of spectral features between BALQSOs and non-BALQSOs: If an accretion disk contributes significantly to the visible and near-UV continuum of QSOs as is now widely believed, then since the observed flux from such a disk is likely to be anisotropic, a flux-limited sample of QSOs may not be a randomly oriented sample.

Before leaving this topic, we mention two other related matters:

(i) Turnshek (1988) presents convincing evidence that properties of the absorption profiles are correlated with properties of the emission profiles in BALQSOs (see his figures 2-5). This suggests that there is a fairly intimate causal connection between the region giving rise to the broad absorption lines and the region giving rise to the high-ionization emission lines. There is a growing suspicion however, that the high ionization and low ionization emission lines come from different regions (cf. Collin-Souffrin et al. 1988, especially figure 1.). It would be interesting to determine, therefore, if a similar correlation exists between the broad absorption line profiles and the profiles of the low ionization emission lines (e.g., MgII).

(ii) Turnshek (1986, 1988) has asserted that there may be very large enhancements of the metal-to-hydrogen ratio in one or two BALQSOs. This is a difficult problem, but one which is obviously of very great interest and importance. More work is required to confirm (or disprove) this result. It would also be of interest to see what reliable constraints on metal enhancement can actually be placed on the broad emission line region with the best models and observations currently available. The possibility that the iron abundance may be significantly enhanced in the emission line region (Netzer 1987) is suggestive, though the models invoking heat deposition in the accretion disk (Collin-Souffrin et al. 1988; see also the discussion by Krolik 1988 at this conference) may avoid the necessity of invoking enhanced iron abundance.

4. Do broad absorption quasars avoid strong radio sources, and if so, how can this be explained?

The evidence continues to become stronger that BALQSOs avoid strong radio sources. We have now examined the spectra of approximately 130 radio-selected QSOs, both some of our own data and that taken by others, which have appropriate signal-to-noise, resolution, and wavelength coverage to allow detection of broad absorption troughs. To date, for only one object, PKS1157+014, is the presence of a broad absorption feature a real possibility.

In making this latter statement, it is important to distinguish between genuine BALQSOs on the one hand, and damped Lyα systems and the "associated" CIV absorption systems (discussed below) on the other. The latter could in some instances be mistaken for BAL troughs. However, one of the working definitions of a BALQSO trough is that it extend at least 5000 km/sec to the blue of the rest frame of the QSO and that there be contiguous absorption over at least 2000 km/sec. This definition serves to separate the BAL troughs from the associated absorption systems. In addition, at moderate resolutions, the associated absorption systems either break up into discrete sharp components or else have both red and blue edges that are quite sharp. Using these precepts, PKS1157+014 is not actually a BALQSO (Weymann et al. 1987). Referring to the middle panel of Figure 8 in Turnshek (1988), the feature marked "24" is definitely what we would classify as "associated absorption", not a BAL trough. The feature marked "22" has the necessary blueward displacement, but is actually narrower than 2000 km/sec when account is taken of the splitting of the CIV doublet.

As in previous discussions of the question of whether BALQSOs avoid strong radio sources, however, we will compute statistics of

(i) 0 or

(ii) not more than 1

BALQSOs occurring in a sample of 130 radio-selected QSOs when the "true" rate (as measured from our survey of optically-selected QSOs) is either

(a) 0.09 (our best estimate), or

(b) 0.06 (one standard deviation below our best estimate).

The following table then summarizes these probabilities:

Table 1. Do BALs "avoid" radio-loud QSOs?

BAL incidence	PKS 1157 a BAL	PKS 1157 not a BAL
0.06	3.0×10^{-3}	3.2×10^{-4}
0.09	6.6×10^{-5}	4.7×10^{-6}

Even if PKS 1157+014 is counted as a BALQSO and the true rate of BALQSOs among optically-selected QSOs is as low as 0.06, a real physical effect seems to be present. What might be the interpretation of this?

Two explanations have often been suggested:

(1) BALQSOs might inhabit only spiral galaxies, while QSOs that are strong radio sources might inhabit only elliptical galaxies. Much more work will be needed to either prove or disprove this suggestion, but in either case it is not what we would consider an "explanation". One would like to know if and how the presence or absence of a gas-rich environment can influence the presence or absence of BAL clouds and compact radio jets if both phenomena arise on scales of only a few parsecs from the QSO nucleus.

(2) Radio jets are beamed; the BALQSO clouds could be ejected along the equatorial regions. Therefore pole-on QSOs would be radio sources, edge-on QSOs could be BALQSOs. However, Malkan, Green, and Hutchings (1987) note that the low-redshift BALQSO 1411+442 does not appear to be in an edge-on galaxy. The main additional difficulty we see with this explanation is that a substantial fraction of the 130 radio-selected QSOs observed not to be BALQSOs have strong extended radio components, and the flux in these sources are thought not to be beamed and therefore to comprise a more-or-less isotropically distributed sample. We have not yet examined in detail the radio properties of the 130 radio QSOs, and the statistics will probably not yet be adequate to establish that BALQSOs avoid both compact and extended strong sources, though our suspicion is that this will turn out to be the case.

If so, this suggests that some physical mechanism selects between the two types of objects. The reason why most QSOs are not powerful radio sources is still not clear. The work of Edelson and Malkan (1986) suggests that a synchrotron source is present but becomes self-absorbed and turns over in the far IR. Edelson and Malkan estimate that the physical size of the region containing the relativistic plasma is comparable to the size expected for the inner portions of an accretion disk. Elsewhere (Weymann et al. 1987), we suggested that whether or not this confined relativistic plasma could break out of the confining region might determine whether the QSO becomes a strong radio source or not. Further, whether such a "break-out" occurs or not might influence, or be influenced by, the amount of energy deposited in the (thermal?) flow giving rise to the BAL phenomenon. It would be useful to model this possibility quantitatively. Our crude estimate suggested that the power in radio jets might be comparable to a putative thermal wind driving the BAL clouds (Weymann et al. 1982).

The power estimate for the BAL outflow made by Weymann et al. (1987) assumed confined clouds embedded in a hot thermal gas under quasi-steady conditions. (See also the discussion in Turnshek 1988). The smooth absorption profiles in BALQSOs like PHL5200 suggest quasi-steady outflow. On the other hand, the objects with complex multiple troughs (see, for example, figure 3 and especially figure 5 in Turnshek 1988) suggest discrete outbursts, and the total energy rather than the power is a more meaningful quantity. There is a potential conflict, however, between the "single outburst" model and recent estimates which put the typical distances of the BAL clouds at a few parsecs (Turnshek 1988). The time for a BAL cloud moving at 30,000 km/sec to cover 10 parsecs or one moving at 10,000 km/sec to cover 3 parsecs is only 300 years. If such complex troughs are truly "one-shot" events then they cannot be detected for much longer than this; otherwise, they would typically be at much larger distances. But since the fraction of all QSOs which are BALQSOs is at least of order 10^{-1}, a lifetime of even 10^3 years for the BAL phenomena would imply a lifetime of only the order of 10^4 years for the (non-BALQSO) progenitors or successors of such objects. On the other hand, recent estimates of ***minimum*** lifetimes for normal QSOs are $> 10^7 - 10^8$ years (e.g., Phinney 1988), several orders of magnitude greater than that consistent with the progenitor/successor lifetime required in the "single outburst" model. Even assuming the objects with complex troughs to be the last phase in an evolutionary development of BALQSOs which start out as PHL5200-type objects does not alleviate the difficulty if the entire BALQSO phase is a one-shot affair. The evolutionary scenario and the small distances for the clouds from the central source lead to a consistent picture if a steady wind is perturbed fairly frequently by strong episodic events.

III. ASSOCIATED CIV ABSORPTION SYSTEMS

We use the term "associated absorption" to refer to absorption at or near the emission line redshift of the quasar. In the work on these systems in which we have been involved, we have defined "at or near" to mean absorption within ±5000 km/sec of the emission line redshift. It should be noted, however, that this same term has also been applied to absorption which is seen at a common redshift in both members of a pair of quasars, as well as to absorption seen in one member of the pair and having about the same redshift as the emission line redshift of the other member of the pair.

Most studies of associated absorption systems have focused on the CIV $\lambda1549$ transition but associated absorption is occasionally seen in the MgII λ2798 transition and, of course, Lyα. In the latter case, however, it is very difficult to disentangle such absorption from the Lyα forest lines.

Recent discussions of the CIV associated absorption systems will be found in Weymann et al. (1987), Anderson et al. (1987), and most recently in Foltz et al. (1988a) in the proceedings of the Baltimore conference. As in the case of Turnshek (1988), the review by Foltz et al. (1988a) covers the significant recent work on this topic and we shall give only a brief synopsis and a few additional comments.

1. Do the CIV Associated Absorption Systems form a class of absorption systems physically distinct from the "intervening systems"?

This question is not very precisely posed, but means roughly the following: If all the CIV absorption systems are ascribed to clouds in the halos of galaxies, then the distribution of these systems should be Poisson. There should be only a very small excess of systems near the emission line redshift of the background QSO due to the correlation of galaxies with the quasar host galaxy. Moreover, the equivalent width distribution of systems at absorption redshifts near the emission redshift should not differ from that at redshifts much smaller than the emission line redshifts. If, on the other hand, as discussed below, one observes significant excess absorption near the emission line redshift in some QSOs, it indicates either that:

(i) The correlation between the putative halos of galaxies and the QSOs are anomalously strong, e.g. the QSOs showing strong associated absorption occur preferentially in clusters of galaxies.

(ii) The clouds are actually intrinsic to the QSO. Here "intrinsic" could mean that they are at galactic distances (kpcs to tens of kpcs), or conceivably even closer (for example, possibly the residue of clouds ejected from the QSO during a prior BALQSO phase).

The initial work of Weymann et al. (1979) led to the conclusion that the frequency of CIV absorption lines near the emission line redshift was larger than expected for purely intervening systems. Subsequently, Young et al. (1982) failed to confirm this result. Foltz et al. (1986) repeated the initial survey of Weymann et al. (1979) using a more suitable observational technique, and went on to suggest that the discrepancy between the initial Weymann et al. (1979) and subsequent Young et al. (1982) results was probably due to a difference between the two samples: The Weymann et al. sample was heavily weighted with powerful radio sources, especially extended, steep spectrum radio sources, while the Young et al. sample had only a few such objects. An additional (albeit small) sample consisting of additional 3C and 3CR sources (Anderson et al. 1987) showed a very high (≈ 50%) incidence of strong associated absorption.

Independently, P. Barthel and collaborators began investigating the statistics of the morphology of radio-selected quasars as a function of redshift (e.g, Barthel and Miley 1988, and references therein). They show that at high redshifts the radio sources are smaller and more distorted than at low redshifts. They interpret this as evidence of a denser inter- or intra-galactic medium which may limit the size to which the radio lobes can grow. They make estimates for the number and physical properties of clouds which could distort the radio jet. They also go on to suggest that these same clouds give rise to the associated absorption.

In a parallel investigation, Barthel and collaborators have therefore been carrying out a program of optical spectroscopy of a large number of steep-spectrum radio sources, especially at high redshifts. Initially, it appeared that there was a very strong correlation between the presence of associated CIV absorption and the presence of bends or "kinks" in the radio morphology. This would appear to support the idea that the same kind of clouds which give rise to the associated absorption are massive enough to deflect the radio jets when the jets encounter such clouds. When a large number of clouds are present, the probability would be high that there would simultaneously be both clouds intersecting the line of sight, and clouds encountered by the jets, hence the correlation. Of course, one would not expect the correlation to be perfect, since it may happen that the jets miss the clouds in cases where the line of sight intersects a cloud, and vice versa.

In fact, the whole question of the nature of the correlation between the presence of associated absorption and the radio properties of the QSOs seems less clear-cut to us than the anti-correlation between radio power and BALQSOs. It still appears that the incidence of strong associated absorption is greater among radio sources than among optically-selected QSOs. However, strong associated absorption is certainly occasionally seen among optically-selected QSOs. Nor is it it yet clear what the relevant radio parameter really is in this correlation, and in fact Moller and Jakobsen (1987) have suggested that the primary parameter may really be optical luminosity (with radio as a secondary parameter). Also, most of the CIV surveys of radio-selected QSOs have been weighted towards steep-spectrum sources. Clarification of these matters must await the completion of the Barthel et al. survey and the survey of Foltz et al. (1988b), and will also benefit from a careful analysis of surveys of optically-selected QSOs (e.g., Steidel and Sargent 1988, Foltz et al. 1987).

2. Assuming that strong associated systems tend to occur in the spectra of strong extended radio sources, what is the origin of these absorption systems?

The two main possibilities for the origin of the associated absorption systems would seem to be: (i) low-velocity remnants of a prior ejection phase of the quasar, perhaps akin to the BAL phenomenon; or, (ii) a gas-rich environment in the neighborhood of the quasar host galaxy itself, or in a cluster in which the quasar resides.

The remnant ejection scenario has been discussed by Briggs et al. (1984). In order to check this notion, Anderson et al. (1988) have undertaken a spectroscopic survey of QSOs known to have strong associated absorption in order to see if there is evidence for any net outflow of the absorbing material. This is done by measuring the systemic velocity of the QSO using the MgII λ2798 (or where possible [OII] λ3727)emission line, since the centroid of the CIV emission line is known not to be a reliable indicator of the "true" systemic velocity (e.g., Gaskell 1982, Wilkes 1986). To date, we have measured velocities for an *a priori* unbiased sample of eleven objects, and find that the mean of the centroids of these associated absorption systems shows a net outflow of 650 km/sec, but with an *rms* dispersion of 970 km/sec. Thus, there is a slight suggestion of net outflow; however, more data (and the systematics of the displacement of MgII with respect to [OII] need to be better understood) before this slight suggestion can be taken seriously. If the existence of net outflow were to be confirmed, it would be a strong argument in favor of some origin similar to the Briggs et al. picture. Even if no outflow is found, however, the Briggs et al. model cannot be ruled out, for one could argue that interaction with the interstellar or intragalactic media has slowed the absorbing material sufficiently that the net outflow velocity has become vanishingly small. By the same token, the cases where there is a clear indication of an "infall" velocity of the absorbing material (Junkarrinen 1987) can be accommodated in the Briggs et al. picture by supposing that the ejected gas has been slowed so that it has nearly the mean velocity of the intracluster medium, but (in these cases) the host QSO galaxy has a peculiar velocity in the cluster which is towards us.

While it seems to us difficult to rule out the remnant ejecta model, our prejudice is that an explanation of the sort advanced by Weymann and Williams (1978)–namely, that the absorption arises in environmental material in a cluster in which the QSO resides is more likely to be correct. Barthel and Miley (1988) have made the interesting suggestion that such absorbers are actually proto-cluster material or debris associated with the formation of the QSO host galaxy. If the estimates of Barthel et al. (1988) about the number and physical properties of the clouds is correct, then *emission* from the clouds, either in CIV, or better still in Lyα should be detectable, as originally suggested by Williams et al. (1975) for the case of 3C191, the prototype associated absorption system object. A program of narrow band Lyα and/or CIV imaging of the associated absorption systems would seem to be a worthwhile undertaking.

Remembering that strong associated absorption systems seem to be found preferentially radio-loud QSOs, the above considerations point to the possibility of an intimate relation between the amount of gas in the QSO environment and the QSO's radio power (cf. also Yee and Green 1987; Hutchings 1987), but still leave us uncertain about the fundamental nature of that relationship.

Further, we again draw attention to the correlations seen in QSOs between their optical spectra and radio properties. For example, Boroson et al. (1985) and Stockton and MacKenty (1987) have found that, in low redshift objects, QSOs with steep radio spectra and extended radio morphology have both resolvable and "nuclear" [OIII] emission that is stronger than that found in radio-quiet QSOs or those with flat radio spectra; and, on the other hand, low redshift radio-quiet QSOs and those having flat radio spectra tend to show stronger nuclear FeII emission than extended, steep-spectrum radio sources. In this connection, Weymann et al. (1987) and Anderson et al. (1987) noted that the optical spectra of high-redshift radio-loud

QSOs seem to have unusually narrow, and yet quite strong, HeII λ1640 emission (as well as a high incidence of strong associated CIV absorption); this may be a high-redshift manifestation of the phenomena discussed by Boroson et al. and Stockton and MacKenty. Inspection of the dataset of Barthel et al. (1988) qualitatively confirms our earlier impression that HeII λ1640 (plus OIII] λ1660) emission is stronger than in optically-selected QSOs, but quantitative confirmation of this must await completion of a survey of radio-quiet QSOs (e.g., a subset of the APM survey) at comparable signal-to-noise to that already available for radio-loud QSOs. Weymann et al. (1987) went on to suggest that the paucity of the BALQSO phenomenon among radio-loud QSOs on the one hand, and the apparent strong FeII emission among BALQSOs on the other hand, is yet further evidence for a high redshift extension of the dichotomy in QSO types proposed by Boroson et al.

Thus, these two apparently disparate types of absorption systems–the BALQSOs and the associated absorption systems–may reflect some more fundamental properties of QSOs, and in particular the circumstances under which the environment allows relativistic plasma to form jets streaming out from the QSO. But we still lack a convincing picture of what these circumstances and fundamental properties are.

Acknowledgements: We thank Drs. Craig Foltz and Peter Barthel for a number of interesting discussions on the topics considered here.

IV. REFERENCES

Anderson, S. F., Weymann, R. J., Foltz, C. B., and Chaffee, F. H., Jr. 1987, *A. J.*, **94**, 278.

Anderson, S. F., et al. 1988, *in preparation.*

Barthel, P. D. and Miley, G. K. 1988, *submitted to Nature.*

Barthel, P. D. et al. 1988, *in preparation.*

Blades, C., Norman, C., and Turnshek, D. (editors) 1988, *Proc. of the STScI Conf. on QSO Abs. Lines (May 1987)*, Cambridge Univ. Press.

Boroson, T. A., Persson, S. E., and Oke, J. B. 1985, *Ap. J.*, **293**, 120.

Briggs, F. H., Turnshek, D. A., and Wolfe, A. M. 1984, *Ap. J.*, **287**, 549.

Collin-Souffrin, S., Dyson, J., McDowell, M., and Perry, J. 1988, *preprint.*

Edelson, R. A. and Malkan, M. A. 1986, *Ap. J.*, **308**, 59.

Foltz, C. B., et al. 1987, *A. J.*, **94**, 1423.

Foltz, C. B., Chaffee, F. H., Jr., Weymann, R. J., and Anderson, S. F. 1988a, to appear in *Proc. of the STScI Conf. on QSO Abs. Lines (May 1987)*, eds. C. Blades, C. Norman, and D. Turnshek, Cambridge Univ. Press.

Foltz, C. B. et al. 1988b, *in preparation.*

Gaskell, C. M. 1982, *Ap. J.*, **263**, 79.

Hartig, G. F. and Baldwin, J. A. 1986, *Ap. J.*, **302**, 64.

Hazard, C., Morton, D. C., Terlevich, R., and McMahon, R. G. 1984, *Ap. J.*, **282**, 33.

Hutchings J. B. 1987, *Ap. J.*, **320**, 122.

Junkkarinen, V. T. 1987, in *QSO Absorption Lines: Probing the Universe, A collection of poster papers from the STScI Conf. on QSO Abs. Lines (May 1987)*, eds. C. Blades, C. Norman, and D. Turnshek, STScI.

Junkkarinen, V. T., Burbidge, E. M., and Smith, H. E. 1987, *Ap. J.*, **317**, 460.

Krolik, J. H. 1988, *proceedings of this conference.*

MacAlpine, G. M. et al. 1988, *proceedings of this conference.*

Malkan, M. A., Green, R. F., and Hutchings, J. B. 1987, *Ap. J.*, **327**, 729.

Miller, J. S. 1988, *proceedings of this conference.*

Netzer, H. 1987, *M.N.R.A.S.*, **225**, 55.

Phinney, S. 1988, *in preparation.*

Steidel, C. and Sargent W. L. W. 1988, *in preparation.*

Stockton, A., and MacKenty, J. W. 1987, *Ap. J.*, **316**, 584.

Surdej, J. and Hutesemekers, D. 1987, *Astr. Ap.*, **177**, 42.

Turnshek, D. A. 1986, in *"Quasars", Proc. of IAU Symp. 119,* eds. G. Swarup, and V. K. Kapahi, Dordrecht: Reidel.

Turnshek, D. A. 1988, to appear in *Proc. of the STScI Conf. on QSO Abs. Lines (May 1987)*, eds. C. Blades, C. Norman, D. Turnshek, Cambridge Univ. Press.

Weymann, R. J. and Williams, R. E. 1978, *Phys. Scripta,* **17**, 217.

Weymann, R. J., Williams, R. E., Peterson B. M., and Turnshck, D. A. 1979, *Ap. J.*, **218**, 619.

Weymann, R. J., Scott, J. S., Schiano, A. V. R., and Christiansen, W. A. 1982, *Ap. J.*, **262**, 497.

Weymann, R. J. and Foltz, C. B. 1983, in *"Quasars and Grav. Lenses", Proc. of the Liege Int. Ap. Colloq. (June 1983)*, p. 538 ff.

Weymann, R. J., Turnshek, D. A., and Christiansen, W. A. 1985, in *"Ap. of Active Galaxies and Quasi-Stellar Objects", Proc. of the Santa Cruz Ap. Workshop (July 1984)*, p. 333 ff.

Weymann, R. J., Anderson, S. F., Foltz, C. B., and Chaffee, F. H., Jr. 1987, *Thirteenth Tex. Symp. Relat. Ap.*, ed. M. Ulmer, Singapore: World Science, p. 297 ff.

Wilkes, B. J. 1986, *M.N.R.A.S.*, **218**, 331.

Williams, R. E., Strittmatter, P. A., Carswell, R. F., and Craine, E. R. 1975, *Ap. J.*, **202**, 296.

Yee, H. K. C. and Green, R. F. 1987, *Ap. J.*, **319**, 28.

Young, P., Sargent, W. L. W., and Boksenberg, A. 1982, *Ap. J. Suppl.*, **48**, 455.

INFRARED SPECTROSCOPY OF NGC 1068

D. L. DePoy
National Optical Astronomy Observatories
P.O.Box 26732, Tucson, AZ 85726

Abstract. Spectroscopy of the nucleus of the nearby Seyfert 2 galaxy NGC 1068 has been obtained using the IRTF and the facility Cooled-Grating Array Spectrometer (CGAS). The wavelengths observed covered the expected wavelengths of the Brγ ($n = 7 \rightarrow 4$), Brα ($n = 5 \rightarrow 4$), and the Pfβ ($n = 7 \rightarrow 5$) hydrogen recombination lines between 2.2 μm and 4.6 μm. The CGAS has a 32-element InSb array multiplexed by a Reticon and was used at a resolution of ~250 km s^{-1}, allowing accurate determination of the line profiles and surrounding continua simultaneously. For example, the S/N at each spectral position in the Brα line spectrum was ~100. The data show that the infrared lines are not more broadened than the optical hydrogen recombination lines, suggesting that the presence of an obscured Seyfert 1-like active nucleus is unlikely unless the visual extinction through any obscuring material present is larger than ~100 mag.

I. INTRODUCTION

Optical spectropolarimetric observations of the nucleus of the Seyfert 2 galaxy NGC 1068 by Antonucci and Miller (1985) have shown the presence of polarized emission lines that have velocity widths much broader than those of unpolarized lines. The polarized spectrum closely resembles the canonical spectrum of Seyfert 1 nuclei, with the Hβ hydrogen recombination line exhibiting a full width at half maximum (FWHM) of ~2500 km s^{-1}. Antonucci and Miller interpret this result as due to electron scattering into the line of sight of emission from a broad-line region (BLR) located within an optically and geometrically obscured disk. They further suggest that this may be a common phenomenon in Seyfert 2 galaxy nuclei.

Observations in the infrared can probe much deeper into a region than optical observations. Therefore, measurements of infrared hydrogen recombination lines should reveal if an obscured Seyfert 1 nucleus exists in NGC 1068. The Brγ ($n = 7 \rightarrow 4$) line has already been measured by several investigators (Hall *et al.* 1981; Thompson, Lebofsky, and Rieke 1978) and shows no evidence for an emission profile broader than the unpolarized optical Hα and Hβ lines (FWHM ~ 1200 km s^{-1}). Therefore, the Brα ($n = 5 \rightarrow 4$) and Pfβ ($n = 7 \rightarrow 5$) lines around 4.05 μm and 4.65 μm, respectively, have been observed. Additionally, Brγ has been measured for comparison.

II. OBSERVATIONS AND RESULTS

Infrared spectroscopy of the nucleus of the nearby Seyfert 2 galaxy NGC 1068 has been obtained using the IRTF and the facility Cooled-Grating Array Spectrometer. The spectrometer has a 32-element InSb array multiplexed by a Reticon and was used at a resolution of ~250 km s^{-1}. The wavelengths observed covered the expected positions of the Brγ, Brα, and the Pfβ hydrogen recombination lines at 2.174 μm, 4.065 μm, and 4.667 μm, respectively. The spectra were ratioed to nearby standard stars or IRC +10216, which have no spectral lines around the wavelengths of the hydrogen lines. Unfortunately, the spectrophotometric accuracy of the Brγ and Pfβ data is poor due to seeing fluctuations. Finally the spectra were smoothed with a 1.5 pixel FWHM Gaussian filter.

The large format of the spectrometer allows accurate determination of the line profiles and surrounding continua simultaneously since the grating is held at a fixed position. The S/N at each spectral position in the Brα line spectrum shown in Figure 1 was ~100; the structure seen in the continuum around the line is residual atmospheric absorption smoothed by the gaussian filter. High S/N was also obtained on the Pfβ and Brγ lines. The Brγ spectrum is similar to the spectrum recorded by Hall *et al.* (1981).

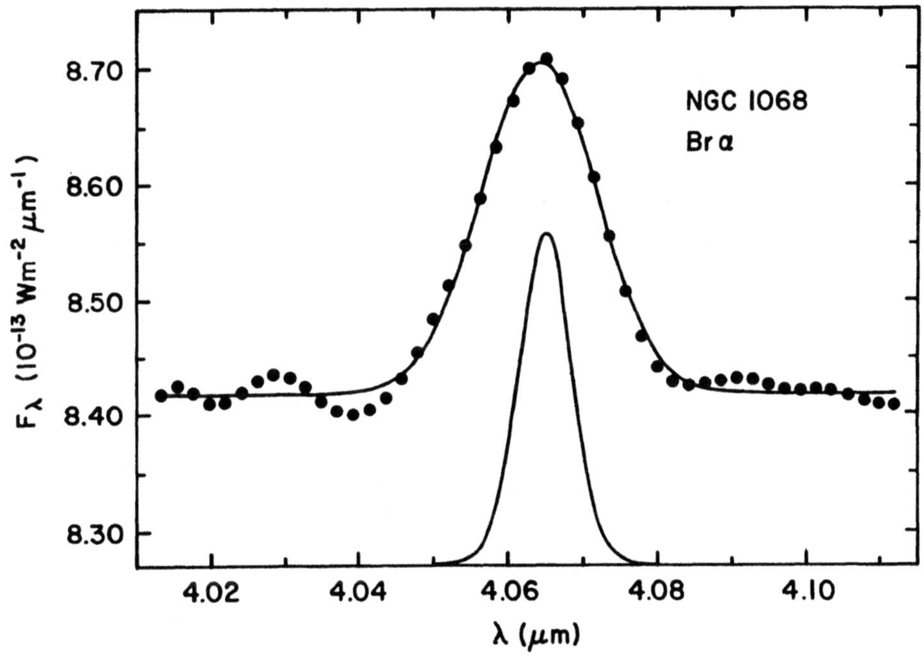

Figure 1: Brα spectrum of NGC 1068.

The data on each spectral line were fit with a linear continuum and a single Gaussian; the minimum χ^2 fit is shown as a solid line in each figure. Comparison of the fits to an unresolved calibration line (shown at the bottom of each spectrum) shows that the lines are resolved; the FWHMs of the fits are very similar, ~1200 km s^{-1}.

III. DISCUSSION

There appears to be no significant increase in the width of the infrared lines compared to the *unpolarized* optical lines. This means that any Seyfert 1 nucleus that is present must be behind a substantial amount of extinction. For example, since the $A_{Pf\beta} \sim 0.01 A_V$ (see Becklin *et al.* 1978), the lack of a Pfβ component as broad as the polarized Hβ line implies that the broad-line-emitting region, if it exists, must be obscured by substantially more than 100 mag of visual extinction. Similar arguments apply to the Brα and Brγ data.

Except for the polarized line observations there is no evidence that an obscured BLR exists. In particular, most observations suggest that the extinction to the nucleus of NGC 1068 is moderate, with $A_V < 10$ mag. For example, extinction estimates based on assumed case B recombination line ratios show that A_V is only ~1.5 mag, with no systematic increase towards longer wavelengths that would indicate the presence of highly obscured emission (Neugebauer *et al.* 1980 combined with the Brα data presented here). If the continuum at 10 μm was observed through dust extinction as thick as implied by the line data upper limits, then the silicate optical depth at 9.7 μm would be much deeper than the $\tau_{Si} \sim 0.5$ observed by Kleinmann, Gillett, and Wright (1976) and Lebofsky, Rieke, and Kemp (1978), especially since most of the 10 μm emission comes from the area around the nucleus (Telesco *et al.* 1984). Infrared continuum polarization studies by Lebofsky, Rieke, and Kemp (1978) and Knacke and Capps (1974) also suggest that the nuclear luminosity source is surrounded by moderate extinction, $A_V < 10$ mag.

The model proposed by Antonucci and Miller (1985) cannot be completely ruled out, however. A very optically thick disk may be present obfuscating even the longer wavelength observations as suggested by Krolik and Begelman (1986). However, other models that explain the broad polarized emission should be investigated. For example, perhaps the polarized emission exhibits broad lines due to reflection off of expanding electron clouds associated with an outflow. This model is especially attractive given the morphological similarities between the radio jets and recent Hα kinematic data (see Cecil, Bland, and Tully in these proceedings). The variations in polarized line ratios may then be due to heating by relativistic particles entrained within the jet (see, for example, Ferland and Mushotsky 1984). In any case, the infrared line observations presented here show that any Seyfert 1-like activity in the nucleus of NGC 1068 must be behind more than 100 mag of visual extinction.

References

Antonucci, R. R. J., and Miller, J. S. 1985, *Ap. J.*, **297**, 621.
Becklin, E. E., Matthews, K., Neugebauer, G., and Willner, S. 1978, *Ap. J.*, **220**, 831.
Ferland, G. J., and Mushotsky, R. F. 1984, *Ap. J.*, **286**, 42.
Hall, D. N. B., Kleinmann, S. G., Scoville, N. Z., and Ridgway, S. T. 1981, *Ap. J.*, **248**, 898.
Kleinmann, D. E., Gillett, F. C., and Wright, E. L. *Ap. J.*, **208**, 42.
Knacke, R. F., and Capps, R. W. 1974, *Ap. J. (Letters)*, **192**, L19.
Krolik, J., and Begelman, M. 1986, *Ap. J. Lett.*, **308**, L55.
Lebofsky, M. J., Rieke, G. H., and Kemp, J. C. 1978, *Ap. J.*, **222**, 95.
Neugebauer, G., *et al.* 1980, *Ap. J.*, **238**, 502.
Telesco, C. M., Becklin, E. E., Wynn-Williams, C. G., and Harper, D. A. 1984, *Ap. J.*, **282**, 427.
Thompson, R. I., Lebofsky, M. J., and Rieke, G. H. 1978, *Ap. J. (Letters)*, **222**, L49.

DUST AND EMISSION-LINE ASYMMETRIES IN ACTIVE GALAXIES

Oved Dahari
Space Telescope Science Institute
Homewood Campus, Baltimore, MD 21218, USA

and

Michael De Robertis
Department of Physics, York University
4700 Keele Street, North York, Ontario M3J 1P3, CANADA

We have compiled data for 268 Seyfert and star-burst galaxies, (including a substantial number of unreported [O III]λ5007 fluxes, widths and asymmetries), and searched for possible correlations between luminosity, extinction, and morphological-related properties. The optical spectra were obtained at the Shane 3-m telescope at Lick Observatory over the past decade by Osterbrock and collaborators using the image-dissector scanner and later the CCD transmission spectrograph. The asymmetry parameter AI20 (as defined by Heckman *et al.* 1981) was measured in 68 galaxies after correcting for broad Hβ and Fe II emission, and for instrumental effects. AI20 values for other 36 galaxies were taken from the literature (Heckman *et al.* 1984; Vrtilek 1985; Whittle 1985). We have also measured narrow Balmer emission line fluxes, and obtained infrared (IRAS), X-ray (primarily from Urry *et al.* 1987), and radio fluxes (primarily from Meurs and Wilson 1984) from the literature. For a complete list of the data, see Dahari and De Robertis (1987).

TABLE 1
Correlation Tests Results

Var1	Var2	N	UL	P(%)	Slope
Seyfert 1-1.5s					
AI20	LIR	57	21	1.8	2.6 ± 1.1
AI20	Eb-v	29	0	11	...
AI20	Rio	56	20	0.5	2.9 ±0.99
Eb-v	LX	22	1	5.7	-0.38 ±0.45
Seyfert 2s					
AI20	LIR	31	4	5.2	3.2 ±1.2
AI20	Eb-v	31	0	0.6	1.1 ±0.51
AI20	Rio	29	4	9.6	2.3 ±1.2
Eb-v	LX	22	8	5.4	0.41 ±0.91

Note: Rio is the logarithm of the flux ratio at 60 to [OIII] 5007. UL are the number of upper limits in the second variable, P is the proability of independance, and LX is the logarithm of the luminosity at the 0.2-4 keV band.

Of interest to this contribution, we tested for correlations between AI20 and "dust signatures," including reddening, far-infrared luminosity (60-100μ, LIR), and the flux ratio F(60μ)/F(λ5007). For data with upper limits, survival analysis techniques were employed (Isobe, Feigelson, and Nelson 1985). The results of the correlation tests are summarized in Table 1. For Seyfert 2s, good correlations are found between AI20 and all three variables (see Figure 1). For Seyfert 1-1.5s, weak correlations are found between AI20 and FIR or F(60μ)/F(λ5007), but no correlation is found with reddening. (The reddening parameter in Seyfert 1s is determined in a different manner from Seyferts with narrow Balmer lines. This could introduce systematic errors into correlations involving Seyfert 1s). These results strongly suggest a direct link between line asymmetry and the presence of dust (*i.e.*, the reddening medium) in the narrow-line region (NLR).

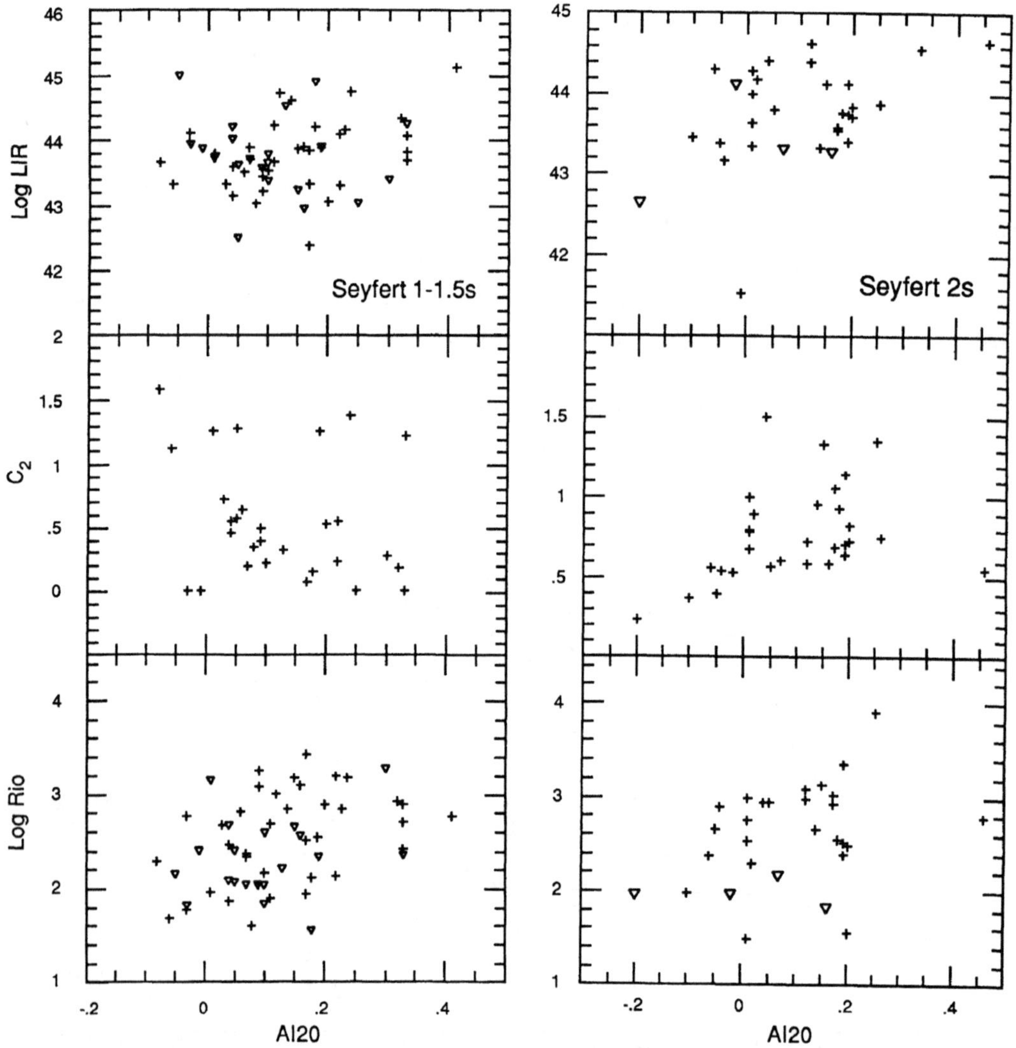

Figure 1: The luminosity at 60-100μ (LIR), reddening indicator $c_2 = 1.38E_{B-V}$, and the flux ratio F(60μ)/F(5007) (Rio), are plotted against the line asymmetry parameter of [O III]λ5007, AI20. Triangles are upper limits.

Under the assumption that the narrow-line clouds are radially outflowing, the predominance of blueward profile asymmetries in active galaxies suggests that the dust responsible for reddening resides in the inter-cloud medium (although see the contribution by De Robertis and Shaw in this volume). If this is true, we would expect emission from the broad-line region and continuum to be reddened by dust in the NLR. Since X-rays above 1 keV are only weakly absorbed by such material, the flux ratio of broad Hβ to 2-keV X-rays should be lower for nuclei with higher NLR reddening. In Figure 2 we plot this ratio versus E_{B-V}, which was obtained from the narrow-line Balmer decrements in Seyfert 1.5s, or from forbidden-line ratios for Seyfert 1s. No correlation is found (the correlation coefficient is 0.04). We conclude that the reddening of the BLR by dust in the inter-cloud medium of the NLR is negligible. This, and the large scatter in the AI20-reddening relations suggest that a third component may be present, perhaps a torus or accretion flow surrounding the BLR. The presence of the torus will lead to a bi-conical emission-line cloud flow and will obscure the inner part of the far cone causing a blueward asymmetry in the narrow-line (*e.g.*, λ5007) profiles. The size of the cone and its inclination angle to the line of sight will determine AI20. The dimensions of the torus and its optical depth (as a function of radius) will determine the degree of reddening in the NLR.

The torus was hypothesized to explain the polarized spectrum of NGC 1068 by Antonucci and Miller (1985). We find further evidence for its existence from the correlation between reddening and soft X-rays in Seyfert 2s (Table 1). This correlation may be explained by X-ray scattering; if dust exists in the inter-cloud medium, the more heavily reddened objects will have a higher X-ray scattering efficiency into the line of sight. Interestingly, the soft X-ray flux *anti*-correlates with reddening among Seyfert 1-1.5s. In these objects the X-ray source should be observed directly so the X-rays must be absorbed by the NLR inter-cloud gas.

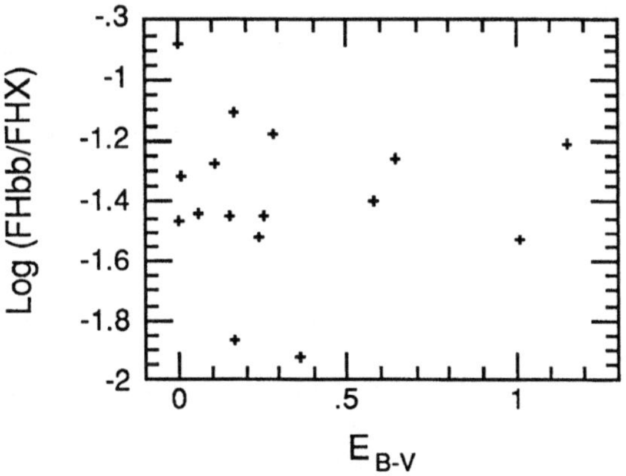

Figure 2: The logarithm of the flux ratio of broad Hβ to 2-keV X-rays (νF_ν), plotted against the reddening in the narrow-line region for Seyfert 1-1.5 galaxies. Both F(Hβ) and E_{B-V} are corrected for Galactic exinction.

References:

Antonucci, R.R.J., and Miller, J.S. 1985, *Ap. J.* **297**, 621.

Dahari, O., and De Robertis, M.M. 1987, *Ap. J. Suppl.* , submitted.

Heckman, T.M., Miley, G.K., and Green, R.F. 1984, *Ap. J.* **281**, 525.

Heckman, T.M., Miley, G.K., van Breugel, W.J.M., and Butcher, H.R 1981, *Ap. J.* **247**, 403.

Isobe, T., Feigelson, E.D., and Nelson, P.I. 1986, *Ap. J.* **306**, 490.

Meurs, E.J.A., and Wilson, A.S. 1984, *Astr. Ap.* **136**, 206.

Urry, C.M., Kruper, J.S., Canizares, C.R., Rohan, M.L., and Oberhardt, M.R 1987, in *Variability of Galactic and Extragalactic X-ray Sources*, ed. A. Traves, preprint.

Vrtilek, J.M. 1985, *Ap. J.* **294**, 121.

Whittle, M. 1985, *M.N.R.A.S.* **213**, 1.

SPECTROPOLARIMETRY AND THE STRUCTURE OF ACTIVE GALACTIC NUCLEI

Joseph S. Miller
Lick Observatory
Board of Studies in Astronomy and Astrophysics
University of California
Santa Cruz, California 95064

ABSTRACT

Previous work on polarized Seyfert galaxies has shown that type 1's have their optical continuum polarization aligned with the nuclear radio structure, while type 2's have a perpendicular relationship between these two features. Radio galaxies also show a bimodal distribution in this relationship, while all low polarization QSOs show a parallel orientation. Polarized Seyfert 2 galaxies tend to have a relatively higher polarization that is wavelength independent. Five out of ten type 2's we have observed have a highly obscured type 1 component that is revealed by the polarization measurements. We interpret this in terms of an obscuring torus which hides our direct view of the type 1 region and continuum source; these are made visible by scattering by electrons located above and below the torus. This is confirmed by the discovery of three off-nuclear regions in NGC 1068 which contain dusty regions that reflect light from the hidden region and allow multidirectional views of the nucleus. These data can be fit by a model consisting of a highly ionized wind flowing in a conical pattern which reflects the type 1 region including the x-ray continuum.

Seyfert 1's generally have low polarizations with varied origins, but the data are consistent with type 1's containing relatively thin disks that do not obscure the central regions. An alternative picture is that type 1's and 2's are the same kind of objects, but differ primarily in viewing aspect; this picture has several problems. Low polarization QSOs appear to be closely related in polarization properties to Seyfert 1's, and the non-variable high polarization QSO OI 287 may be an example of thin disk that is seen nearly edge on.

I. INTRODUCTION

Polarization measurements of extragalactic objects are among the least frequent type of observation made by optical astronomers. In fact our group at Lick Observatory may be the only ones in the world who regularly and routinely (almost every month) make spectrophotometric observations of active galaxies. The rarity of such observations is not hard to understand, because they require a large telescope and very specialized equipment. This type of observation also requires a real commitment, because the polarizations of extragalactic objects are generally low and the objects are faint. In spite of the difficulties, I feel the results are worth the effort, as I hope this review will show.

Because polarizations of objects such as Seyfert galaxies and normal QSOs (i.e., those not in the optically violently variable class) are typically less than 2%, instrumentation for spectropolarimetry must be very efficient, have low instrumental sources of error, and provide excellent sky subtraction. The latter is especially important at Lick Observatory, because the night sky light is dominated by highly polarized scattered light of nearby San Jose and other cities. For maximum accuracy, polarimeters have usually used some kind of analyzing device that rapidly modulates the light in polarization. For a number of years at Lick we used a spectropolarimeter employing a Pockels cell acting as a reversible quarter-wave plate. This instrument (Miller, Robinson and Schmidt 1980) used the image-tube, image-dissector scanner as a detector and typically modulated the incoming light at a frequency of 0.2 Hz. This rate effectively cancelled variations in seeing, guiding, cathode sensitivity, etc., and produced highly dependable results. However, with the installation of a new CCD spectrograph on the 3-m telescope, it quickly became obvious that large speed gains could be achieved if the new spectrograph were converted to a spectropolarimeter. The device we finally constructed and steps leading to it are described in detail elsewhere (Miller, Robinson, and Goodrich 1988). In fact we achieved a speed gain of nearly a factor of 20, and the device serves as an imaging polarimeter as well. The spectropolarimeter uses a rotatable achromatic half-wave retarder as a modulator. The modulation frequency is very slow (many minutes), but because it is a double beam instrument (both senses of polarization being detected simultaneously) and the CCD is very stable, it is a quite accurate instrument. New results with the CCD spectropolarimeter will be presented below.

Polarization of celestial objects can arise by three main processes. First it can be a direct result of the emission process, as in synchrotion emission. Second it can be produced in two ways by scattering of light by dust grains. The scattered light itself is generally polarized and much bluer than that of the source, which usually causes an increase to the blue in polarized flux. Alternatively we observe the light transmitted through aligned dust grains, which produces a polarization that changes slowly with wavelength. The third polarization process of concern here is scattering by electrons, which produces wavelength-independent polarization. It is often difficult to determine with certainty what kind of polarizing mechanism is at work, but at least if polarized emission-lines are observed, it is clear that some kind of scattering is taking place for their light.

In this review I will concentrate on our spectrophotometric results for three classes of active objects. Section II discusses Seyfert galaxies, Section III radio galaxies, and Section IV quasars. The last section gives a summary and overview.

II. SEYFERT GALAXIES

Seyfert galaxies can be classified into two main types. The type 1's have broad permitted lines, narrow forbidden lines, and a relatively low ratio of [O III]/Hβ; permitted Fe II emission is also common. Type 2's have permitted and forbidden lines that have the same, narrow profiles and [O III] considerably stronger than Hβ. Fe II is not present in type 2's. Osterbrock and his co-workers have classified Seyferts having composite Hβ profiles, with both narrow and broad components, as type 1.x's, where x can range up to 9. As will be discussed below, we will consider the 1.x's as subclasses of the 1's, not intermediate objects.

a) Seyfert 2 galaxies

i) NGC 1068

Studies of type 2's will be the most extensively described in this review. Optical polarization for the very bright Seyfert 2 galaxy NGC 1068 was discovered by Walker (1968). The first spectropolarimetry of this object was obtained by Visvanathan and Oke (1968), who found that the polarization increased significantly in the UV. By making the assumption that the polarization of the polarized component was wavelength-independent, they interpreted the data as being the composite of a relatively flat, powerlaw continuum plus a substantial amount of light from unpolarized, late-type stars longward of 4000Å. They concluded that the polarized featureless continuum was probably produced by the synchrotron process, and the apparent increase in polarization toward the blue was a result of decreasing dilution by unpolarized starlight.

Angel *et al.* (1976) made spectropolarimetric observations of NGC 1068 at higher spectral resolution and found what appeared to be a remarkable result. The strong lines of [O III] were polarized much less than and at a different angle from the nearby continuum, while Hβ

had the same polarization as the continuum. This appeared to contradict the standard interpretation of Seyfert 2 galaxies, as the different observed polarizations implied that Hβ and [O III] were formed in different regions in spite of their having identical profiles. They concluded that the rise in polarization toward the blue and the similarity of the polarizations of Hβ and the nearby continuum argued for a common polarizing agent: scattering in an asymmetrical dust distribution.

Our work on NGC 1068 to date is contained in three papers. Paper I (Miller and Antonucci 1983) presented spectropolarimetric data for the nucleus obtained at higher spectral resolution than earlier studies. We confirmed in general the results of the earlier studies: the polarization rises steeply in the UV, and [O III] is apparently polarized differently from Hβ. However, our higher quality data allowed us to establish clearly an important additional result: there is substantial dilution of the polarized featureless continuum by unpolarized starlight. When the starlight contribution is removed, in itself not a straightforward procedure, we found that the residual continuum has wavelength-independent polarization and a powerlaw distribution of index approximately -2. Furthermore, the dilution correction removed the apparent similarity of the polarization of Hβ and the nearby continuum. We felt these data eliminated direct evidence for dust scattering. We proposed that these data could be understood if the nucleus of NGC 1068 contained a disk of material that restricted our direct view of the central continuum source. Light emerging above the disk near its axis encounters electrons, which scatter the light into our line of sight. This hypothesis would also help to explain the apparent low power of the central ionizing source in comparison to the power radiated in the emission lines: from our viewing direction the central source is largely blocked.

We also noted that the direction of ejection of the central radio source (Ulvestad and Wilson 1984) was very close to being perpendicular to the polarization position angle of the optical continuum. If the radio axis indeed lies close to the axis of the central obscuring disk, scattering material above and below the disk would produce polarization at right angles to the axis of the disk. Antonucci (1983) has shown that in general polarized type 2 Seyferts have their inner radio structure perpendicular to the optical polarization direction, while type 1's show a parallel relationship. This is an extremely important result that will be discussed further below.

These results were further confirmed and extended by the material presented in Paper II (Antonucci and Miller 1985). There we showed that not only does the obscuring material hide the source of continuum emission, but there is also a hidden broad-line region revealed only in the polarized flux. Figure 1 shows the total flux and polarized flux spectra of the nucleus obtained with the new CCD spectropolarimeter. Since the polarized flux spectrum suppresses all unpolarized light, such as that of the stars, the difference is striking. In fact the polarized flux spectrum looks very similar to a Seyfert type 1 galaxy. We see broad permitted lines plus the Fe II emission characteristic of a type 1. With the obscuring disk or torus picture presented above, these results would imply that if we observed NGC 1068 from a direction roughly aligned with the axis of the disk, we would call it a type 1 Seyfert galaxy and infer for it a much higher optical luminosity than we derive from our viewing direction. We also pointed out in Paper II that the ratio of scattered broad Hβ flux to x-ray flux was consistent with the x-rays also being scattered by the same electron-scattering medium. Also, careful treatment of the polarization of Hβ showed that the polarization of narrow Hβ was consistent with that of [O III].

Further confirmation of this picture comes from work using the new CCD imaging polarimeter. Imaging studies done in the violet show that there are three regions in the vicinity of the nucleus that are substantially polarized: (1) the nucleus itself, (2) a relatively narrow strip of light extending several seconds from the nucleus toward the north and south, and (3) a small region located about $5''$ to the north-east of the nucleus ("the north-east knot"). Spectropolarimetry of the regions outside the nucleus show that the polarization rises rapidly in the UV, indicating they are dust-scattering regions, and the position angles are all at right angles to the direction to the nucleus, suggesting that it is scattered nuclear light itself that is being detected. This is confirmed by the polarized flux spectra, which indeed show the same Seyfert type 1 spectrum observed in the polarized flux of the nucleus. Figure 2 shows the flux and polarized flux spectra of the north-east knot, to be compared with those in Figure 1. Note, however, that Hβ is not as broad in the type 1 spectrum reflected from the knot as it is seen to be in the nuclear polarized flux.

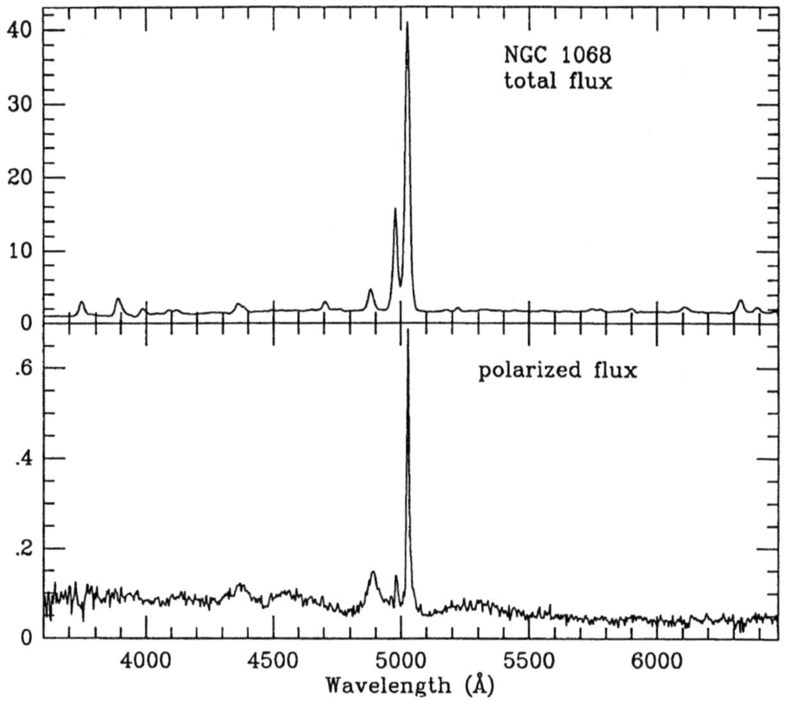

Figure 1. The total flux spectrum and polarized flux spectrum of the nucleus of NGC 1068.

Our third paper on NGC 1068 (Miller, Goodrich, and Mathews 1988) describes in detail how these results can be interpreted in terms of the obscuring torus picture. The north-south strip is produced by light emerging along the unobscured axis of the torus and undergoing scattering into our line of sight by dust grains in the interstellar medium of the galaxy. The north-east knot is a particularly dense cloud which has a direct view of the nucleus and can reflect light from it. It is interesting to note that the knot is situated near the tip of the radio jet emerging from the nucleus, and it is possible that the jet itself has cleared out material between the nucleus and the knot. The greater Hβ width in the nuclear polarized flux spectrum than that seen reflected by the north-east knot is understandable in terms of the electron scattering hypothesis for the polarization of the nucleus: electron scattering broadens the nuclear light while dust scattering in the knot has a negligible affect on the line width. These data then provide constraints for model calculations of the nuclear scattering region.

Mathews has performed an extensive series of calculations for Paper III. Briefly, he considered two basic types of models based on outward flow in a conical pattern along the axis of the torus. One involves a wind model such that the density falls off as r^2, while the second consists of a constant density scattering medium. The models must be in accord with five constraints: (1) they must explain the amount of polarization in the nuclear continuum (2) they must reproduce the line profile of scattered Hβ, (3) the scattering region must be optically thin to x-rays, (4) the broad line clouds must have the right ionization conditions even though they are buried inside the torus, and (5) the overall energy budget must be acceptable. Condition (3) stems from the result that the x-ray radiation from the nucleus is apparently scattered in the same manner as the optical light, and thus the scattering region must be transparent to x-rays.

Both the constant density and wind models require a relatively narrow cone (opening angle $\leq 30°$) and a line-of-sight viewing angle that is slightly bigger than the cone opening angle. In order to have the scattering medium transparent to x-rays, the ionization parameter must be extremely high. The constraint provided by the data on the total luminosity from x-rays to the infrared indicates that the scattering region should be small and close to the central source. Size scales of less than a parsec are required to keep the luminosity within reasonable bounds, and even so it appears that significant amounts of energy must be radiated from the nucleus in a non-isotropic distribution. While these simple models provide reasonable consistency with the data, it is clear that the real situation is quite complex.

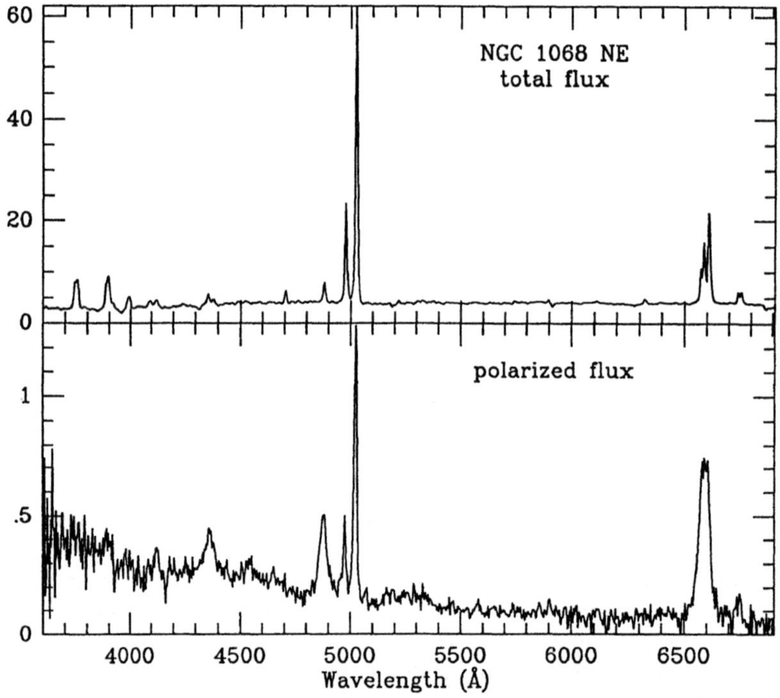

Figure 2. The total flux spectrum and polarized flux spectrum of the "north-east knot" in NGC 1068.

ii) Other Seyfert 2 galaxies

We have carried out spectropolarimetry on nine polarized Seyfert 2's in addition to NGC 1068. Two of these, Mrk 1066 and NGC 2992, appear to be polarized by dust transmission in the host galaxy. Our data for two others that are weakly polarized, Mrk 78 and NGC 7672, are inconclusive as to the nature of the polarizing mechanism. The other five – Mrk 3, Mrk 348, Mrk 463E, NGC 591, and NGC 7674 – all show a hidden type 1 region in the polarized flux, though the data for NGC 7674 provide only marginal evidence for this. These results indicate that at least half of our sample of polarized Seyfert 2's are similar to NGC 1068 in having an obscured high density emitting region, and thus NGC 1068 cannot be considered a fluke. The data on the other Seyferts are being prepared for publication, but it is clear that they show a variety of hidden type 1 regions in terms of line widths, Fe II strengths, and broad-line Balmer decrements. The question of the relationship between type 2 and type 1 Seyfert galaxies in the light of these results will be discussed below.

b) Seyfert 1 galaxies

There is much less data available on the characteristics of polarization of type 1 Seyferts than on type 2's. Type 1's typically have low polarizations (<1%), and the polarization mechanism is often not clear. For example, in our spectropolarimetric study of NGC 4151 (Schmidt and Miller 1980) we found the data indicated a uniformly polarized continuum, unpolarized broad permitted lines, and forbidden lines polarized much less than and at different angles from the continuum. The origin of the continuum polarization is uncertain; it could arise from either the synchrotron process or from electron scattering. However, the continuum polarization position direction is parallel to the nuclear radio source elongation. This is the same relation that was found for quasars (Stockman, Angel, and Miley 1978), which are of course also "type 1" objects. As was mentioned above, Seyfert 1 and 2 galaxies appear to separate cleanly in this relationship between polarization angle and radio structure.

Further study of polarized Seyfert 1's by Schmidt and Miller (1985) shows the situation is by no means simple. Some are clearly polarized by dust scattering, while others do not show any direct evidence for a dust-scattering origin of the polarization.

The bimodel distribution of polarization angle and radio structure correlation for type 1's and type 2's led us in Paper I to suggest that, if type 2's have thick disks, type 1's could be understood if they had thin disks. The scattering would then take place along the surface of the disk, producing the parallel relationship. Further studies of radio galaxies and quasars (see below) are consistent with this picture.

c) One type of Seyfert?

The discovery that Seyfert 2 galaxies can contain highly obscured type 1 regions raises the following question: can type 1's and 2's be the same thing seen from different viewing directions? If NGC 1068 were viewed from a direction which had no obscuration of the broad line region, it would indeed be called a type 1 Seyfert. Below are some of the arguments for and against one type of Seyfert galaxy:

For:

1. NGC 1068 and at least some other Seyfert 2's would be called 1's if they were seen from other directions.
2. There may be four times as many Seyfert 2's as Seyfert 1's (see the paper by Osterbrock in this volume), which is roughly consistent with the obscuring torus picture if the opening angle is around 30°.
3. The narrow line regions are the same in both types.

Against:

1. There appears to be a difference in the radio luminosities of 1's and 2's (Meurs and Wilson 1984). It has been argued this could be a selection effect (Lawrence 1987).
2. If the broad line region is blocked in 2's, so should be the featureless continuum. But Koski (1978) and others have shown that a featureless continuum is detectable in most Seyfert 2's, and most of them have not been reported to be polarized.
3. Virtually all QSOs are type 1's. Their polarization properties are similar to Seyfert 1's. If 1's and 2's are the same thing, where are the type 2 QSOs?

Neither set of arguments above are very compelling, but I lean toward there being two physically distinct types and make the following remarks:

a) At least some fraction of type 1's are really 2's or objects like NGC 1068. Perhaps the far infrared luminosity would be a clue as to which 1's were like NGC 1068, because it might be expected that type 1's surrounded by thick tori would show large conversions of energy to infrared radiation. See also Goodrich's paper in this volume.

b) I suspect most 1's are not unobscured 2's. A careful study of the featureless continuum of type 2's, including polarization measurements, would be important for investigating whether type 2's generally have obscured continuum-emitting regions. Since all QSOs are type 1's and are more luminous than Seyfert 1's, which in turn are more luminous than Seyfert 2's, it could be luminosity is an important factor in determining type.

c) Seyferts classified 1.x by Osterbrock, where x is between 1 and 9, are polarimetrically similar to type 1's. Furthermore, many exhibit variability of the broad line component, changing their "x" value considerably. Occasionally the broad component may disappear entirely. When this happens, the featureless continuum also disappears and the continuum polarization goes to zero; they do not look like type 2's in these respects. I feel the 1.x's are all subclasses of the 1's, not intermediate objects

III. RADIO GALAXIES

The major polarization study of radio galaxies is that of Antonucci (1984). In a sample of about 60 objects he found 22 to be polarized. Of these, six were definitely variable and six were probably variable, but these objects tended to have a preferred polarization angle. A very important result of this study was the discovery that the polarized radio galaxies could be divided into two classes: those with their optical continuum polarization perpendicular to inner radio structure and those showing a parallel relationship. Furthermore, there were other systematic differences between the two classes. The perpendicular objects tended to be N galaxies, have higher polarizations (up to 15%), be non-variable, and have less dominant radio cores. Also, there was some indication of the lines being polarized the same as the continuum, favoring a scattering origin for the polarization. The parallel objects tended to be E galaxies, have low polarizations ($\sim$ 1%), be variable, and more dominant radio cores. It is less clear what is causing the polarization in these objects, since they tend to show different polarizations for the lines and the continua.

The polarization behavior of the radio galaxies shows some obvious similarities to the Seyfert galaxies. As a dramatic demonstration of this, our (Miller and Goodrich, in preparation) recent observations of 3C 234, a rather narrow-lined radio galaxy in the perpendicular class, reveal the existence of a broad permitted line region of polarization similar to that of the continuum ($\sim$ 15%). We feel the central object has a structure similar to that of NGC 1068, which indicates that similar central objects can be in elliptical galaxies and spiral galaxies.

At present Antonucci and I are investigating a new sample of radio galaxies selected for having low expected galactic interstellar polarization. We feel these objects show similarities both to quasars and Seyfert galaxies, and a large sample at higher signal-to-noise could be of considerable value.

IV. LOW POLARIZATION QUASARS

Our information about the nature of the polarization of quasars is necessarily more limited than that about the above classes of objects because they tend to be fainter. We exclude from this discussion the objects ascribed to the "blazar" class, as they show high and highly variable polarization and are clearly physically very different. Also, the radio quiet "broad absorption line" QSOs (BALQSOs) are a distinct class which can show polarizations up to 5%, but relatively little is known about the polarization behavior of these objects and they will not be discussed here.

The low polarization quasars have several things in common. They have polarizations less than 2%, and the polarizations apparently are not highly variable. The polarization position angles generally are aligned with the radio structure (Stockman, Angel, and Miley 1979), analogous to Seyfert 1 galaxies. Our recent results for several quasars (Miller, Goodrich, and Antonucci, in preparation) show that the lines are not polarized or at least are polarized much less than the continuum. Also the polarization appears to be wavelength-independent over a wide range of wavelengths, except that the "3000 Å bump" is not polarized. This latter result is consistent with the bump to a large extent being made of Balmer emission, since the Balmer lines themselves are not polarized.

The simplest interpretation of these quasars follows that given for Seyfert 1 galaxies: the central objects contain thin disks which do not obscure our direct view of the central continuum source. Scattering of central source light by electrons near the surface of the disk produces the polarized light aligned with the disk and radio axis. The polarization is low because of strong dilution of the scattered light by the direct, unpolarized emission of the central source. Our (Goodrich and Miller 1988) recent results for the quasar O I 287 are consistent with this picture. This object is unique among the non-variable, non-BAL QSOs in showing high polarization ($\sim$ 8%). The polarization is constant, independent of wavelength, and parallel to the radio structure. Furthermore Hβ is polarized the same as the continuum, while the forbidden lines are unpolarized. We interpret O I 287 as that rare case of a thin disk that is nearly edge-on, so that the broad line region and central continuum source are both covered while the narrow line region is not. The polarization of the broad lines and continuum is a result of electron scattering occurring near the surface of the disk.

We are completing a survey of a few of the brighter low-polarization quasars, but more detailed results await the construction of larger telescopes. For the present the data suggest that low polarization quasars may have central objects similar to those found in Seyfert 1 galaxies.

V. SUMMARY REMARKS

The material presented in this paper illustrates some insights, many unique, that can be provided by spectropolarimetry of active galaxies. While the observations require large amounts of time on a large telescope, they give direct clues about the structures of central objects that are not obtainable by conventional techniques.

One of the main conclusions is that there are clear, systematic relationships among the characteristics of the line spectrum, the continuum and line polarization, and the radio structure. This means that physical processes occurring in one object on size scales ranging from a few light days to many kpc are intimately related. Several years ago Miley and I (Miley and Miller 1979) showed that the structure of radio sources of quasars was related to the profile of the Hβ emission line and the presence of Fe II emission in the nuclear spectrum. In this paper I have outlined the results which show a correlation between the axis of the radio structure, the angle of optical continuum polarization, and the type of emission line spectrum in the nucleus. The broad line objects – Seyfert 1 galaxies, broad line radio galaxies, and quasars – show a parallel relationship in polarization direction while the narrow permitted line objects – Seyfert 2 galaxies and narrow line radio galaxies – show a perpendicular relationship. At face value this would suggest that the central regions of broad line objects are similar to one another over a large range of luminosity and in both spiral and elliptical galaxies. The same can be said for the narrow line objects. While some of the classification of objects into these two types clearly depends on viewing angle, I believe there are real physical differences between them in general.

We have proposed that the data for the narrow line objects can be understood if they tend to contain rather thick tori that can obscure our direct view of the inner regions nucleus, while the broad line objects have relatively thin disks that allow direct views of the broad line region and the central continuum source. Since we have shown that in fact some so-called "narrow line" objects harbor a broad line or, better, a high density region, the distinction based on line width is clearly not ideal. If there are physically two types of central objects, then clearly some can have their classification influenced by aspect effects.

We do not yet have a physical theory which ties all this material together. But the establishment of systematic relationship between a variety of phenomena provides meaningful constraints on modelling of the central source, and we look forward to the day when a models are available that treat the emission line region, radio structure, continuum source, and central energy generation in a comprehensive, uniform way.

I wish to acknowledge and thank Robert Antonucci, Robert Goodrich, and Gary Schmidt for their large and significant contribution to the research described in this paper. This research was supported in part by NSF Grant AST 84-06843.

References

Angel, J.R.P., Stockman, H.J.S., Woolf, N.J., Beaver, E.A., and Martin, P.G. 1976, *Ap. J. (Letters)*, **206**, L5.

Antonucci, R.R.J. 1983, *Nature*, **303**, 158.

________. 1984, *Ap. J.*, **278**, 499.

Antonucci, R.R.J., and Miller, J.S. 1985, *Ap. J.,*, **297**, 621 (Paper II).

Goodrich, R.A., and Miller, J.S. 1988, *Ap. J.,* in press.

Koski, A.T. 1978, *Ap. J.*, **223**, 56.

Lawrence, A. 1987, *Pub. A.S.P.*, **99**, 309.

Meurs, E.J.A., and Wilson, A.S. 1984, *Astron. Ap.,* **136**, 206.

Miley, G.K., and Miller, J.S. 1979, *Ap. J. (Letters)*, **228**, L55.

Miller, J.S., and Antonucci, R.R.J. 1983, *Ap. J. (Letters)*, **271**, L7 (Paper I).

Miller, J.S., Goodrich, R.A., and Mathews, W.G. 1988, in preparation (Paper III).

Miller, J.S., Robinson, L.B., and Goodrich, R.A. 1988, in *Instrumentation for Ground-Based Optical Astronomy: Present and Future* (Springer Verlag: New York), in press.

Miller, J.S., Robinson, L. B., and Schmidt, G.D. 1980, *Pub. A.S.P.*, **92**, 702.

Schmidt, G.D., and Miller, J.S. 1980, *Ap. J.,* **240**, 759.

________. 1985, *Ap. J.*, **290**, 517.

Stockman, H.S., Angel, J.R.P., and Miley, G.K. 1979, *Ap. J. (Letters)*, **227**, L55.

Ulvestad, J.S., and Wilson, A.S. 1984, *Ap. J.,* **285**, 439.

Visvanathan, N., and Oke, J.B. 1968, *Ap. J. (Letters)*, **152**, L165.

Walker, M.F. 1968, *Ap. J.,* **151**, 71.

SPECTROPOLARIMETRY OF "NARROW-LINE SEYFERT 1s"

R. W. Goodrich
Lick Observatory, Santa Cruz, California, U.S.A.

The use of spectropolarimetry as a diagnostic tool in AGN research has proven to be quite powerful. In fact the discovery by Joe Miller and his collaborators of hidden broad-line regions (BLRs) in several Seyfert 2s threatens to overthrow our perception of the true differences between Seyfert 1s and 2s. To further clarify the interrelationships between different Seyfert classes we have observed several "complete" samples of what might be called intermediate or transition classes, one of which, the narrow-line Seyfert 1s ("NL 1s"), is discussed in this contribution.

The term "narrow-line Seyfert 1" was introduced by Osterbrock and Pogge (1985). The definition of the class includes that the line widths of the permitted and forbidden lines are similar, and lower than in "normal" Seyfert 1s. Also the [O III]/Hβ ratio is < 3, and Fe II, [Fe VII], and/or [Fe X] are often present. Because of the line ratio characteristics of the spectra it appears that there exist "BLRs" in these objects which have relatively narrow lines (typically < 2000 km s^{-1}). (Hence we should probably not talk about "*broad*-line regions," but rather high-density regions where the forbidden lines are supressed.) Are the NL 1s then Seyfert 1s with unusually narrow "BLR" widths, or are they Seyfert 2s with unusually high ionization and strong permitted lines?

Spectropolarimetry can be used to attack this problem in several ways. First, Antonucci (1984) showed in his thesis that the optical polarization position angles (p.a.s) are almost always perpendicular to the radio structure p.a.s for Seyfert 2s — for Seyfert 1s they are almost always parallel to each other. VLA observations by Jim Ulvestad, Ski Antonucci, and myself are currently awaiting reduction to test the behaviour of the NL 1s, although p.a.s for Mrk 766 and Mrk 1126 exist in the literature. Secondly, spectropolarimetry can provide information on the kinematics and geometry of the both the emitting gas and the polarizing medium.

Table 1 contains a summary of the data as of October 1987. Three objects (Mrks 766 and 1239 and IRAS 1509 − 211) have high polarizations and provide information on the line profiles in polarized light. Four more objects have moderate polarization and line ratio information in polarized light may be obtained, while two more have the potential for a useful comparison between radio and optical p.a.s.

TABLE 1

THE SAMPLE

Object	Optical Pol. % @ p.a.	Object	Optical Pol. % @ p.a.
Mrk 957	0.6 @ 65	NGC 4748	0.1 @ 87
0111.9-0132	...	Mrk 783	0.3 @ 178
0112.9-0147	...	Mrk 684	0.2 @ 86
Mrk 359	0.5 @ 112	Mrk 1388	1.0 @ 80
Mrk 1044	0.5 @ 141	IRAS 1509	4.3 @ 62
0449-184	...	Mrk 493	0.2 @ 86
0919+518	...	1747.3+6836	0.0 @ 92
Mrk 1239	2.9 @ 130	Mrk 507	0.8 @ 8
PG 1016	0.1 @ 89	Akn 564	0.4 @ 92
Mrk 42	0.3 @ 29	Mrk 1126	0.5 @ 165
Mrk 766	2.4 @ 90	1E2348.6	0.5 @ 75
1228.7+1219	0.1 @ 57		

In Mrk 766 and IRAS 1509 − 211 the lines in polarized flux are slightly broader than they are in integrated light, but typically only by ~ 1000 km s^{-1}. So there appears to be no evidence for a hidden, "normal" (i.e. high-velocity) BLR in these objects. All seven objects with line ratio information show higher polarization of the permitted lines than of the forbidden lines, indicating that the two sets of lines originate from spatially distinct regions. In Mrks 766 and 1126 the optical polarization p.a. is more nearly perpendicular to the published radio p.a.s, indicating a similarity to Seyfert 2s in this respect. Finally, in Mrk 1239 the polarized Balmer lines show a red asymmetry, which in analogy to NGC 1068 (see Miller, this volume) may indicate the presence of an outflowing mirror of hot electrons. (Alternatively there may be a second component at a higher velocity which may be independent of the main part of the profile and simply added to it.)

We have seen that the "BLR" and "NLR" are spatially distinct in these objects, and further that the "BLRs" really do have small velocities! Fe II is seen in 84% of the objects, which is more or less normal for BLRs in Seyfert 1s. After the revelations by Miller and collaborators, however, it may turn out to be rather meaningless to ask whether these are really Seyfert 1s or 2s — the distinction between the classes may be an artificial one due to aspect! In any case the "BLRs" of the NL 1s seem to be a continuous extension of more normal BLRs towards lower velocity widths. The fraction of NL 1s which are moderate to highly polarized (7/19) is rather higher than the fraction of Seyfert 1s but lower than the fraction of Seyfert 2s with intrinsic polarizations. Clearly more work will need to be done to determine whether there are any other peculiarities of the NL 1s when taken as an extension of the class of Seyfert 1s.

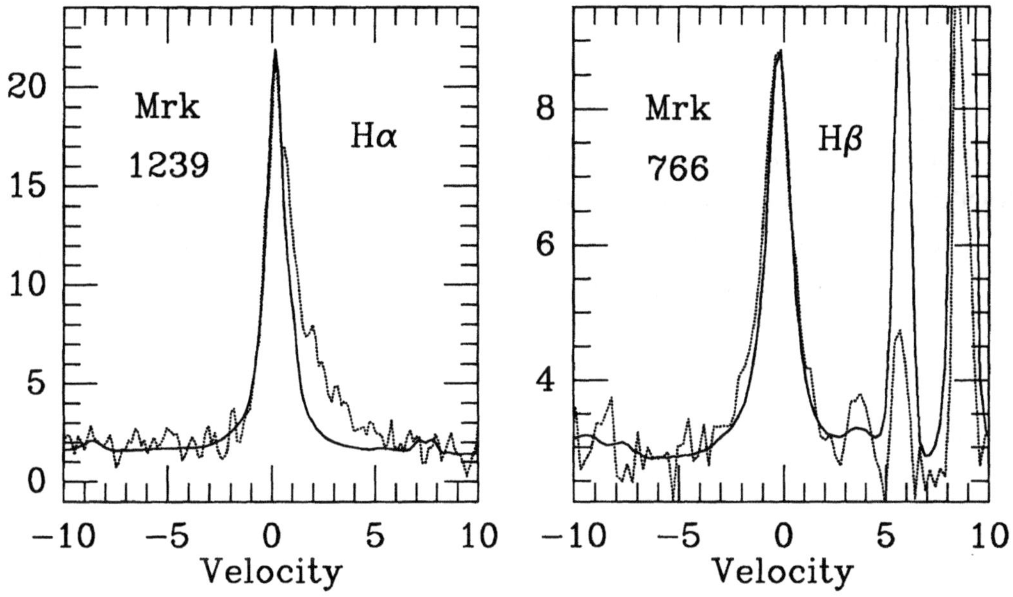

Figure 1. Polarized flux (dotted lines) plotted on top of the integrated-light spectrum (solid line) for the Mrk 1239 Hα and Mrk 766 Hβ emission lines. The vertical scale is F_λ and the horizontal velocity scale is in units of 1000 km s^{-1}. Note the red wing in Mrk 1239, also seen at Hβ.

REFERENCES

Antonucci, R. R. J. 1984, *Ap. J.*, **278**, 499.

Osterbrock, D. E., and Pogge, R. W. 1985, *Ap. J.*, **297**, 166.

MUSHOTZKY: Are there any peculiarities in line ratios? Are they peculiar in their colors or luminosity distributions?

GOODRICH: The line ratios, optical luminosities, and infrared luminosities are all more or less normal. In the *IRAS* infrared color-color diagram the narrow-line Seyfert 1s are found in a narrow band of $25-60\mu$m colors, although they have a wide range of $60-100\mu$m colors. I'm not sure what this means, or whether it is significant.

GASKELL: A couple of years ago (Gaskell 1985, *Ap. J.*, **291**, 112) I showed that these narrow-line Seyfert 1s had much higher Fe II/Hβ ratios. One of the misconceptions about this result is that many people think I was saying that the Fe II was strange. It isn't. It is the Hβ which is weaker in these narrow-line Seyfert 1s.

ANTONUCCI: Don Osterbrock said that narrow-line Seyfert 1s are probably not a distinct class, different from normal Seyfert 1s. But do your results show that statistically they are more likely to have polarized permitted lines?

GOODRICH: I think we can say that they are more likely to have high polarizations. The statistics on the *permitted line* polarization specifically for Seyfert 1s is very poor, and I wouldn't hazard a comment on that.

EVIDENCE FOR THERMAL EMISSION COMPONENTS IN HIGHLY POLARIZED QUASARS

Paul S. Smith and Richard Elston
Steward Observatory, University of Arizona
Tucson, Arizona 85721

I. Introduction and Observations

It has been found through simultaneous, broad-band optical and near-infrared polarimetry (Smith *et al.* 1986) and multi-frequency flux measurements (Bregman *et al.* 1986) that the highly polarized quasar (HPQ) 3C 345 has a broad continuum "bump" at optical and near-UV wavelengths. Both groups of investigators suggest that this spectral feature is optically thick thermal emission from an accretion disk. Optical polarimetry can detect this continuum component because the thermal emission dilutes the polarized flux of the synchrotron emission. Because the "blue bump" contributes less to the total flux at longer wavelengths, the degree of linear polarization (P) increases toward longer wavelengths. Our observational program is designed to find possible accretion disk components in HPQs by finding those quasars that exhibit increasing polarization with wavelength.

Broad-band optical polarimetry and photometry of HPQs was done using the "Two-Holer" polarimeter. Most of the observations were made using the Steward 2.3 m telescope. Some of the observations of 0420−014, 1510−089, and 3C 454.3 were made using the UCSD/University of Minnesota 1.5 m telescope on Mt. Lemmon, Arizona. Our sample of highly polarized QSOs was chosen mainly from the list of Moore and Stockman (1981).

We find that 0420−014, 1156+295 and 3C 454.3 clearly show increasing P with wavelength. Marginal evidence is found for wavelength-dependent P in the same sense in 1510−089 and 2345−167. The HPQs 0716+332 and 3C 216 exhibited decreasing P with wavelength (BL Lacertae objects often show this type of wavelength dependence in P; Smith, Elston, and Balonek 1986). Our observations of 0420−014, 1156+295, and 3C 454.3 are shown in Figure 1.

II. Modeling the Continua of HPQs

We construct simple synthetic spectra to explain wavelength-dependent polarization in the following way:

a) We assume that the emission from the broad-line region (BLR) is unpolarized and subtract the BLR contribution from the total flux in the *UBVRI* bands using the "3000Å bump" models of Wills, Netzer, and Wills (1985). The remaining flux is modeled using two continuum components.

b) The synchrotron component is approximated by a power-law and its polarization is assumed to be independent of wavelength.

c) The accretion disk component is approximated with a flat spectrum over the limited wavelength range of our observations (Malkan 1983). Like the BLR, the emission from this component is assumed to be unpolarized.

The model has four free parameters: brightness, spectral index, and intrinsic polarization of the synchrotron component, and the brightness of the optically thick thermal component. The best-fitting model parameters for five HPQs are given in Table 1. Approximate upper limits on the brightness of possible accretion disk components are also given for five other HPQs.

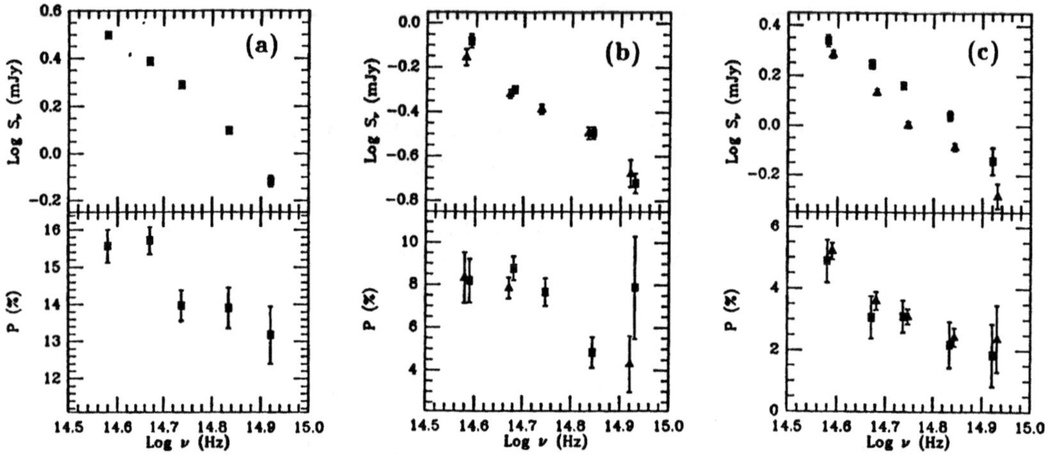

Figure 1: (*a*) Observations of PKS 0420−014 on 1986 November 27. The degree of linear polarization is clearly seen (at the 4σ level) to increase with wavelength. The very weak dependence of P on wavelength is expected since the object was very bright (B=16.5; near its historical photometric maximum) and the synchrotron emission was dominant. (*b*) Observations of 1156+295 (Ton 599) during 1986 November (squares) and 1987 June (triangles). This quasar was faint (B=17.7) during these observations. When bright, 1156+295 looks like a BL Lac object in its polarization characteristics (P is often observed to *decrease* with wavelength; Smith, Elston, and Balonek 1986). (*c*) Observations of 3C 454.3 on 1986 December 23 (squares). Our observations of 1987 June 23 and 24 are averaged and denoted by triangles. The polarization measurements at I show the highest optical polarization reported for this object.

Table 1. Thermal Emission Components in HPQs.

Object	α_{syn} [a]	P_0 (%)	S_{syn} [b]	S_{th} [b]	B [c]
0420−014	−1.9	16.9	2.33	0.36	18.6
1156+295	−2.2	11.1	0.27	0.14	19.2
1510−089	−1.8	15.3	0.30	0.60	17.8
3C 345[d]	−1.9	37.8	3.10	0.53	17.7
3C 454.3	−2.7	6.2	0.66	0.53	18.1
0336−019				< 0.15	> 19.5
1958−179				< 0.29	> 18.4
2201+171				< 0.31	> 18.6
3C 446				< 0.14	> 19.9
2345−167				< 0.17	> 18.9

Notes to the table

[a] Spectral index of the synchrotron component ($S_\nu \propto \nu^{\alpha_{syn}}$).

[b] Rest frame flux density of the synchrotron component (S_{syn}) and accretion disk component (S_{th}) at 2500Å in mJy (10^{-26} ergs cm^{-2} s^{-1} Hz^{-1}).

[c] Approximate B magnitude in the observer's frame of the accretion disk component.

[d] The values listed for P_0 and S_{syn} for 3C 345 are for the 1983 February data reported in Smith *et al.* (1986). During 1984 June P_0 = 10.8% and S_{syn} = 0.7 mJy.

III. Luminosity of the Accretion Disk Components

We find a range of at least a factor of 10 in luminosity for the HPQ thermal components. 3C 454.3 has the most luminous thermal component in our sample (more than twice as luminous as the thermal component of 3C 345). In comparison, the thermal component of 3C 345 is at least three times more luminous than the luminosity upper limit found for 2345−167.

The near-ultraviolet luminosities of the HPQ thermal components fall close to the average near-UV luminosity of the quasars in the PG survey (Neugebauer *et al.* 1987). We assume that $H_0 = 75$ km/sec/Mpc, $q_0 = 0$, and

$$L_{2500} = 4\pi \left(\frac{cz}{H_0}\right)^2 \left(1 + \frac{z}{2}\right)^2 S_{2500},$$

where L_{2500} is the monochromatic luminosity at 2500Å and S_{2500} is the rest frame flux density at 2500Å.

$\mathrm{Log}(L_{2500})$	average	range
PG quasars	30.4	32.8 – 28.5
HPQ thermal components	30.8	31.2 – <30.2

Upper limits on the luminosity of possible accretion disk components for a few BL Lac objects can also be estimated from the data of Smith *et al.* (1987). We find $\log(L_{2500}) < 29.9$ for AO 0235+164, < 29.8 for OJ 287, and < 28.5 for BL Lac.

The present data do not allow us to determine if some HPQs and BL Lac objects have accretion disk components substantially less luminous than the faintest PG quasars. Such faint components would give rise to very weak wavelength dependence in their optical polarization. However, it appears that only the presence of a bright synchrotron component distinguishes some HPQs from low-polarization, optically selected QSOs at optical and near-UV wavelengths.

References

Bregman, J.N., *et al.* 1986, Ap. J., **301**, 708.

Malkan, M.A. 1983, Ap. J., **268**, 582.

Moore, R.L. and Stockman, H.S. 1981, Ap. J., **243**, 60.

Neugebauer, G., Green, R.F., Matthews, K., Schmidt, M., Soifer, B.T., and Bennett, J. 1987, Ap. J. Suppl., **63**, 615.

Smith, P.S., Balonek, T.J., Elston, R., and Heckert, P.A. 1987, Ap. J. Suppl., **64**, 459.

Smith, P.S., Balonek, T.J., Heckert, P.A., and Elston, R. 1986, Ap. J., **305**, 484.

Smith, P.S., Elston, R., and Balonek, T.J. 1986, Continuum Emission in Active Galactic Nuclei, ed. M.L. Sitko (Tucson: NOAO), p. 171.

Wills, B.J., Netzer, H., and Wills, D. 1985, Ap. J., **288**, 94.

THE SHAPE OF THE ULTRAVIOLET CONTINUUM OF QUASARS

Anuradha P. Koratkar
Astronomy Program, State University of New York, Stony Brook, NY 11794
Fu-Hua Cheng
Lick Observatory, University of California, Santa Cruz, CA 95064
C. Martin Gaskell
Astronomy Department, University of Michigan, Ann Arbor, MI 48109

ABSTRACT

We investigate the dependence of the UV spectral index (α) on luminosity, redshift, and radio properties of the quasar. We find that α is remarkably independent of luminosity. We confirm earlier results that there is a slight dependence of α on the radio properties and redshift, but believe that the latter is perhaps an optical selection effect. We find no evidence for intergalactic reddening.

INTRODUCTION

It has been known for sometime that the optical to X-ray spectral index of quasars is a function of the optical luminosity (Zamorani *et al.* 1981 and Reichert *et al.* 1982). In this paper we ask what is happening between the optical and the X-ray regions in the ultraviolet. If the "blue bump" in the UV is thermal in origin and the result of accretion onto a blackhole, the temperature associated with it is expected to vary with luminosity. It might then be expected that the UV slope is luminosity dependent. It is well known that quasars evolve in luminosity and it has been reported previously that α flattens with redshift (O'Brien, Gondhalekar and Wilson 1986 and Cheng and Fang 1987), and, also that the radio-loud quasars have systematically steeper slopes than the radio-quiet ones (Neugebauer *et al.* 1979 and Moore and Stockman 1981). We therefore address the following questions: Is there an "universal" spectral slope in the UV?, Is this slope a function of luminosity?, Does it evolve with redshift?, Does it depend on the radio properties? and, Is there any evidence for intergalactic reddening?

DATA SET

The data set used consists of quasar samples obtained by different observers. The high redshift optically selected quasars are from Osmer and Smith (OS) (1977, 1978). The intermediate flat radio spectrum quasars are from Gaskell (1981). The low redshift quasars are from Kinney *et al.* (1985) and Cheng and Fang (1987). The Seyfert galaxies are from Wu, Boggess and Gull (1983). The spectral indices were all measured in the *same rest frame* for all the quasars (λ1216 – λ2200). The average error in the spectral index for the OS sample is $\pm$ 0.27, for the Gaskell sample is $\pm$ 0.21, for the Kinney *et al.* sample it is $\pm$ 0.23, for the Cheng and Fang sample it is $\pm$ 0.27, and for the Wu *et al.* sample it is $\pm$ 0.31. All the values have been corrected for local galactic reddening. Full details of the measuring procedure are in Cheng, Gaskell and Koratkar (1988).

Absolute luminosities (L_{1450} ergs s^{-1} Hz^{-1}) have been calculated from the estimated UV continuum flux at $\lambda1450$ assuming H_0 of 50 km s^{-1} Mpc^{-1} and q_0 of 1.0. Our criterion of whether a quasar was radio-loud or radio-quiet was simply whether or not entries existed for radio fluxes in the Veron and Veron (1987) ESO catalogue.

ANALYSIS

We performed a multivariate linear regression analysis on the data as we were interested in the dependence of α (the spectral index) on many factors. The form of the regression curve was

$$\alpha = \alpha_o + A_1\, Type + A_2\, log\frac{L_{1450}}{< L_{1450} >} + A_3\, log\frac{1+z}{< 1+z >}$$

where the convention Type = 0.0 is for radio-quiet quasars and Type = 1.0 for radio-loud quasars. Table 1. gives the results of our analysis.

TABLE 1

REGRESSION COEFFICIENTS AND ERRORS

Sample	α_o	A_1	A_2	A_3
all objects	0.85±0.06	0.0*	-0.2 ± 0.05	0.0*
all objects	0.88±0.06	0.0*	-0.06 ± 0.08	-0.82 ± 0.35
excluding Osmer & Smith	0.93±0.07	0.0*	-0.07 ± 0.08	-0.50 ± 0.43
all objects	0.81±0.08	0.1±0.1	-0.08 ± 0.08	-0.84 ± 0.34
excluding Osmer & Smith	0.93±0.16	-0.001 ± 0.16	-0.07 ± 0.08	-0.51 ± 0.50

* These coefficients have been set to zero.

From line 1 of table 1 we can see that there is an apparent decrease in α as the luminosity increases, but when the effect of redshift is considered this dependence vanishes (see line 2) and the redshift effect dominates (98% significance). From figure 1 it can be seen that the OS quasars dominate the red shift correlation although it is weakly present (75% significance) without them (see line 3). The reasons why the OS sample could be influencing the correlation in addition to the possibility of a real evolutionary effect are; (1) the quasars are optically selected and objective prism surveys bias the sample towards quasars with a high uv excess, (2) the quasars are radio-quiet and thus tend to have flatter slopes. Line 4 shows that when the radio dependence is included there is small dependence on radio type but this does not change the dependence on redshift. Line 5 shows that the radio-quietness of the OS sample alone is not the cause of the apparent evolutionary effect, so we suspect an optical selection effect.

Most of the scatter about the correlation is due to the measuring errors and there is little intrinsic scatter. Evidence for this is the lack of correlation of Lyα with α. The variation in the Lyα equivalent width is probably due to the variation of the covering factor. In the spectral region we see no signs of a luminosity/temperature effect.

Wright (1981) has suggested that a tentative earlier result by Richstone and Schmidt (1980) claiming a steepening of α with redshift at the highest redshift could be due to intergalactic dust. We do not confirm this result, which we believe was a consequence of not measuring at the same rest wavelength. We therefore feel that our results exclude intergalactic reddening being the cause of a redshift cutoff for quasars (Ostriker and Heisler 1984).

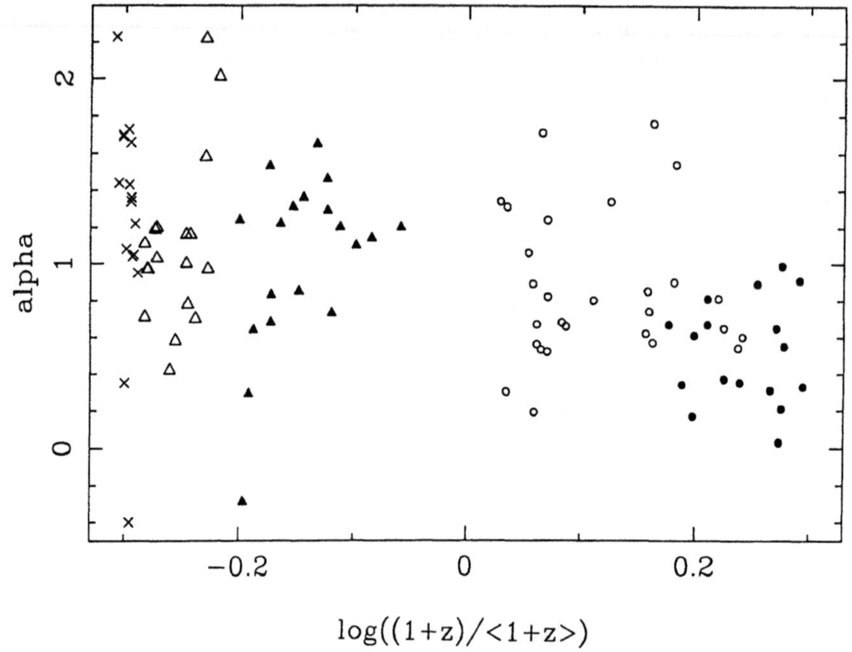

Fig. 1. This graph shows the dependence of α on the redshift. The average redshift is 2.04. The different samples are indicated by different symbols. × Wu, Boggess and Gull (1983), △ Cheng and Fang (1987), ▲ Kinney *et al.* (1985), ○ Gaskell (1981), ● Osmer and Smith (1976,1977). Note that the Osmer and Smith quasars have a large weight in the correlation of α and z.

RESULTS

Our main result is that α is remarkably independent of luminosity over a range of 4 orders of magnitude. It depends slightly on the redshift but we suspect that most of the correlation is due to the optical selection of the Osmer and Smith quasars. As has previously been reported there is a very small dependence of the spectral index on the radio properties. The variation in the Lyα equivalent width is probably due to variation of the covering factor and not the intrinsic variation of α. We do not see intergalactic reddening in our data.

REFERENCES

Cheng, F.H. and Fang, L.Z. 1987, *M.N.R.A.S.*, **226**, 485.

Cheng, F.H., Gaskell, C.M. and Koratkar, A.P. 1988 in preparation.

Gaskell, C.M. 1981, Ph.D thesis, University of California at Santa Cruz.

Kinney, A.L., Huggins, P.L., Bergman, J.N. and Glassgold, A.E. 1985, *Ap. J.*, **291**, 128.

Moore, R.L. and Stockman, H.S. 1981, *Ap. J.*, **243**, 558.

Neugebauer, G., Oke, J.B., Becklin, E.E. and Matthews, K. 1979, *Ap. J.*, **230**, 79.

O'Brien, P.T., Gondhalekar, P.M. and Wilson, R. 1986 in *New Insights in Astrophysics*, ESA SP-263, p 649.

Osmer, P.S. and Smith, M.G. 1976, *Ap. J.*, **210**, 267.

Osmer, P.S. and Smith, M.G. 1977, *Ap. J.*, **213**, 607.

Ostriker, J.P., and Heisler, J. 1984, *Ap. J.*, **278**, 1.

Reichert, G.A., Mason, K.O., Thorstensen, J.R. and Bowyer, C.S. 1982 *Ap. J.*, **264**, 575.

Wright, E.L. 1981, *Ap. J.*, **250**, 1.

Wu, C.C., Boggess, A. and Gull, T.R. 1983, *Ap.J.*, **266**, 28.

Zamorani, G., Henry, J.P., Maccocaro, T., Tananbaum, H.D., Soltan, A., Avni, Y., Liebert, J.W., Stocke, J.T., Strittmatter, P.A., Weymann, R.J., Smith, M.G. and Condon, J.J. 1981 *Ap. J.*, **245**, 357.

A SEARCH FOR IRAS AGN

Charlene Anne Heisler[1,2] and J. Patricia Vader[1]

1) Dept. of Astronomy , Yale University Box 6666 , New Haven , CT. , 06511

2) Guest Investigator, NOAO, Cerro Tololo Inter-American Observatory, Operated by Associated Universities for Research in Astronomy, Inc.

The infrared astronomical satellite (IRAS) has made it possible to study the far infrared properties of a large number of active galactic nuclei (AGN). The presence of a host galaxy in which a QSO is embedded is commonly accepted to be the determinant of the relative importance of thermal to non-thermal radiation. Thus the far infrared properties are critical in helping to determine whether the infrared radiation is primarily non-thermal or thermal dust emission.

So far only two previously unidentified quasars have been discovered in the IRAS data base (Beichman et al. 1986, Vader and Simon 1987). One of these quasars (00275-2859) was discovered during a redshift survey investigation to determine the optical luminosity function of IRAS galaxies. The discovery of these infrared-luminous quasars, and the possibility that many other yet unidentified quasars may exist in the IRAS data base has been the motivation for carrying out an unbiased survey of IRAS quasar candidates.

QSOs detected by IRAS seem to be preferentially strong FeII emitters. (Fig. 3) Since the host systems of many QSOs show signs of *interaction* (e.g. Hutchings et.al. 1984), this may fit into the interpretation according to which FeII emission is associated with thick accretion disks resulting from accretion rates above the Eddington limit. This in turn favors large far-infrared luminosities by thermal reradiation by dust. Thus the role of interactions may be crucial to the AGN phenomenon.

Examination of the properties of the IRAS AGN in Table I seems to support the above hypothesis. IRAS 00275-2859 ranks among both the strongest FeII and far-infrared emitters, and it is embedded in an interacting system (Vader et.al. 1987). IRAS 02580-1136 is a double nucleus object (Fig. 1) with 'warm' IRAS colours in which the central component has a UV excess and a spectrum characteristic of a Seyfert 2 (Fig. 2). This system also appears to show tidal tails.

Thus far in our search for AGN in the IRAS data base, a narrow emission line galaxy (NELG) and an HII region-like galaxy have been found. IRAS 22584+2319 is a narrow emission line galaxy (NELG) which has evidence for a smaller companion (separated by a distance of 10 kpc if assumed to be at the same redshift). IRAS 01053+2417 is an HII region-like galaxy with a peculiar morphology - it has wisps or tidal arms extending from the compact nuclear component. If the powering source of FIR luminous HII region-like galaxies is non-thermal in origin (Shu F. 1987), then this latter object may also be linked to AGN.

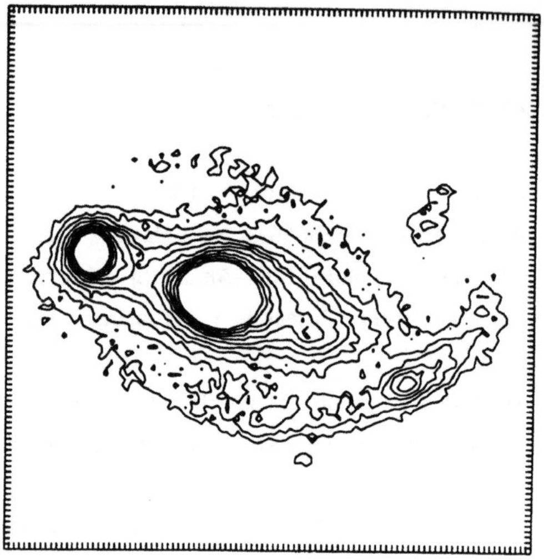

Fig.1: 50" by 50" contour plot of IRAS 02580-1136 from median of three 600s exposures in B. The tick marks indicate pixel size (1 pixel =0.5"). Lowest contour corresponds to B=21.64 mag/arcsec^2 and contour spacing is 0.075 mag.

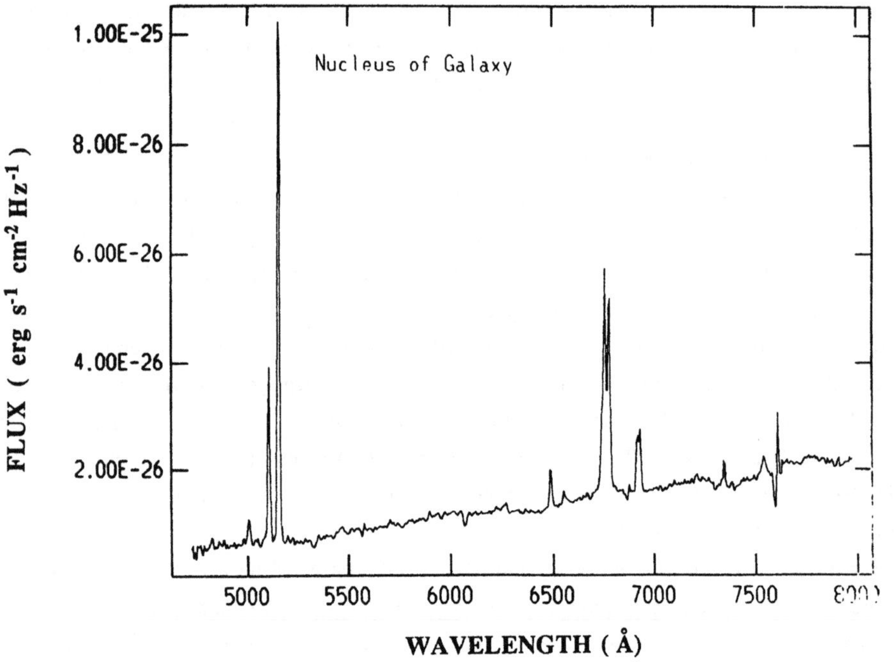

Fig. 2: Optical spectrum of central nucleus of IRAS 02580-1136. This 300 second exposure (RC spectrograph/CTIO 4m telescope) was kindly provided by Jay Frogel. Rest frame line widths are 230 km/s for [OIII]5007 and 330 km/s for the Balmer lines.

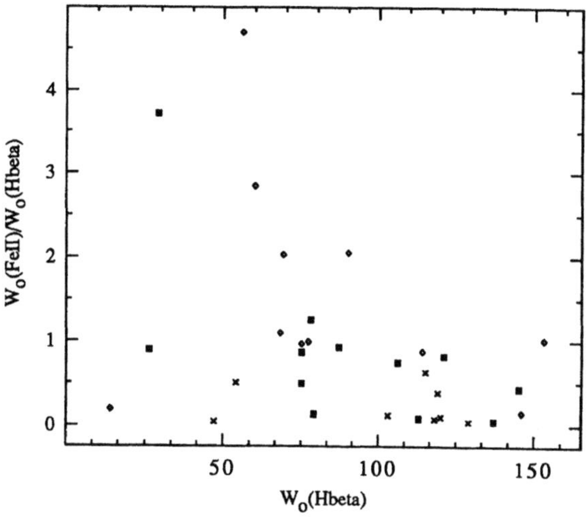

Fig. 3: Ratio of rest frame equivalent width of FeII (λ4570 + λ5190 +λ5320) and Hβ against Hβ equivalent width. (from Steiner J. 1981, Grandi S. 1981). Diamonds: QSOs in IRAS Point Source Catalog(1985). Squares: QSOs detected in at least one wavelength band by IRAS during pointed observations; crosses: QSOs not detected by IRAS during pointed observations (Neugebauer et al. 1986).

Table I: IRAS CHARACTERISTICS

Object	z	f(60μ)	α(60,25)	α(100,60)	L(60μ) / L_O*	type
00275-2859	0.28	0.69	-1.16	-1.10	1.5×10^{12}	QSO
02580-1136	0.03	0.56	-0.15	-0.62	1.3×10^{10}	SY 2
22584+2319	0.05	0.65	<1.05	-2.35	4.2×10^{10}	NELG
01053+2147	0.05	1.66	-1.68	+0.12	1.1×10^{11}	HII

* assuming $H_o = 75$ km s^{-1} Mpc^{-1} and $q_o = 0.5$

This research was supported in part by grant JPL 958027.

REFERENCES

Beichman C. A., Soifer B. T., Helou G., Chester T. J., Neugebauer G. , Gillett F. C., and Low F. J. , (1986), Astrophys.J. Lett. **308**,L1

Grandi S., (1981), Astrophys. J., **251**,451

Hutchings J.B., Crampton D., and Campbell B., (1984), Astrophys. J. **280**, 41

IRAS Point Source Catalog, Version 2, (1985), Joint IRAS Science Working Group, (Washington D.C.,: U.S. Government Printing Office)

Miley G. K., Neugebauer G., and Soifer B. T., (1985), Astrophys. J. Lett., **293**, L11

Neugbauer G., Miley G.K., Soifer B.T., and Clegg P.E., (1986), Astrophys. J., **250**, 469

Osterbrock D.E., (1977), Astrophys. J. , **215**, 733

Shu F. H.,(1987) in Star Formation in Galaxies, ed. C.J. Persson, in press

Vader J.P. and Simon M., (1987), Nature, **327**, 304

Vader J. P., Da Costa G.S., Frogel J.A., Heisler C.A. and Simon M., (1987), Astron. J. , **94**, 847

NUCLEAR ACTIVITY IN WARM IRAS GALAXIES*

J. Patricia Vader
Yale university, Department of Astronomy
Box 6666, New Haven CT

Jay A. Frogel & Fred C. Gillett
Kitt Peak National Observatory
Box 26732, 950 North Cherry Avenue, Tucson AZ

*based on observations made at CTIO and KPNO, operated by Associated Universities for Research in Astronomy, Inc. under contract with the National Science Foundation.

Quasars have the flattest, i.e., warmest far-infrared (FIR) energy distributions of all extra-galactic sources, followed by Seyfert 1 and Seyfert 2 galaxies. The discovery of the unusually warm galaxy IRAS 0052-7054, whose properties are indicative of a dust-embedded active nucleus (Frogel and Elias 1987), prompted a systematic study of IRAS sources a FIR energy distribution peaking at 60 μm. Selection of all sources at galactic latitude |b| > 10° from the IRAS Point Source Catalog Working Survey Data Base, with good quality detections at 25 and 60 μm and flux density ratios $f_{60}/f_{100} > 1$, $4 > f_{60}/f_{25} > 1$, yielded a majority of planetary nebulae and smaller numbers of stars and galaxies, including the FIR ultraluminous galaxy Mrk 231 (Sanders et al. 1987). The 22 previously unidentified galaxies among these sources, including 0052-7054 and the akin object 2306+0505 (Hill et al. 1987), are the subject of this study. Table 1 gives the redshift, 60 μm luminosity, Hα to Hβ flux ratios (uncorrected for reddening and starlight), and the rest frame FWHM's velocity widths (km/s) of [OIII] and Hα for these galaxies, as well as for the double nucleus galaxy IRAS 0258-1136 discussed by Heisler and Vader (1988). Near-IR data have been obtained in three or more bands from 1.2 to 10 μm. In case of two or three optical candidates within the IRAS position error ellipse, the optical counterpart has been taken to be that with the reddest H-K color. Optical spectra with 11 A resolution were taken at the CTIO 4m. Fig. 1 displays the optical continuum as determined from the available spectra, along with the smallest aperture IR (4.7 to 11.7") and the IRAS measurements. The spectral energy distributions (SED) all peak around 25 - 60 μm, but otherwise show a wide variety. The three objects with unusually 'warm' SED's (0438, 0518, and 1345), similar to those of 0052 and 2306, have the broadest emission lines and high excitation line strength ratios, classifying them as

TABLE 1

GALAXY	z	$\log L_{60}$	log Hα/Hβ	FWHM [OIII]	FWHM Hα
0016-0719B	0.018	10.16			
0019-7926A	0.079	11.70	1.01	317	
0034-3349B	0.021	10.84	0.92		
0052-7054	0.069	11.01	1.13	370	546
0105+2147	0.055	11.10	1.17		360
0107-0348	0.054	10.89	0.84	529	494
0425-0440	0.015	10.37			
0438-0828	0.015	10.21		889	441
0518-2524	0.042	11.79		786	790
0800-6600A	0.041	11.20	1.05		
0817-1126A	0.054	11.11	0.92		
1056-3323	0.059	11.27	0.94		
1345-4155A	0.039	10.85	0.97	409	339
1416-7236	0.026	10.51			
0258-1136A	0.030	10.08	0.97		
1638-8120					
1931+4054					
1937-0131					
2021+1121A					
2025-8152					
2306+0505	0.174	11.96			
2344+1519					
2354+1914A					

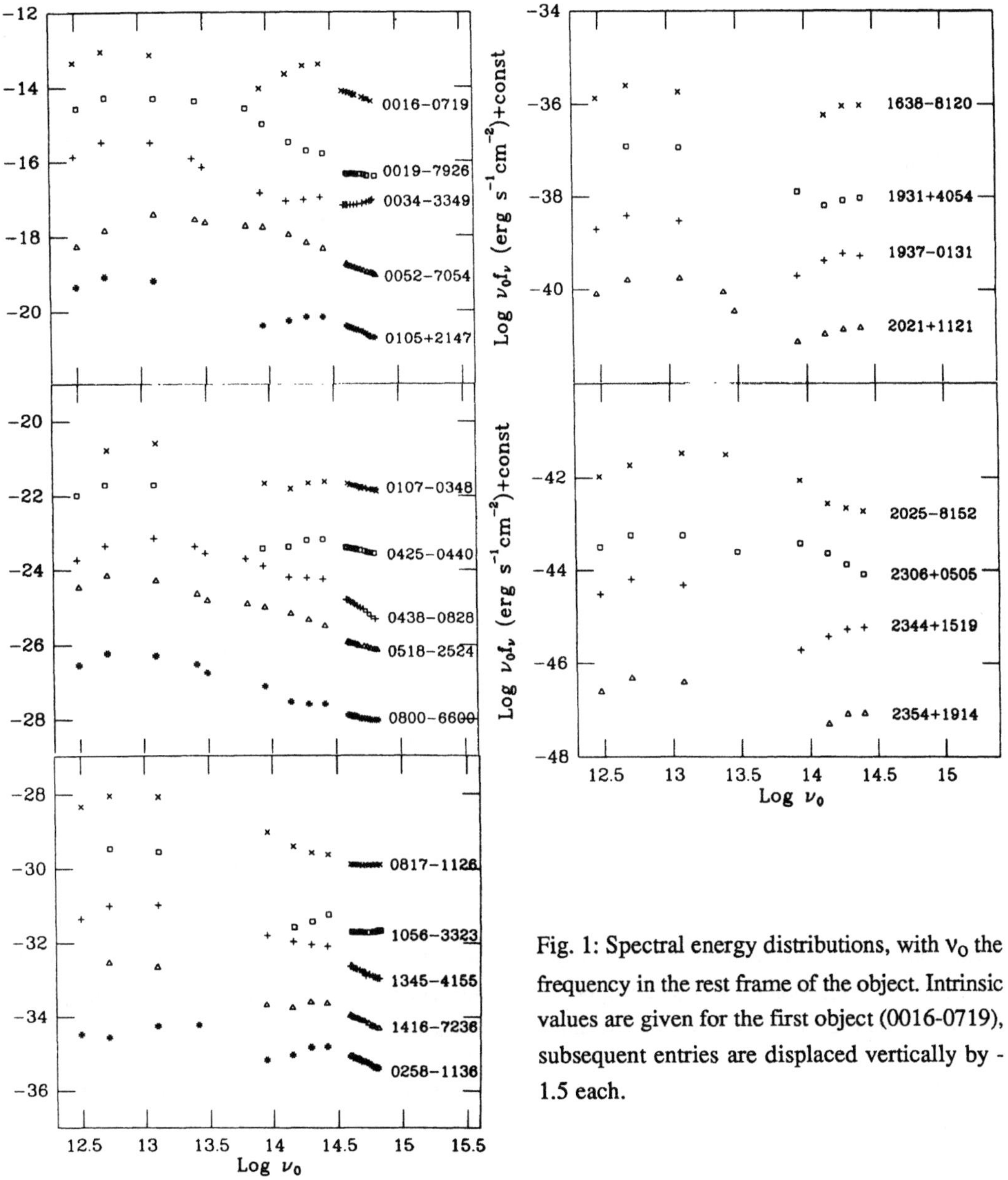

Fig. 1: Spectral energy distributions, with ν_0 the frequency in the rest frame of the object. Intrinsic values are given for the first object (0016-0719), subsequent entries are displaced vertically by -1.5 each.

Seyfert 2's with a dust-obscured AGN. A second class of objects with relatively larger FIR than near-IR excesses, i.e., more cold dust, comprises 0019, 0800, 0817, and 2025. All of the above occupy the upper right corner of the (J-K) vs (H-K) diagram (Fig. 2) where objects dominated by reddened starlight are ruled out. The remaining objects lie closer to the locus of normal spirals in Fig.2 and their SED's seem more typical of starburst galaxies. Among these 0107 stands out with broad permitted and forbidden lines. The large Hα to Hβ ratios are indicative of significant internal reddening and/or stellar absorption. Stellar absorption features are clearly present in many of the spectra, indiscriminately among the three SED classes defined above. Emission line ratios of [OI]6300, [SII]6716+6731, and [OIII]5007 to Hα are

Fig. 2: near-IR color-color plot. Dots: galaxies in our sample; crosses: galaxies from Sanders et al. (1987), 5" aperture diameter measurements. The solid line indicates the locus of 'normal' spiral galaxies.

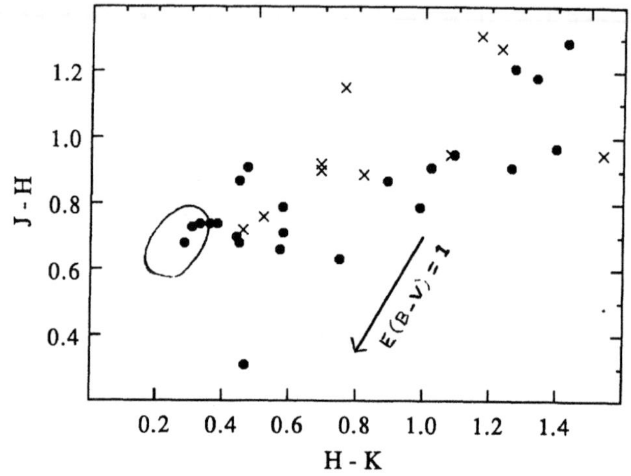

Fig. 3: optical spectrum of the warm galaxy IRAS 0438-0838

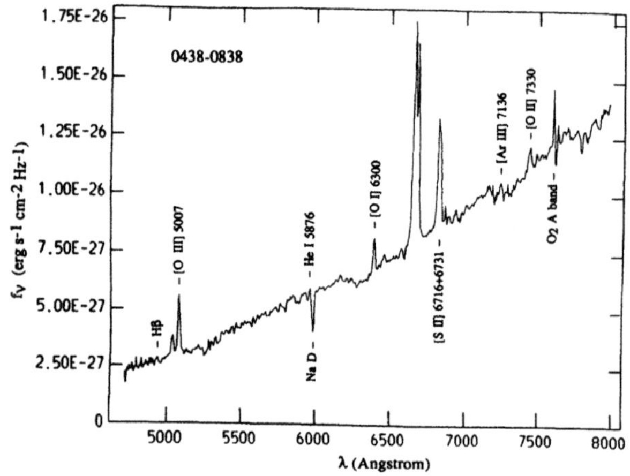

mostly indicative of nuclear activity. CCD imaging shows that most galaxies are disturbed or interacting. A conspicuous exception is 0438 whose optical image appears to be a lenticular or elliptical galaxy with axial ratio 2. However, it has an asymmetrical nucleus, suggesting the presence of dust.

The galaxies in our sample show various amounts of star formation, non-thermal activity, and thermal emission of dust with a wide range in temperature. A more detailed account of this research is in preparation.

JPV acknowledges partial financial support by NASA grant JPL 958027. FG is supported in part by IPAC.

REFERENCES

Frogel, J.A. and Elias J.H. 1987, Ap. J. (Letters) **313**, L53.

Heisler, C.A. and Vader, J.P. 1988, this volume.

Hill, G.J., Wynn-Williams, C.G., and Becklin, E.E. 1987, Ap.J. (Letters) **316**, L11.

Sanders, D.B., Soifer, B.T., Elias, J.H., Madore, B.F., Matthews, K., Neugebauer, G., and Scoville, N.Z. 1988, Ap.J., in press.

An Optical and Radio Survey of Southern Radio Galaxies.

S. Caganoff[1,2], R.D. Ekers[1], G.V. Bicknell[2] and D. Carter[2]

1. NRAO, Socorro N.M., U.S.A.

2. Mount Stromlo Observatory, A.C.T., Australia.

Abstract: We present results from a subset of our survey of southern elliptical radio galaxies. Using VLA snapshot data and high sensitivity spectra from the AAT Faint Object Red Spectrograph we find a strong correlation between the luminosity in the Hα+[NII] emission lines and the total radio power from the source, but a poor correlation between Hα+[NII] luminosity and core radio power. A possible interpretation of these data is discussed.

I. Introduction:

Our sample of Parkes radio sources has been defined by the following criteria:

1. $-35^\circ \leq$ declination $\leq -10^\circ$
2. flux at 2.7GHz ≥ 0.5 Jy
3. optical identification with $14 \leq m_B \leq 18$.

These selection criteria bias the sample toward nearby ($z \simeq 0.05$–0.3) radio sources lying around the Fanaroff-Riley (F-R) class I/class II break at $P_{20cm} = 10^{25}\mathrm{WHz}^{-1}$ (see Bridle 1984). An optical and radio survey of a sample in this medium power range will complement previous studies of lower power radio sources (e.g. Phillips *et al.* 1986) and the numerous surveys of the higher powered 3C radio sources.

II. Results:

The results presented here are from a subset of approximately 50% of our final sample. This subset comprises galaxies for which we already have optical spectra using the Anglo-Australian Telescope. Although the subsample is not statistically complete, our selection of these objects has not been significantly biased and we expect it to be representative of the whole sample.

Redshifts are obtained from blue spectra of resolution $\simeq$ 4Å around the Ca h and k absorption lines using the RGO spectrograph. The Faint Object Red Spectrograph (FORS) yields low resolution ($\simeq$ 25Å) spectra over the range 4500Å to 8500Å. The FORS has the high sensitivity needed for a survey of faint galaxies such as ours.

Complementary to the AAT spectrographic survey is a snapshot survey using the VLA. Snapshots of most of the sample have been made at 6cm and 20cm in two scaled array configurations. Accurate core fluxes are obtained by filtering out the low spatial frequency information in the map making process. This suppresses confusion of the core with extended radio structure. Virtually all our sources exhibit a core down to a sensitivity of a few mJy.

Our sample contains radio galaxies with nuclear emission lines spanning a wide range of ionisation level from LINER-like to seyfert-like spectra. All but one are narrow line radio galaxies. Figure 1a shows a graph of the luminosity in the $H\alpha$+[NII] emission lines as a function of the total radio power at 20cm. Using a modified version of the Kendall τ test to take into account upper limits (Isobe *et al.* 1986) we find a good correlation, with the null hypothesis rejected at a significance level of 98.6%. Note that this method of dealing with upper limits avoids detecting spurious correlations in the luminosity domain. The dashed line is a line of best fit with $L_{H\alpha+[NII]} \propto P_{total}^{1.2\pm0.4}$. Figure 1b shows the relation between $H\alpha$+[NII] luminosity and the core radio power at 6cm. This is, at best, a poor correlation with a significance level of 84%.

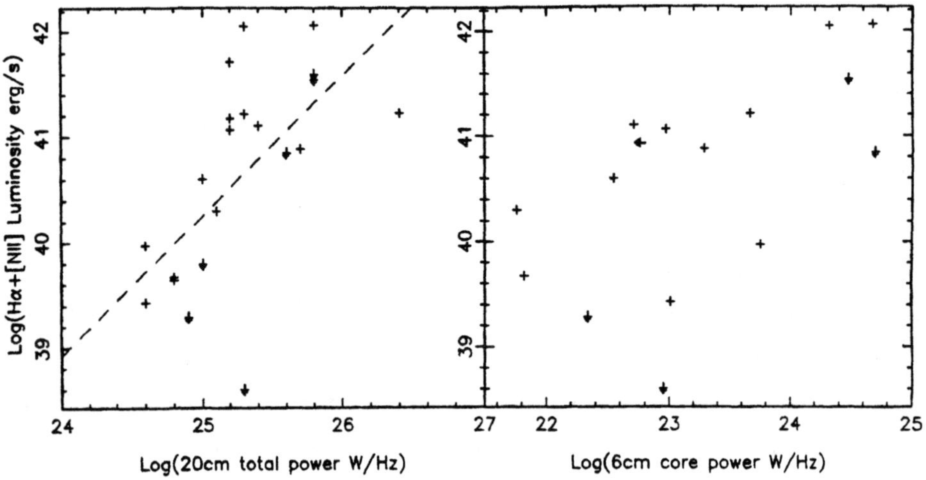

Figure 1a: $H\alpha$ + [NII] luminosity vs. the total radio power at 20cm. The correlation is significant at the 98.6% level.

Figure 1b: $H\alpha$ + [NII] liminosity vs. the core radio power at 6cm. Arrows represent upper limits in both diagrams.

III. Discussion:

Clearly the emission line strength correlates better with the *total* radio power than with the *core* radio power. Assuming the power in the radio core is closely related to the power of the active nucleus, one possible hypothesis is that both the emission line region and the extended radio source are influenced by a common set of fundamental parameters. While the active nucleus may play a fundamental role in powering the radio source and photo-ionising the emission line clouds, we might also expect the hot interstellar medium which confines the emission line clouds (Krolik, McKee and Tartar 1981) to influence the collimation of the jet which powers the extended radio source.

Independant evidence for the effects of a hot interstellar medium on radio sources comes from the results of Fabbiano *et al.* (1987) and Miller *et al.* (1985). They show that while F-R class I sources appear to be confined by an extended, X-ray emitting gas, the more powerful F-R class II sources exhibit little extended X-ray emission, being dominated instead by non-thermal emission from the active nucleus. The class II sources are probably not confined by a hot interstellar medium.

It is possible that a radio galaxy with a low density interstellar medium will yield strong optical emission lines from low density clouds with a high ionisation parameter (ratio of ionising photon density to electron density). A low density medium may also be less disruptive to radio jets. Conversely, radio jets which power the extended radio emission may be significantly disrupted (via entrainment or plasma instabilities) by a dense interstellar medium containing dense clouds with correspondingly low ionisation parameter and hence, weaker emission lines. Such a scenario could yield a correlation between optical emission line luminosity and total radio power. This effect may break down in the most powerful radio galaxies where the intrinsic power in the active nucleus and the jets may dominate the effects of the interstellar medium.

IV. Conclusion:

We argue that not only the central engine, but also the hot interstellar medium can significantly influence both the emission line clouds and the extended radio emission in medium powered radio sources. This may explain the correlation we find between the total radio power and the Hα+[NII] emission line luminosity. We intend to carry out further observations combined with theoretical modelling to discern the role of these parameters.

References:

Bridle, A.H., 1984, Astron. J., **89**, 979.

Fabbiano G., *et al.*, 1987, Ap. J., **312**, 111.

Isobe T., Feigelson E.D., and Nelson P.I., 1986, Ap. J., **306**, 490.

Krolik J.H., McKee C.F., and Tartar C.B., 1981, Ap. J., **249**, 422.

Miller L., *et al.*, 1985, M.N.R.A.S., **215**, 799.

Phillips *et al.*, 1986, Astron. J., **91**, 1062.

A study of Sérsic-Pastoriza galaxies

D.J. Saikia[1,2], G.J. Yates[1], A. Pedlar[1], D.J. Axon[1], J. van Gorkom[3], R.D. Wolstencroft[4], S.W. Unger[5]

[1] University of Manchester, Nuffield Radio Astronomy Laboratories, Jodrell Bank, Macclesfield, Cheshire SK11 9DL, England

[2] Tata Institute of Fundamental Research, Post Box 1234, Bangalore 560 012, India

[3] National Radio Astronomy Observatory, Post Box 0, Socorro, New-Mexico 87801, USA

[4] Royal Observatory, Blackford Hill, Edinburgh, England

[5] Royal Greenwich Observatory, Herstmonceux Castle, Hailsham, East Sussex, England

1 Introduction

In his paper on the classification of galaxies, Morgan (1958) noticed that the nuclear regions of some spiral galaxies consist of clusters of bright spots which he termed nuclear "hot-spots". Spectroscopic observations soon showed that the hot-spots are complexes of HII regions. Some years after Morgan's work, Sérsic & Pastoriza (1965, 1967), who were studying morphologically peculiar nuclei, pointed out that the observed nuclear structures exhibit at least two extreme forms. These are the hot-spot systems and the ones with "amorphous nuclei". The latter consist of spherical formations surrounded by diffuse and asymmetric structures. Sérsic's (1973) definition of a peculiar nucleus as one in which there is a change in the slope of the luminosity profile and evidence of some structure, includes the conventional hot-spot systems as well as the ones with amorphous nuclei. Following Osmer, Smith & Weedman (1974) we shall refer to these spirals with morphologically peculiar nuclei as Sérsic-Pastoriza or S-P galaxies.

Although many of the hot-spot systems have been studied in some detail, a systematic study of the entire range of S-P galaxies is essential in order to understand their formation, evolution and possible relationships between the different forms of morphologically peculiar nuclei. Prabhu (1980) made a systematic, photographic study of a large sample of S-P galaxies and suggested a possible classification scheme depending on the relative brightness of the nuclear and perinuclear formation. He also noted the similarity between Sersic's criterion of a change in the slope of the luminosity profile, and the nucleus-shoulder-arms structure of Seyfert galaxies (Morgan, Walborn & Tapscott 1971). A number of S-P galaxies are known to exhibit both Seyfert and starburst activity and it is important to clarify any link between these two forms of activity. With some of these problems in mind we have begun a radio, optical and infrared study of S-P galaxies. In this contribution we first present radio maps of a few of the sources we have observed, and discuss a possible relationship between nuclear morphology and radio emission.

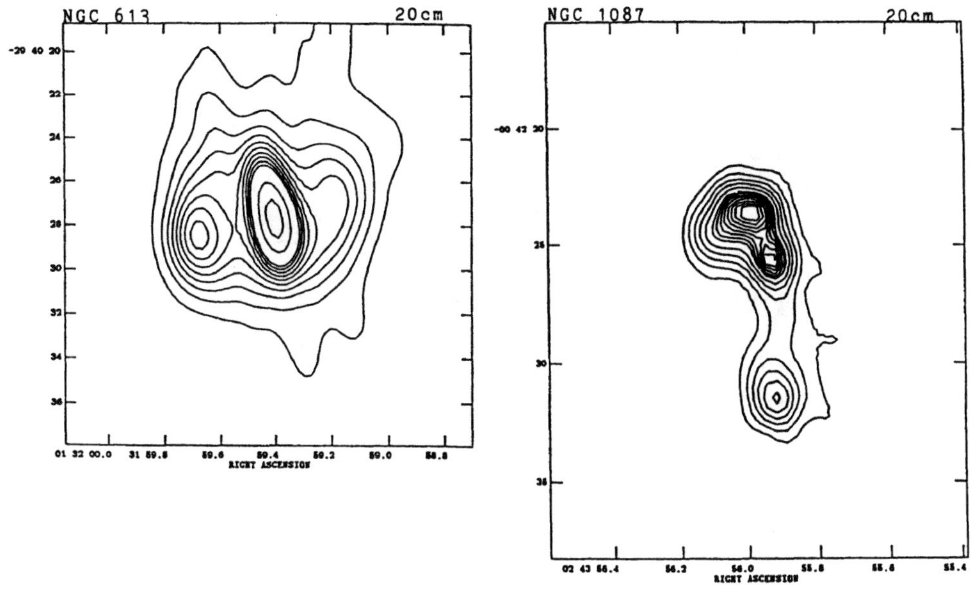

Figure 1: VLA A-array maps of NGC 613 and 1087 at 20cm.

2 Observational results

The galaxy NGC 613, which is being studied in detail by Hummel *et al.* (1987 and in preparation), is of morphological class σ according to Prabhu's classification scheme. Galaxies in this group have a bright nucleus and also bright, fairly distinct hot-spots. Our radio map with the VLA A-array at 20 cm (Figure 1) shows a central component with weaker components on either side of it. The central feature appears extended along a PA of ~10°. A higher-resolution radio map shows that this feature has a jet-like extension (Hummel *et al.* 1987). Figure 1 also shows our map of NGC 1087 which is of morphological class $\sigma\iota$. This group consists of galaxies which have a somewhat fainter nucleus and a perinuclear formation which is bar-like and generally gets resolved into hot-spots. The bar-like structure in the optical image of this galaxy is along a PA ~135° (Prabhu 1980), while the radio structure is almost north-south. At 6cm, we find evidence for a weak component about 1 arcmin to the west.

The giant barred spiral galaxy in the Fornax cluster, NGC 1365, has been well studied by a number of authors. Sandqvist, Jörsäter & Lindblad (1982) have presented early VLA maps of the source at 20 and 6 cm. They, however, did not detect any radio emission from the Seyfert nucleus of the galaxy. Our VLA A-array 20-cm map (Figure 2) shows a clear ring-like structure and a radio component within about an arcsec of the position of the nucleus listed by Sandqvist *et al.* (1982). The central regions of NGC 1365 as well as of NGC 1808, have been classified to be of type σ. Long-slit spectra of the central regions of NGC 1808 show that although most of the bright spots are luminous HII regions, there is also a weak Seyfert-like nucleus (Véron-Cetty & Véron 1985). Our VLA A-array map at 20 cm is shown in Figure 2.

A detailed discussion of these sources along with our optical data, which is currently being

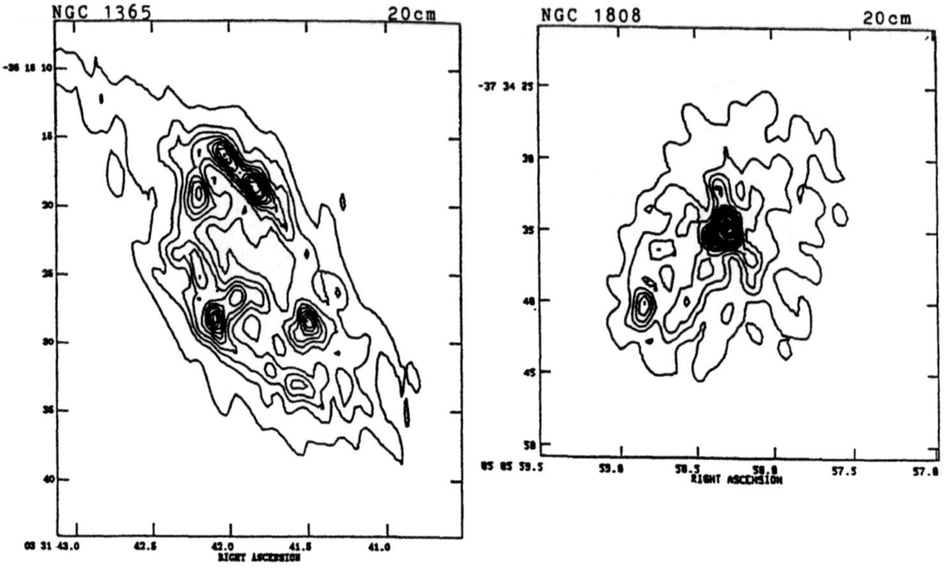

Figure 2: VLA A-array maps of NGC 1365 and 1808 at 20cm

processed, will be presented later. We, however, note here that of the 20 galaxies observed in our first observing session of S-P galaxies with the VLA A-array at 20 cm, 13 are radio loud while the remaining 7 are at best marginally detected. Galaxies whose nuclear morphology have been classified to be of type σ are among the stronger radio sources, while all the three type ι systems, which are galaxies with very faint or no visible nucleus (*cf.* Prabhu 1980), have no significant radio emission in spite of being hot-spot galaxies. This possible relationship between nuclear morphology and the incidence of radio emission is being investigated using a larger sample of S-P galaxies which we have observed with the VLA.

References

Hummel, E., Jörsäter, S., Lindblad, P.O., Sandqvist, A. 1987. *Astron. Astrophys.*, **172**, 51.

Morgan, W.W. 1958. *Publ. astr. Soc. Pacific*, **70**, 364.

Morgan, W.W., Walborn, N.R. & Tapscott, J.W. 1971. *Nuclei of Galaxies*, ed O'Connell, D.J.K., North Holland Publishing Co., Amsterdam, p. 27..

Osmer, P.S., Smith, M.G. & Weedman, D.W. 1974. *Astrophys. J.*, **192**, 279.

Prabhu, T.P. 1980. *J. Astrophys. astr.*, **1**, 129.

Sandqvist, A., Jörsäter, S. & Lindblad, P.O. 1982. *Astr. Astrophys.*, **110**, 336.

Sérsic, J.L. 1973. *Publ. astr. Soc. Pacific*, **85**, 103.

Sérsic, J.L. & Pastoriza, M.G. 1965. *Publ. astr. Soc. Pacific*, **77**, 287.

Sérsic, J.L. & Pastoriza, M.G. 1967. *Publ. astr. Soc. Pacific*, **79**, 152.

Véron-Cetty, M.-P. & Véron, P. 1985. *Astr. Astrophys*, **145**, 425.

INFRARED EMISSION AND TIDAL INTERACTIONS OF SPIRAL GALAXIES

G. G. Byrd
Department of Physics and Astronomy, University of Alabama
Tuscaloosa, Alabama 35487, USA
(Summer ASEE visitor, Marshall Space Flight Center, Huntsville, AL)

B. Sundelius
Department of Astronomy, University of Gothenburg
41296 Gothenburg, Sweden

M. Thomasson
Onsala Space Observatory, Sweden

M. Valtonen
Turku University Observatory, 20520 Turku, Finland

ABSTRACT. We use computer simulations of tidal interactions of spiral galaxies and find that the stronger the tidal perturbation by a companion, the more disk gas clouds are thrown into nucleus-crossing orbits and the greater their collision velocities in the disk. More intense and hotter IR emission would result, matching observations by Telesco, Wolstencroft and Done of greater and hotter IR emission for spiral galaxies with closer and/or more massive companions. Disturbances propagate inward from the edges to cause nuclear inflow. In Telesco et al.'s IR pairs, the larger members are the more IR luminous. Our simulations indicate that mass and size/separation effects cause this pattern.

INTRODUCTION, SIMULATIONS AND DISCUSSION.

Telesco et al. (1987) used IRAS data and a sample of interacting pairs of galaxies from Arp and Madore (1987) to study the ratio, R, of their fluxes at 60 and 100 micrometers, defining R as the 60 flux/100 micrometer flux. For R> 0.5, the galaxy is "hot", smaller "cool". An R of 0.5 is ~ 60° K >> the 5° K typical of galactic molecular clouds. Telesco et al. found that IR temperature and intensity were correlated with the degree of tidal perturbation. Dahari (1984) previously found a correlation between tidal perturbation and Seyfert activity. Telesco et al. also found the larger member of a pair was either the more IR luminous or the only detectable member.

We use a two-dimensional n-body program by Miller (1974, 1978) to simulate the self-gravitating disk of a spiral galaxy. The disk is represented by 60,000 particles. The halo and disk are equal in mass and the rotation curve flat to the disk edge.

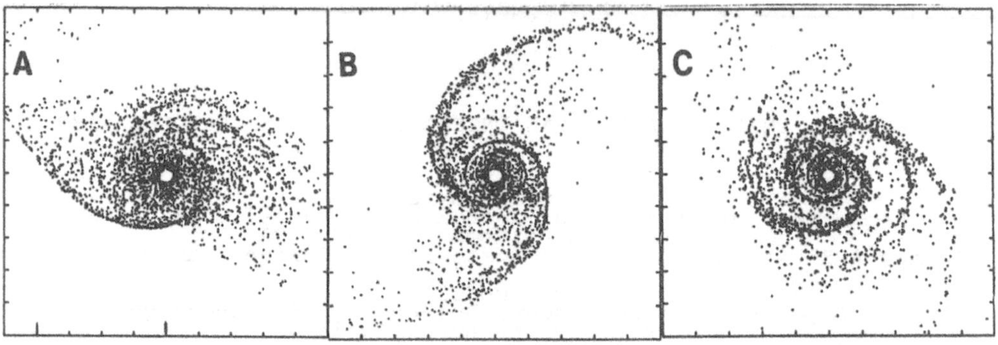

Figure 1- Shapes (A,B,C) during a simulation. A 0.2 galaxy mass disturber starts from right of frames passing counter-clockwise to 1.5 radii above the disk at a time before frame A, then retreating to left. Nuclear inflow starts at B when disturbances reach center.

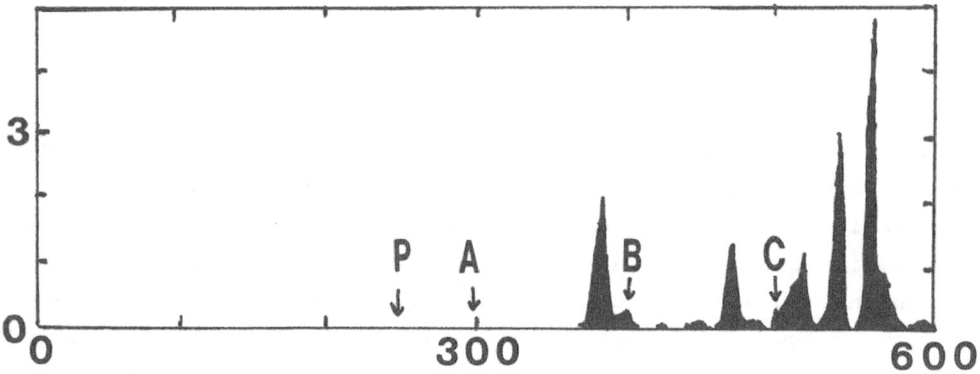

Figure 2- Nuclear inflow rate during Fig. 1 encounter in solar masses/year versus time in 10^6 yr units. Perigalactic point P and points A,B,C from Fig. 1 are marked.

In our current and earlier simulations of tidal effects on spiral galaxies (Byrd et al. 1986, 1987), stronger tidal perturbation results in greater inflow of disk gas clouds into the nuclear regions which could cause IR emission. This result agrees with the tidal perturbation-IR correlation found by Telesco et al.

Our stronger perturbation runs also show IR emission can also arise in the disk. Radial velocity deviations are much larger than those in weaker perturbation runs with large inward and outward velocities in the same areas. Collisions would occur at these locations with relative velocities sufficient to excite IR directly (>100 km/s, Harwit 1987). Afterward, massive star formation would quickly produce indirect IR. So IR emission should be seen from the disk of the galaxy as well as the nuclear regions. Stronger perturbations would cause greater velocity differences, more energy release in collisions, greater grain temperatures and hotter IR temperature emission as observed.

We find maximum inflow occurs well after perigalactic passage with self-gravitation being crucial (Fig. 1). Velocity and morphological deviations appear first at the edge of the disk where local tidal effects are strongest. They then appear to "avalanche" inward to the nuclear regions with the inflow starting upon arrival. Inflow continues even after the perturber has receeded (Fig. 2). This delay could explain "paradoxical" observations of Keel, Kennicutt, Hummel and van der Hulst (1985) and Bushouse (1986, 1987) of some disturbed galaxies with no past or present nuclear activity.

Telesco et al.'s "largest most IR luminous" rule may be due to a number of factors. Tidally equivalent runs with smaller, closer perturber runs versus more massive, distant runs show that this rule appears in our simulations. We see higher peak rates of mass inflow (solar masses/yr) in the primary compared to the secondary. The peak rates of inflow also appear to be sharper in time. Major causes of these effects are the greater mass of gas in the primary's disk and the greater strength of the local tidal field on the inside edge of the primary at close approach.

The above rule applies to encounters with orbital sense the same as the galaxy disk spin. If the orbit is direct relative to the small companion disk and retrograde to the primary, the secondary could show more activity. Spin effects plus episodic inflow may cause the "larger more IR active" rule not to be universal. The M81-M82 system with the IR luminous but small M82 is a counter-example.

ACKNOWLEDGEMENTS

We express our appreciation for computer time to NASA/Marshall Space Flight Center and Onsala Space Observatory (operated by Chalmers University of Technology with support from the Swedish Natural Science Research Council SNFR).

REFERENCES

Arp, H.C. and Madore, B.F. 1987, A Catalogue of Southern Peculiar Galaxies and Associations (Cambridge Univ. Press: New York, N.Y.)
Bushouse, H.A. 1987, Ap. J. *320*, 49.
Bushouse, H.A. 1986, Ap. J. *91*, 255.
Byrd, G.G., Valtonen, M.J., Sundelius, B. and Valtaoja, L. 1986, Astron. Astrophys. *166*, 75.
Byrd, G.G., Sundelius, B. and Valtonen, M. 1987, Ast. Ap. *171*, 16.
Dahari, O. 1984, A. J. *89*, 966.
Harwit, M., Houck, J.R., Soifer, B.T., Palumbo, G.G.C. 1987, Ap. J., *315*, 28.
Keel, W.C., Kennicutt, R.C., Hummel, E. and van der Hulst, J.M. 1985, A. J. *90*, 708.
Miller, R.H. 1974, Ap. J. *190*, 539.
Miller, R.H. 1978, Ap. J. *224*, 32.
Telesco, C.M., Wolstencroft, R.D. and Done, C. 1987 (in press).

The Optical Variability of Selected Blazars

H. Richard Miller
Department of Physics and Astronomy
Georgia State University

Abstract

The optical variability of a small number of BL Lacertae objects and OVV quasars have been monitored from 1972 to the present time. Major long term trends in their brightness have been observed with total amplitudes sometimes as large as 5.0 magnitudes detected as in the case of OJ 287. These variations have been observed to occur on the timescales of years. The nature of shorter term variations with amplitudes from 0.5 - 2.0 magnitudes on timescales of days to months have also been observed and investigated. An extensive effort has been made to investigate the possibility of significant optical variations which might occur on timescales much shorter than one day. Observations have been obtained which conclusively demonstrate that short-term, small amplitude variations occur frequently for BL Lacertae. An investigation of a number of other members of this class of objects has also revealed this type of optical activity to be present.

I. Introduction

Blazars, a term generally reserved to identify the class of objects which includes BL Lacertae objects and OVV quasars are of fundamental importance to understanding the underlying energy mechanism in active galactic nuclei (AGNs). The rapid variations observed at all wavelengths, in both the luminosity and polarization, for members of this class of objects are identifying characteristics of Blazars (Stein, O'Dell, and Strittmatter, 1976). The optical variations which have been observed for Blazars suggest that the tremendous energy output must be generated in regions with dimensions of the order of light-days unless there is significant relativistic enhancement due to relativistic bulk motion of material in the central region (Miller, 1975). The problems associated with producing extreme luminosities in such compact sources have provided a significant impetus for the detailed development of relativistic jet models. In this paper, I will present the results of a long term study of the nature and the character of the optical variability of a selected sample of Blazars and the constraints that this study may place on the energy generation processes.

This program was initiated fifteen years ago. The first five years of the program consisted of a photographic monitoring program of selected objects using the 30-inch Beck Telescope at Agnes Scott College. This program allowed one to track the long-term, large-amplitude variations which occurred for these objects and had the advantage that observations were obtained several times a month on program objects since the telescope was dedicated to this program during "dark time." However, photographic photometry has a typical uncertainty of 0.10 magnitude, and therefore smaller amplitude variations could not be accurately followed.

In 1976, photoelectric photometry of a number of these objects was begun using observing time obtained on the 0.9 meter and 1.3 meter telescopes at KPNO and the 1.05 meter telescope at Lowell Observatory. These observations provided UBV magnitudes and colors which complemented the photographic observations, and which had accuracies of 0.01-0.03 magnitudes, depending on the brightness of the object of interest at the time the observations were obtained.

II. Long Term Variability Studies

The results of the long term studies of the optical variability for five of the more intensively monitored blazars are presented in the following sections.

Markarian 421

One of the objects which we studied was Markarian 421 which has z = 0.0308. The results of this study of the optical variability for Markarian 421 are shown in Figure 1. This galaxy has exhibited a range of approximately 2.0 magnitudes over the past 12 years. An outburst of 0.75 magnitude was observed from early 1981 to April, 1982 when Markarian 421 reached its a maximum brightness of 12.63. The B-V colors appeared somewhat bluer during this outburst but no significant change was detected for the U-B color index. This suggests that there were no major changes in the spectral index associated with the luminosity changes of the non thermal central source, but that the observed color variations were due to changing contributions from the non thermal central source and the galaxy in which this source is embedded.

The overall pattern of variability for Markarian 421 is one of erratic variations in which the long-term and short-term variations are of comparable amplitude. Extremely rapid

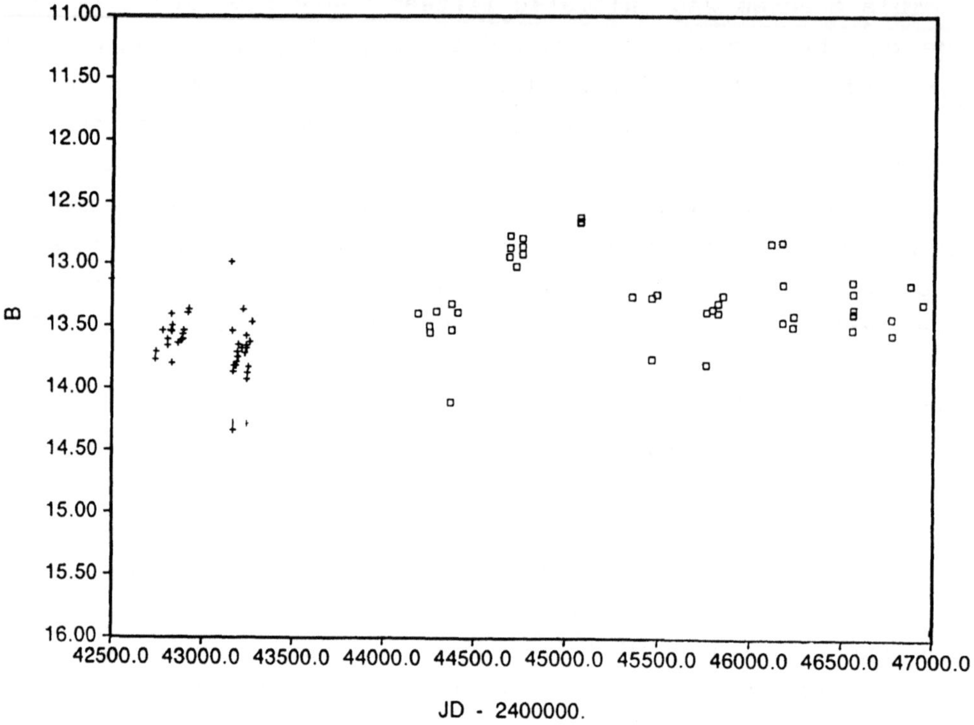

Figure 1. The optical variability of Markarian 421 is displayed from 1975 to 1987.

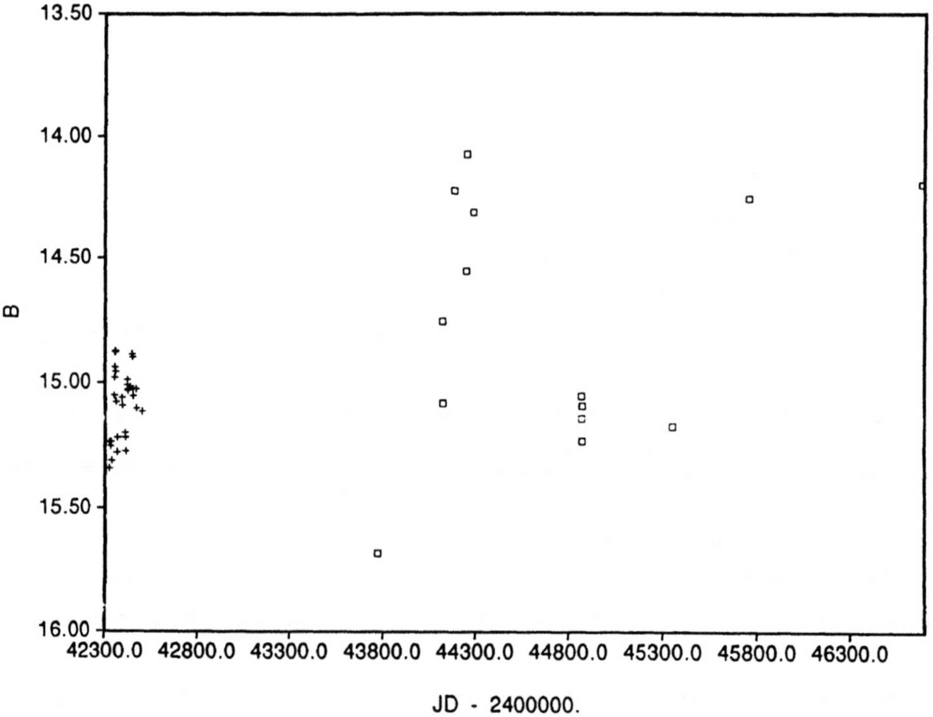

Figure 2. The long term optical light curve for 3C 66A is shown from 1973 to 1987.

variations of the order of 0.3 - 0.4 magnitude/day have also been detected on several occasions (e.g. May 15-16, 1980 (0.2 mag); May 9-10, 1986 (0.38 mag); and May 13-15, 1980 (.79 mag.)). Variation s on these time scales imply a source size of 10^{-3} pc, assuming no relativistic bulk motion.

3C 66A

The optical variability of the BL Lac object, 3C 66A (z = 0.44) was first reported by Folsom et al (1976). We have continued to monitor the long term variability of this object and the results are shown in Figure 2. Photographically, a flare of 0.50 magnitude. was observed in 1981. This 0.5 magnitude event had a duration of approximately a month. A major outburst was also observed in early 1980 when this source reached a B = 14.08, which is the brightest level at which it has ever been observed. Extremely rapid variations were observed to accompany this outburst with the source rising 0.48 magnitudes from 1980 January 16 to January 17. This rapid variability implies a source size of approximately 10^{-3} pc. 3C 66A was also observed in bright phases in 1984 (14.26) and 1986 (14.20). However, the paucity of observations precludes any detailed discussions of the nature of these outbursts. In addition, no significant color variations were found to be present during any of these variations.

PKS 0735 + 178

Pks 0735 + 178 has long been identified as a BL Lac object with z = 0.424. It is well known to exhibit significant optical variability. The results of our study of the optical variability of this object are shown in Figure 3. Pks 0735 + 178 has exhibited episodic outbursts with long term trends of up to 5.0 magnitudes with shorter term variations of amplitude of 1-2 magnitudes superposed on the long term trends. No color variation were detected to be associated with either the long term or short term variations. However when the object went through a minimum in 1979-80 of 18.34 (Jan. 12, 1980), the character of the optical variations changed from a "bursty"-type behavior to long term changes with smaller term events superimposed on the long term changes. Rapid variations have been observed for this object with a change of 0.49 mag. observed from 1982 April 19-20 (15.97-15.48). Similar studies of the nature of the optical variability for Pks 0735 + 178 (Webb et al 1988) are in excellent agreement with the observations presented

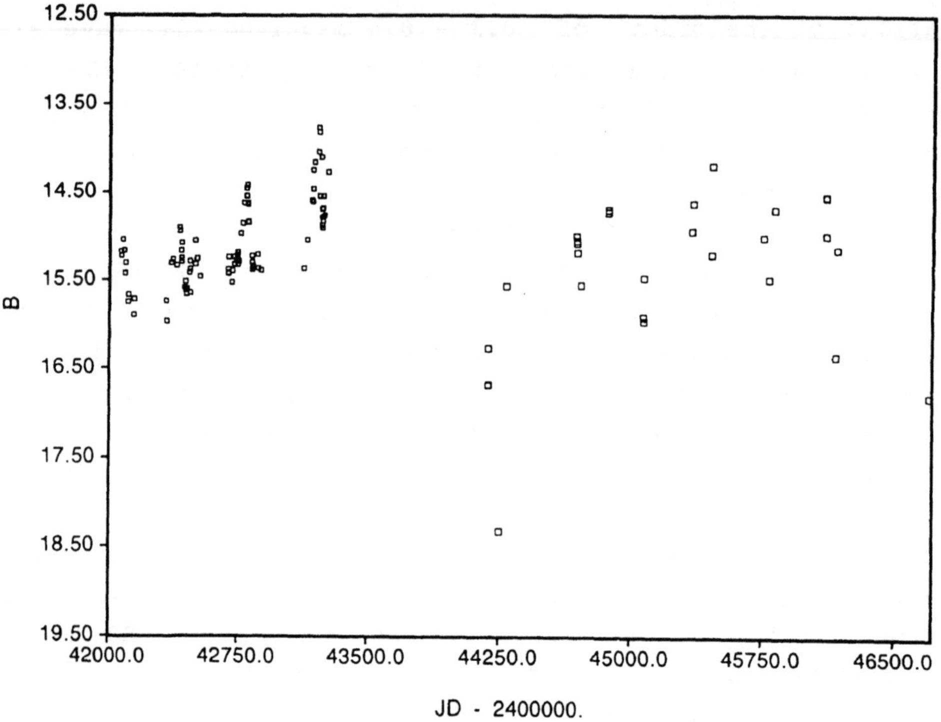

Figure 3. The long term light curve for PKS 0735 + 178 is shown for the period 1973 to 1987. The dramatic change in the nature is clearly visible following the minimum detected in 1980.

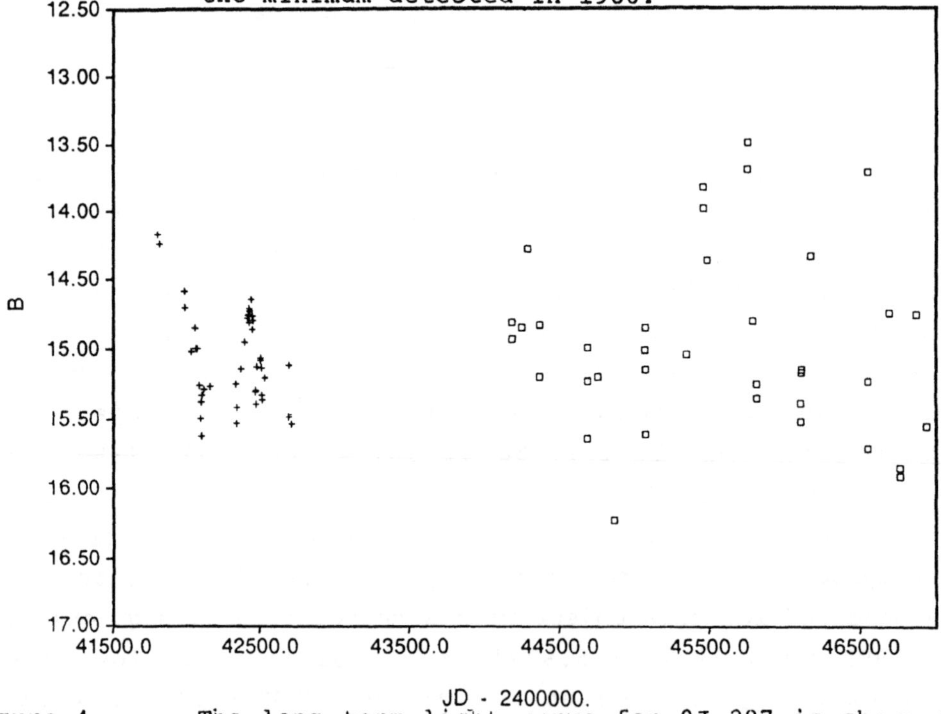

Figure 4. The long term light curve for OJ 287 is shown for the period 1973 to 1987.

here. This once again indicates variations on timescales of a day and implies source sizes of 10^{-3} pc, which is similar to that predicted for the first two objects discussed.

OJ 287

OJ 287 (z=0.306) has long been known to exhibit long term trends in its brightness with large scale, short term events being present as well. The results of our study are shown in Figure 4. In the early 1970s, OJ 287 is observed to be decaying in brightness following a major outburst. A minimum is reached in late 1974 followed by a lower amplitude outburst in early 1975. Short-term, daily excursions are found, based on the photographic studies, to be superposed on the longer term events detected from 1973 to 1977. Photoelectric observations have indicated a continuation of this type of rapid variability. There is a slight suggestion that the colors of OJ 287 appear somewhat bluer at maximum light than at minimum light. This is particularly true for a set of observations obtained as shown below:

Date	V	B-V
1986 May 10	13.72	0.47
1986 May 11	15.24	0.52
1986 May 12	15.72	0.59

This set of observations also indicates one of the most dramatic variations detected for OJ 287 in which the source faded by 1.52 magnitudes in a single day followed by another 0.48 magnitude decline on the following day. This again suggests that the size of the source is on the order of 10^{-3} pc.

BL Lacertae

BL Lacertae (z = 0.069), the proto type for this class of objects, has long been known to be variable since the time it was mistakenly identified as a variable star. We began monitoring this object in 1973 (Figure 5). Erratic variations with an amplitude of the order of 2.0 magnitudes were detected in our photographic monitoring program from 1973 to 1977. No significant long term trends were detected. Our program of photoelectric photometry of BL Lac was begun in 1977 and continues to the present. Bl Lac was found to exhibit the same type of variations as seen in the photographic study except that the total range was found into be in excess of 3.0 magnitudes. No significant color variations were found to accompany the large scale variations in brightness. There are several instances when

BL Lacertae has been observed to vary by more than a magnitude on adjacent days (e.g. 1981 June 6-7: 14.56 - 15.87). This again implies source sizes of less than a 10^{-3} pc. Recently, 1987 September 19, BL Lacertae was observed near V = 16.5. This magnitude corresponds very closely to the expected magnitude of the underlying galaxy. This represents a situation, similar to that reported for PKS 2201 + 044 by Carini et al (1988) at this meeting, in which the non thermal central source has become very weak. If one assumes that the accretion process is responsible for the radiation, this suggests that the accretion process has been significantly reduced.

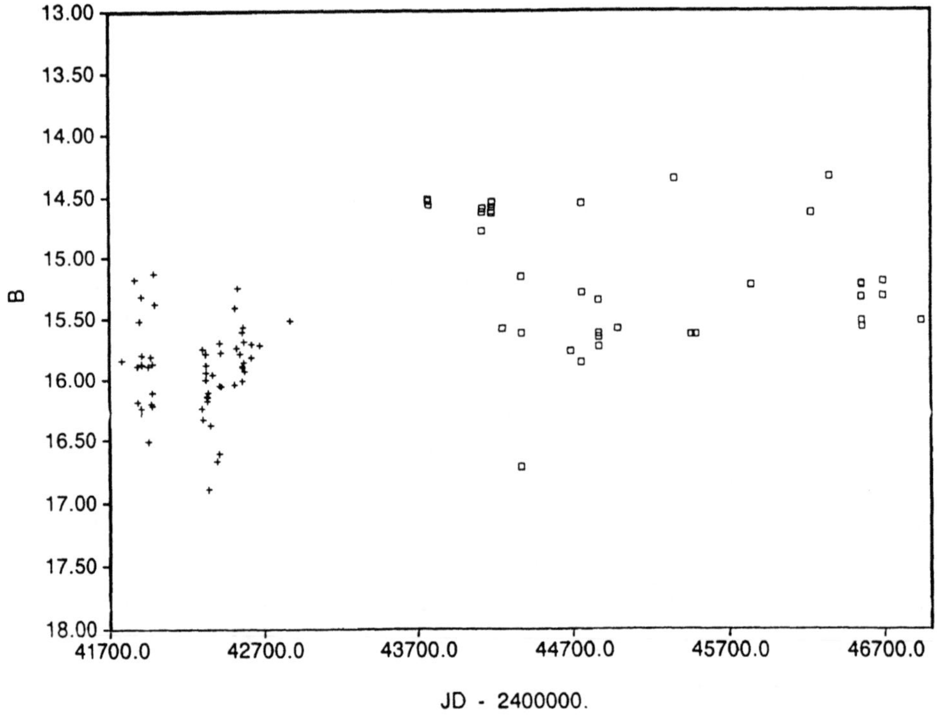

Figure 5. The long term light curve for BL Lacertae is shown for the period 1973 to 1987.

III. Short Term Variability Studies

Note that the size for the source resions are all essentially the same for the five selected BL Lacs, which are very different kinds of members of this class i.e. Mkn 421 is embedded in a galaxy; 3C 66A, is found in a cluster of galaxies; Pks 0735 + 178 exhibits absorption-lines in its spectrum; OJ 287, exhibits long term trends in its brightness and variations; BL Lacertae exhibits short term and long term trends which are of comparable amplitude. They all show evidence, at least some of the time, for variations on timescales shorter than one day. If the timescales and the nature of the variations can be more precisely defined by the observations, this would provide much more stringent constraints for models of the source region.

Rapid, intraday variations have been reported for a number of Blazars. Grauer (1984) and Wills et al (1983) reported short-term changes for 1156 + 295 with timescales of less than an hour while it was near a peak of a major outburst. Observations which I obtained on May 6, 1981 of B2 1308 + 326 during a major outburst shows an increase of 0.10 mag. in less than three hours.

OJ 287 also has a history of suspected variations on the timescale of minutes. Visvanathan and Elliot (1973) detected optical variations with a period of 40 minutes which were later confirmed by Frohlich et al (1974). However, Kiplinger (1974) failed to detect the 40-minute period. More recently, Carrasco et al (1985) have reported a 23-minute period for OJ 287 with an amplitude of 0.03 magnitued. Observationally, it has been difficult to conclusively demonstrate the reality of such small amplitude, short-term variations. Some of the factors which have contributed to this situation are: 1) non simultaneity of observations of comparison stars and the object of interest, and 2) if the observations are made with 2-star photometers, the maintenance of a constant gain difference between the two channels.

Recently, observations have been obtained for OJ 287 using the RCA CCD direct camera on the 0.9 meter telescope at KPNO. Short exposures were obtained and the variable object was reduced differentially using at least four different comparison stars. The comparison stars are also reduced differentially relative to each other. The comparison stars showed no evidence of variability with a scatter of 0.01 magnitude. A total increase in brightness of 0.08 magnitude was detected in 1.5 hours for OJ 287. Although there was no evidence for the 40-minute period of Visvanathan and Elliot and the 23-minute period (amplitude = 0.03 mag.) of Carrasco et al, a possible 28-minute period with an amplitude of 0.03 magnitude could not be ruled out. OJ 287 was near a maximum in brightness when observed by Carrasco et al. However, OJ 287 was moderately faint, near V = 15.4, when the present observations were obtained.

The Bolometric luminosity for OJ 287 is $L = 10^{46}$ ergs/sec at this brightness level. Carrasco et al interpreted the short period, i.e. 23-minute or 40-minute, variations in terms of an accretion disk which contains one or more "hotspots" which are spiralling into a black hole. For a Schwarzschild black hole, thin accretion disks have their inner most stable orbit at 6 GM/c, and thick disks reduce this value to 4 GM/c. The radius of the event horizon of a Kerr black hole of mass M and angular momentum parameter a is given by (Wiita 1985) as

$$r = 1.48 \times 10^5 \ (1 + \sqrt{1-a^2}) \ M/M_\odot \quad \text{cm}$$

For an extreme Kerr black hole (a=1), both the innermost bound and stable orbits coincide at

$$r = GM/c$$

Therefore periodic phenomenon could exist down to timescales of

$$p = \frac{2\pi r}{c} = 3.1 \times 10^{-5} \ M/M_\odot$$

The 28-min period implies a mass for the central black hole of $5.4 \times 10^7 \ M/M_\odot$ and an Eddington luminosity of 9×10^{45} ergs/sec. These numbers are too small to explain the object's bolometric luminosity. However, thick accretion disk models allow for actual luminosities that exceed L_E. In addition, the effects of relativistic bulk motion must also be considered. In this case, the observed luminosity is pumped up by γ^2 (the Lorentz factor) and the total energy required (in the rest frame of the source) is reduced by γ^{-4}. The conflict is removed if the apparent luminosity $L = 10L_E$ is produced by a thick disc coupled with bulk relativistic motion of $\gamma = 2$.

Observations have also been obtained for BL Lacertae. This object was observed for approximately 1.5 hours on 1986 November 9 (Figure 6). BL Lac exhibited a total variation of 0.14 mag. during this interval. BL Lac was observed again on 1986 November 11 for approximately 3 hours. (Figure 7). A much more

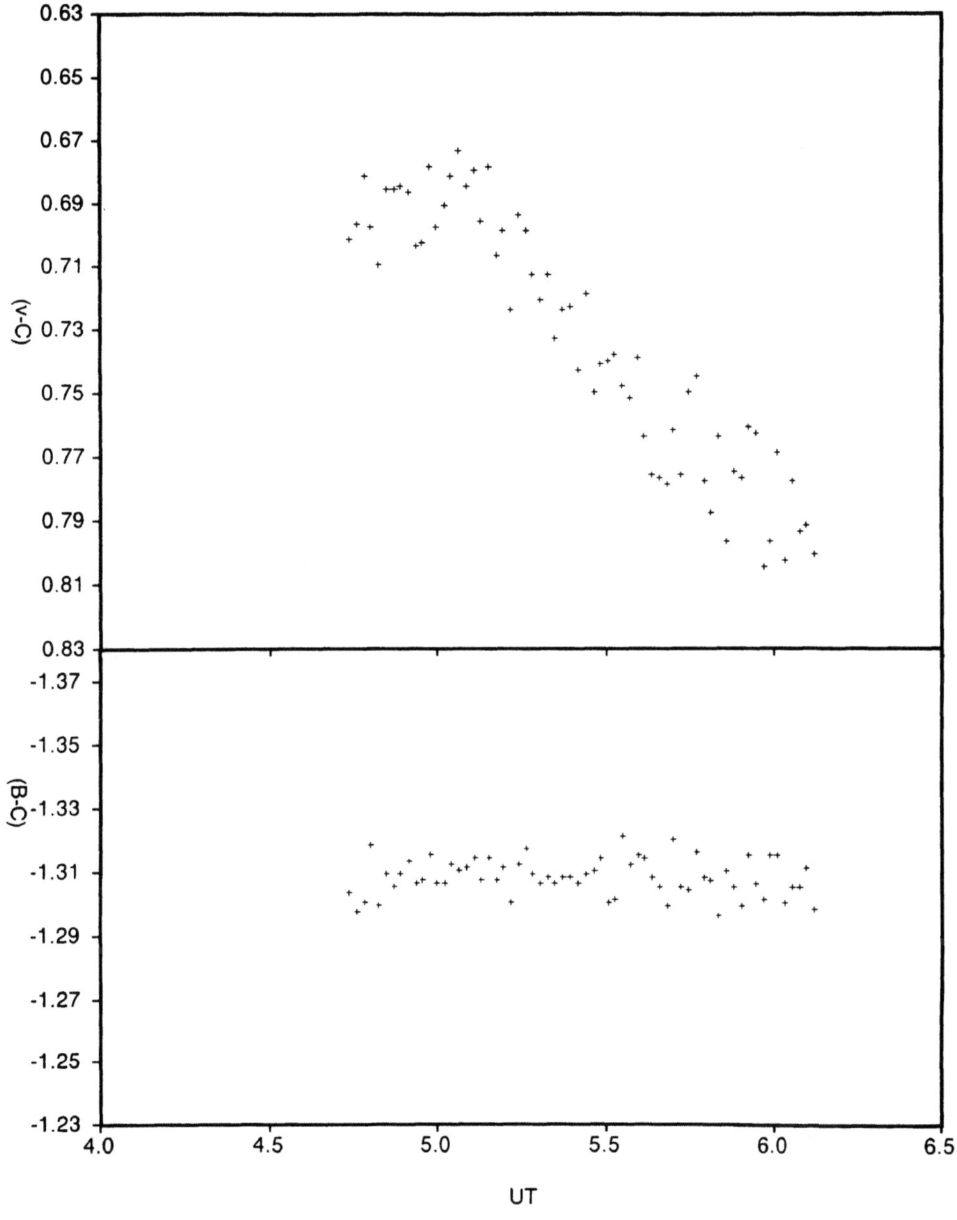

Figure 6. The observed variability of BL Lacertae on 1986 November 9.

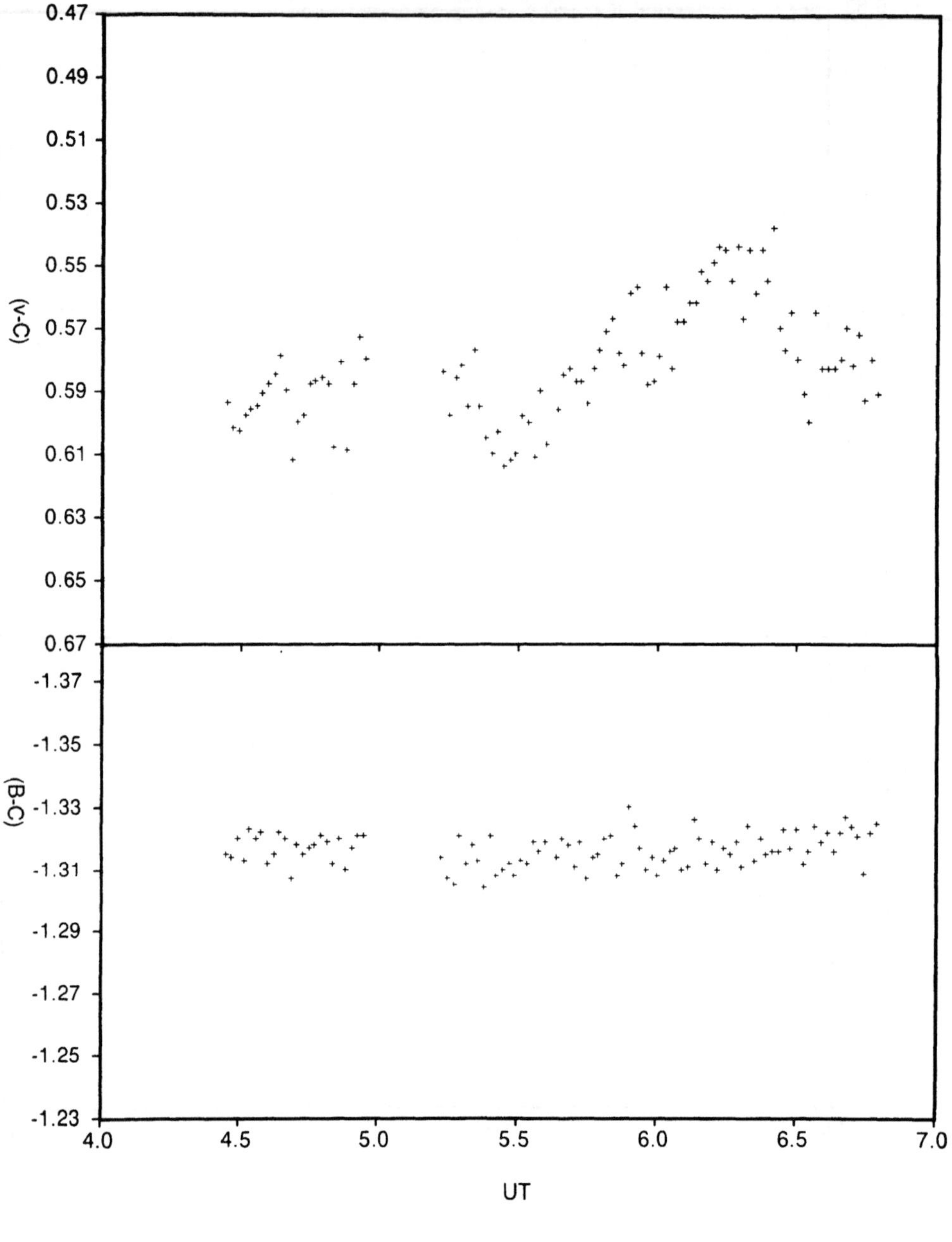

Figure 7. The observed variability of BL Lacertae on 1986 November 12.

complicated structure for the variations were observed. Low amplitude variations were observed during the first 1.5 hours. However, an outburst is observed between 6:00 - 6:30 U.T. with an amplitude of approximately 0.08 magnitude and a duration of 30 minutes. BL Lac is near V = 14.90 on 1986 November 9 and near V = 14.70 on 1986 November 11.

No evidence was found for the presence of any short term periodicities, to be present but one would have to look for such effects superposed on the longer term non periodic changes in brightness. These longer term changes could be due to variations in the accretion rate, thermal instabilities in the relativistic flow, or variations in the injection rate for the radiating particles. Clearly a great deal of additional work needs to be done with both the observationist and theorist providing a significant exchange of information and ideas.

The author thanks Lowell Observatory and Kitt Peak National Observatory for generous allocations of observing time. This work was supported in part by grants from the Vice President's Research Fund at Georgia State University.

References

Carini, M.T., Miller, H.R, and Wilson, J. W. 1988, Active Galactic Nuclei, Ed. by H. R. Miler and P. J. Wiita (In Press).
Carrasco, L., Dultzin-Haagen, D., Cruz-Gonzalez, I. 1985, Nature, 318, 416.
Folsom, G.H., Miller, H.R., Wingert, D.W., and Williamon, R.M., 1976, A. J., 81, 145.
Frohlick, A., Goldsmith, S. and Weistig, D. 1974, M.N.R.A.S., 168 417.
Grauer, A.D. 1984, Ap. J., 277, 77.
Kiplinger, A. L., 1974, Ap.J., 191, L110.
Miller, H.R. 1975, Ap.J., 201, L109.
Stein, W.A., O'Dell, S.L., and Strittmatter, P.A. 1976, Ann. Rev A. and Ap., 14, 173.
Visvanathan, N. and Elliot, J.L., 1973, Ap.J., 179, 721.
Webb, J.R., Smith, A. G., Leacock, R.J. Fitzgibbons, G.L., Ganbola, P.P., and Shepard, D.W. 1988, A. J., 95, 37.
Wiita, P.J., 1985, Phys. Rep. 123, 117.
Wills, B.J., et. al, 1983, Ap.J., 274, 62.

Long-Term Monitoring of a Large Sample of Active Galactic Nuclei

Alex G. Smith, R.J. Leacock, and J.R. Webb
Department of Astronomy
University of Florida

An uninterrupted program of monitoring Active Galactic Nuclei has been carried on at the Rosemary Hill Observatory since 1968 (see, for example, Pollock et al. 1979 and Pica et al. 1980). The current observing list numbers 223 objects, including 156 classical quasars, 35 BL Lac objects, 25 active galaxies, and 6 sources with dubious identifications. Of the foregoing objects, 19 are radio-quiet QSOs, 55 are strong x-ray sources, and 16 are very high redshift objects.

We have classified 22 of our objects as "optically violent variables" or OVVs (Webb et al. 1988), indicating that they show variations of a magnitude or more on timescales of days or weeks. A more recent term for such sources is "blazars". Since we now have as many as 19 years of internally consistent data for some of these objects, an excellent opportunity exists to search for long-term trends. Fortuitiously, our blazar sample divides almost evenly into BL Lac objects and quasars, making it possible to compare the two groups (one blazar, 3C 120, is classified as a Seyfert).

Linear Trends. An obvious question is whether the sources show long-term trends in luminosity underlying the short-term fluctuations. Table 1 shows the results of applying the linear regression routine of Bevington (1969). The quantity L is the slope of the linear regression expressed in magnitudes/yr. Positive values denote increasing magnitudes or decreasing brightness. It is evident that both groups of objects show similar ranges in L, with positive and negative values fairly evenly divided.

Table 1. Linear Trends L in magnitudes per year.

BL Lacs			Quasars		
Object	Years Monitored	L in Mag./yr.	Object	Years Monitored	L in Mag./yr.
1308 + 32	11	+0.162	1156 + 295	7	+0.157
OJ 287	16	+ .146	2345 - 16	15	+ .044
OE 110	12	+ .124	0906 + 01	18	+ .035
0735 + 17	15	+ .038	OX 074	16	+ .017
0235 + 164	12	+ .029	NRAO 512	18	+ .004
BL Lac	16	+ .004	3C 345	15	- .005
0215 + 01	5	.000	3C 454.3	14	- .014
3C 371	19	- .001	0420 - 01	18	- .093
3C 446	15	- .025	NRAO 530	14	- .095
1514-24	15	- .029	NRAO 190	18	- .097
ON 231	15	- .067			

If we ignore the signs and sum absolute values of L, the mean slope for the BL Lacs is 0.057 mag./yr., while for the quasars it is a virtually identical 0.056 mag./yr. It may be significant, however, that if we take sign into account, positive slopes (decreasing brightness) predominate for the BL Lacs by $L^+ = 0.084$ to $L^- = 0.031$, a nearly 3:1 ratio. For the quasars, the mean positive and negative slopes are nearly equal: $L^+ = 0.051$ and $L^- = 0.061$. Is it possible that many of the BL Lacs are fading from some recent event?

Timescale Analysis. A second obvious question is whether any of the blazar data suggest periodicities, or at least characteristic timescales. This is, of course, a question concerning AGNs that has been examined repeatedly in the literature. Well-known difficulties are introduced by relatively short runs of data, by unequally spaced observations, and by spurious periodicities introduced by solar and lunar modulations of the times of observation.

After removal of the linear trends displayed in Table 1, the light curves of our 22 blazars were analyzed by the unequal-interval Fourier transform technique of Deeming (1975). This method yields a power spectrum of the putative periodicities in the data, as well as a "spectral window" that aids in identifying spurious aliases that may contaminate the power spectrum.

Table 2 lists for each object the strongest periods found in the power spectrum in order of decreasing period. In each case an asterisk denotes the strongest of the several periods. If we average the periods P_1, P_2, P_3, we find that the quasars tend toward longer timescales than the BL Lacs. However, in the BL Lacs the longest period, P_1, tends to be most prominent, whereas in the quasars more of the stronger timescales are found in P_2 and P_3.

Table 2. Strongest periods found in the power spectra of the light curves of the 22 OVVs.

BL Lacs				Quasars			
Object	P_1	P_2	P_3	Object	P_1	P_2	P_3
0215+01	2.17*	0.73		0420–01	13.04*	4.64	2.46
0235+164	2.79*	1.53	1.29	NRAO 190	3.65	0.88*	
OE 110	2.88	0.29*		0906+01	3.70*	2.79*	
0735+17	4.89*	1.20		1156+295	3.38	1.41*	0.52
OJ 287	11.90*	5.30	2.79	NRAO 512	2.46	0.66	
ON 231	3.91*	1.94	0.78	3C 345	11.40	5.60	
1308+32	5.59*	1.59	1.28	NRAO 530	4.80	1.71	1.03*
1514-24	1.98*	0.70		OX 074	5.07*	2.46	0.48
3C 371	2.38	0.99		3C 454.3	6.39	2.97	0.83*
BL Lac	0.88	0.60	0.31*	2345–16	5.07*	3.34	1.06
3C 446	4.70*	2.40	1.29	Avg.	5.9	2.6	1.1
Avg.	4.1	1.6	1.3				

The uncertainties inherent in this kind of analysis can be demonstrated by looking at the results for the quasar 3C 345. Table 3 compares the Rosemary Hill analysis with the results of several earlier analyses. It can be seen that there is little correlation between any of the analyses, a contributing factor doubtless being that they represent different intervals of time. Further study of the timescales of 3C 345 was undertaken by adding to the RHO data observations collected from the literature by Pollock (1982). The resulting data set, consisting of 1052 observations from 1965 through 1986, yielded the periods shown on the second line of Table 4. This 21-year run of data was finally split into two segments, one preceding and the other following the major flare of 1971, with the results shown on lines 3 and 4 of Table 4. The most conspicuous feature of Table 4 is the persistence in all four data sets of a period of about 4 years, and a similar period appeared in the analysis of Barbieri et al. It must be admitted, however, that such a period is not conspicuous in a visual inspection of the light curve.

Table 3. Periods found in 3C 345 data by various investigators.

Authors	Interval Covered by Data	Periods Found, Yrs. P_1	P_2	P_3
Smyth & Wolstencroft (1970)	1965-1969	2.80	1.52	0.45
Barbieri et al. (1977)	1964-1976	4.38	2.19	0.38
Babadzhayants & Belokon (1984)	1965-1983	15	1	0.03
Rosemary Hill B-band data	1971-1986	11.4	5.6	4

Table 4. Periods found in 3C 345 by adding historical data to RHO B-band data and subdividing at peak of great flare of 1971.

Data Set	Interval Covered by Data	Periods Found, Yrs. P_1	P_2	P_3
Rosemary Hill B-band data	1971-1986	11.4	5.6	4
RHO + Historical (Pollock, 1982)	5/65-4/86	11.4	3.9	2.16
1st 6 yr., RHO + Historical	5/65-5/71	4.03	1.98	
Last 15 yr., RHO + Historical	5/71-4/86	10.14	3.9	5.47

Using the methods of Barbieri et al. (1977), the frequencies derived from the power spectra can be combined, with appropriate amplitudes, to model the light curves. Figure 1 is an illustration of such modelling, again applied to 3C 345. Much of the deviation from the model is due to rapid events with timescales too short to be resolved by the Fourier analysis.

Variability Subclasses. In keeping with earlier work, we have assigned so-called "variability subclasses" to each of the 22 light curves to convey at least a rough impression of the appearance of the curve. Subclass I displays rapid "flickering" without significant long-term trends. Conversely, subclass II shows slow, long-term changes that dominate any brief variations.

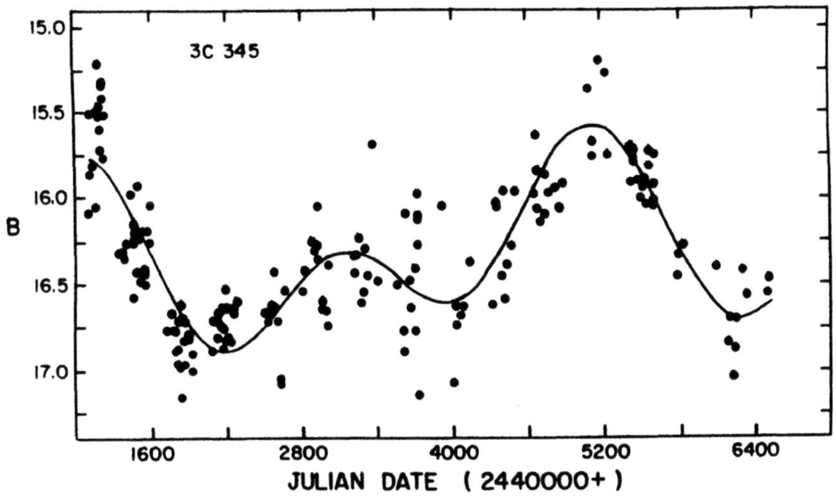

Figure 1. Model light curve (solid line) derived from Fourier analysis of RHO data for 3C 345, 1970-1986. Filled circles are actual data points.

Subclass III is a mix of short and long-term fluctuations of comparable amplitudes. Finally, subclass IV is episodic, with outbursts punctuating long periods of inactivity. An example of each subclass is illustrated in Fig. 2. Table 5 shows the relative numbers of BL Lacs and quasars in each subclass. It is noteworthy that the BL Lacs and quasars are distributed almost equally in subclasses I and III (flickering and mixed flickering and long-term trends). The only object in subclass II (dominant long-term trends) is the BL Lac OJ 287. None of the BL Lacs is in the episodic subclass IV, implying that they are all continuously active in one way or another.

Subclass I is reminiscent of the random fluctuations produced in electronic circuits by thermal noise and the shot and flicker effects. In AGNs it may well point to stochastic processes such as turbulence, in which case scale lengths and other physical parameters may be derivable. However, random noise cannot produce the varied structures seen in subclasses II, III, and IV. Here it is evident that the "noise" is superimposed on larger-scale changes in the rotating accretion disk, or in our mode of viewing it.

Lensing. In 1985 Ostriker and Vietri hypothesized that BL Lacs are "normal" OVVs that are being gravitationally lensed by an intervening galaxy. They theorized that the BL Lac characteristics are produced by so-called "mini-lensing" by individual point objects in the galaxy, such as stars or stellar remnants, rather than by the galaxy as a whole. Lawrence Molnar of CFA (1986) called this work to our attention and suggested that we examine our BL Lac records

Table 5. Break-down of 21 BL Lacs and quasars into variability subclasses.

Subclass	I	II	III	IV
BL Lacs	4	1	6	0
Quasars	3	0	4	3

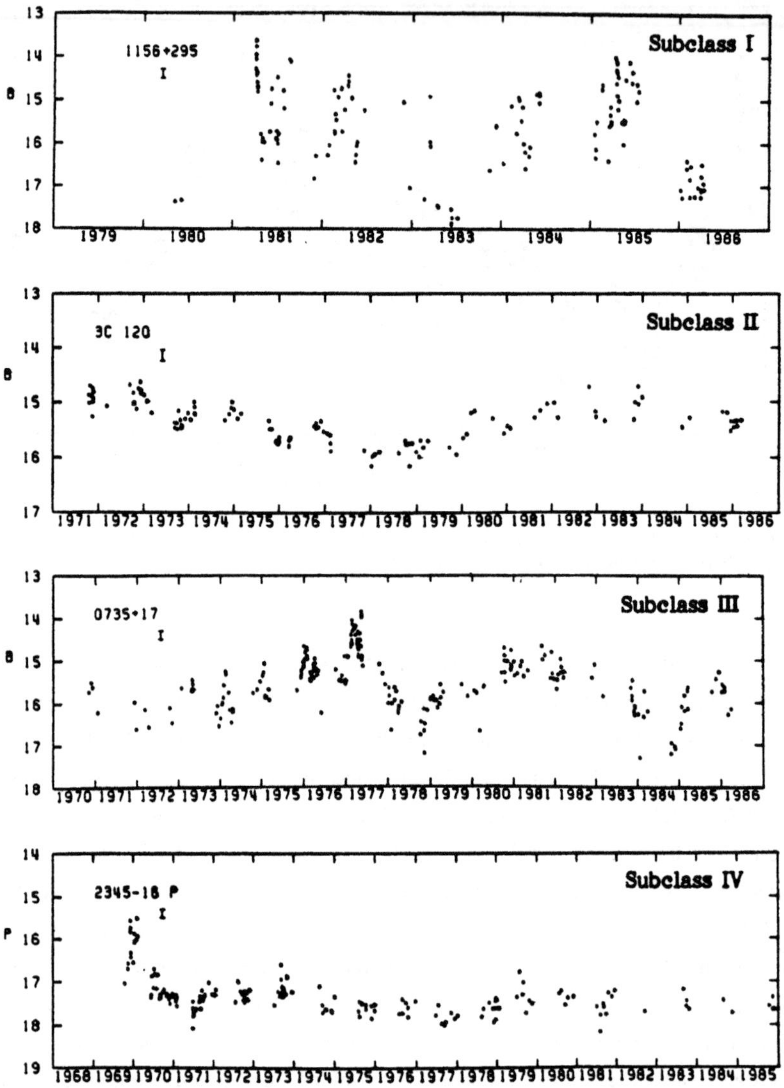

Figure 2. RHO observations representative of the four "variability subclasses".

for evidence of brightenings on timescales of several years due to mini-lensing. Ostriker and Vietri found, for example, that occultation of a small quasar by a moving one solar mass star should produce a flux variation of the order of 5 years with an amplification of about 8, or roughly 2 magnitudes.

If we study the timescales found in the power spectra of our 22 blazars (Table 2), both the quasars and the BL Lacs seem to produce components in the suggested range of several years. The relatively uniform distribution of quasars and BL Lacs into subclasses I and III (Table 5) similarly does not seem to suggest totally different mechanisms. If one were looking for relatively isolated events caused by infrequent occultations in the fringes of a galaxy, one might expect to find some BL Lacs in subclass IV, but as Table 5 shows, it is the quasars that fall in this subclass. Finally, of course, one should simply examine the light curves for events with the expected lensing signature. While this is subjective, 8 of our blazars appear to show multiple long-term events, each of which lasts 2.5 to 5 years. Four of these sources are BL Lacs and 4 are quasars, so again we fail to support lensing as unique to BL Lacs.

It should be remarked that in some cases the morphology of the light curve may change with time. An example is BL Lac itself (Fig. 3). Prior to 1980 the source displayed large-amplitude flickering about a seemingly constant mean; one could say it was a pure subclass I object. However, 1980 marked the beginning of a long-term swing of about 1.5 magnitudes in the mean, and the amplitude of the flickering seems to have declined markedly. The later segment of the light curve is more representative of subclass III. In the early 1970's 0906-01, NRAO 512, and 2345-16 displayed periods of violent activity, but subsequently they became relatively quiescent; 2345-16, in particular, has shown almost no activity for 13 years.

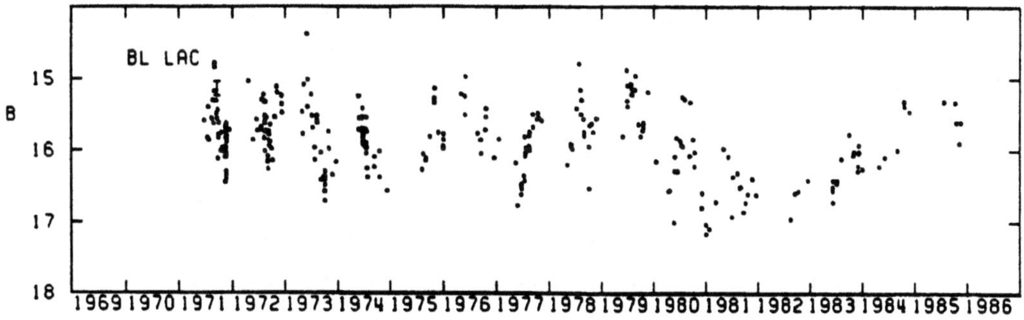

Figure 3. RHO B-band data for BL Lacertae.

We are currently in the final stages of analyzing 145 sources that previously have not been grouped with our blazars. Of these, 24 are classified as BL Lacs by Hewitt and Burbidge (1987). Since it is already evident that some of these 145 objects will have to be reclassified as blazars, we should soon be able to add significantly to the statistics presented in the present paper.

Acknowledgements. We are grateful to the National Science Foundation for the continuing support that has made this research possible; the present grant is AST-8516269. As usual, we must acknowledge the indispensable contributions of numerous workers who have assisted in the making and reduction of thousands of observations since 1968.

REFERENCES

Babadzhanyants, M.K. and Beloken, E.T. (1984). Astrofizika 21, 231.

Barbieri, C., Romano, G., di Serigio, S., and Zambon, M. (1977). Astron. and Astrophys. 59, 419.

Bevington, P.R. (1969). Data Reduction and Error Analysis for the Physical Sciences (McGraw-Hill).

Pollock, J.T., Pica, A.J., Smith, A.G., Leacock, R.J., Edwards, P.L. and Scott, R.L. (1979). Astron. J. 84, 1658.

Smyth, M.J. and Wollstencroft, R.D. (1970). Astrophys. and Space Sci. 8, 471.

Webb, J.R., Smith, A.G., Leacock, R.J., Fitzgibbons, G.L., Gombola, P., and Shepherd, D.W. (1988). In the press, Astron., J., Feb. 1988.

Discussion

A. E. Wehrle: Would you describe the error bars for individual points in your slide of the 3C 345 light curve?

Ans: To avoid excessive clutter, we have adopted the convention of placing a single error bar representing $\pm 0^{m}.1$ on each plot. Our experience indicates this is representative of the errors typical of photographic photometry.

R. A. Shaw: Looking at your light curve for BL Lac, it seems we might be looking at a large number of periods beating against each other prior to 1980, followed by reenforcement in later years. Is this reasonable, and would it be worthwhile to monitor these light curves for 50 or more years?

Ans: The pre-1980 data do strongly suggest a stochastic process such as multi-scale turbulence. Because marked changes in behaviour are seen in our 19-year light curves, as well as in historical records, we clearly gain much more information about a source by very long-term studies. In particular, efforts to determine whether a source is variable or not by making spot checks at 2 or 3 isolated epochs clearly can be highly deceptive.

W. Stein: The problems of sky brightness and observing seasons can be eliminated by observing at 3 or 10 microns.

Ans: Yes, we have followed flaring QSOs into moonlight by working at I-band; even V-band gives a considerable gain over U or B because of the reduced sky background. However, to work at several microns requires a different technology than photographic photometry.

J. S. Miller: Among your new objects to be assigned to the blazar class, are any of them radio-quiet?

Ans: We hope to have a more definitive answer in a few months. The 19 RQQSOs selected for specific study in our program tend to have a smaller amplitude of variability than the group as a whole. Bonoli et al. reported that the incidence of OVVs among a sample of radio-quiets they studied was 10 times less than that of the radio-louds.

D. E. Osterbrock: The long-term linear trends in your data are interesting. It would be interesting to divide a 20-year run of data, say, into two 10-year sets and see if the same trend exists, or if different segments might show opposite trends, indicating that in different epochs either increases or decreases might exist for the same object.

Ans: I agree; and in fact visual inspection suggests that a given object may for a while show a long-term decline that then reverses and becomes a long-term brightening. (We might note that removal of the overall linear trend was a necessary preliminary to the Fourier analysis for periodicities.)

THE LONG TERM OPTICAL VARIABILITY OF PKS 2201 + 044

M. T. Carini, H. R. Miller, and J. W. Wilson
Department of Physics and Astronomy, Georgia State University, Atlanta, Georgia 30303

BL Lacertae objects have been characterized by rapid and large amplitude optical variability, by a highly variable and polarized optical continuum, and by a featureless optical continuum or one which exhibits only low contrast discrete features. In the present investigation, the BL Lacertae object with the strongest galaxy component relative to the nonthermal component detected to date, PKS 2201 + 044, has been studied.

The optical counterpart of the radio source PKS 2201 + 044 was originally identified with a 15.0 mag. galaxy by Clarke, Bolton and Shimmins (1966). Wills and Wills (1976) obtained a spectrum of this galaxy and have determined a redshift of z = 0.0281. M.-H. Ulrich (private communication) suggested that this object might be related to the BL Lacertae class of objects based on the presence of a very blue continuum. PKS 2201 + 044 was included in a photometric survey of BL Lacertae objects by Kinman (1976) in which he determined V = 15.2, B-V = +1.00 and U-B = +0.22 for this object. Miller (1977) obtained a long term light curve for PKS 2201 + 044 based on archival plates at the Harvard College Observatory. The object exhibited a total range of 1.5 mag. in a month. The purpose of the present investigation is to present the results of a decade of photometric monitoring of this BL Lacertae galaxy.

Multiple aperture observations were obtained on several nights. These were used to derive an aperture correction via the method outlined by Sandage (1973), which we applied to all the observations. Sandage assumes a two component model for the source: an unresolved central nonthermal source and an underlying giant E galaxy. We found the galaxy magnitude and colors in the 15.42" aperture to be: V = 16.57, B-V = 1.58, U-B = 0.55. These colors are not inconsistent with the colors of a giant E galaxy. Correcting the galaxy magnitude to a standard aperture(Sandage,1973) we find V = 15.55.

PKS 2201 + 044 is observed to vary erratically from V = 14.44 to V = 16.34 over this ten year period. These observations, when compared with the historical light curve for

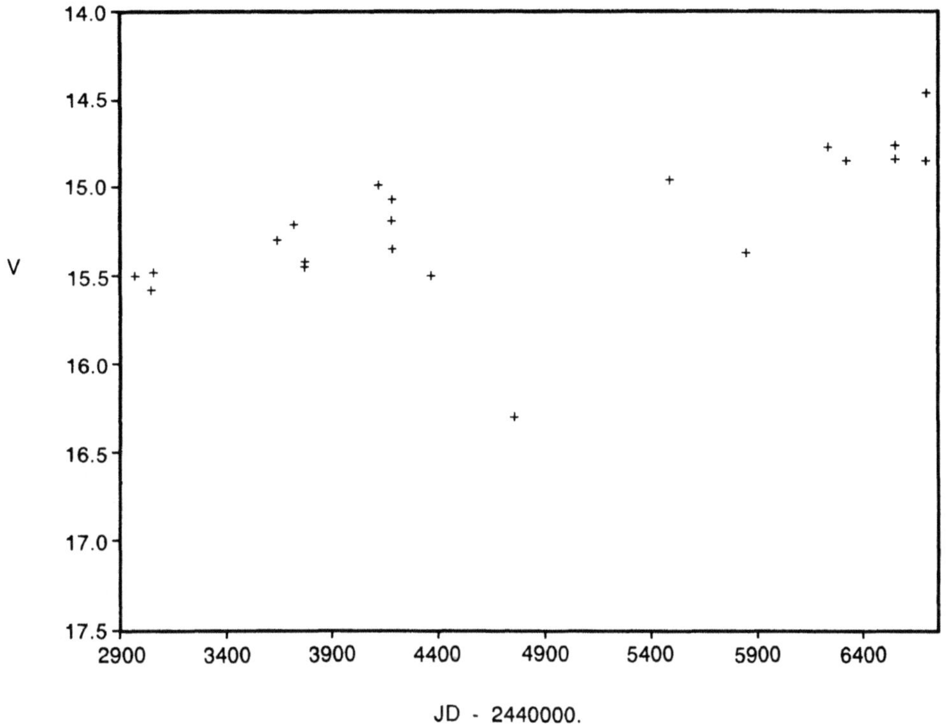

Figure 1. The long-term optical variability of PKS 2201 + 044 is shown from 1977 to 1987.

this object, indicates a total range of 2.2 mag. which is somewhat greater than the reported previously (Miller, 1977). Mild color variations were found to accompany the long term changes in brightness with B-V varying from 0.78 to 1.18 and U-B ranging from 0.12 to 0.68. In general, there does not appear to be a high degree of correlation between the brightness of the object and the colors. However, the U-B colors of 0.6f8 are the reddest detected and occur when the source was at its faintest detected level on 1981 June 7. When the source attained its maximum detected brightness of V = 14.44 on 1986 September 30, U-B = 0.12, which was the second bluest color measured during this ten year period. This suggests that there were no major color variations associated with luminosity changes of the nonthermal central source, but that the observed color variations were due to changing contributions from the non thermal central source and the galaxy in which this source is embedded.

Figure 1 is a plot of V vs Julian Day. Over the ten year period of the observations, we see a general brightening trend, with variations on timescales of a few years superimposed. A major exception to this is the observation of 7 June 1981. The magnitude measured here is V = 16.31, very close to the expected magnitude of the underlying galaxy. This indicates that the nonthermal component has almost disappeared. If an accretion process is responsible for the nonthermal component, then this minimum which is observed suggests either the accretion has ceased or that the nonthermal central source has been obscured by a cloud of material. This is similar to recent behavior the authors have observed in BL Lac itself, which in September 1987 was observed at a magnitude of V = 16.50, very close to the expected magnitude of its underlying galaxy.

Although this object exhibits a long-term brightening trend, on at least two occasions significant fading in brightness is observed to take place. The first of these is noted in 1978 September 24 and 26 when the object is observed to have V = 15.46 and V = 15.43 respectively, compared to V = 15.22 on 1978 August 6. No significant color variations were found to accompany this decline. The second of these is observed on 1984 May 29. PKS 2201 + 044 is 0.41 mag. fainter than the observations obtained 1983 June 6 and is 0.60 mag. fainter than the observations obtained 1985 June 21. Again, no significant color variations were present. The absence of significant color variations accompanying these changes indicates that the spectral index of the nonthermal continuum source must exhibit sufficient variations in order to compensate for the varying dilution of the nonthermal central source and the underlying galaxy. Both of these events are typical of the shorter-term erratic variations which are found superposed on the long term brightening which was observed.

The most rapid variation which was observed was the 0.26 mag. change which was detected between 1979 November 13 and 1979 November 14. This variation places an upper limit on the size of the emitting region of 8.39DEX(-4) pc., based on light-travel time arguments and assuming that no relativistic bulk motion of material is present in the source.

The authors wish to thank KPNO, CTIO and Lowell Observatory for providing allocations of observing time to this project. This work has been supported in part by grants from the National Science Foundation and the Vice President's Research Grant Program at Georgia State University.

References

Clarke, M. E., Bolton, J. G. and Shimmins, A. J. 1966, *Australilan J. Physics*, *19*, 375.
Kinman, T. D. 1976, *Ap. J.*, *205*, 1.
Miller, H. R. 1977, *Ap. J.*, *212*, 34.
Wills, D. and Wills, B. J. 1976, *Ap. J. Suppl.*, *31*, 143.

A PHOTOMETRIC INVESTIGATION OF THE OPTICAL VARIABILITY OF MARKARIAN 501

B.Q. McGimsey, H.R. Miller, M.T. Carini, and J.W. Wilson
Department of Physics and Astronomy
Georgia State University
Atlanta, GA 30303

ABSTRACT

The optical variability of the BL Lacertae object, Markarian 501, has been monitored from 1976-1987. Long-term optical variations have been detected with an amplitude of 0.65 mag. The long-term small amplitude, color variations which have been observed are found to be weakly correlated with the long-term trends in brightness in the sense that the object appears bluer when it is brighter. No evidence has been found for variability on a time scale of days or less for this object.

I. Introduction

The object Markarian 501 (B2 1652+39) is a member of a class of objects similar to BL Lacertae and AP Librae which contain strong nonthermal sources imbedded within the nucleus of a luminous galaxy (Ulrich et al. 1975). Little evidence of varibility has been reported in the literature to date (Kondo et al. 1981, Mufson et al. 1984). We report here the results of eleven years of photoelectric observations giving evidence of long-term changes in brightness with some evidence of variations of over one half magnitude within one month.

II. Observations

Photometric observations were obtained for Markarian 501 at two observatories. At Kitt Peak National Observatory the 1.3 meter and 0.9 meter telescopes were used. They were equipped with the computer photometer and the automated filter photometer respectively. Each of these photometers was equipped with a 1p21 photomultiplier tube and used with standard Johnson UBV filters. The 1.05 meter telescope at

Lowell Observatory was used with the Bowell two-channel photometer. Both channels were equipped with an S-20 photomultiplier. V and R filters were used in the first channel and U and B filters were used in the second channel.

Transformation coefficients were determined independently for each observing run using the Landolt (1973) UBV standard stars. The extinction coefficients were determined each night. No K correction was applied due to the high galactic latitude of the object.

Multiple aperture observations were obtained on August 25, 1976 and March 29, 1981. These were used to derive an aperture correction to the V magnitudes via the method outlined by Sandage (1973). Assuming that Markarian 501 is a non-thermal source embedded within a first rank giant elliptical galaxy, we applied this method to correct all observations to an aperture of 15.5 seconds. In this manner we corrected for changes in brightness due to differing aperture rather than intrinsic variability. These aperture corrections were small (on the order of 0.05 mag. or less) so that no correction was applied to the colors.

III. Discussion

The results of the observations are shown in table 1. Comparison of the corrected results from the multiple aperture observations on 1976 August 25 and 1981 March 29 indicates that the V brightnesses measured in differing apertures were substantially identical, indicating the correction methodology to be accurate. The light curve in V is plotted in figure 1. Observations by Kondo et al. and Mufson et al. are also indicated. The light curve indicates continuing fluctuations on the order of 0.3 magnitudes with an indication of one rapid increase in brightness of 0.3 magnitude from the pervious average followed within three months by a 0.68 magnitude decline (0.38 magnitude below the average). The decline was observed on three dates at two different telescopes and so seems certain to be real. Following this episode, the object's average brightness appears to have increased about 0.2 magnitudes, lending credence to the reality of the sudden brightening mentioned above. In summary, this appears to be typical behavior of a BL Lacertae object although at a somewhat lower amplitude than is observed for some of the objects.

Assuming from the variability in V that the variable part of the emitting region is on the order of three light months in size, it can be shown that that region at the distance indicated by $z = .0337$ is approximately .078 milli arcseconds in size. This is much smaller than the 2.2 mas derived by Mufson et al. for their synchrotron self-Compton model, which assumes that the emitting region is homogeneous.

Mufson et al. indicate that they themselves prefer an inhomogeneous jet model for Markarian 501. Although this model does not explicitly give the source size, the current results are consistent with variability in some portion of the jet.

The colors shown in table 1 are weakly correlated with brightness in that B - V changes from an average of 0.73 before the proposed event in 1984 to an average of 0.68 afterward when the object is brighter. This is consistent with an increase in brightness of the bluer non-thermal source.

REFERENCES

Kondo, Y. et al., 1981, Ap.J., 243, 690.

Landolt, A.U. 1973, A.J., 78, 959.

Mufson, S.L. et al. 1984, Ap.J., 285, 571.

Sandage, A. 1973, Ap.J., 180, 687.

Ulrich, M.-H., Kinman, T.D., Lynds, C.R., Rieke, G.H., Ekers, R.D. 1975, Ap.J., 198, 261.

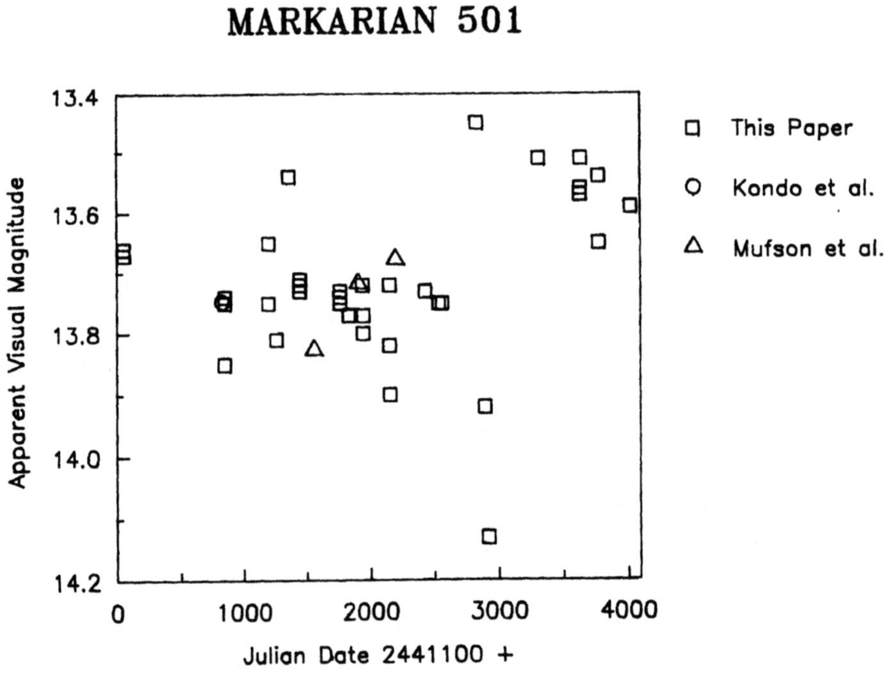

Figure 1. The long-term light curve for Markarian 501

THE OPTICAL VARIABILITY OF ARAKELIAN 120, 1977-1987

J. W. Wilson, H. R. Miller, and M. T. Carini
Department of Physics and Astronomy, Georgia State University
Atlanta, GA 30303

Arakelian 120 was first identified as a possible Seyfert galaxy in a survey of high surface brightness galaxies by Arakelian (1975) and Arakelian, Dibai, and Esipou (1975). This identification was confirmed by Osterbrock and Phillips (1977). Miller (1979a) demonstrated that Arakelian 120 was variable with a range B = 2.1 mag. based on a search of archival plates investigated at the Harvard College Observatory. Miller (1979b) also reported a more recent optical outburst which occurred for this galaxy in 1977 and 1978. More recently Peterson el al (1983, 1985) have discovered that the broad Hβ line exhibits both significant variations in its equivalent width and line profile on timescales as short as a month or less. The results presented in this paper are based on a decade of photometric observations of Arakelian 120 obtained between 1977 and 1987 and are shown in Figure 1.

The long term light curve indicates that Arakelian 120 has undergone four significant outbursts between 1977 and 1987 while exhibiting a total range of V=0.96 mag. and B=1.06 mag.

The first outburst reached a maximum observed brightness on 1978 March 4 of V_C=13.33. The rise to this maximum brightness is gradual without any significant short term flickering detected. However, a rapid decline in brightness of 0.39 mag. occurred in four days immediately following this maximum. No significant color variations were detected during this outburst.

A second smaller outburst of 0.22 mag. was observed on 1979 April 4. Following this outburst, the galaxy declined over 0.5 mag. to V_C=14.04 on 1979 September 13 which is the second faintest brightness level at which the galaxy was observed during this 10-year period.

The galaxy brightened steadily from this minimum to the brightest value it reached of V_C=13.12 on 1980 February 23. It was observed at essentially the same brightness over a year later on 1981 March 30 when it was at V_C=13.13. Unfortunately no observations were obtained in the intervening 13 months and the detailed structure of this outburst is not known. No significant color variation were observed for B-V near the maximum of this outburst, but U-B was approximately 0.1 mag. bluer. As the brightness of Arakelian 120 declined, the U-B color slowly returned to its pre-outburst value.

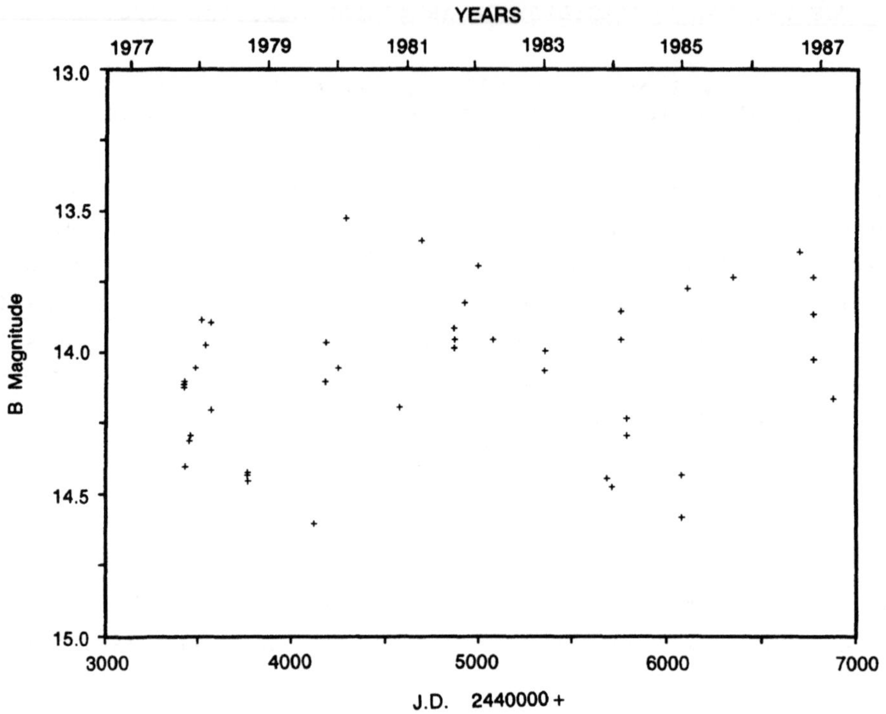

Figure 1. The long-term light curve for Arakelian 120 is shown from 1977 to 1987.

The galaxy, continued to decline until it reached its minimum observed brightness of V_C=14.08 on 1984 January . It then rebounded sharply in brightness on 1984 March 1 to V_C=13.51 with U-B = 0.79, once again approximately 0.10 mag. bluer than the value of U-B = -0.71 when it was observed at its minimum.

Following this outburst, Arakelian 120 has slowly declined in brightness reaching V_C=13.99 on 1986 December 13. During this decline, the U-B colors have appeared somewhat redder with U-B = -0.50.

The U-B color variations which were observed when Arakelian 120 was at the extremes in its brightness variations are consistent with a Sandage-type model for this Seyfert galaxy in which there is a mix of the galaxy component and the nonthermal component. Variations of the nonthermal component, which is blue, contributes more significantly to the total luminosity of the composite system when the object is bright, while the galaxy-component dominates when the object is faint. Therefore one expects more of an ultraviolet excess to be present near maximum than near minimum brightness levels which is what is observed.

However, the smaller amplitude outbursts discussed above did not exhibit any significant color variations. This implies that another mechanism must be present which is producing a compensating change in color. Two possibilities exist: (1) the spectral index of the nonthermal continuum may be varying, at least for short term small-amplitude events, in just such a way so as to compensate for the changing contributions from the two components or (2) the variations in color and luminosity are due to variability in both the broad emissions lines which are present and in the nonthermal continuum.

The authors gratefully acknowledge partial support from the Vice President's Research Fund at Georgia State University. The authors also thank KPNO, CTIO, Lowell Observatory, and Agnes Scott College for granting observing time for this program.

References

Arakelian, M. A. (1975). Publ. Byurakan Obs. 47, 3.
Arakelian, M. A., Dibae, E. A., and Esipov, V. F. (1975). Astiofizika, 11, 377.
Landolt, A. U. (1973). Astron. J. 78, 959.
Miller, H. R. (1979a). Astron. Astrophys. Suppl. 35, 386.
Miller, H. R. (1979b). Publ. Astron. Soc. Pac. 91, 624.
Miller, H. R. (1979c). Astrophys. J. 227, 52.
Osterbrock, D. E. and Phillips, M. M. (1977). Publ. Astron. Soc. Pac. 89, 251.
Peterson, B. M., Foltz, C. B., Miller, H. R., Wagner, R. M., Crenshaw, D. M., Meyers, K. A., and Byard, P. L. (1983). Astron. J. 88, 926.
Peterson, B. M., Myers, K. A., Capriotti, E. R., Foltz, C. B., Wilkes, B. J., and Miller, H. R. (1985). Astrophys. J. 292, 164.

OBSERVATIONS OF THE 1987 OUTBURST OF AO 0235+164

James R. Webb and Alex G. Smith
Department of Astronomy
University of Florida

I. Introduction

The BL Lacertae object AO 0235+164 (AO) has been observed from Rosemary Hill observatory (RHO) since 1975. It has exhibited optically violent variations with a brightness range of 5.5 magnitudes. Exceptionally rapid, high-amplitude outbursts were seen in 1975 (O'Dell et al. 1977) and 1979 (Pica, Smith and Pollock 1980). Since the 1979 event, no major outbursts were detected from RHO until the early 1987 event reported here.

II. Observations

The photographic observations of the 1987 outburst of AO are listed in Table 1. The magnitudes given are standard Johnson U, B, and V magnitudes and the errors are estimated from the scatter of 20 comparison stars around a parabolic calibration curve calculated for each exposure. Figure 1 shows these data plotted versus Julian date. As the outburst faded, the proximity of the Sun to the AO field prevented further observations. Since AO became accessible again to ground-based observations in August, one exposure imaged the source at B = 20.30 magnitudes.

TABLE 1. Optical Observations

U. T. Date	U	B	V
10/ 2/86		18.84 (0.19)	
11/24/86		17.38 (0.15)	
12/29/86		16.14 (0.18)	
1/ 6/87		15.77 (0.11)	
1/23/87	16.99 (0.04)*	16.82 (0.10)*	16.08 (0.07)
1/24/87	16.98 (0.06)*	16.80 (0.06)*	
1/26/87	16.86 (0.12)	16.95 (0.06)*	15.53 (0.04)
1/28/87	16.41 (0.05)*	16.35 (0.09)*	15.50 (0.05)*
1/29/87	16.05 (0.07)*	15.60 (0.11)*	
1/30/87	16.96 (0.10)	16.70 (0.10)*	15.78 (0.09)
2/ 1/87	17.12 (0.10)	17.11 (0.08)*	16.18 (0.05)
3/21/87		17.41 (0.19)*	
3/22/87		17.41 (0.31)	

* denotes the RMS weighted mean of 2 or more exposures.

III. Analysis

We analyzed the energetics of the 1987 outburst of AO assuming a Friedman cosmology with q_o = 0.0 and H_o = 55.0 Km/sec/Mpc. The redshift was assumed to be 0.851 (a value since revised by Hewitt and Burbidge to 0.94). The maximum luminosity of AO during the 1987 event was 2.7×10^{49} ergs/sec on J. D. 2446824. The total energy released during the outburst was 1.36×10^{56} ergs. This value assumed a base luminosity of 1.07×10^{48} ergs/sec, which corresponds to B = 18.84 magnitudes, and a duration of 197 days in the observer's frame.

The long-term light curve of AO is shown in Figure 2. The 1975 observations are from O'Dell et al. (1977) and the remainder of the data are Rosemary Hill observations. Fourier transform analysis of the light curve revealed characteristic timescales of 2.8 and 1.5 years in the observer's frame. We assumed energy is stored in a disk over these timescales and applied the Shields and Wheeler (1976) model of flares on magnetized accretion disks. In the "real time" scenario, the total outburst energy must be generated during the outburst. A magnetic accretion disk around a massive 10^8 solar mass black hole cannot generate the necessary 10^{56} ergs of energy in 197 days (106.4 days in source rest frame). The second scenario assumes the energy released in the outburst is stored during periods of relative inactivity between outbursts. If we use the timescales identified in the Fourier analysis of the long-term light curve as storage timescales, we find that the outburst energy cannot be stored outside 3 gravitational radii (3Rg). The energy released in the 1987 outburst must either have been stored over timescales longer than 2.8 years (1.5 years in source rest frame) or relativistic beaming is present. No major outburst was observed in 1984 so we are free to speculate that the energy released in the 1987 outburst was stored over a longer time period. Similar calculations for the even more energetic 1979 outburst place similar energy storage constraints on the accretion models. The presence of the 1975 outburst prevents us from extending the storage timescales for the 1979 event. If we insist the Shields and Wheeler model gives reasonable results, then we must find a way to reduce the energy requirements of the outbursts. The most natural method is to invoke relativistic beaming. The presence of relativistic beaming in BL Lacs has been suggested by several authors (Hutter and Mufson 1986, Urry 1986, and Worrall 1986). They invoke beaming with Doppler factors of 1.0 to 10.0 to explain the multifrequency spectra of BL Lacs in terms of synchrotron-self-Compton models. A Doppler factor of 1.4 is needed to reconcile the AO observations and the Shields and Wheeler model. This provides independent evidence that relativistic bulk motion is probably present in some BL Lac objects, particularly in AO 0235+164.

IV. References

Hutter, D. J., and Mufson, S. L., 1986, Astrophys. J., 301, 50.

O'Dell, S. L., Puschell, J. J., Stein, W. A., and Warner, J. W., 1977, Astrophys. J., (Letters), 214, L105.

Pica, A. J., Smith, A. G., and Pollock, J. T., 1980, Astrophys. J., 236, 84.

Shields, G. A. and Wheeler, J. C., 1976, Astrophys. J., Lett., 17, L69.

Urry, C. M., 1986, Proceedings of the Workshop on Accretion onto Compact Objects, ed. K. O. Mason, M. G. Watson and N. E. White, (New York: Springer-Verlag Co.), pg. 357.

Worrall, D. M., 1986, Continuum Emission in Active Galactic Nuclei, ed. M. Sitko, workshop sponsored by NOAO, pg. 97.

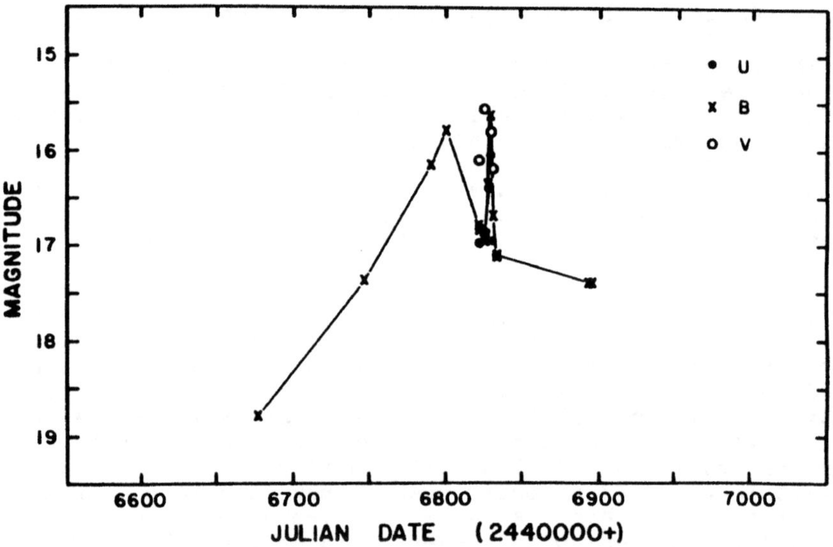

Figure 1. The 1987 outburst of AO 0235+164 is shown in three colors. The x's represent B magnitudes and are connected by a solid line. The U magnitudes are denoted by closed circles and the open circles denote V magnitudes.

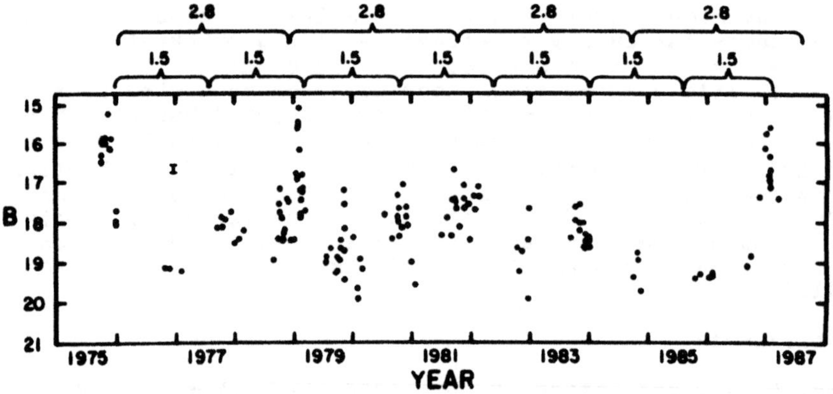

Figure 2. The long term light curve of AO 0235+154. The brackets represent timescales derived from the Fourier analysis.

A COMPLETE SPECTRAL ANALYSIS OF THE FLARE OF THE QUASAR 3C 273

Alberto C. Sadun
Bradley Observatory, Agnes Scott College
Decatur, Georgia 30030

Observations spanning six decades of frequency have been compiled from many sources of the 1983 outburst of the quasar 3C 273 (Robson et al., 1983; Aller et al., 1985; Corso et al., 1985; Courvoisier and Ulrich, 1985; Sadun, 1985; Smith et al., 1986; Barvainis, 1987; Cohen et al., 1987). The flare has been monitored for many months from radio frequencies through the ultraviolet. The data are presented in Fig. 1 in terms of spectra for different points in time, with day 1 being defined as January 1, 1983. We see in the evolution of the flare that the peak frequency shifts to lower frequencies at later times. We interpret this in light of a model of a cloud of gas, expanding and moving towards us at a relativistic velocity, radiating doppler boosted synchrotron emission. VLBI studies of superluminal motion in this source identify the cloud responsible for the enhanced emission.

This phenomenon describes a region that becomes optically thin at progressively longer wavelengths. This is consistent with identifying the area of frequency less than the peak frequency as the self absorbed optically thick part of the spectrum.

Upon careful examination of Fig. 1 we see a persistent hump in the optical and UV region, whose peak seems to slowly decrease in frequency with time. One also notices a power law in the millimeter part of the spectrum. There is also a transient IR feature that appears at around day 80 and seems to disappear around 45 days later.

We consider the following model to describe the features observed in this outburst. The quasar undergoes a disturbance of some kind which manifests itself as a rapidly expanding hot spot on its accretion disc. More importantly, the quasar ejects an expanding cloud of gas that has highly relativistic bulk motion with a Lorentz factor of around 10, as inferred from superluminal studies (see Cohen et al., 1987). The expanding cloud radiates via the synchrotron process, and the hot spot via black body emission. The transient IR feature is to be explained as a black body emission resulting from the heating of a nearby static cloud of dust of radius 4×10^{16} cm, that is being pumped with energy from the optical emission to heat it to a temperature of approximately 500 degrees Kelvin (see Edelson and Malkan, 1986).

As the cloud expands and its surface area increases, the flux increases. The cloud, initially optically thick to most frequencies, becomes optically thin to lower and lower frequencies as it expands; the emission spectrum evolves with the peak frequency slowly shifting to lower values as shown in Fig. 1, and with the cloud experiencing self absorption at frequencies less than the peak frequency.

One notices from Fig. 1, however, that the low frequency spectrum does not follow the classical positive slope of 2.5 but rather there appears to be a shoulder near the peak frequency. This is because 3C 273 is still settling down from its previous outburst. That increase in flux has passed for the higher frequencies, but since the low frequency emission lags behind the higher frequency, during the period in which one outburst is showing up in the optical or IR, the previous outburst is still influencing the radio portion of the spectrum. Hence, instead of a steep self absorption drop off in radio, we see a shoulder. In fact, one can see the "wave" of the previous outburst move through the radio. A radius of 10^{16} cm may be calculated by using the criterion for synchrotron self absorption.

The optical and UV bump is consistent with there being a black body hot spot of a temperature of around 8,000 degrees Kelvin on the accretion disc surrounding the compact object. This hot spot is a region of radius nearly 10^{17} cm and mass of 10^7 solar masses. It is expanding at a mildly relativistic speed, namely at 0.13 c, too quickly for adiabatic cooling.

The observations imply that an expanding cloud of gas was released at a highly relativistic speed nearly along our line of sight. This gas then radiated in synchrotron that was doppler shifted and boosted in accordance to its relativistic motion. In addition, there seemed to have been a persistent hot spot on the accretion disc surrounding the core, emitting black body radiation. Finally, during the travel and evolution of the gas cloud, a nearby region of dust was heated and it radiated as a black body for a time, peaking in the IR. It is hoped that a careful analysis of this system may lead to a better understanding of active galactic nuclei in general.

Aller, H.D., Aller, M.F., Latimer, G.E., and Hodge, P.E.: 1985, *Astrophys. J. Suppl.* **59**, 513

Barvainis, R.E.: 1987 (private communication)

Cohen, M.H., Zensus, J.A., Biretta, J.A., Comoretto, G., Kaufmann, P., and Abraham, Z.: 1987, *Astrophys. J.* **315**, L89

Corso, G.J., Schultz, J., Purcell, B., Garino, G., and Dey, A.: 1985, *Publ. Astron. Soc. Pacific* **97**, 118

Courvoisier, T.J.-L., and Ulrich, M.H.: 1985, *Nature* **316**, 524

Edelson, R.A., and Malkan, M.A.: 1986, *Astrophys. J.* **308**, 59

Robson, E.I., Gear, W.K., Clegg, P.E., Ade, P.A.R., Smith, M.G., Griffin, M.J., Nolt, I.G., Radostitz, J.V., and Howard, R.J.: 1983, *Nature* **305**, 194

Sadun, A.: 1985, *Publ. Astron. Soc. Pacific* **97**, 395

Smith, P.S., Balonek, T.J., Elston, R., and Heckert, P.A.: 1987, *Astrophys. J. Suppl.* **64**, 459

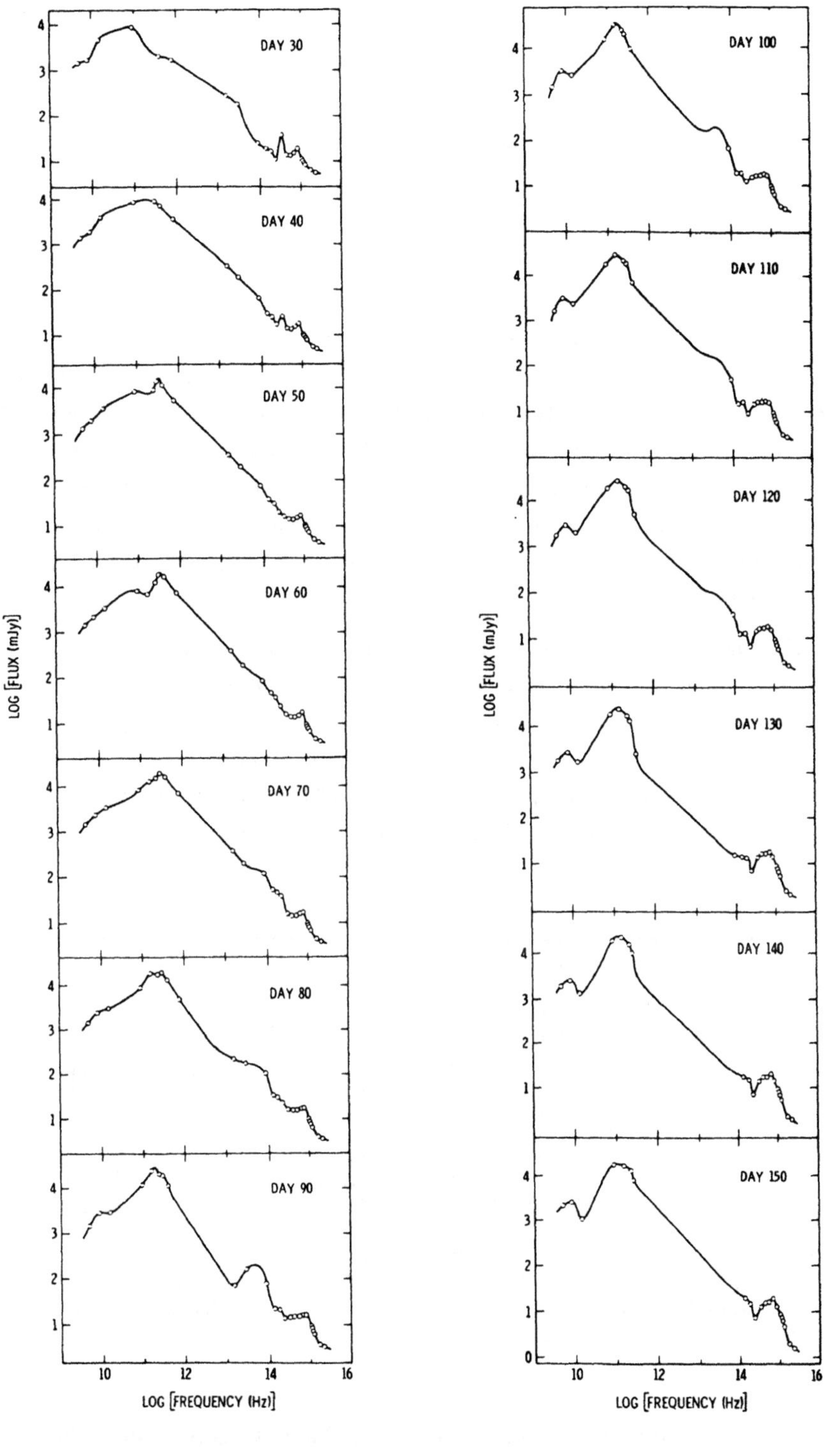

Figure 1. The changing spectrum of the outburst of 3C 273.

Broad Line Variations of Seyfert Galaxies

Edward I. Rosenblatt and Matthew A. Malkan
University of California, Los Angeles

The flux and profile variations of the broad emission lines in active galactic nuclei can determine, in principle, the structure and kinematics of the broad-line region (BLR). Since variations cannot occur on a timescale much shorter than the size of the emitting region, line flux changes place an upper limit on the BLR size. In addition, variations in line profile parameters such as asymmetry and width can place constraints on kinematics. For instance, a systematic red or blue asymmetry may imply radial motion of BLR clouds. However, an increase in line width during a continuum outburst suggests rotational motion whereas little or no change in width implies chaotic bulk motion.

Clearly, line variations represent an important diagnostic tool for understanding BLR structure. We are thus measuring and analyzing variations of 20 bright Seyfert galaxies to determine characteristic timescales and amplitudes of flux and profile changes. One object in our sample is the Seyfert 1.5 galaxy NGC 5548. Thirteen observations were made with the 60" Palomar telescope and a SIT vidicon spectrograph from 1979 to 1982. Figure 1 shows two spectra taken less than four months apart that differ in Hβ line flux by 65%.

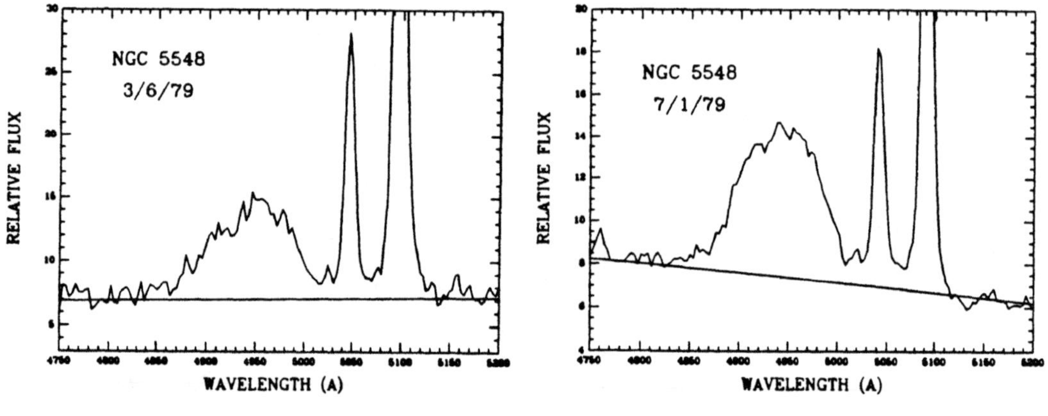

Figure 1. A change in Hβ line flux of 65% (relative to O[III] λ4959,5007) is readily apparent in these two 60" Palomar spectra taken less than four months apart.

Further significant flux changes occurred throughout the the four year time period (also see Peterson 1987). These changes are apparent in Figure 2a which shows the equivalent width of Hβ to O[III] λ5007, which is proportional to Hβ line flux assuming constant O[III]. Figure 2b shows the equivalent width of O[III] λ5007 which is inversely proportional to the non-stellar continuum level. Note that the Hβ line flux doubled and the continuum tripled over this time interval and that these changes were positively correlated as one would expect from photoionization by a central continuum source. Figure 3 also shows this effect quite well.

In contrast to ealier studies which made qualitative claims that the Balmer lines in NGC 5548 were "quite asymmetric" (cf. Anderson 1971), we find Hβ to be only marginally blue asymmetric. Our measurements of three different types of asymmetry parameters $A(\%)$, AI_{20}, and α_β (as defined by Whittle 1985, Heckman *et al.* 1981, and De Robertis 1985 respectively) yield a systematic blue asymmetry of only 10% or less after correcting for Fe II and narrow Hβ contamination. It is important to emphasize that this slight asymmetry, or rather this overall line *symmetry*, persisted without significant deviations for the entire four year period. Additionally, line widths at several levels (measured from the peak) also remained constant, generally to within 5%. For instance, we measured an average Hβ FWHM of 5725 ± 285 km/s with no individual measurement deviating more than 2σ.

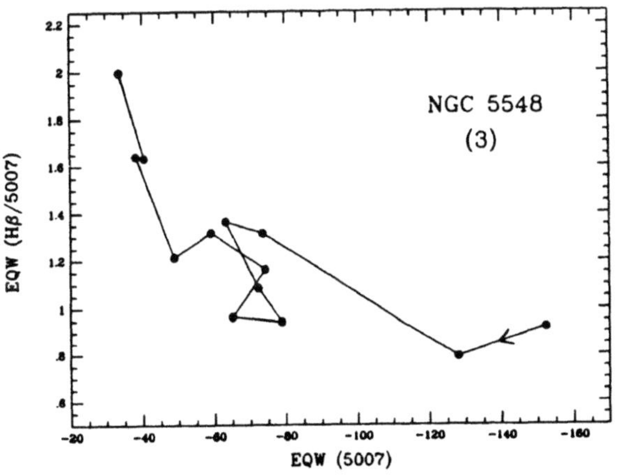

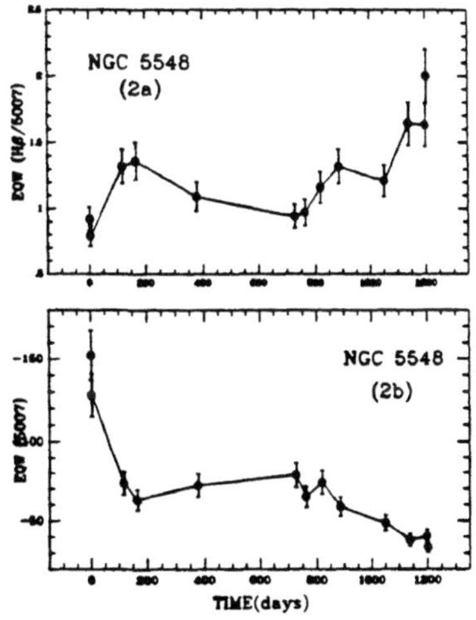

Figures 2. and 3. (2) Light curves for NGC 5548 beginning March 1979 and ending July 1982. Assuming that O[III] λ5007 is constant, equivalent width (Hβ/5007) represents the Hβ line intensity and the reciprocal of equivalent width (5007) represents the non-stellar continuum level. Note the temporal correlation between Hβ line strength, which doubled over this time period, and continuum flux which tripled. 1σ errors are 10%. (3) The correlation between the non-stellar continuum and Hβ line flux is shown. The continuum and Hβ fluxes both start in a low state at the lower right-hand corner of the figure, and then increase together, with the direction of time indicated by the arrow. This increase partially reversed itself during a 1.5 year period that occurred near the middle part of the observing program.

As described by Ulrich *et al.* (1984), if the velocities of most of the BLR clouds are everywhere distributed randomly (i.e. chaotic motion), we would expect to observe line flux variations *without* significant profile changes following a continuum outburst. Indeed, since our observations show that Hβ varied dramatically in intensity while maintaining a similar profile, we tentatively conclude that chaotic bulk motions dominate in the BLR of NGC 5548. Further measurements and observations, however, will be required to confirm this hypothesis.

REFERENCES

Anderson, K. S. 1971, *Ap. J.*, 169:449.

De Robertis, M. 1985, *Ap. J.*, 289:67.

Heckman, T. M., Miley, G. K., and van Breugel, W. J. M. 1981, *Ap. J.*, 247:403.

Peterson, B. M. 1987, *Ap. J.*, 312:79.

Ulrich, M. H. *et al.* 1984, *M.N.R.A.S.*, 206:221.

Whittle, M. 1985, *M.N.R.A.S.*, 213:1.

EFFECT OF PULSED VARIATIONS OF THE IONIZING CONTINUUM ON THE NARROW LINE REGION

Luc Binette
Canadian Institute of Theoretical Astrophysics
60 St. George Street, Toronto, Ontario M5S 1A1

1. INTRODUCTION

As the continuum emission of many Active Galactic Nuclei is known to be variable on time scales of months to years, it is not unreasonable to consider the possibility that many of them may also be variable on much larger time scales. Long term effects of the ionizing continuum variability on the Narrow Line Region (NLR) can be expected to be more subtle than a simple proportional change in the gas excitation (as represented by the ionization parameter U) if this variability is characterized by time scales shorter than the time scale for the NLR photoionized clouds to readjust to steady state ionization and temperature equilibrium conditions. Using the time-dependent multipurpose code *MAPPINGS*[1], preliminary results are given of the effect of periodic ionizing continuum "flashes" [significantly shorter than the recombination time scale] on the emission spectrum of a single cloud.

2. RELEVANT TIME SCALES

Assuming that we have a photoionized cloud in steady state ionization and temperature equilibrium, the interesting time scales in the context of the problem defined above are found to be the following:

$$\tau_{NLR} > 10^2$$
$$\tau_{rec} \approx 42$$
$$\tau_{cooling} \approx 5$$
$$\tau_{light-cross} \approx 2$$
$$\tau_{heating} \approx 0.1 + \tau_{light-cross}$$

where τ_{NLR} is the light crossing time within the Narrow Line Region, τ_{rec} and $\tau_{cooling}$ the recombination and cooling time scales if the ionizing continuum was instantaneously turned off, $\tau_{light-cross}$ the light crossing time scale through the photoionized layer of the cloudlet and $\tau_{heating}$ the heating time scale if the ionizing continuum was to increase dramatically in intensity. The time units are in DSU[2] which can be converted to *sec* (but not τ_{NLR}) when the cloudlet density n_H is specified: $1\, DSU = 1.03/n_H \times 10^{11}\, s$. The above quantities were estimated assuming $n_H = 100 cm^{-3}$, an ionization parameter $U \simeq 10^{-3}$, solar abundances[3] and an energy distribution for the ionizing continuum corresponding to a black body[3,4] of temperature 130 000 K.

3. CONTINUUM BEHAVIOUR

To exacerbate the effect of continuum variations, let us assume that the ionizing continuum behaves in a pulsed fashion as a way of mimicking a series of high and low activity levels in AGN. This is obviously an idealized way of stating the problem but it remains useful insofar as it allows one to explore second order effects such as departure from steady state line ratios due to continuum variability. In the calculations presented below, the ionizing continuum is in the high state 25% of the cycle duration and the jump in continuum luminosity between low and high states corresponds to a factor 5. Let us assume a density

Table 1: Calculated emission line intensities (relative to $H\beta$) [plane parallel slab[a] geometry]

Ion ID.	λ Å	$I_{equil.}$ (steady state)	$I^b_{cycle\,1}$ time average	$I^b_{cycle\,2}$ time average
$\log < U >$		-2.7	-2.7	-2.7
$L_{H\beta}/10^{-3}\, erg.s^{-1}.cm^{-2}$		2.27	2.67	2.40
$L_{cooling}\, erg.s^{-1}.cm^{-2}$		0.146	0.130	0.113
$T_{H^+}/10^4\, K$		0.967	0.897	0.920
$T_{O^{++}}/10^4\, K$		0.953	0.964	0.993
$d_{H^+}/10^{17}\, cm$		7.7	8.7	7.1
$d_{O^{++}}/10^{17}\, cm$		6.6	5.7	4.7
$m_{H^+}/10^{-4}\, g.cm^{-2}$		2.8	3.1	2.8
$\log \tau_H$		243	277	377
(1)	(2)	(3)	(4)	(5)
H I	6563	1.0	1.0	1.0
H I	4861	2.87	2.91	2.95
He I	5876	0.11	0.11	0.11
He II	4686	0.34	0.31	0.31
[N I]	5199	0.04	0.04	0.03
[N II]	6583	1.59	1.33	1.31
[O I]	6300	0.24	0.13	0.12
[O II]	3727	6.11	4.28	4.34
[O III]	5007	13.3	9.04	8.64
[O III]	4363	0.087	0.071	0.077
[Ne III]	3869	0.98	0.66	0.65
[S II]	6731	0.14	0.22	0.19
[S III]	9532	2.39	1.55	1.54

[a] density $n_H = 100 cm^{-3}$; ionizing continuum: $T_{bb} = 130\,000$ K black body

[b] $\log U_{min} = -3.0$, $\log U_{max} = -2.3$; 25% of *cycle duration* at U_{max}

period of cycle $= 2.0 \times 10^{10}\, s \approx 0.46 \tau_{rec}$

$n_H = 100cm^{-3}$, an ionization parameter $U_{min} = 10^{-3}$ for the low activity level (and $U_{max} = 10^{-2.3}$ for the high level) and suppose that the succession of high (25% or $4.9DSU$) and low (75% or $14.6DSU$) levels (or full cycle duration) repeats itself every $2 \times 10^{10}\,s$ (or $19.4DSU$, which is significantly shorter than τ_{rec}).

4. RESULTS TAKING INTO ACCOUNT NLR LIGHT CROSSING TIME EFFECTS

The time-dependent photoionization calculations cannot be directly used to make predictions that can be observationally tested because of light travel time effects[5] drastically affecting the integrated spectrum of an ensemble of clouds which the Narrow Line Region is believed to consist of. The simplest way to circumvent entirely this difficulty is to integrate the spectral response of a single cloud over the duration of the complete cycle (complete phase coverage) and see how the resulting *time average* spectrum differs from the steady state spectrum at the *average* ionization parameter value $U_{average} = 10^{-2.7}$. Table 1, presents the results and allows such a comparison to be made. Note that cycle 1 and cycle 2 give very similar line ratios and shows us that the first cycle already produces a *time average* spectrum which remains fairly similar in the following cycles. [The initial ionization conditions assumed before beginning the time-dependent calculations (*i.e.*,before the first cycle) were those corresponding to steady state calculations at $U = 10^{-2.7}$.] Although the spectrum of a single cloud at any given time along the time axis can be shown to be considerably different from that given by steady state calculations, the fact that the *time average* line ratios of column 4 (or 5) are very similar to those of column 3 (steady state) indicates that there is little remaining trace of this evolution. To what extent this would remain so under very different variability characteristics ha yet to be verified. An interesting exception which, however, is apparent in Table 1 is the ratio [S II]/[O I] which increases by as much as $0.3dex$ in the case of the time-dependent calculations. Although this ratio is sensitive to n_H, the actual amount of increase in [S II]/[O I] [going from steady state to time-dependent calculations] can be expected to remain of the same order of magnitude under very different initial conditions from those assumed above.

REFERENCES

1. Binette, L., Dopita, M.A., Tuohy, I.R.: 1985, *Ap. J.* **297**, 476
2. Binette, L. and Robinson, A.: 1987, *Astronomy & Astrophysics*, **177**, 11
3. Binette, L., Robinson, A., Courvoisier, T.J.-L.: 1988, *Astron & Astrophys* (in press)
4. Stasińska, G.: 1984, *A & A* **135**, 341
5. Robinson A. and Binette, L.: 1986, "Workshop on Model Nebulae" (Meudon, July 8-20 1985) ed. D. Péquignot, publication de l'observatoire de Paris, p. 140

OBSERVATIONS AND INTERPRETATION OF THE MULTIFREQUENCY CONTINUA OF AGN AND QSOs

Wayne A. Stein
University of Minnesota
and
University of California, San Diego

Abstract. The continuum distribution of AGN and QSOs extends from radio through x-ray frequencies to γ-ray energies in a few cases. Strongly active nonthermal sources (blazar) spectral-distributions seem to be adequately interpreted in terms of synchrotron self-Compton emission of nonthermal electrons although many interesting physical questions remain. The less strongly active emission line objects exhibit more complicated multifrequency spectral-distributions with a large blue bump at ultraviolet wavelengths as an energetically important and observationally prominent feature. Models are being explored that explain the large blue bump as synchrotron emission from secondary e^+e^- pairs radiating in the high magnetic field region of a supermassive compact object. As a natural consequence of secondary pair production we also anticipate a high energy x-ray component due to Compton scattering from the large blue bump, a nonthermal infrared component due to pairs radiating in the low field of the broad line region, a Compton scattered x-ray component from the infrared synchrotron emission and π° decay γ-rays with energies greater than 100 Mev. The sum of these components closely resembles the spectral distribution of 3C273 for 10^9 Hz < ν < 10^{23} Hz.

1. Introduction

Understanding the origin of the emission components contributing to the radio to γ-ray continuum of AGN and QSOs is a problem of prime importance. This continuum contains nearly the entire luminosity radiated by these objects, except for the line emission, and it may range up to 10^{48} erg sec^{-1}. It will be the purpose of this review to discuss the observational evidence and interpretations that may impose constraints on the physical processes generating these enormous luminosities. Most of this discussion will focus on the radio to ultraviolet continuum in order to avoid overlap with other discussions on high energy photons. However, some comments about the x-ray and γ-ray energy regimes will be included as necessary implications of models.

The multifrequency continuum of 'blazars' - BL Lac objects and OVV QSOs - will not be discussed in any depth. There are many details about these objects that we still do not understand. However, the general description of the continuum in terms of the synchrotron self-Compton theory seems generally accepted. There are various forms of this theory (Jones et al. 1974, Marscher 1977, Konigl 1981), but by comparison with the state of our understanding of the continua of emission line QSOs, further details involving particular geometries in blazar emission do not seem warranted. A summary of characteristics and interpretations of blazars was given by Stein and O'Dell (1985). Major remaining questions regarding blazar continua are the relationship of observed phenomena from this class of object to those from emission line QSOs (both radio quiet and radio loud), a better understanding of the relativistic particle acceleration mechanism, a determination of the evolution of short lived visual-infrared outbursts to those observed at radio frequencies (see Mufson et al. 1985) and interpretation of the observed polarization characteristics (Sitko, Schmidt and Stein 1985).

This discussion of the continua of emission line objects will center on observed characteristics of 3C273. This object, the brightest QSO in the sky, has been observed more extensively than any other at all wavelengths. Thus, we are given an opportunity to propose and test models with the broadest possible wavelength base. In this approach there is the danger of overinterpretation of some characteristics that may not be typical of other objects. However, the matter of the relationship of models to other objects will be discussed separately. The ultimate goal ought to be to find a coordinated series of models that explains the wide variety of observed phenomena seen in many objects as special cases.

2. The Radio to γ-Ray Continuum of 3C273

A. The Spectral-Flux Distribution

A number of investigators have reviewed the multifrequency spectral distribution of 3C273 (Worrall et al. 1980, Ulrich 1981, Ramaty and Lingenfelter 1982, Bezler et al. 1984, Perry, Ward and Jones 1987). This is the only QSO observed at $E_\gamma \sim 100$ Mev photon energies and the spectral distribution of this object must extend from $\sim 10^8$ Hz to 10^{24}Hz. It has been widely noted that from roughly the millimeter band to γ-rays the spectral-flux distribution exhibits approximately equal energy per decade. Thus, in order to simplify the problem of plotting this distribution over many orders of magnitude investigators have plotted the quantity log νF_ν vs. log ν. A plot of this type for 3C273 is shown in the above references.

It has been popular, because of the rough equipartition of emission over the many orders of magnitude of photon energy, to refer to the $F(\nu) \propto \nu^{-1}$ continuum of these objects. This statement can be easily misinterpreted as implying literally a $F(\nu) \propto \nu^{-1}$ continuum. It should be apparent to anyone that the multifrequency continuum of 3C273 is not simply $F(\nu) \propto \nu^{-1}$ (or $\nu F_\nu \alpha$ constant). It is a gross oversimplification to interpret the continuum in terms of models that result in νF_ν constant (see Perry et al. 1987, also). The observed curvature of the luminosity distribution in the radio and infrared is well outside of errors (Neugebauer et al. 1985, Robson et al. 1983), the visual-ultraviolet continuum is much flatter than $F(\nu) \propto \nu^{-1}$ (Puetter et al. 1982) and the x-ray distribution is clearly not $F(\nu) \propto \nu^{-1}$ (Worrall et al. 1979). In arriving at an acceptable model for the continuum of 3C273 the bumps, wiggles and inflections in the spectral distribution must be explained.

B. Variability

3C273 was known historically as a visual wavelength variable at a time not long after its discovery (Smith 1965). In recent years it has been only weakly variable (Cutri et al. 1985). This may mean that the physical processes giving rise to activity have been subdued or that the volume of emission at visual and infrared wavelengths is large. We would have to conclude on the basis of recent visual-infrared variability that the constraint on the region of emission at visual wavelengths is $R < c\tau \sim 1$ parsec.

Variability of ultraviolet emission has been observed from IUE by Couvousier and Ulrich (1985). Comparison of these results with those of Cutri et al. (1985) show that the $\lambda \approx 1275$Å and 1675Å flux changes have been in phase with those at $\lambda \approx 10{,}000$Å. The time scale of these changes ($\tau \sim 1$ year) imply a size limit on the radiating region of $R < c\tau \sim 1$ pc. It is difficult to separate the emission of the Balmer continuum (small blue bump) from the flat photoionizing continuum (large blue bump) because of the possible effects of large optical depth in the emission line gas (Puetter and LeVan 1982). A significant fraction of the ultraviolet emission may arise from emission line clouds within the 1 parsec volume and not from the central source.

The spectral distribution and variability of 0.75-4.5 kev x-rays from 3C273 have been reported by Bradt et al. (1979), White and Ricketts

(1979), Marshall et al. 1981, and Petre et al. (1984). A time scale of variability of these x-rays has generally been established at about one year. However, Marshall et al. (1981) indicate a significant change on a timescale of one day. At higher energies (up to $E_\gamma \sim 100$ kev) variability has also been claimed (Bezler et al. 1984). We must conclude that for the 1-100 kev region the emission volume is limited to $R_x \leq 1$ pc. This linear size is consistent with the VLBI dimension for this source (Kellermann 1971, Worrall et al. 1979). This dimension for the x-ray emission from 3C273 is consistent with a synchrotron self-Compton origin (Jones, O'Dell and Stein 1974). However, a significant one-day flux change would imply another mechanism for at least some fraction of the 2-10 kev luminosity.

C. γ-Rays

3C273 is the only extragalactic source observed in $E_\gamma \sim 100$ Mev γ-rays (Swanenburg et al. 1978). The spectral shape of these hard photons has been determined to be $N_\gamma(E) \propto E^{-2.5}$ (Bignami et al. 1981).

Adopting a cosmological distance of 630 Mpc ($H_o \sim 75$ km sec^{-1} Mpc^{-1}) we can determine the total rate of γ-ray emission from 3C273 ($z = 0.158$) assuming that is unlikely that we are at a special place in the universe and thus the γ-rays are emitted isotropically. We find that 10^{50} γ-rays sec^{-1} are radiated with a luminosity of $L_\gamma \sim 2\times10^{46}$ erg sec^{-1}. This is a luminosity comparable to that in essentially all other photon energy regimes.

It has been argued that the inner region of 3C273 should be opaque to 10^2 Mev γ-rays because of photon-photon interactions (Bignami et al. 1981). The threshold for this process is of the order of $\varepsilon_1\varepsilon_2 \sim (m_e c^2)^2$ and the cross section is $\sigma_{\gamma\gamma} \sim 2\times10^{-26}$ cm (Gould and Rephaeli 1978). In order for photon-photon interactions involving $\varepsilon_1 \sim 10^2$ Mev γ-rays to be significant the region of emission must have a very high density of photons with energies $\varepsilon_2 > 10^{-2}$ Mev = 1 kev. If the 2-10 kev emission region is $R < 3\times10^{15}$ cm the $E > 100$ Mev gamma rays must originate in a larger volume. This may be the case for either π° decay or second-order Compton scattering (Jones 1979) hypotheses. If the x-rays are attributed to Compton scattered radio frequency photons they arise in a volume of size $R \sim 1$ pc. Alternatively, the optical depth due to the γ-rays themselves can easily be shown to be $\tau_{\gamma\gamma} \sim 10^{-4}$ for $R \sim 10^{16}$ cm.

3. A Nonthermal Model for the Radio to γ-Ray Emission of 3C273

A. The Large Blue Bump and the Blue Bump Compton Component

The γ-ray emission of 3C273 strongly suggests the overwhelming importance of nonthermal processes in this object. A model involving pair production in this source and the various observable consequences will be summarized and compared with observations of the multifrequency continuum exhibited (references cited previously).

Since the implications of γ-ray emission are suggestive of nonthermal mechanisms, we assume that relativistic particles are produced in the innermost region of the object near a supermassive compact object either by accretion or by magnetohydrodynamics. Since electrons lose energy so easily, we assume that the primary result of the acceleration is the isotropic emission from the immediate volume surrounding the massive object of a spectrum of relativistic protons $N_p(E) \propto E^{-\alpha}$ where typically $\alpha \leq 3$ as for galactic cosmic rays (Burbidge 1956). Thus, we have a source of cosmic ray protons imbedded in ambient thermal gas that eventually (for $R > 10^{17}$ cm) becomes the gas normally described as the broad emission line region. The density of the thermal gas in this innermost region is assumed to have a particle density of 10^{10} cm^{-3} although all results can be scaled for other values.

Cosmic ray protons will interact with the ambient thermal gas with a cross section $\sigma_{p+p} \simeq 2\times10^{-26}$ cm^2 above threshold of about 300 Mev (Ramaty and Lingenfelter 1966). Thus a significant fraction of the accelerated protons will interact with ambient gas on a length scale of $R \sim 10^{16}$ cm (for $n_{gas} \sim 10^{10}$ cm^{-3}). As a result of pp interactions $\Pi^{\pm}$ and Π° will be produced that will decay to e^+e^- pairs and γ-rays with a spectrum of the following type:

$$N_e \propto \begin{cases} \text{const} & \gamma_e < 300 \\ E^{-3} & \gamma_e > 300 \end{cases}$$

A number of analyses have demonstrated this type of $e^{\pm}$ spectrum with differences that depend on initial assumptions (Kazanas 1984, Eilek et al. 1986, Zdziarski 1986).

O'Dell, Scott and Stein (1987) have proposed that the e^+e^- pairs generated with the above characteristic spectrum radiate synchrotron radiation in the strong fields near the supermassive object and produce what has been referred to as the large blue bump. The extension of this large blue bump beyond the particle energy spectral break above $\gamma_e \sim 300$ may extrapolate to soft x-ray energies (Jones and Stein 1987).

The parameters that are required are as follows:

$$\nu_b = 4\times10^6\gamma_e^2 B = 3\times10^{15} \text{ Hz}$$

for $\gamma_e = 300$ (the break energy) if $B \simeq 10^4$ gauss. This field strength is suggested in some QSO models (Layzer 1965, Ozernoy 1966, Morrison 1969, Rees 1984), although it is not appropriate to discuss models for the central object here.

The synchrotron radiation spectrum from secondary electrons will have a spectrum of the following type:

$$F(\nu) \propto \begin{cases} \nu^{-1/2} & \nu < \nu_b \sim 3\times10^{15} \text{ Hz} \\ \nu^{-3/2} & \nu > \nu_b \end{cases}$$

where the power-law index above ν_b is somewhat uncertain because of different results of e^+e^- cascade calculations and the assumption of the power-law index of the incident protons. A spectrum of this type fits observations of many QSOs well over the frequency range 10^{14} Hz < $\nu < 10^{17}$ Hz (Green et al. 1980, Bechtold et al. 1984, Elvis and Lawrence 1985). We display this component in Figure 1 in terms of $\log \nu F_\nu$ versus $\log \nu$.

It is a natural extension of the synchrotron blue bump model that there should also be a hard photon component due to synchrotron self-Compton of nonthermal electrons scattering on ultraviolet photons. This "blue bump Compton" component will extend over the frequency range given by $h\nu' = \gamma_e^2 h\nu$ which in this case covers 10^{18} Hz < ν < 10^{23} Hz with a spectrum reflecting the shape of the nonthermal large blue bump. It is also shown in Figure 1 under the assumption that there is energetic equipartition between the synchrotron and Compton components.

B. Infrared Emission and Infrared Compton Components

Let us pursue the other natural implications of our secondary electron synchrotron model of 3C273. We have discussed e^+e^- production and its consequences in the innermost region ($R < 10^{16}$ cm) of the object. γ-rays and protons interact with thermal ambient gas. We have calculated that the total observed gamma ray emission to be 10^{50} photons sec^{-1}. The rate of production of electrons resulting from interactions may then be

$$N_e = \tau_e \frac{dN_\gamma}{dt} = 10^{50}$$

with a lifetime 3×10^8 seconds for $B = 7\times10^{-2}$ gauss (Jones et al. 1974). Thus, we estimate from the γ-ray observation that 3×10^{58} electrons may exist with a spectrum of approximately $N_e(E) \propto E^{-2.5}$. The power observed at infrared wavelengths of $\nu F_\nu \simeq 3\times10^{-14}$ ωm^{-2} at $\nu \sim 10^{12}$ Hz would imply $\sim10^{58}$ e's for $B = 10^{-1}$ gauss and $\gamma \sim 3\times10^2$ if interpreted in terms of synchrotron radiation. In addition, an electron spectrum $N_e(\gamma) \propto \gamma^{-2.5}$ would imply a synchrotron radiation spectrum of $F(\nu) \propto \nu^{-0.75}$ at $\nu \gtrsim 4\times10^6\gamma^2$ BHz $\sim 3\times10^{10}$ Hz in conformity with the observed infrared spectral index of 3C273.

Our model seems to imply the existence of a synchrotron emission component at millimeter to infrared frequencies with the observed slope that originates in the broad line region R ~ 1-10 pc of 3C273. This is in agreement with observed parameters at high radio frequencies (Jones et al. 1974). This is in contrast to the suggestion of Zdziarski (1986) that the infrared emission arises in a small ($R \sim 10^{15}$ cm) strong field region ($B > 100$ gauss) which would imply synchrotron self absorption at $\nu < 10^{13}$ Hz (see Section 4 for details) and thus not allow for the observed extension to radio frequencies. The hypothesized nonthermal infrared emission spectrum from secondary electrons is shown in Figure 1 as the nonthermal radio-IR component.

An additional implication of our secondary electron infrared model is the natural existence of a high frequency component due to synchrotron self-Compton that would be observed over the frequency range 10^{16} Hz $< \nu < 10^{21}$ Hz. It is labeled Infrared-Compton in Figure 1. This component would be the likely explanation of observed x-ray emission of 3C273. Note that it would originate on a size scale of parsecs in the radiating volume and therefore would not be expected to have rapid fluctuations on timescales <1 year. This is in agreement with observations reviewed earlier. (However, recall the question of short-term variability.)

C. The Extended Multifrequency Spectral Distribution

The various components of synchrotron and Compton emission expected in our secondary electron model of 3C273 are displayed in Figure 1. The various components have been arbitrarily normalized to have the same peak νF_ν in this figure. For more generality we include a potential dust radiation component as well, although we do not suggest that dust emission is observed from 3C273. The total spectral distribution $\log \nu F_\nu$ as a function of $\log \nu$ would be the sum of the various components and it can be seen that this would result in an extended multifrequency distribution that would approximate $F_\nu \propto \nu^{-1}$ energetically. However, note that we now can explain observed bumps, wiggles and inflections in the spectral distribution. A comparison of Figure 1 with the observed distribution shown in various compilations cited previously reveals a close similarity. Application of this hypothetical spectrum to other sources will be discussed in the following section.

4. Observations and Hypotheses Regarding Other AGN

Multifrequency spectral-flux distributions of a number of extragalactic sources have been recently summarized by Edelson and Malkan (1986). In general we see that, energetically, the spectral-flux distributions of many of these sources, correspond to equipartition of energy over all frequencies. However, it would be incorrect to describe these spectral distributions as $F(\nu) \propto \nu^{-1}$ - the bumps and wiggles are real and must be explained by any theory.

An especially interesting set of spectral-flux distributions has been shown by Neugebauer et al. (1985) that include results in the 10μm-100μm wavelength range as observed from Infrared Astronomical Satellite (IRAS). The comparison is made of the 10^9 Hz $< \nu <$ 10^{16} Hz results of 3C48, 3C273, 3C345 and PG1351+640. This is an interesting combination of objects because 3C48 has been hypothesized by Neugebauer et al. (1985) to exhibit a $\lambda \sim$ 100μm dust bump from a host galaxy, 3C345 is a well known OVV quasar (also a strong flat spectrum radio source), PG1351+640 is an optically selected QSO (Schmidt and Green 1983) with a strong, large blue bump that is relatively radio quiet (Condon et al. 1981) and 3C273 is a strong radio source with a large blue bump.

Our model may be adapted to sources of various types through the following parameters that may be different from one source to another as follows:

(1) Density of gas as a function of distance will determine the degree of compactness of the nonthermal source. Smaller sources radiating in stronger B fields will become synchrotron self-absorbed at higher frequencies. An example of a series of models of this type may be that of Zdziarski (1986). This may be the explanation of the radio loud-radio quiet phenomenon (Condon et al. 1981).

(2) The ratio of the peak νF_ν for the various nonthermal components may vary from one source to another. A more diffuse synchrotron source, having a smaller synchrotron radiation energy density, may have a lower ratio of νF_ν)Compton to νF_ν) synchrotron. Thus, we may

explain sources that have stronger or weaker x-ray emission relative to synchrotron emission.

(3) The ratio of nonthermal blue bump emission to nonthermal infrared emission (as well as the Compton components associated with each) may vary from one object to another depending on the magnetic field distribution within each.

(4) Dust, if it exists in the galaxy surrounding an active nucleus, will produce a spectral emission bump over the 10^{11} Hz < ν < 10^{14} Hz. frequency range. In some cases the existence of dust may be detectable by visual reddening, however, geometry may prevent such detection merely by orientation of the object relative to the line of sight. Dust observed in emission is not so strongly geometry sensitive.

We will discuss these effects more quantitatively in the following paragraphs. Synchrotron self-absorption becomes important at frequencies given by

$$\frac{\nu}{10^8 \text{ Hz}} = \frac{(4\times10^{-3})^{4/5}(F_{-26})^{2/5}B^{1/5}}{\theta^{4/5}\text{ (arc sec)}}$$

in which $F_{-26} = F_{obs} \times 10^{26}$ erg cm^{-1} sec^{-1} Hz^{-} and [B] = gauss (e.g., Tucker 1975). For parameters relevant to 3C273 θ_d - 10^{-3} arc seconds, $F \simeq 10^{-25}$ watt m^{-2} Hz^{-1} at $\nu \sim 10^{11}$ Hz, $B \sim 10^{-1}$ gauss (Jones et al. 1974) we find $\nu_{SA} < 10^{11}$ Hz. However, in the models of Zdziarski (1986) $B \sim 10^2$ gauss and $R \sim 10^{15}$ cm ($\theta \sim 10^{-7}$ radian), therefore $\nu_{SA} \le 6\times10^{13}$ Hz and it would not be possible to explain the infrared to radio frequency portion of the spectral distribution. A larger source dimension with lower B than those hypothesized by Zdziarski are required to explain radio loud objects. However, the smaller regions and higher fields may explain radio quiet objects as was also discussed by Condon et al. (1981).

The ratio of Compton scattered photon luminosity to its associated synchrotron component will depend on magnetic field and synchrotron energy density

$$\frac{\nu\, F_\nu)s}{\nu\, F_\nu)c} = \frac{B^2/8\Pi}{U_{rad}} \; .$$

Frequently it is found that extragalactic sources exhibit a ratio $\nu\, F_\nu)s/\nu\, F_\nu)c \simeq 1$ although this may not necessarily be the case for all objects. We may find situations in which the synchrotron emission component dominates although it is rare to find more luminosity associated with the Compton component than the synchrotron component.

The ratio of radio-infrared synchrotron emission to blue bump non-thermal emission is given by

$$\frac{\nu F_\nu)_{IR}}{\nu F_\nu)_{BB}} = \frac{\gamma_{IR}^2 B_{IR}^2 N_{IR}}{\gamma_{BB}^2 B_{BB}^2 N_{BB}} .$$

If both infrared and blue bump components are attributed to secondary electrons a typical γ for both is $\gamma_e \sim 300$. Therefore, the ratio of emission in the two frequency ranges is determined by the magnetic field distribution and the numbers of electron-positron pairs in the high and low field regions. Here we are oversimplifying in reference to high and low B field regions since in reality there is probably a continuous distribution of B within the AGN. Models of such will be generated in later work.

Finally we may expect that some AGN will exhibit a dust emission peak. We have called attention earlier to this component, shown in Figure 1, that probably does not apply to an interpretation of 3C273, but could be significant in 3C48 (Neugebauer et al. 1985). If dust is present in a galaxy we may expect that the galaxy will exhibit a dust bump, similar to that shown, that contains a luminosity of $>10^{11}$ L_o based just on absorption of starlight. This would not apply to dust free ellipticals, however. Thus, we may expect that in galaxies containing dust, for which lack of angular resolution prevents resolving only the nucleus (e.g., IRAS), $\nu F_\nu)_{dust} > 10^{-14}$ watt m^{-2} at $\nu \sim 3\times10^{12}$Hz. Any short wavelength luminosity attributable to nonthermal emission from the nucleus would add to this minimum expected value based on stars alone. In the IRAS measurements reported by Neugebauer et al. PG1351+650 exhibits a 100μm emission of $\nu F_\nu \sim 3\times10^{-14}$ wm^{-2}. 3C48 exhibits $\nu F_\nu \sim 3\times10^{-14}$ wm^{-2} also.

5. Summary

We have presented a model for the entire radio to γ-ray emission of active galactic nuclei and QSOs based on acceleration of relativistic particles and consequences of secondary electron production. The total emission as a function of frequency is in close agreement with the spectral-flux distribution of 3C273. By changing physical parameters within the source emission at various frequencies can be modified to conform with the observed luminosity versus frequency of other sources.

In the case of 3C273 relativistic secondary electrons produce the spectral distribution of the large blue bump in a high field of $B \sim 10^4$ gauss near a supermassive compact object. The infrared emission is produced by secondary electrons resulting from γ-ray pair production

in a field B < 0.1 gauss in conformity with previous estimates based on high radio frequency observations. Compton scattering of the synchrotron photons result in the x-ray spectrum. We anticipate further applications of the model to other sources in future investigations.
The author wishes to acknowledge helpful discussions with T.W. Jones and L. Rudnick.

References

Bechtold, J. et al.: 1984, Astrophys. J. 281, 76.
Bethe, H. A. and Heitler, W.: 1934, Proc. Roy. Soc. London A 146, 83.
Bezler, M., Kenziorra, E., Staubert, R., Hasinger, G., Pietsch, W., Reppin, C., Trumper, J. and Voges, W.: 1984, Astron. Astrophys. 136, 351.
Bignami, G. F. et al.: 1981, Astron. Astrophys. 93, 71.
Bradt, H. V., Doxsey, R. E., Johnston, M. D., Schwartz, D. A., Burkhead, M. S., Dent, W. A., Liller, W. and Smith, A. G.: 1979, Astrophys. J. (Letters) 230, 416.
Burbidge, G. R.: 1956, Astrophys. J. 124, 416.
Condon, J. J., O'Dell, S. L., Puschell, J. J. and Stein, W. A.: 1981, Astrophys. J. 246, 624.
Courvoisier, T. J.-L. and Ulrich, M. H.: 1985, Nature 316, 524.
Cutri, R. M., Wisniewski, W. A., Rieke, G. H. and Lebofsky, M. J.: 1985, Astrophys. J. 296, 423.
Edelson, R. A. and Malkan, M. A.: 1986, Astrophys. J. 308, 59.
Eilek, J. A., Caroff, L. J., Noerdlinger, P. D. and Dove, M. E.: 1986, Proceedings of NOAO Workshop Continuum Emission of Active Galactic Nuclei, ed. M.L. Sitko (Tucson: Arizona), p. 134.
Elvis, M. and Lawrence, A.: 1985, in Astrophysics of Active Galaxies and Quasi-Stellar Objects, ed. J. S. Miller (Mill Valley: California University Science Books), p. 289.
Green, R. F. et al.: 1980, Astrophys. J. 239, 483.
Jones, T. W.: 1979, Astrophys. J. 233, 796.
Jones, T. W., O'Dell, S. L. and Stein, W. A.: 1974, Astrophys. J. 192, 261.
Jones, T. W. and Stein, W. A.: 1987, Astrophys. J. (Letters), 320, L1.
Kazanas, D.: 1984, Astrophys. J. 287, 112.
Kellermann, K. I. et al.: 1971, Astrophys. J. 169, 1.
Königl. A.: 1981, Astrophys. J. 243, 700.
Layzer, D.: 1965, Astrophys. J. 141, 837.
Marscher, A. P.: 1977, Astrophys. J. 216, 244.
Marshall, N., Warwick, R. S. and Pounds, K. A.: 1981, Monthly Not. Roy. Astron. Soc. 194, 987.
Morrison, P.: 1969, Astrophys. J. (Letters) 157, L73.
Mufson, S. L. et al.: 1985, Astrophys. J. 288, 718.
Neugebauer, G., Soifer, B. T. and Miley, G. K.: 1985, Astrophys. J. (Letters) 295, L27.
O'Dell, S. L., Scott, H. A. and Stein, W. A.: 1987, Astrophys. J. 313, 164.
Ozernoy, L. M.: 1966, Astr. Zh. 43, 300.
Perry, J. L., Ward, M. J. and Jones, M.: 1987, Monthly Not. Roy. Astron. Soc. 228, 623.
Petre, R., Mushotzky, R. F., Krolik, J. H. and Holt, S. S.: 1984, Astrophys. J. 280, 499.
Puetter, R. C., Burbidge, E. M., Smith, H. E. and Stein, W. A.: 1982, Astrophys. J. 257, 487.
Puetter, R. C. and LeVan, P. D.: 1982, Astrophys. J. 260, 44.
Ramaty, R. and Lingenfelter, R. E.: 1966, J. Geophys. Res. 71, 3687.

Ramaty, R. and Lingenfelter, R. E.: 1982, Ann. Rev. Nucl. Part. Sci. 32, 235.
Rees, M. J.: 1984, Ann. Rev. Astr. Ap. 22, 471.
Robson, E. I. et al.: 1983, Nature 305, 194.
Sitko, M. L., Schmidt, G. D. and Stein, W. A.: 1985, Astrophys. J. Suppl. 59, 323.
Smith, H. J.: 1964, Quasi-Stellar Sources and Gravitational Collapse, eds. I. Robinson, H. Schild and E. L. Schuicking (University Chicago Press), p. 221.
Stein, W. A. and O'Dell, S. L.: 1985, in Astrophysics of Active Galaxies and Quasi-Stellar Objects, ed. J. S. Miller (University Science Books: Mill Valley), p. 381.
Swanenburg, B. N. et al.: 1978, Nature 275, 298.
Tucker, W. H.: 1975, Radiation Processes in Astrophysics (MIT Press: Cambridge, Massachusetts).
Ulrich, M.-H.: 1981, Space Sci. Rev. 28, 89.
White, G. J. and Ricketts, M. J.: 1979, Monthly Not. Roy Astron. Soc. 187, 757.
Worrall, D. M., Boldt, E. A., Holt, S. S. and Sermitsos, P. J.: 1980, Astrophys. J. 240, 421.
Worrall, D. M., Mushotzky, R. F., Boldt, E. A., Holt, S. S. and Serlemitsos, P. J.: 1979, Astrophys. J. 232, 683.
Zdziarski, A. A.: 1986, Astrophys. J. 305, 45.

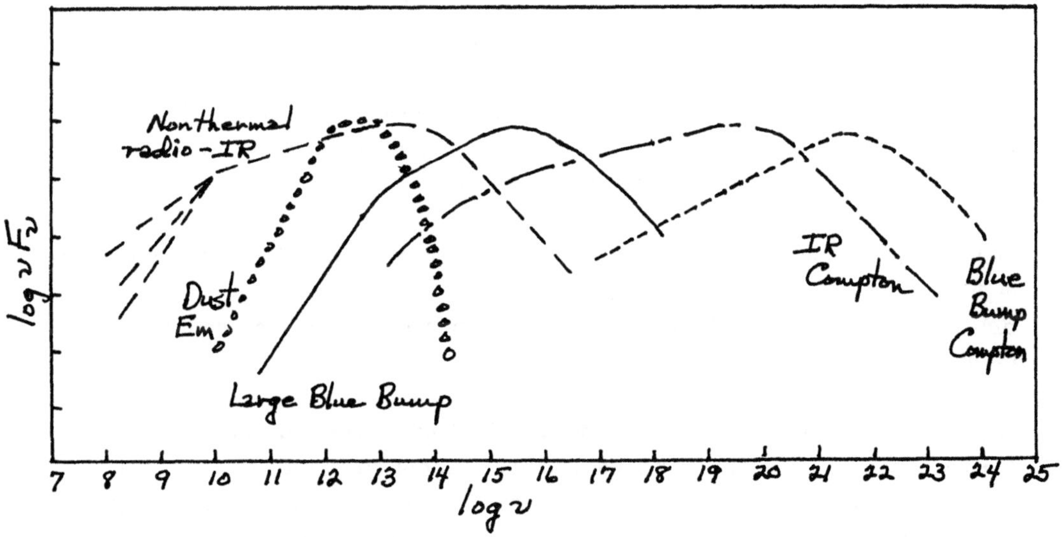

Figure 1 - Schematic representation of secondary electron synchrotron self-Compton model.

QUESTIONS AND COMMENTS

A.C. Sadun: The quasar 3C 273 is the only quasar known with > 100 MeV γ-ray spectrum. COS-B observations have shown this to be a constant source while the emission varies considerably at all other wavelength regimes. This has led many to believe that the γ-ray emission comes in fact from the jet (Morrison, *et al.* *Ap. J.*, **280**, 483; 1984). In what way, with your model, can you have a constant source of γ-rays in the core, while you have a great deal of variability at other energy regimes?

W.A. Stein: The E > 100 MeV γ-rays may be attributabe to π^o decay or to Compton scattering. To the extent that Compton γ-rays may dominate, they would arise in a region $R \gtrsim 1$ pc and thus may not vary on short time scales. π^o's may also be generated throughout a large region.

Y. Tomozawa: What is the energy spectrum of the protons which creates the electron-positron pairs in your model for the inner core of 3C 273? How sensitive is the assumption for a specific form of the energy spectrum in explaining the experimental data?

W.A. Stein: We have adopted a standard cosmic ray power law spectrum (*e.g.* $N(E) \propto E^{-2.5}$ or E^{-3}.

Y. Tomozawa: In a model of quantum mechanical black holes which I presented here, one can obtain emission of protons with the E^{-3} energy spectrum, which is identical with that of high energy cosmic rays.

R. Hackney: Would you expect any distinctive observable effects of annihilation of the pairs?

W.A. Stein: If the pairs thermalize and cool then there may be annihilation radiation at ~0.5 MeV.

C.M. Gaskell: First a historical comment: the idea of the blue bump being of thermal origin goes back a long way, actually being first proposed in 1964 by Johnson and Low (*Ap. J.*). My question is: In our paper (Koratkar, Cheng and Gaskell, this volume) we show that the UV slopes have very little object to object variation, and in particular, that there is absolutely no luminosity dependence for the UV slope. Would your theory predict this constancy of shape?

W.A. Stein: A spectral shape that is flat (*e.g.* $F(\nu) \propto \nu^{-0.5}$) followed by a break to a steeper slope is a natural consequence. The break frequency depends on B and this parameter could be different in various sources.

M. Ward: It appears that your model predicts a somewhat steep "ionizing continuum". Is this consistent with the observed emission line strengths?

G. MacAlpine: The shape of the ionizing radiation spectrum does have some important impacts on theoretical predictions for photoionization models. One example is the He II λ1640/ C IV λ1549 emission-line ratio, which provides a measure of $h\nu \geqslant 4$ Ry photon input compared with overall radiative energy input to the gas. Another notable example involves He II line equivalent widths, which provided the inital realization that intrinsic far UV quasar spectra ("big blue bumps") are much more energetic than implied by IUE observations.

D. Kazanas: A comment first on the question of photoionization: if your model reproduces the observed continuum as you have shown then the line emission should be that observed and accounted for by the models. The question is: granted your one assumption about the emission mechanism, what is the number of parameters one needs to fit the spectrum?
W.A. Stein: The most important one is the magnetic field, B(r).

C.M. Urry: As you pointed out, SSC models (particularly tapered models) reproduce nicely the very smooth broad-band spectra of blazars. It seems to me that if you adjust your model to fit quasar spectra, which, in contrast to blazars, are full of bumps and wiggles, you will have to postulate physically distinct emitting regions, with each synchrotron self-absorption peak tailored to a particular bump in the spectrum. I would expect tapered models to be too smooth. Could you comment on this and on the physical consequences?
W.A. Stein: This is exactly the matter that we are currently investigating numerically.

THEORY OF AGN CONTINUUM RADIATION

Mitchell C. Begelman
Joint Institute for Laboratory Astrophysics
University of Colorado and National Bureau of Standards
Boulder, CO 80309-0440

I. INTRODUCTION

The theory of the continuum radiation from AGNs is in a very primitive state. We don't really know what the principal radiation mechanisms are, or even whether the spectrum in a particular frequency range is produced by thermal or nonthermal processes. A possible breakthrough has been the identification of the "UV bump" with thermal radiation from the inner part of an accretion disk (Malkan 1983, Malkan and Sargent 1982). This identification is both plausible and encouraging (particularly for theorists), although Guilbert and Rees (1987) point out that non-disk-like configurations involving thermal gas are possible, such as cool filaments embedded in a hot substrate. Unfortunately, much of the UV bump radiation may lie at frequencies which are not readily observed, so we do not have a good handle on the shape of the spectrum. Clear signatures of disk geometry (e.g., atomic spectral features, aspect effects) have yet to be found.

Still less is known about the origin of radiation in other parts of the spectrum. The uncertainties have permitted a wide range of theoretical constructs to flourish, and although there are fads and fashions, it is not clear whether any meaningful convergence of the models is occurring. One approach to modeling the continuum proceeds from "pure thought," e.g., one tries to model the likely physical properties of accretion flows onto massive black holes, then to deduce the radiative properties of such flows. Aspects of this approach have been discussed in reviews by Begelman (1985) and Rees (1984). While this approach leads to some interesting arguments concerning the plausibility of various radiation mechanisms, it is weak on predicting the details, particularly where nonthermal radiation mechanisms are involved.[1] A second approach focuses on a particular set of radiation mechanisms,

---

[1]This approach can give much cleaner results if one considers only a restricted set of thermal radiation mechanisms and geometries. A familiar example is standard thin accretion disk theory; less familiar ones include crude models for the spectra of radiation tori (Begelman 1985) and thermal pair production models for the AGN continuum (e.g., Dermer, these proceedings).

with the details of the environment regarded as secondary. Such models are usually aimed at "explaining" particular spectral features, hopefully more than one, and are better suited to making predictions about future observations. A particularly important component of the latter class of models is an economy of assumptions about the geometry and spatial stratification of the emitting region. As in the simple models for the broad line region (which are now coming under attack), there is a premium on the "one-zone model," a single, homogeneous or simply stratified emitting region which can explain everything. Unfortunately, given what we think we know about accretion flows and jets, such an assumption is probably unrealistic. Perhaps more comprehensive multi-spectral data on variability will help to indicate which spectral bands are closely interrelated.

I have generally operated in the "pure thought" mode, so I will try a slightly different tack this time. I will try to say as little as possible about my prejudices concerning accretion onto black holes, and instead allow my argument to be guided by what the observers tell us needs explaining. Of course, it is impossible to resist checking whether my prejudices about accretion flows are bolstered or refuted, so I will allow myself a few a posteriori comparisons. I will not comment on the origin of the "UV bump," although I will consider it as a possible source of seed photons for other parts of the spectrum. And, unlike most recent treatments of the "nonthermal" continuum, which concentrate on the X-ray part of the spectrum, I will launch my discussion by referring to a recent series of observations in the infrared, which I think may hold important clues to the physics behind the continuum radiation.

II. THE FAR INFRARED PHOTOSPHERE

Since most quasars and virtually all Seyferts are "radio-quiet," it has long been recognized that the optical-near IR spectrum must "turn over" before reaching radio wavelengths. We simply did not know where the turnover occurred. Edelson (1986), Edelson and Malkan (1986) and Edelson, Malkan and Rieke (1987) have recently shown that the turnover typically occurs at frequencies in the range $\nu_t \sim 10^{12}$ - 10^{13} Hz, in optically selected AGNs without strong evidence for dust. Between the start of the turnover and frequencies $\sim 3 \times 10^{14}$ Hz, the spectra of these "bare AGNs" appear to have a standard "nonthermal" shape, with a spectral index $\alpha \sim 1$ (Ward et al. 1987, Carleton et al. 1987). Both the turnover and the spectrum shortward of the turnover are readily explained in terms of self-absorbed synchrotron radiation from relativistic electrons with a distribution of Lorentz factors $N_\gamma \propto \gamma^{-3}$, or by Comptonized, self-absorbed thermal cyclotron radiation from a mildly relativistic electron plasma. A straightforward application of the condition for self-absorption (basically the Rayleigh-Jeans law with the correct expression inserted for the brightness temperature) immediately leads to the surface area A_t of the "far infrared photosphere," or equivalently to its characteristic radius

$$R_t = (A_t/4\pi)^{1/2} \sim 1.6 \times 10^{15} L_{45}^{1/2} B_4^{1/4} \nu_{13}^{-7/4} \times \begin{matrix} 1 & \text{synchrotron} \\ 3.6 & \text{cyclotron} \end{matrix} \text{ cm} . \quad (1)$$

Here $L_{45} = (\nu L_\nu)_t/10^{45}$ ergs s^{-1} (measured just shortward of the turnover), $B_4 = B/10^4$ Gauss is the magnetic field, and $\nu_{13} = \nu_t/10^{13}$ Hz. The magnetic field is normalized to a value thought to be typical of the inner parts of accretion flows (e.g., Begelman 1985), but note that R_t is insensitive to the value of B. Both L and ν_t are observables. The estimated value of R_t is comparable to the radii believed to characterize the "central engines" of luminous AGNs. This agreement would be remarkable if it were completely fortuitous, because R_t is quite sensitive to the value of ν_t.

Note that we are still presented with a choice between two versions of the emission mechanism, one thermal and one nonthermal. It is not entirely clear which one applies, or whether both are relevant. Explaining the $\alpha \sim 1$ power-law by unsaturated thermal Comptonization does seem to require some fine-tuning of the temperature and optical depth in the plasma, and geometric effects (e.g., partial covering) must be invoked to explain why the same gas does not strongly Comptonize the UV bump. Neither of these points condemns the Comptonized cyclotron interpretation, but for brevity I will henceforth assume that the dominant mechanism is nonthermal synchrotron radiation. (Some of the following arguments also apply to the cyclotron case, with minor modification.)

It is straightforward to associate a dynamical timescale with the radius R_t. Such a timescale might be a light travel time, or a Keplerian time (in which case it is necessary to assume a value for the central mass). For typical values of the parameters, such a timescale is unlikely to exceed a few weeks. This is comfortably shorter than the observed variability timescales in the infrared. In fact, some people are worried that the accommodation is <u>too</u> comfortable, since IRAS observations have failed to detect variability on timescales longer than a year (Edelson 1987). Note that the distance of the far-IR photosphere from the central engine can be increased only by making the covering factor, f, small: the radial scale would vary as $f^{-1/2}$. The dynamical time is orders of magnitude longer than the synchrotron cooling time, which is $\sim 0.5\ B_4^{-3/2}\ \nu_{13}^{-1/2}$ seconds for electrons radiating close to the turnover frequency, and even shorter for higher energy electrons. Thus, electrons must be continually accelerated <u>in situ</u>, and the overall patterns of variability are presumably conditioned by macroscopic (hydrodynamic, hydromagnetic) properties of the flow. The latter are notoriously difficult to model.

What determines the radius of the far-IR photosphere? One possibility is that it is related to a structural feature in the accretion flow; in this case, we would expect R_t to be some modest multiple (of order 10-100) of the gravitational radius of the black hole, $R_g = GM/c^2$. The luminosity in the far-IR band (L^{FIR}) is typically a few per cent of the inferred luminosity in the UV bump. If the latter represents the bulk of the accretion luminosity, and if the power locally available from accretion scales with radius as R^{-1}, then it is hard to see how the far-IR flux could be powered by the local release of gravitational binding energy unless $R_t \lesssim 100\ R_g$. But from eq. (1), we infer that $R_t \gg 100\ R_g$ for $\nu_t \sim 3 \times 10^{12}$ Hz, unless L^{FIR} is much smaller than the Eddington limit, $L^{FIR}/L_E < 0.01$. Note that the far-IR turnover represents the innermost self-absorbed region of the flow. Self-absorbed regions at larger radii would have turnovers at lower frequencies ($\nu_t \propto R^{-1}$, if the luminosity is a fixed fraction of the accretion power; see §V). Hence if R_t were very far from the black hole, then we would have to conclude that for some reason a strong synchrotron source is not established at smaller radii; otherwise the turnover would be at higher frequencies than observed. Of course the energy for the far-IR photosphere could be transported to large radii by a jet or wind, in which case the constraint imposed by local gravitational binding energy does not apply.

Regardless of its global relation to a possible accretion flow, we can say something about the local energetics of the far-IR photosphere. The mean energy density in relativistic electrons,

$$\bar{U}_{rel} \approx 0.03\ L_{45}^{-1/2}\ B_4^{-9/4}\ \nu_{13}^{5/4}\ \text{ergs cm}^{-3}, \tag{2}$$

is much smaller than the magnetic energy density, $U_B = 4 \times 10^6\ B_4^2$ ergs cm^{-3}. Thus, the relativistic electrons are either far from equipartition with the magnetic field, and/or they are found only in highly localized regions of the flow, e.g., close to shock fronts or sites of magnetic reconnection. The radiation energy density is

$$U_{rad} \sim 10^3 (L^{tot}/L^{FIR})\ B_4^{-1/2}\ \nu_{13}^{7/2}\ \text{ergs cm}^{-3}, \tag{3}$$

where L^{tot} is the total luminosity passing through the far-IR photosphere. U_{rad} is smaller than U_B if

$$B > 360\ (L^{tot}/L^{FIR})^{2/5}\ \nu_{13}^{7/5}\ \text{Gauss}. \tag{4}$$

If condition (4) is satisfied, then synchrotron radiation is a more important loss mechanism than Compton scattering for the relativistic electrons at R_t. There is some weak observational support for this conclusion, since the same relativistic electrons responsible for some of the infrared synchrotron continuum have Lorentz factors such that they would Comptonize UV bump photons into the ~2 keV X-ray band. In the composite spectrum for radio-quiet quasars constructed by Kriss

(1988), the ~2 keV luminosity is smaller than (or at best comparable to) the IR luminosity. If the UV bump is the main source of seed photons for Comptonization at R_t, this implies that synchrotron cooling is as important or more important than Compton cooling.

III. PARTICLE ACCELERATION AND PAIR CASCADES: THE INFRARED CONTINUUM

If we do indeed have a self-absorbed synchrotron source at R_t, where do the relativistic electrons come from? To obtain a spectral index $\alpha \sim 1$, we need an electron energy distribution $N_\gamma \underset{\sim}{\propto} \gamma^{-3}$. If these electrons are "injected" with some energy distribution S_γ, then the solution of the steady-state equation for energy diffusion (for optically thin radiative cooling) is

$$N_\gamma \propto \frac{1}{\gamma^2} \int_\gamma^\infty S_\gamma d\gamma \; . \tag{5}$$

Given a power-law injection function, $S_\gamma \propto \gamma^{-s}$, with some high-energy cutoff at γ_{max}, there are two characteristic behaviors of N_γ at $\gamma \ll \gamma_{max}$. If $s < 1$, then $N_\gamma \propto \gamma^{-2}$ and $\alpha \sim 0.5$. If $s > 1$, then $N_\gamma \propto \gamma^{-1-s}$ and $\alpha \sim s/2$. Thus, to obtain $\alpha \sim 1$, we need injection with $s = 2$. Unfortunately, we know little about how relativistic electrons are created in AGNs. Some proposed particle acceleration processes, such as magnetic reconnection, are not very well understood. Others are in better shape, such as the shock acceleration mechanism which has been applied to particle acceleration in the interplanetary medium and to the production of cosmic rays (Blandford and Eichler 1987). In the latter model, a strong shock with a density jump of 4 naturally injects relativistic particles with an index $s = 2$. While it seems that this mechanism can channel a sizable fraction of the shock energy into relativistic protons, there are doubts about its efficiency at accelerating electrons. In order to account for the large luminosity in the nonthermal parts of the spectrum, the energy transfer to relativistic electrons has to be rather efficient.

Observations of "universal" power laws in AGN spectra have led researchers to search for methods of producing power-law electron distributions which are insensitive to the primary injection mechanism. The mechanism which has received the greatest attention of late is the electron-positron pair cascade, in which part of the energy radiated by relativistic pairs is transformed into rest mass of pairs at lower energies. The concept of a "pair cascade" was introduced by Bonometto and Rees (1971), although the first self-consistent models have been worked out only recently (e.g., by Zdziarski and Lightman 1985, Fabian et al. 1986, Svennson 1987, and Ghisellini 1987). The basic idea is illustrated in Fig. 1. At the i-th stage in the cascade, gamma-rays with energy $h\nu_i \gg m_e c^2$ interact with soft photons of energy $(m_e c^2)^2/h\nu_i$ (since the cross section is maximized for center-of-mass energy

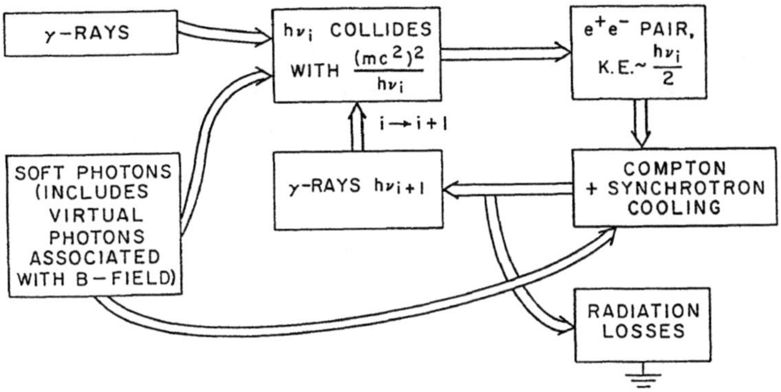

Fig. 1. Schematic diagram of a pair cascade. Note the roles played by external sources of soft photons and γ-rays, as well as radiation losses. Details are described in text.

$\sim 2m_ec^2$), to produce an electron and a positron, each with energy $\sim h\nu_i/2$. These in turn cool by synchrotron radiation and Compton scattering, with a certain fraction of the radiation carried by gamma-rays of energy $h\nu_{i+1}$. If the soft photon field is sufficiently optically thick to these new photons, then they too will produce pairs and the cycle repeats. The characteristic time scale for one cycle in the cascade is the cooling time scale, $t_{cool} \propto \gamma^{-1}$. Now suppose that all of the electrons are initially injected with high Lorentz factors, so that the energy distribution at lower γ is governed entirely by cooling processes. If all of the energy radiated by the pairs goes into producing new pairs, then the total energy of the system is conserved as the pairs cool, and the energy diffusion equation implies $\gamma^2 N_\gamma / t_{cool} \sim$ constant, or $N_\gamma \propto \gamma^{-3}$. On the other hand, if much of the radiated energy escapes without producing pairs, then it is rest mass which is conserved, and we obtain the old result $N_\gamma \propto \gamma^{-2}$. Thus, a pair cascade process can result in spectral indices between 0.5 and 1, depending on how efficiently the radiation from the cooling pairs is converted into new pairs. This efficiency determines the level of "saturation" of the pair cascade.

Two conditions must be met in order for a pair cascade to operate efficiently at a given pair energy γm_ec^2. First, the background soft photon field must be optically thick to gamma-rays with energies comparable to the pair energy. For an $\alpha \sim 1$ power-law radiation spectrum, this implies that the "compactness parameter," $\ell = (L/R)\,\sigma_T/m_ec^3$, must exceed $10/\gamma$. The compactness parameter at the far-IR photosphere is

$$\ell(R_t) \sim 17\, L_{45}^{1/2}\, B_4^{-1/4}\, \nu_{13}^{7/4}\,, \qquad (6)$$

so we see that a pair cascade is certainly possible in the region producing the infrared radiation. Once the compactness condition is met, the degree of saturation of the cascade depends on the energy of the injected electrons and the relative importance of synchrotron and Compton cooling, which determine the fraction of radiation emitted with photon energy above the pair production threshold.

To explain the $\alpha \sim 1$ spectrum, we need a highly saturated pair cascade. Previous applications of the pair cascade have concentrated on producing the "universal" $\alpha \sim 0.7$ spectrum at hard X-ray energies, which requires a moderate level of saturation. Such cascades are possible if electrons are injected with Lorentz factors in the range $10^5 - 10^6$ and if synchrotron cooling is not too important. This is because synchrotron photons produced by such electrons tend to be rather soft, in the magnetic fields believed to be present. We have already argued that synchrotron radiation may well dominate over inverse Compton cooling at R_t, so we would require much higher injected electron energies even if our sole aim were to explain the $\alpha \sim 0.7$ X-ray spectrum. The situation is even worse if we want to obtain a spectrum extending down to the far infrared with $\alpha \sim 1$.

Lorentz factors of order $10^6 B_4^{-1/2}$ are the largest that can be produced directly by electron acceleration mechanisms involving scattering off magnetic inhomogeneities. In such processes, the characteristic acceleration time scale exceeds the light travel time across the Larmor radius. Acceleration stops when the cooling time scale becomes shorter than the acceleration time (or the gyroradius exceeds the shock size). Synchrotron photons produced by magnetically accelerated electrons are limited to energies $\lesssim \alpha_f^{-1} m_e c^2 = 70$ MeV, where α_f is the fine structure constant (Phinney 1983, Guilbert, Fabian and Rees 1983); this is somewhat above the pair production threshold but not high enough to lead to a saturated cascade. We (Sikora et al. 1987) have suggested an indirect way of injecting electrons with extremely high Lorentz factors. In our picture, no direct acceleration of electrons is required; all that is needed is the efficient injection of relativistic <u>protons</u> with $s < 2$ (i.e., most of the energy concentrated in protons with the highest Lorentz factors). Although the protons are not subject to significant radiative losses, they are cooled very effectively by interactions with soft photons, and cannot be accelerated to arbitrarily high Lorentz factors. The main reactions are direct pair production by a photon in the Coulomb field of the proton and photo-production of (mainly neutral) pions. The latter process dominates for a soft photon background with $\alpha \sim 1$, and leads to the creation of pairs with Lorentz factors as large as

$$\gamma_{max} \sim 10^{10} \ell^{-1/4} \left(\frac{U_B}{U_{rad}}\right)^{1/4} . \qquad (7)$$

These electrons are able to initiate a highly saturated pair cascade, even if synchrotron radiation is the dominant cooling process. Such a cascade would produce an underlying $\alpha \sim 1$ continuum extending from the gamma-rays down to the far-IR turnover. The Sikora et al. process has the additional feature that it avoids the difficulty of accelerating electrons from the thermal background, which is inherent in current versions of the shock acceleration theory. Note that inelastic proton-proton collisions may also lead to the creation of relativistic electron-positron pairs, as has been pointed out by Kazanas and Ellison (1986) and Zdziarski (1986). This process is energetically important only if the proton injection index is larger than 2 (i.e., most energy concentrated at low Lorentz factors) and it tends to be less efficient than the Sikora et al. process. Proton-proton collisions can, however, create a supply of pairs with Lorentz factors of a few hundred, which can then be accelerated further in shocks.

IV. X-RAY CONTINUUM

The main observational features of the X-ray continuum are: 1) a steep soft X-ray spectrum, which may be the (Comptonized?) tail of the UV bump (Wilkes and Elvis 1987); 2) a ~ 2 keV flux which fits roughly on the $\alpha \sim 1$ extrapolation of the IR flux, suggesting a single synchrotron spectrum extending over many decades in frequency (Malkan and Sargent 1982, but see Kriss 1988 for some refinements to this view); and 3) a flattening of the spectrum at $h\nu >$ a few keV, to a "universal" $\alpha \sim 0.7$ power-law (Mushotzky 1984). It is the latter feature that has excited the greatest interest among theorists, and there are several models now extant to explain both the flattening and the universality of the power law. Band and Grindlay (1986) have suggested that the $\alpha \sim 0.7$ power-law results from Comptonization of the UV bump by a nonthermal distribution of electrons with $N_\gamma \propto \gamma^{-2.4}$ (injection index 1.4). They do not specify why the electron distribution should have this unique form, but several researchers (Zdziarski and Lightman 1985, Fabian et al. 1986, Svensson 1987) have shown that a spectrum with approximately this slope results from a partially saturated pair cascade. Note that such a cascade must be spatially distinct from a fully saturated ($\alpha \sim 1$) cascade, like the one described in §III.

Another possiblity is that the flattening results from unsaturated thermal Comptonization of the UV bump (e.g., Maraschi, Roasio and Treves 1982, Ipser and Price 1983). This explanation suffers from the same need for "fine-tuning" as does the thermal model for the $\alpha \sim 1$ infrared spectrum (§II), i.e., the Compton y-parameter [$y \sim \max(\tau, \tau^2)\, 4kT/m_ec^2$] must have a unique value if the 0.7 spectral slope is truly "universal." Note that the existence of a pair cascade automatically places a lower limit on the optical depth in thermal plasma. Generally, pairs are expected to cool to nonrelativistic energies and to thermalize before annihilating. The optical depth in the thermal pair plasma depends on the compactness, $\tau \sim \ell^{1/2}$, and is larger than one for an efficient pair cascade. This thermal plasma is unavoidably mixed with the relativistic pairs responsible for the nonthermal spectrum. If the y-parameter of the thermal plasma is too large (i.e., temperature too large for a given optical depth), then Comptonization by the thermal plasma will alter the slope of the nonthermal spectrum. Note also that if the $\alpha \sim 0.7$ slope is due to thermal Comptonization of UV bump photons, then the geometry must be arranged so that only part of the UV bump flux passes through the Comptonizing region -- otherwise the bump will lose its identity due to scattering effects.

V. RADIO-LOUD VERSUS RADIO-QUIET AGNs

One of the venerable puzzles for AGN theorists is, why are some AGNs radio loud, while others are radio quiet? Flat-spectrum radio sources are produced on scales smaller than a parsec, so the answer would seem to lie in the nature of the flow close to the central black hole. Even more puzzling is the apparent correlation between radio loudness and Hubble type. Virtually all radio galaxies are ellipticals, while Seyfert galaxies are radio-quiet spirals. It is not yet known whether radio-loud and radio-quiet quasars fit this pattern, but there is anecdotal evidence suggesting that they do. How could the central engine of an AGN know what the Hubble type of the host galaxy is?

Sparke and Shu (1980) were the first to suggest that the large-scale gas dynamics could affect the formation of extended radio structure. Since gas accreting through an elliptical galaxy would have relatively low angular momentum (compared to the Keplerian value) it would tend to form a narrow funnel capable of confining and collimating a jet. However, in order to explain the small-scale radio structure one would have to believe that this behavior extends deep into the potential well of the black hole. Rees et al. (1982) broached this possibility by suggesting that two distinct thermal states of accretion were possible at a given accretion rate: a thin accretion disk or a hot "ion torus." The mode of accretion chosen could depend on boundary conditions far from the black hole (Begelman 1986). Thus, in a spiral galaxy, material is fed into the central engine in a thin disk, so it remains a thin disk, whereas in an elliptical, the inflow is presumably hotter and more quasispherical (e.g., a cooling flow), hence the bulk of the flow remains hot all the way into the center.

How would these different flow patterns show up in the AGN spectrum? Surely the two types of flow pattern are not mutually exclusive, in the sense that the thin disk presumably forms a hot corona where much of the nonthermal spectrum formation takes place, while the ion torus may develop cool filaments or a thin disk component as a result of thermal instabilities. However, we might expect the ion torus to be characterized by a more extensive nonthermal corona, perhaps with accretion energy being dissipated over a large range of radii. If we consider the spherical superposition of self-absorbed synchrotron sources with $L(>R) \propto R^{-1}$ (i.e., with a fixed fraction of the accretion energy going into nonthermal radiation at each R), then we obtain the spectrum $F_\nu \propto \nu^{1/6}$ for uniform B, and $F_\nu \propto \nu^{1/28}$ for $B \propto R^{-5/4}$ (the expected scaling for a spherical accretion flow). Thus, flat radio-submillimeter spectra are produced naturally by a quasispherical inflow with distributed non-thermal energy deposition. Alternatively, the self-absorbed sources could be produced in a collimated outflow (jet), the formation of which is associated with the existence of a quasispherical accretion flow.

What about connections with other spectral bands? Wilkes and Elvis (1987) claim that the presence of an "extra" flat component of the X-ray spectrum is correlated with radio loudness. They suggest that the X-rays are produced by the same relativisitic electrons which produce the radio emission, e.g., by a synchrotron-self-Compton process. One might test this hypothesis by looking for correlated radio/X-ray variability. The X-rays could also be produced by an independent mechanism which happens to operate in the extended inflow. For example, a pair cascade operating over a range of radii might produce a synchrotron-dominated $\alpha \sim 1$ spectrum in the inner regions, and a flatter, Compton-dominated spectrum further out.

VI. A NEW MECHANISM FOR THE BLAZAR CONTINUUM

Lurking behind my cheerful recital of physical mechanisms which may be operating in AGNs is the suspicion that an understanding of the spatial stratification and geometric structure of these objects is crucial to explaining their spectra. Of course, we are dealing with angular scales that are far too small for direct imaging in the foreseeable future. In describing "quiescent" AGNs, geometric effects are often disregarded, simply because we have little evidence of what they are. However, in the cases of blazars the role of beaming by relativistic jets is usually regarded as they key to understanding the continuum radiation (Blandford and Rees 1978, Blandford and Königl 1979). Because it is highly polarized, the optical and IR emission is usually ascribed to synchrotron radiation. Measurements of spectral curvature and complex time and wavelength-dependence of the flux and polarization have led researchers to propose increasingly complex models including multiple radiating components and spatial stratification within the jet. Unfortunately, there has been little evidence to tie these models to the physics of either the accretion process or the jet formation mechanism.

Marek Sikora and I (1987) have proposed a new mechanism for producing beamed, polarized continuum from a relativistic jet, which depends strongly on the jet formation process and the underlying accretion flow. If jets are really accelerated to relativistic speeds within a few gravitational radii of the black hole, then they must traverse the background radiation field produced by the accretion flow. In a frame of reference which is stationary with respect to the black hole, this radiation may be roughly isotropic and weakly polarized. Viewed from the relativistic jet frame, however, the ambient radiation will be concentrated into a narrow cone about the apex of bulk particle motion. Therefore, after scattering by cold electrons in the jet, it will be highly polarized, and in the observer frame, strongly beamed. We have performed detailed calculations to illustrate the angular distributions of polarization and intensity of radiation scattered by jets having a variety of velocity and density profiles. As a result of relativistic beaming, the observed quantities are sensitive to small asymmetries and fluctuations in jet properties; this sensitivity could give rise to the high degree of variability that is observed. We have also found that the mean direction of polarization can serve as a diagnostic of jet structure, if the mean jet axis is known (e.g., from observations of a VLBI jet). "Pencil beam" jets, i.e., jets with velocity peaking on the axis of symmetry, always produce perpendicular polarization. On the other hand, "hollow cone" jets, i.e., jets with velocity peaking off-axis, will generally be observed to yield polarization parallel to the projected jet axis. A number of observational studies (Ulrich 1984, Rusk and Seaquist 1985, Smith et al. 1986, Brindle et al. 1986) indicate that parallel polarization is favored on average, suggesting that the hollow cone description is more apt. Such a configuration might arise naturally if jets originate as centrifugally-driven hydromagnetic winds (Phinney 1983, 1987).

The spectrum predicted in this model reflects the spectrum of the underlying accretion flow. Thus, it is not surprising that the mean spectral slope is similar to that in non-blazar AGNs (Malkan 1984). Curvature in the spectrum could arise from Doppler-boosting (and broadening) of the far-IR turnover, and as a result of uncertain geometric details in the jet structure and ambient radiation fields. The latter could also account for some of the observed wavelength-dependence of polarization. Doppler boosting could also "drown out" the UV bump, which is not prominent in blazars. In principle the drag associated with the inverse Compton scattering could be the most important factor in determining the terminal speed of the jet, and the scattered radiation flux could exceed the kinetic energy flux in the jet.

This is not to say that intrinsic radiation mechanisms (such as synchrotron and synchrotron-self-Compton) do not also make important or perhaps dominant contributions to blazar spectra. However, inverse Compton scattering of the ambient radiation <u>must</u> occur if the jets reach relativistic speeds (exceeding the average random speed of an electron) close to the black hole. I am encouraged by this new connection between the accretion flow and the jets, because it suggests that we may yet find ways to unravel the geometric complexities of AGN central engines by studying their continuum spectra.

Much of the original work described in this article is the result of my extensive collaboration with Marek Sikora over the past two years. This work was supported in part by NSF Grant AST 83-51997 (Presidential Young Investigator Award), NASA Grant NAGW-766 (NASA Astrophysical Theory Program), and grants from the Alfred P. Sloan Foundation, Rockwell International Corporation, and Ball Aerospace Systems Division.

REFERENCES

Band, D. and Grindlay, J. 1986, Ap. J., 308, 576.

Begelman, M. C. 1985, in Astrophysics of Active Galaxies and Quasi-Stellar Objects, ed. J. S. Miller (Mill Valley, CA; University Science Books), p. 411.

Begelman, M. C. 1986, Nature, 322, 614.

Begelman, M. C. and Sikora, M. 1987, Ap. J., 322, 650.

Blandford, R. D. and Eichler, D. 1987, Physics Reports, 154, 1.

Blandford, R. D. and Königl 1979, Ap. J., 232, 34.

Blandford, R. D. and Rees, M. J. 1978, in Pittsburgh Conference on BL Lac Objects, ed. A. M. Wolfe (Pittsburgh; University of Pittsburgh), p. 328.

Bonometto, S. and Rees, M. J. 1971, M. N. R. A. S., 152, 21.

Brindle, C., Hough, J. H., Bailey, J. A., Axon, D. J., and Hyland, A. R. 1986, M. N. R. A. S., 221, 739.

Carleton, N. P., Elvis, M., Fabbiano, G., Willner, S. P., Lawrence, A., and Ward, M. 1987, Ap. J., 318, 595.

Edelson, R. A. 1986, Ap. J. (Letters), 309, L69.

Edelson, R. A. 1987, Ap. J., in press.

Edelson, R. A. and Malkan, M. A. 1986, Ap. J., 308, 59.

Edelson, R. A., Malkan, M. A., and Rieke, G. H. 1987, Ap. J., 321, 233.

Fabian, A. C., Blandford, R. D., Guilbert, P. W., Phinney, E. S., and Cuellar, L. 1986, M. N. R. A. S., 221, 931.

Ghisellini, G. 1987, M. N. R. A. S., 224, 1.

Guilbert, P. W., Fabian, A. C., and Rees, M. J. 1983, M. N. R. A. S., 205, 593.

Guilbert, P. W. and Rees, M. J. 1987, M. N. R. A. S., submitted.

Ipser, J. R. and Price, R. H. 1983, Ap. J., 267, 371.

Kazanas, D. and Ellison, D. C. 1986, Ap. J., 304, 178.

Kriss, G. A. 1988, Ap. J., in press.

Malkan, M. A. 1983, Ap. J., 268, 582.

Malkan, M. A. 1984, in X-Ray and UV Emission from Active Galactic Nuclei, ed. W. Brinkmann and J. Trumper (Garching, West Germany; MPE Rept. 184), p. 104.

Malkan, M. A. and Sargent, W. L. W. 1982, Ap. J., 254, 22.

Maraschi. L., Roasio, R., and Treves, A. 1982, Ap. J., 253, 312.

Mushotzky, R. F. 1984, Adv. Space Res., 3, 157.

Phinney, E. S. 1983, Ph. D. thesis, University of Cambridge.

Phinney, E. S. 1987, in Supermassive Black Holes, ed. M. Kafatos (in press).

Rees, M. J. 1984, Ann. Rev. Astr. Ap., 22, 471.

Rees, M. J., Begelman, M. C., Blandford, R. D., and Phinney, E. S. 1982, Nature, 295, 17.

Rusk, R. and Seaquist, E. R. 1985, A. J., 90, 30.

Sikora, M., Kirk, J. G., Begelman, M. C., and Schneider, P. 1987, Ap. J. (Letters), 320, L81.

Smith, P. S., Balonek, T. J., Heckert, P. A., and Elston, R. 1986, Ap. J., 305, 484.

Sparke, L. S. and Shu, F. H. 1980, Ap. J. (Letters), 241, L65.

Svennson, R. 1987, M. N. R. A. S., 227, 403.

Ulrich, M.-H. 1984, in IAU Symposium 110, VLBI and Compact Radio Sources, ed. R. Fanti, K. Kellermann, and G. Setti (Dordrecht; Reidel), p. 73.

Ward, M., Elvis, M., Fabbiano, G., Carleton, N. P., Willner, S. P., and Lawrence, A. 1987 Ap. J., 315, 74.

Wilkes, B. J. and Elvis, M. 1987, Ap. J., in press.

Zdziarski, A. A. 1986, Ap. J., 305, 45.

Zdziarski, A. A. and Lightman, A. P. 1985, Ap. J. (Letters), 294, L79.

QUESTIONS:

P. J. Wiita:

What model do you prefer for producing hollow cone jets starting very close to the black hole?

M. C. Begelman:

If you believe Phinney's black hole flywheel model, the jet starts out as a centrifugally driven wind, and is later collimated by Maxwell stresses (a group in Japan have recently done 2-D numerical simulations demonstrating this). Therefore it would seem natural for most of the jet momentum to be concentrated off-axis, in a hollow cone.

S. P. Reynolds:

In your model of blazar jets, they predominantly Compton scatter photons from the ambient thermal radiation field rather than jet synchrotron photons, demanding that in the jet, the synchrotron radiation energy density be less than that in ambient photons. Does this provide a useful constraint? This also means that any synchrotron radiation from the jet can never be observable.

M. C. Begelman:

Our model does not preclude the jet from putting out a comparable or larger flux of synchrotron radiation and Comptonized synchrotron radiation than scattered ambient radiation. In order for our mechanism to contribute significantly to the observed spectrum, the intensity of scattered ambient radiation at a given frequency (and in a given direction) must be competitive with the intensity in other forms of radiation. They only physical requirement is that the bulk Lorentz factor exceed the random Lorentz factor for an average electron, so that Comptonization by the bulk flow makes sense. Even if the jet were a strong source of self-Comptonized synchrotron radiation, it is likely that only a small fraction of the electrons would have large Lorentz factors at any one time.

C. D. Dermer:

Explain how your model accounts for the absence of time variability observed in the infrared continua of these sources. Also, the observations show that there is roughly equal energy per decade in the rands $\sim 10^{12}$ - $\sim 10^{20}$ Hz. You assume that protons have injection spectra with equal energy per decade (or even an enhancement of energy toward higher energies). They then reradiate their energy at lower energies. Are you assuming the answer?

M. C. Begelman:

My model does not account for the absence of time variability in the infrared. The apparent steadiness of the IR flux is puzzling. In answer to your second question, I need to assume that the proton injection spectrum has most of the energy concentrated at the highest Lorentz factors. The exact form of the injection spectrum does not matter, since it is the pair cascade that creates the final electron energy distribution.

W. van Breugel:

You propose that radio power might be related to the local environment of the central engine, with spirals having perhaps a cold, flat accretion disk, and ellipticals having more spherical, dense coronae originating in hot cooling flows. However, powerful radio galaxies are often (30-60%) associated with very peculiar elliptical-like galaxies (Heckman et al. 1986, Ap. J.; Hutchings 1987, Ap. J.), suggesting tidal interactions and merging of parent (expectant?) radio galaxies with cold, gas-rich companions. Could it not be that the fueling mechanism, and amount of material, is the more important factor in triggering AGN activity, and that this might be more efficient in more massive (elliptical) galaxies because of their larger and deeper potential wells (and perhaps bigger central black holes as well)?

M. C. Begelman:

The differences between radio-loud and radio-quiet AGNs are manifested on scales smaller than a parsec. It is hard to understand how the depth of the galactic potential well could have much to do with this. You might want to argue that M_{BH} and $\dot{M}/\dot{M}_{EDD}$ differs systematically with Hubble type (I used the latter argument in my 1986 Nature paper), but bear in mind that the luminosity functions of radio-loud and radio-quiet AGNs are quite broad. I am simply suggesting that differences in the outer boundary conditions of the accretion flow might affect the properties of the flow close to the black hole. The argument does not have to be cast in terms of thin accretion disks vs. laminar cooling flows.

R. Antonucci:

To clarify Wil van Breugel's comment that radio galaxies don't have normal elliptical hosts, everyone agrees that the hosts of the optically-dull Fanaroff-Riley Class I radio galaxies are basically normal ellipticals.

MODEL FOR THE CONTINUUM EMISSION OF ACTIVE GALACTIC NUCLEI

Charles D. Dermer
Lawrence Livermore National Laboratory
P. O. Box 808, L-297, Livermore, CA 94550 USA

I. Introduction. We propose a hot proton model for the production of the continuum emission of AGN. A dissipative process operating in the vicinity of collapsed objects is assumed to heat protons to temperatures $T_p \sim 100$ MeV, near the virial value. Nonthermal pion-decay electrons and positrons and proton-heated thermal electrons and pairs radiate a large fraction of the energy originally stored in the protons. We calculate time-averaged emergent spectra following impulsive proton injection and discuss observable features of the model. This work advances previous treatments of relativistic two-temperature plasmas [1,2,3] by including effects of both pion production and magnetic fields, and differs from earlier thermal Comptonization models [1,4] by directly specifying the electron-heating mechanism and the origin of the soft photons.

II. Model and Analysis. In the hot proton model, protons produce pions in proton-proton collisions and transfer energy to a thermal electron-pair component (the "pair plasma") through Coulomb energy exchange. Neutral pions decay into γ rays with photon energy $\varepsilon \sim 70$ MeV, and charged pions decay into nonthermal e^+ and e^- which produce soft synchrotron emission, Comptonize photons of the radiation field, and become part of the pair plasma if they survive annihilation in flight. The pair plasma is assumed to be in pair and energy balance. The most important heating mechanism of the pair plasma is Coulomb energy exchange with the protons. The pair plasma loses energy through bremsstrahlung and annihilation losses, Compton cooling, and the emission of soft cyclo-synchrotron photons. Thermal Comptonization by the pair plasma is primarily responsible for the formation of the continuum between the low energy absorption cutoff and $\varepsilon \approx T_e$, where T_e is the temperature of the pair plasma.

We assume that protons are impulsively heated to $T_p = 100$ MeV, and calculate the electron temperature, pair density and time-averaged emission spectrum of the system [cf. 3] in a homogeneous, spherically-symmetric framework. For the rates and luminosities of the various channels of pion production in thermal proton collisions, we fit the exact calculations [5]. For the Coulomb energy exchange rate between thermal protons and the pair plasma, we use equation (11) of [5], valid for nonrelativistic protons and electrons at arbitrary temperatures. Bremsstrahlung and annihilation loss rates for the pair plasma are taken from [6]. The thermal synchrotron absorption cutoff energy ε_{co} is determined by solving the implicit equation (7) of [7] at unity optical depth. The thermal synchrotron energy loss rate is approximated by integrating the Rayleigh-Jeans spectrum up to $\varepsilon = \varepsilon_{co}$.

The Compton cooling rate of the pair plasma is given by

$$L_{Comp}\left[\frac{\text{MeV}}{\text{cm}^3\text{-s}}\right] = (1+2z)\, n_p\, c \int_0^\infty d\varepsilon\, \varepsilon\, n(\varepsilon)\,[A(\varepsilon,\theta) - 1]\, \sigma_C(\varepsilon,\theta) \ , \qquad (1)$$

where z is the pair fraction, $n(\varepsilon)$ is the differential photon density, n_p is the proton density, $\theta = T_e/m_e c^2$, A is the energy amplification factor, and σ_C is the thermal-averaged Compton cross section (see [2,8] for graphs of A and σ_C). We approximate $A \cong A_o/(1+A_o x/\theta)$ and $\sigma_C \cong \sigma_{KN}(\varepsilon)/(1+0.67\theta x^{1/2})$ at $x \le \theta$, where x =

ε/m_ec^2, $A_o = 1+4\theta+16\theta^2$, and σ_{KN} is the exact Compton cross section. The integrand is set equal to zero at $\varepsilon \geq T_e$, as Compton heating is negligible compared to Coulomb heating.

For the calculation of the spectra, we approximate thermal bremsstrahlung by the expression $dN/dVdtd\varepsilon \propto \varepsilon^{-1.1} \exp(-\varepsilon/T_e)$, normalized to the luminosities [6] of e-p, e-e, and e^+-e^- thermal bremsstrahlung. Thermal annihlation radiation is approximated by a Gaussian function normalized to the thermal annihilation rate with peak energy $\varepsilon_p = 0.511$ MeV + $3T_e/4$ and FWHM = $1.18\sqrt{T_e}$, $\varepsilon \leq 0.12$ MeV; FWHM = $2.74\, T_e^{0.9}$, $\varepsilon > 0.12$ MeV [9]. These give rough approximations to the exact thermal bremsstrahlung [5] and thermal annihilation [10,11] spectra. Thermal cyclo-synchrotron radiation is approximated by a Rayleigh-Jeans spectrum at $\varepsilon \leq \varepsilon_{co}$. At $\varepsilon > \varepsilon_{co}$, the soft nonthermal synchrotron spectrum from pion-decay e^+ and e^- is approximated by the expression $dN/dVdtd\varepsilon \propto \varepsilon^{-1.5} \exp(-\varepsilon/\varepsilon^*)$ expected from Compton-synchrotron losses, normalized to the nonthermal synchrotron luminosity. We take $\varepsilon^* = 1.2\times10^{-10}$ B MeV, where B is the magnetic field in Gauss, corresponding to a typical nonthermal e^+ or e^- Lorentz factor ~ 100. For nonthermal Comptonization, we numerically solve a simple integral given through a delta-function approximation. Pair absorption and photon-photon pair production are numerically calculated on the basis of the exact formalism [12]. Pair cascade effects are neglected. Thermal Comptonization follows the approximate treatment of [8]. All Compton-scattered photons with $\varepsilon \geq T_e$ are assumed to become part of the Wien component. Comparison with exact Monte-Carlo simulations [13,14] show that this method is accurate at $\varepsilon \ll T_e$, but somewhat inaccurate at $\varepsilon \approx T_e$. Its advantage is that it is numerically very fast.

III. Discussion and Speculation. In Figure 1 we show numerical calculations of spectra averaged over 30 Thomson times t_p ($= 1/n_p\sigma_Tc$), beginning with the initial injection of protons at 100 MeV. During this period ~80% of the original proton energy is radiated, giving average bolometric luminosities $\sim 8\times10^{43}$ erg s^{-1}. The spectra display a roughly power law behavior between $\varepsilon \approx \varepsilon_{co}$ and $\varepsilon \approx T_e$. Some structure is evident in the continuum of the B = 10 G case partly due to the discrete nature of the approximations and partly due to the well-known behavior of relativistic thermal Comptonization [15]. Average pair temperatures for B = 10, 100, and 1000 G are 300, 200, and 150 keV, respectively. The simulations show evidence of a pair annihilation feature. A pion signature is present at $\varepsilon \gg T_e$ due to nonthermal Comptonization and neutral pion decay.

This model may account for the X-ray continuum of AGN but suggests a thermal rollover at $\varepsilon \approx$ few hundred keV. This does not conflict with data of NGC 4151 during certain periods of observation, but may conflict with data of MCG 8-11-11 and Cen A, which show a turnover at higher energies (the location of the break in 3C 273 is uncertain) . The strongest support for this model would be the discovery that most Sy 1s exhibit a thermal rollover at $\varepsilon \approx$ few hundred keV. Indeed, a typical Sy spectrum softer, on average, than has been observed in NGC 4151 or MCG 8-11-11 is suggested by observations of the diffuse γ-ray background [16]. The model predicts correlated X- and gamma-ray variability, but the infrared excess observed in the canonical Sy 1 is evidently of different origin as is, of course, the "blue bump." Although the spectrum is self-absorbed in the radio regime, there may be some associated radio emission if the energetic pion-decay e^+ and e^- leak out of the system to a region of lower field strength (see below). We now speculate on various features of this model which warrant further study.

(a) X-ray pulsations. Our simulations show that the pair plasma temperature does not always approach a monotonic value, but may exhibit time-variable behavior. The physical origin of this behavior is clear: The electron temperature initially rises in response to the hot protons because the photon energy density u_{ph} [and therefore the Compton cooling rate; see equation (1)] is small at early times. The amplification of the soft photon field by the warm pair plasma increases u_{ph} and the Compton cooling rate, thus causing the pair plasma to cool. This lower electron temperature leads to a smaller amplification factor A and therefore a declining u_{ph}, permitting the electron temperature once again to rise. Under certain choices of parameters, this cycle may repeat and produce X-ray pulsations. Whether this is an artifact of the spherically-symmetric approach requires further study.

(b) Radio-to-X-ray intensity correlation. Exact calculations of pion production [5] in thermal proton plasmas show that the nonthermal electron luminosity increases by over three orders of magnitude when the proton temperature increases from 20 to 100 MeV. If the initial proton temperatures in the central engines of AGN were distributed in this range, and if some fraction of the pion-decay e^+ and e^- escapes from the central region to produce radio emission, then a small change in the X-ray intensity would be reflected in a large change in the radio intensity.

(c) Variation of the X-ray spectral index. The continuum spectral index α deduced from studies of thermal Comptonization of a soft photon source is a function of $\tau_p(1+2z)$ and T_e [e.g., 14]. Suppose we make the *ansatz* that the magnetic field energy density is in equipartition with the gas pressure. The equipartition magnetic field strenth $B_{eq} \approx 800\,(\tau_p T_p/R_{14})^{1/2}$ G, where $R_{14} = R/10^{14}$ cm. Preliminary calculations show that α decreases from ~1 to ~0.5 as τ_p increases from 0.1 to 1, nearly independently of R and $T_p^{init} > 10$ MeV (the compactness $\propto \tau_p^2$ and is independent of R in this model). Further work is required to see if this leads to a range of α in accord with observations (see [17] and Mushotzky; this volume).

Acknowledgements. I thank David Band, Edison Liang and Jon Arons for useful discussions.

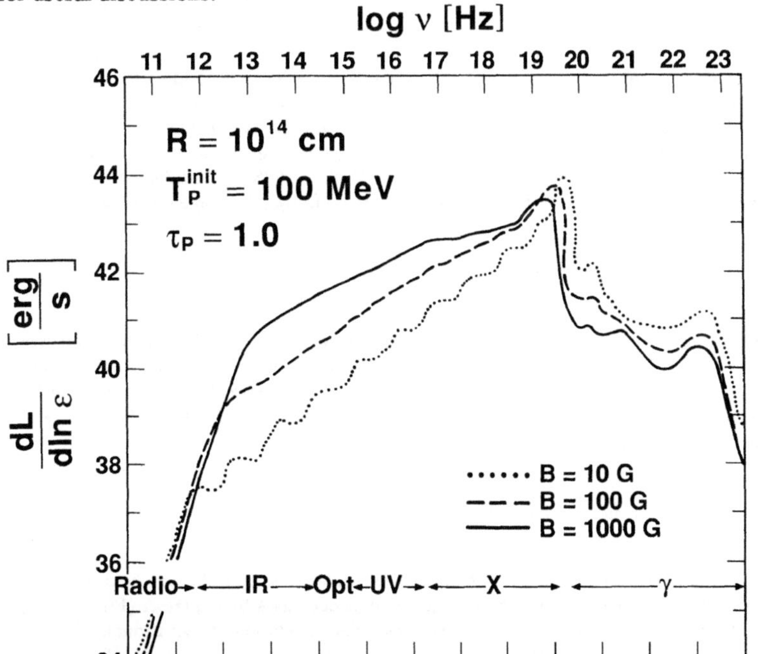

Figure 1. Numerical calculations of emitted spectra for various values of the magnetic field B. The spectra are averaged over 30 Thomson times following the initial injection of protons with temperature $T_p^{init} = 100$ MeV and density 1.5×10^{10} cm^{-3} into a region of radius $R = 10^{14}$ cm (proton optical depth $\tau_p = n_p\sigma_T R = 1$).

References.

[1] Shapiro, S. et al. 1976, ApJ., **204**, 187.
[2] Sikora, M. and Zbyszewska, M. 1985, Acta Astr., **36**, 255.
[3] Guilbert, P. W. and Stepney, S. 1985, MNRAS, **212**, 523.
[4] Katz, J. I. 1976, Ap. J., **206**, 910.
[5] Dermer, C. D. 1986, Ap. J., **307**, 47.
[6] Svensson, R. 1982, Ap. J., **258**, 321.
[7] Kusunose, M. and Takahara, F. 1985, Prog. Theor. Phys., **73**, 41.
[8] Dermer, C. D. 1984, Ph. D. Thesis, UCSD (NASA TM 86154).
[9] Ramaty, R. and Meszaros, P. 1981, Ap. J., **250**, 384.
[10] Svensson, R. 1983, Ap. J., **270**, 300.
[11] Dermer, C. D. 1984, Ap. J., **280**, 328.
[12] Gould, R. J. and Schreder, G. P. 1967, Phys. Rev., **155**, 1404.
[13] Zdziarski, A. A. 1984, Ap. J., **283**, 842.
[14] Zdziarski, A. A. 1986, Ap. J., **303**, 94.
[15] Pozdnyakov, L. A. et al. 1977, Sov. Astr., **21**, 708.
[16] Bassani, L. and Dean, A. J. 1983, Space Sci. Rev., **35**, 367.
[17] Lightman, A. P. and Zdziarski, A. A. 1987, Ap. J., **319**, 643.

FITTING MULTI-WAVELENGTH CONTINUUM OF AGN WITH IMPROVED ACCRETION DISK MODELS

Wei-Hsin Sun (NASA/GSFC) and Matthew A. Malkan (UCLA)

ABSTRACT

We present fits to the UV-Optical-IR spectra of 60 quasars and active galactic nuclei using improved black hole accretion disk models. For both Kerr and Schwarzschild black holes, we calculated the effects of inclination, including Doppler boosting, gravitational focusing, and gravitational redshift, on the observed disk spectra. In the Kerr case, a simple inverse relation was found between the inferred black hole mass and the assumed inclination angle, while the inferred accretion rate ($M_\odot$/yr) remains constant. Disk spectra calculated with non-LTE stellar atmospheres instead of blackbodies do not reproduce the observations as well. Low-redshift Seyfert galaxies have relatively low accretion rates, only a few percent of their Eddington luminosities (L_{Edd}), while the most luminous quasars are accreting nearer their Eddington limits.

IMPROVED MODELS OF ACCRETION DISK SPECTRA

Thermal radiation from a geometrically thin and optically thick accretion disk surrounding a supermassive black hole is the most promising explanation for the strong optical and ultraviolet fluxes observed in many quasars and active galactic nuclei (Malkan 1983). Therefore we calculated the spectra of disk models, for which the effective surface temperature at each radius is proportional to $[E/\sigma]^{1/4}$ (proportional to $r^{-3/4}$), from energy conservation. The integrated spectrum was then derived by summing the locally emerging spectra from each radius. No information about the internal vertical disk structure or viscosity is required. We have recently (Sun and Malkan 1986,1987) included two important effects: 1) To account for the effects of electron scattering opacity, we described the locally emerging flux spectra with non-LTE stellar atmospheres, rather than blackbodies; 2) We included the effects of inclination, including Doppler boosting, gravitational focusing, and gravitational redshift, on the observed disk spectra. A total of 14 quasars were fitted with face-on Kerr disks (Sun and Malkan 1988). The inferred black hole mass falls roughly in a range between 10^8 and 10^9 $M_\odot$ with the luminosity ratio (L/L_{Edd}) between 10% and 50%.

MODEL FITTING RESULTS

The multi-wavelength spectra of a sample of 60 QSOs and AGNs were fitted with Kerr blackbody disks, for five assumed inclination angles (cos i = 1/0.75/0.5/0.25/0). Since the fits are very good in all cases, the uncertainties in the fitting parameters are less than 20%. Two typical examples are illustrated in Figure 1. A subsample of 24 QSOs and AGNs, selected to have the best measured spectra, were fitted with Schwarzschild blackbody disks, and with both Kerr and Schwarzschild stellar-atmospheres disks, for all five inclination angles.

i) Kerr Geometry with Blackbody Disks

A simple relation between inferred black hole mass and the assumed disk angles was found for the Kerr disk fits. The inferred mass doubles when the cosine of the assumed disk inclination angle decreases by a quarter (Fig. 2): Log $(M/M_{face-on}) = -1.2\times\cos i + 1.2$. Observing a disk at higher inclination Doppler-boosts and gravitationally focuses the harder radiation from the innermost radii, and consequently makes the apparent disk spectrum harder. Increasing the black hole mass acts in the opposite sense. Since the effective temperature at a given number of gravitational radii decreases with black hole mass ($T \sim M^{-1/4}$), the larger the mass, the softer the disk spectrum appears. We can therefore obtain equally good fits at different inclination angles with different masses, at a given total luminosity, i.e., a fixed accretion rate. This simple relation reduces the number of fitting parameters for an accretion disk from three (M, M, cosi) to two (M, M(cosi)).

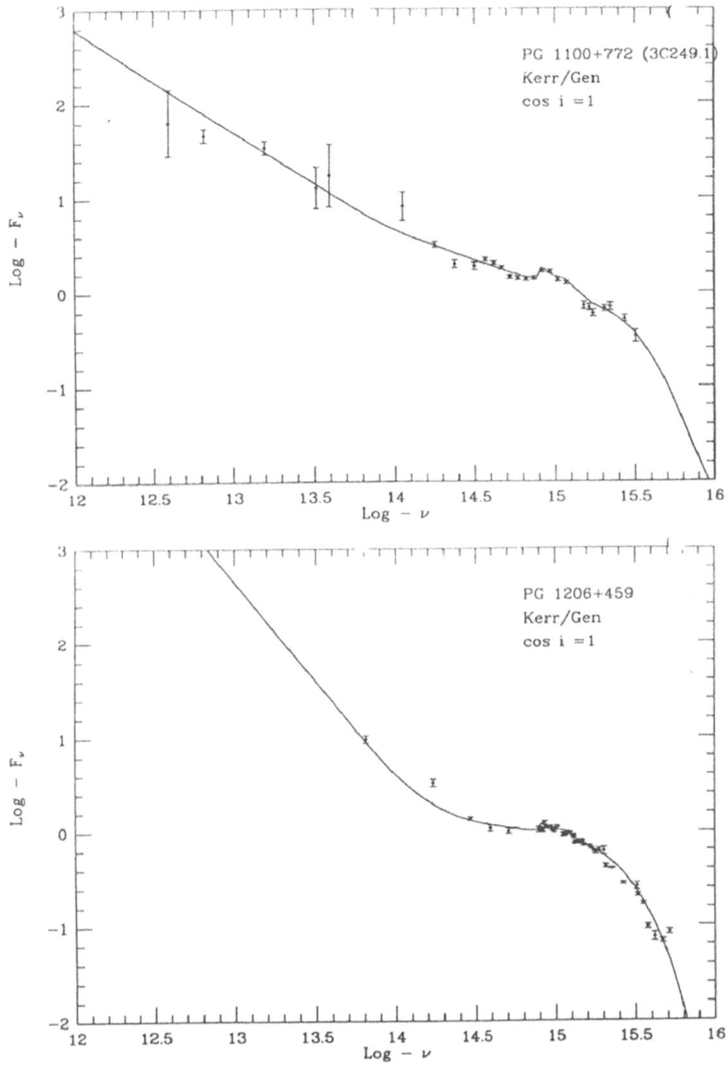

Fig. 1 Model fits of two AGN in f_ν *vs* ν.

The inferred accretion rate ($M_\odot$/yr) was found to be extremely well defined for each object, varying by 5-15% from face-on to edge-on fits (Fig. 3). The estimated black hole mass varies as the assumed inclination angle changes. We are therefore able to obtain an independent measurement of the accretion rate, which is the total energy generated from the disk, even though we cannot determine the mass and the disk inclination independently. Below we quote the face-on disk parameters for simplicity.

About three-quarters of the sample objects have fitted masses between 3×10^8 and $3\times10^9 M_\odot$, with a very weak positive dependence on the redshift. We caution that these masses are lower limits, because they were derived assuming face-on disks. If some disks are actually inclined, their black holes may be as massive as $10^{10} M_\odot$.

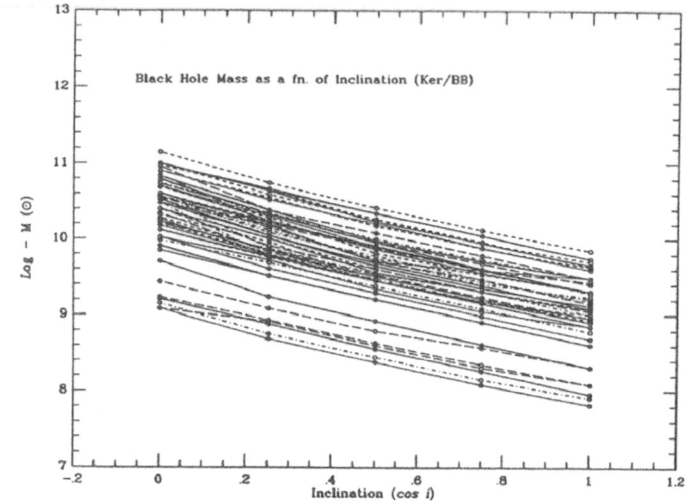

Fig. 2 The inferred black hole doubles each time the cosine of the inclination angle decreases by a quarter, in the Kerr case.

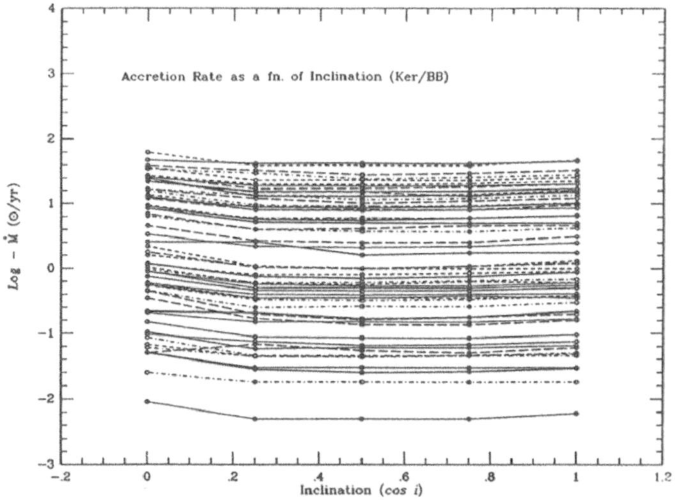

Fig. 3 The inferred accretion rate is very well-defined regardless of the inclination angle in the Kerr case.

ii) Schwarzschild Geometry with Blackbody Disks

As we fit a spectrum with Schwarzschild disks of increasing inclination angles, the inferred black hole mass and accretion rate both increase gradually. The ratio of luminoisity to the Eddington limit (L/L_{Edd}, which is proportional to $\dot{M}/M$) remains roughly constant, independent of assumed inclination (Fig. 4). As the disk tilts to higher inclinations, the observer sees some Doppler-boosted and gravitationally-focused hard radiation from the disk center (but not as strong as in the Kerr case) and, at the same time, loses the softer radiation from the outer disk due to the geometric inclination effect. Increasing the black hole mass properly will cancel the former effect (as in the Kerr case). However, to make up the lost energy due to the latter effect (which was negligible in the Kerr case) requires an increase in the accretion rate.

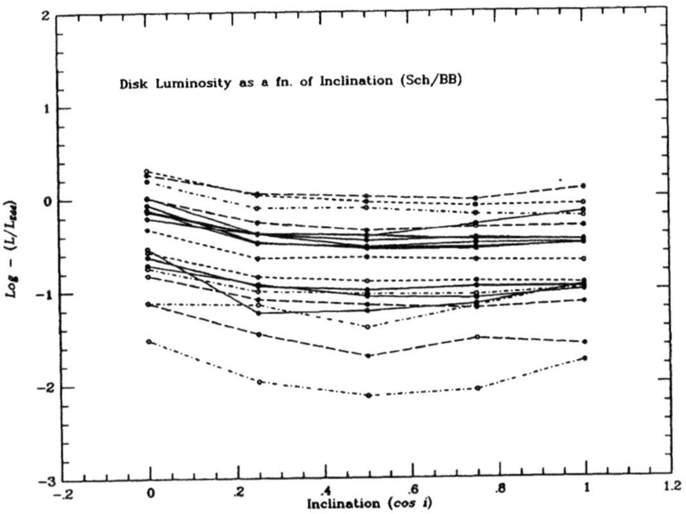

Fig. 4 L/L_{Edd} remains roughly constant, while inclination angle varies, in the Schwarzschild case.

The estimated accretion rate for a given quasar in the Schwarzschild case is 2.5 times higher than that in the Kerr metric. The inferred black hole mass in the Schwarzschild disk, however, is 2.5 times smaller. Thus Schwarzschild fits give a higher (approximately 6 times) M/M ratio. Since the efficiency of a Schwarzschild disk (0.057) is roughly 6 times smaller than that of a Kerr disk (0.32), the luminosity ratio (L/L_{Edd}) is about the same in both cases.

iii) Stellar-Atmosphere Disks

The stellar-atmosphere disk spectra do not fit most of the quasar spectra as well as the blackbody disks. The largest discrepancy appears at the Lyman edge, where the model spectra have a Lyman limit absorption edge with strength depending on the inclination (strongest in the face-on case). Most observed spectra do not show such a strong Lyman edge. However, this need not imply that modeling the disks with atmospheres is inappropriate. The emergent fluxes around the Lyman limit are known to be very sensitive to the assumed surface gravity. The strength of the Lyman limit absorption in the disk spectrum may have been overestimated by using stellar-atmosphere models with unrealistically high surface gravities. Further uncertainty might have been introduced by the necessity of using a mixture of models from different sources whose predictions are not always consistent with one another. Finally, the assumption of a stellar atmosphere, that all energy is produced at infinite optical depth, may not apply to accretion disks.

CONCLUSION

The most important implication of the fitting results is that the low-redshift Seyfert galaxies are accreting at a significantly lower rate (a few percent of their Eddington limits), while intermediate- to high-redshift quasars ($z \geq 1$) are accreting at rates close to their Eddington limits (Fig. 5). This may suggest that Seyfert galaxies were once luminous quasars which currently have central engines nearly starved for fuel. Is this suggesting some kind of luminosity evolution? Naively pure luminosity evolution would predict that low-redshift AGNs would have black holes at least as massive as those in high-redshift quasars. Our fitting results do show that the inferred black hole masses for both Seyferts and quasars are similar, to within an order of magnitude. However, almost all the Seyfert galaxies have slightly smaller black holes than the quasars (Fig. 5). This positive dependence of hole mass on redshift is significant due to the sample size. This might not be fatal to the pure luminosity evolution hypothesis if we are not actually comparing the same kind of objects. Perhaps those predecessors of today's Seyfert galaxies were much fainter (with smaller black holes) than those rare, extremely luminous quasars we actually observe at high redshifts. Answering this question will require Space Telescope observations of the ultraviolet spectra of faint high-redshift quasars, and more extensive disk-fitting of luminous Seyfert galaxies, which we are pursuing.

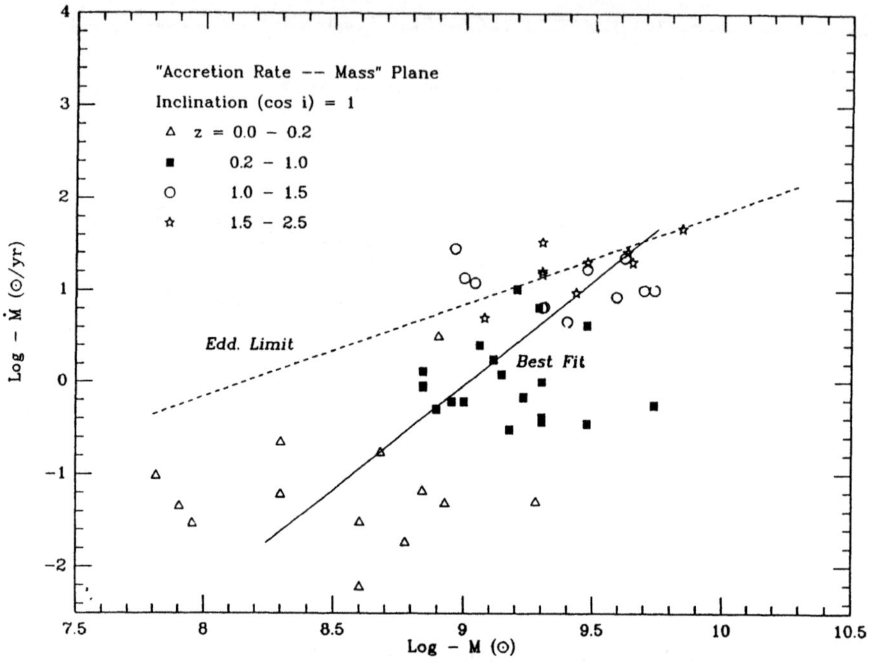

Fig. 5 Seyfert galaxies are accreting at very low accretion rate while the luminous quasars are accreting at rates near their Eddington limits.

REFERENCE

1. Malkan, M.A., 1983, Ap.J. 268, 582.

2. Sun, W.-H. and Malkan, M.A., in "New Insights in Astrophysics: 8 years of UV Astronomy with IUE", 1986, SP 263, P.641, ESA Pub., Noordwijk, The Netherlands.

3. Sun, W.-H. and Malkan, M.A., in "Astrophysical Jets and Their Engines", 1987, p.125, ed. Kundt, W., Reidel.

4. Sun, W.-H. and Malkan, M.A., in "Supermassive Blackhole Accretion", 1988, in press, ed. Kafatos, M., Reidel.

ACCRETION-DISK MODELLING

OF THE OPTICAL-UV SPECTRUM OF QUASARS

Amri Wandel and Vahé Petrosian
Center of Space Science and Astrophysics
Stanford University, ERL, Stanford, CA 94305 USA

Abstract

The thin accretion-disk spectrum, including the effect of electron scattering , is calculated for a grid of the accretion parameters (black hole mass and accretion rate). It is shown that the luminosity and spectral slope in the UV and optical bands are uniquely determined by these parameters and depend only weakly on the viscosity parameter. The model calculations are compared with data from IUE and ground based observations for a sample of ~70 quasars and Seyfert 1 nuclei, with UV spectra and fluxes in the 912-1450A band. It is found that the black hole mass is in the range $10^8 - 10^{9.5} M_\odot$ for the quasars, and $10^{7.5} - 10^{8.5} M_\odot$ for the Seyferts in the sample, and that the ratio of UV to Eddington luminosities varies from ~ 1 for luminous, high redshift quasars, to < 0.01 for low luminosity objects. The optical (4200-7500A) band gives similar results. The correlation between the spectral slope and luminosity of quasars in the UV and optical bands can be explained by evolution along curves of constant black hole mass and decreasing accretion rate. The relation between the spectrum and the accretion parameters can be used to constrain the cosmological evolution of the objects.

1. Introduction

Although the interpretation of the "Big UV Bump" in quasar specctra as a signature of an accretion disk is not unique (see e.g. O'Dell, Scott and Stein 1986 for a different explanation), it is attractive, because it may enable a determination of the black hole mass and the accretion rate from the observed continuum spectrum. Previous workers have found that AGN UV spectra turn over at a surprisingly uniform frequency of $\sim$ 1200A, corresponding to a maximum disk temperature of $\sim$ 30000K (Malkan 1983). If the turnover represents the Wien cutoff, one could determine the black hole mass, but this interpretation disagrees with more recent IUE observations of high redshift quasars which show strong emission down to at least 600A (Bechtold *et. al.* 1984). Czerny and Elvis (1987) show that the decrease in the UV observed in quasar spectra must therefore not be interpreted as a Wien cutoff; rather, it reflects the domain where electron scattering becomes important. We try to relate the parameters of the accretion disk model to the observed properties of the continuum spectrum (luminosity and spectral index) in the optical and UV bands. As a byproduct we infer possible paths of a cosmological evolution of the quasars and active galactic nuclei.

2. The Thin Disk Spectrum

The disk structure is calculated following the formulae of Shakura and Sunyaev (1973). The disc structure and the emitted radiation spectrum are determined by the black hole mass M, the accretion rate $\dot{M}$, the viscosity parameter α and the opacity. The surface temperature T_s at a radial distance R from the black hole is found from the condition

$$\pi \int_0^\infty I_\nu(T_s) d\nu = \frac{3GM\dot{M}}{8\pi R^3} \phi(R) \tag{1}$$

where I_ν is the emitted radiation flux. The RHS is the local energy release in the disk, $\phi(R) = 1 - (R/6R_s)^{-1/2}$ is the modification at the inner disk boundary, and R_s is the Schwarzschild radius.

While in the outer part of the disk a black body spectrum is a good approximation and $I_\nu \approx B_\nu$ (B_ν being the Planck function), in the inner part the spectrum is modified by electron scattering , becomming optically thin and Comptonization. It can be shown (Wandel and Petrosian 1987) that the spectrum in the UV band considered here (1000-1500A) will originate mainly from the regions of the disk which are radiation-pressure dominated, the opacity is dominated by electron scattering , but the disk is effectively optically thick. Also, the modification due to Compton scattering is shown to be small.

We calculate the spectrum by integrating numerically the local spectra over the whole disk, taking into account the effects of the optically thin region, Comptonization and electron scattering . The surface temperature is self consistently determined from eq. (1), with a spectrum of the form

$$I_\nu = \left(\frac{2(1-A_\nu)(1-e^{-\tau_*})}{1+(1+\kappa_{es}/\kappa_a)^{1/2}} + C \right) B_\nu, \tag{2}$$

where τ_* is the effectve optical depth of the disk, κ_a and κ_{es} are the opacities due to absorption and electron scattering respectively, and A_ν and C are determined by the Comptonization. Fig. (1) shows the dependence of the spectrum on the viscosity parameter α and on the dimensionless accretion rate, defined as

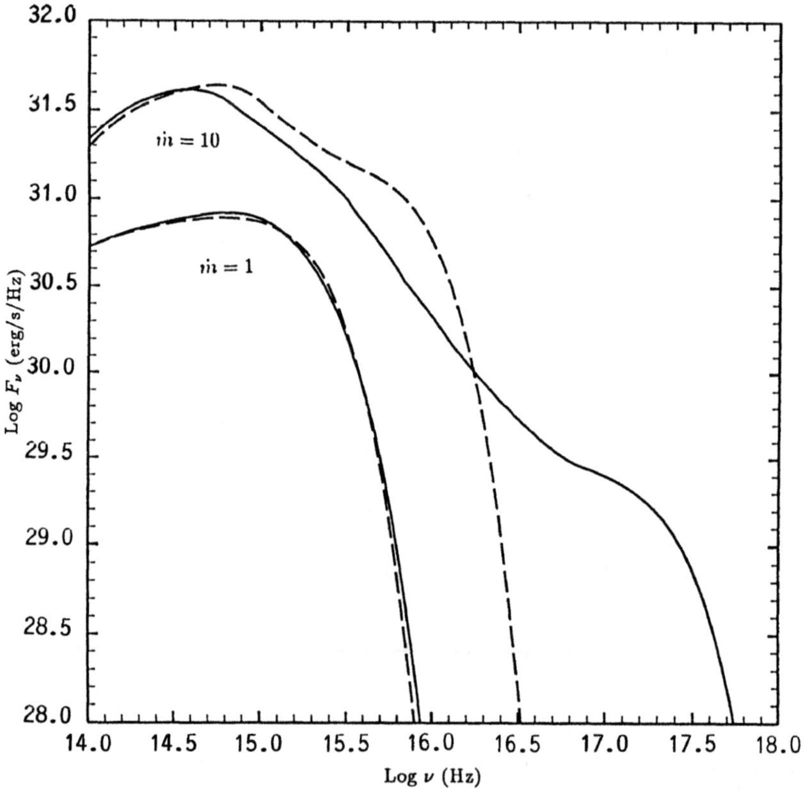

Fig 1. The calculated spectrum of a disk around a black hole of $M = 10^9 M_\odot$ for two accretion rates ($\dot{m} = 1, 10$). Continuous curves are for a viscosity paremeter of $\alpha = 1$ and dashed curves are for $\alpha = 0.1$.

$$\dot{m} = \dot{M}c^2/L_{Edd} = 4.4\ \dot{M}/(M_\odot \mathrm{yr}^{-1})(M/10^8 M_\odot)^{-1}. \tag{3}$$

The spectrum is sensitive to the viscosity parameter only at high accretion rates, and also then, only at frequencies higher than the UV band (see fig. 1).

3. Comparison with the Data

We concentrate on the UV (1000-1500A) band which has many advantages: there exist a fairly large body of data in this band (IUE and high redshift quasars), it is relatively free from obvious contamination by other components (which may dominate the spectrum in other wavelength ranges) such as the infrared-optical power-law, emission from dust (Edelson and Malkan 1986), starlight and emission lines (including the "small" 3000A bump).

We have calculated the ultraviolet luminosity L_{UV} and spectral index α_{UV} for a grid of face-on disk models, spanning the $M - \dot{M} - \alpha$ parameter space. The results are only weakly dependent on the viscosity parameter α or the opacity coefficients; Fig 2. shows the $\alpha_{UV} - L_{UV}$ relation for models with constant mass (solid lines) and constant accretion rate parameter (dashed lines). Also shown in Fig. 2 are the data points for a sample of $\sim$ 70 quasars and Seyfert nuclei. We find that most quasars are well fit by masses in the range $M = 10^{8-9.5} M_\odot$ and accretion rates $\dot{m} \sim 0.3 - 30$. Seyfert 1 nuclei have typically $M \sim 10^8 M_\odot$ and $\dot{m} \sim 0.1 - 1$. Note that luminous quasars with $\alpha_{UV} < 0.4$ cannot be fitted.

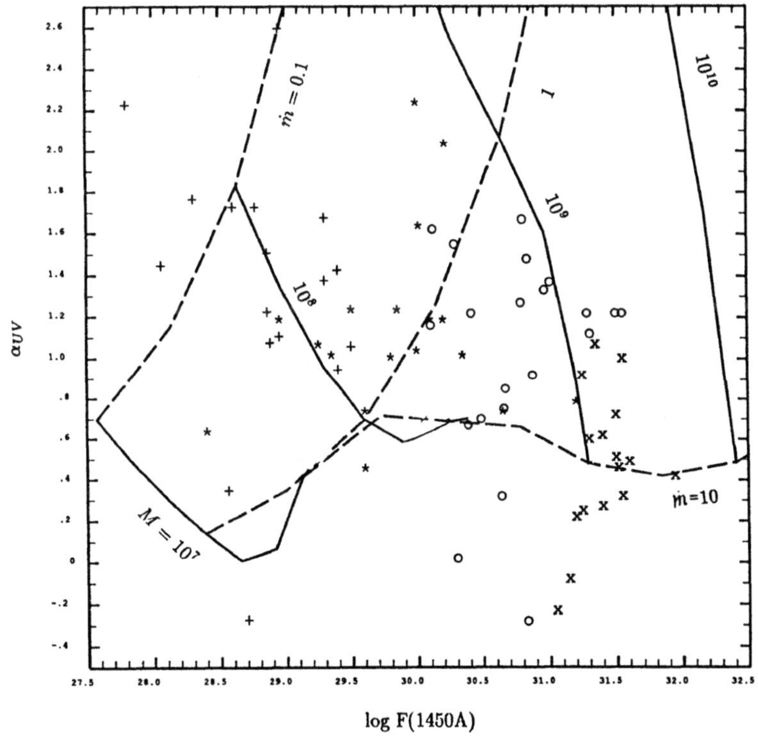

Fig 2. Mapping of the accretion disk parameter space ($M - \dot{m}$) onto the observable UV plane (spectral index in the 1000-1450A band vs. UV flux F(1450A)). Continous lines are curves for constant black hole mass, dashed lines are for constant accretion rate ($\dot{m}$). The viscosity parameter is $\alpha = 0.1$. The data shown are Seyfert 1 nuclei from Wu (+), low redshift quasars (Cheng and Fang 1987 (*), and Kinney et al 1985 (o)), and high redshift quasars (x) from Osmer and Smith (1977 and 1980).

We do a similar mapping in the optical regime. We calculate the spectral index α_{rb} between the red (7500A)and blue (4200A). We compare the model calculations with the data from the complete sample of Neugebauer *et. al.* (1987).

The theoretical $M - \dot{m}$ grid (modified for the presence of a power law component of slope 1.4) together with the optical data are shown in Fig. 3. The inferred accretion parameters are similar to what we have found for the UV sample and the masses are slightly higher. Note that the optical spectral index of objects with $L < 10^{45}\,\mathrm{erg\ s^{-1}}$ is probably contaminated by starlight from the host galaxy (cf. Wandel 1987), so their true spectral index and luminosity would be lower than actually observed, placing them in a lower mass regime.

4. Implications and Evolution

The black hole mass range found for quasars in both samples roughly agrees with estimates based on emission-line analysis (Wandel and Yahil 1985) and X-ray variability (Wandel and Mushotzky 1986). It is of interest to find the Eddington ratio in the observed bands. Curves of constant $\ell_{UV} = L_{UV}/L_{Edd}$ can be calculated from Fig. 2. Approximately, they follow the $\dot{m}$ curves, with $\ell_{UV} \sim 0.03\dot{m}$. We find a range of $\ell_{UV} \sim 0.01 - 1$, with most high-redshift luminous quasars requiring Eddington ratios of order unity. The Eddington ratio for the optical band

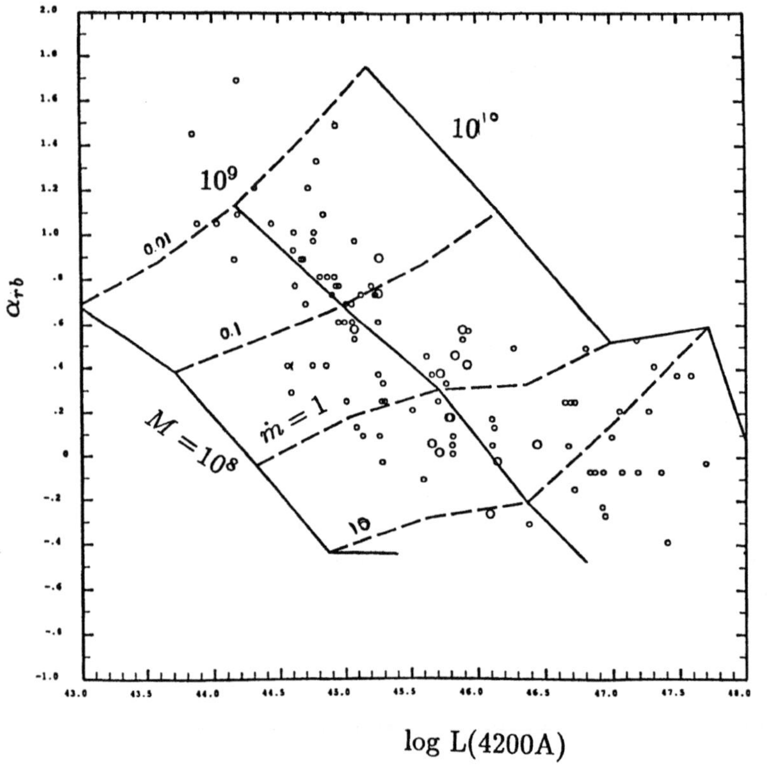

Fig 3. Mapping of the accretion-parameter space onto the plane of optical spectral index (between rest frame wavelengths 4200 and 7500A) vs blue luminosity L(4200A). Continuous lines are curves of constant black hole mass; dashed lines are constant accretion rate ($\dot{m}$). The theoretical curves are corrected for a contribution from a $\nu^{-1.4}$ power law normalized so that it is equal to the disk component at frequency $\log\nu_o = 14.3$. The data are from Neugebauer et al (1987); larger circles denote radio loud quasars.

(Fig. 3) is lower than in the UV (for most quasars in the sample $L_{opt}/L_{Edd} \sim 0.001 - 0.1$. For the low luminosity objects this is in rough agreement with Wandel and Yahil (1985). The high redshift, very luminous quasars, (which were not included in their sample), the Eddington ratio is significantly higher.

Previous workers noted that for large enough samples of quasars, the luminosity is anti-correlated with the spectral index in the UV (Cheng and Fang 1987) and optical (Wandel 1987) bands. In other words, the spectra become flatter for more luminous (or higher redshift) objects. Because of the strong correlation between luminosity and redshift in magnitude-limited samples, it is impossible to determine if the primary correlation of the spectral index is with luminosity or redshift, but in either case this effect can be explained as evolution along curves of approximately constant mass in figures 2 and 3 (however, for large values of $\dot{m}$ the mass will change).

The dependence of the luminosity and spectrum on the mass and the accretion rate may be used to place a constraint on the luminosity evolution of quasars. The exact path of evolution of the luminosity and spectrum of an accretion disk will depend on the initial mass of the black hole M and the rate of accretion $\dot{M}$ throughout its lifetime. The mass will be increasing monotonically with an e-folding time

$$\tau = (\dot{M}/M)^{-1} = 5 \times 10^8 \dot{m}^{-1} \mathrm{yr}, \tag{4}$$

so that a significant change in mass will take place over a cosmological time span of a few billion years if $\dot{m} > 1$. In addition, we need to know how $\dot{m}$ (or $\dot{M}$) evolves so that an evolution path through figure 1 or 2 can be determined. The accretion rate and its evolution depends on the unknown environment of the accreting black hole. The above qualitative analysis shows that the expected evolutionary path for an accretion disk around a black hole is quite different from what is expected from pure luminosity evolution (requiering the luminosity of different objects to evolve in the same way); for example, it is very difficult to see how a luminous quasar with the requisite high mass will evolve into a low luminosity object (Seyfert galaxy or quasar).

References

Bechtold, J., Czerny, B., Elvis, M., Fabbiano, G., and Green, R. 1986, *Ap. J.,* **314**, 699.

Cheng, F.H., and Fang, L.Z. 1987, preprint.

Czerney, B., and Elvis, M. 1987, *Ap. J.,* , in press.

Edelson, R.A. and Malkan, M.A. 1987, *Ap. J.,* **308**, 59.

Kinney, A.L., Huggins, P.J., Bregman, J.N. and Glassgold, A.E., 1985, *Ap. J.,* **291**, 128.

Malkan, M.A. 1983, *Ap. J.,* **268**, 582.

Neugebauer, G., Green, R.J., Matthews, K., Schmidt, M., Soifer, B.T., and Bennett, J. 1987, *Ap. J. Sup.,* **63**, 615.

O'Dell, S.L., Scott, H.A., and Stein, W.A. 1987 *Ap. J.,* ,

Osmer, P.S. and Smith, M.G. 1977, *Ap. J.,* **213**, 607.

——— 1980, *Ap. J. Sup.,* **42**, 333.

Sakura, N.I. and Sunyaev, R.A. 1973, *Astr. Ap.,* **24**, 337.

Wandel, A. 1987, *Ap. J. (Letters),* **316**, L55.

Wandel, A., and Yahil, A. 1985, *Ap. J. (Letters),* **295**, L1.

Wandel, A., and Mushotzky, R.F. 1986, *Ap. J. (Letters),* **306**, L61.

Wandel, A., and Petrosian 1987, *Ap. J. (Letters),* , L(in press).

Supported by NASA grant NCC 2-322 and NGR 05-020-668

SYNTHESIS OF ACCRETION DISK AND NONTHERMAL SOURCE MODELS FOR AGN

DAVID L. BAND[1] AND MATTHEW A. MALKAN[2]
[1]*IGPP-LLNL, L-413, Livermore, CA 94550*
[2]*UCLA, Department of Astronomy, Los Angeles, CA 90024*

A scenario for the central engine of AGN has been developed consisting of a massive black hole (MBH) onto which gas accretes through an accretion disk. The accretion disk radiates the observed optical and ultraviolet continua. Surrounding the MBH is a nonthermal source which produces the infrared and soft X-ray continua by synchrotron emission, and the X-ray spectrum by inverse Compton scattering of the optical-ultraviolet photons from the accretion disk. Previously we modeled the accretion disk (M.A.M.) and nonthermal source (D.L.B.) separately, and here we combine the two models to form a unified description of the AGN engine. This combined model can be inverted to determine source parameters from observed spectra. A group of AGN for which multiband observations exist can then be modeled to: demonstrate the validity of the combined model for a large number of objects; establish the range of parameter values that describe the source; and search for any correlations between source description and type.

An accretion disk around a MBH is the most promising origin of the optical-ultraviolet "big blue bump." The methodology has been developed by Sun and Malkan (1987a,b) and is described by Sun in the present proceedings. The spectrum is modeled by assuming that the accretion disk is geometrically thin and optically thick. At each radius either a black body or a low surface-gravity non-LTE stellar atmosphere radiating the necessary energy flux is used to calculate the spectrum from that radius. The stellar atmospheres are less efficient radiators than a black body, and consequently they emit at a higher frequency than would a black body. Gravitational defocusing (light that passes close to the black hole is bent) and Doppler beaming (the light from the rapidly rotating material at the center of the disk is beamed along the disk) is included, providing a strong inclination dependence. In particular, observers near the plane of the disk see a significantly harder spectrum than those who view the disk face on. Both stationary Schwartzschild and maximally rotating Kerr black holes can be modeled.

The nonthermal model is a variant of the Synchrotron-Self Compton (SSC) model and differs from it in two respects: a) the soft photons which are scattered into the X-ray band originate in the "big blue bump" and not the synchrotron source; and b) the electron distribution is assumed to consist of low energy flat and high energy steep components (Band and Grindlay 1985, 1986 and Band 1987). The nonthermal source is modeled as a sphere of radius R surrounding the accretion disk (for the sake of simplicity the accretion disk is assumed to be a point source at the center). The source is permeated by a tangled magnetic field B (allowing the field to be treated as a scalar). Within the source is a homogeneous "broken" electron density: $n_e = n\gamma^{-p}$, $1 \leq \gamma \leq \gamma_b$ and $n_e = n\gamma_b\gamma^{-p-1}$, $\gamma_b \leq \gamma_b$, where $\gamma = E/mc^2$. This electron distribution could arise due to injection $\propto \gamma^{-p}$, with radiative losses above γ_b steepening the distribution. The resulting spectrum is:

- Synchrotron self-absorbed below ν_T
- A flat synchrotron power law between ν_T and ν_b with $L_\nu \propto \nu^{-\alpha}$ where $\alpha = (p-1)/2$.
- A steep synchrotron power law above ν_b with $L_\nu \propto \nu^{-\alpha-1/2}$.
- A flat X-ray power law due to inverse Compton scattering by the flat component of the electron distribution. The flat scattered component extends to $\gamma_b^2 \nu_c$, where ν_c is the characteristic frequency of the soft photon source.
- The gamma ray spectrum is affected by the break in the electron distribution, the decrease in the Klein-Nishina cross section, and absorption due to pair creation.

The equations for the various spectral features can be inverted to derive the source parameters from the observed spectrum. The observed spectrum from the far-infrared through X-rays is fit with the following components (see example of such a fit in the Figure below):

- Balmer jump and blended Fe II lines due to emission in the Broad Line Region
- Flux normalization of starlight from host galaxy
- Absorption in both the Milky Way and the host galaxy
- Accretion disk defined by: M_B—black hole mass; $\dot{m}$—accretion rate; and μ_i—cosine of disk inclination angle to the line of sight.
- Synchrotron spectrum defined by: α'—spectral index of steep component; $L_\nu(\nu_{fid})$—normalization of steep component at fiducial ν_{fid}; ν_b—break frequency between flat and steep components; and ν_T—turnover frequency due to synchrotron self-absorption.

This spectral decomposition gives the parameters for the accretion disk model, e.g., M_B and $\dot{m}$. The synchrotron turnover frequency ν_T, break frequency ν_b, and normalization $L_\nu(\nu_{fid})$ are used in conjunction with an X-ray flux L_x and a moment of the accretion disk luminosity to derive the parameters of the non-thermal source: the normalization of the electron distribution n, the magnetic field B, and the source radius R.

Preliminary fits to five objects are given in the Table below. In all cases the accretion disk was assumed to be face on. The sources show remarkably similar parameters: $n \sim 10^7$ cm^3, $B \sim 1000$ gauss, $R \sim 2 \times 10^{15}$ cm or about 30 Schwarzschild radii, black hole masses of a few times $10^8 M_\odot$, and total luminosities of a large fraction of the Eddington limit. The break energy $\gamma_b \sim 50 - 90$ is perhaps an indication of the microphysics within the source. Finally, the magnetic and photon fields are in approximate equipartition, while the relativistic electron energy density is much smaller. The compactness $l = (L_{acc}/R)(\sigma_T/mc^3)$ is quite large. However, L_{acc} is the ultraviolet accretion disk luminosity, whereas it is the usually unobserved luminosity at $\nu \sim mc^2/h$ that is relevant to the production of electron-positron pairs (Band 1987).

This work is supported by the U.S. Department of Energy, under contract No. W-7405-ENG-48 to the Lawrence Livermore National Laboratory (D.L.B) and the NSF, under grant AST-86-14510 (M.A.M.).

References

Band, D. L., and Grindlay, J. E. 1985, *Ap. J.*, **298**, 128.
——— 1986, *Ap. J.*, **308**, 576.
Band, D. L. 1987, *Ap. J.*, **321**, 80.

Sun, W.H., and Malkan, M.A. 1987a, in *Supermassive Black Holes*, M. Kafatos, ed. (Cambridge: Cambridge University Press).
——— 1987b, in *Active Galactic Nuclei, Neutron Stars, and Jets: Astrophysics of the Central Engines*, W. Kundt, ed. (Dordrecht: Reidel), p. 124.

Table—Preliminary Fits

	Mkn 841	II Zw 136	3C249.1	Mkn 1383	3C273
Fitted Parameters					
$n_6 = n/10^6$ cm^{-3}	19.2	12.4	7.2	19.3	9.4
$B_2 = B/100$ gauss	13.3	17.0	8.9	8.2	12.0
$R_{15} = R/10^{15}$ cm	0.8	1.2	1.9	1.1	5.2
γ_b	56.	53.	83.	89.	82.
$M_8 = M_B/10^8\, M_\odot$	0.7	1.1	8.0	1.6	5.0
$\dot{m} = \dot{M}/(M_\odot\text{-yr}^{-1})$	0.15	0.41	1.0	0.56	14.0
Underlying α	0.73	0.80	0.62	0.65	0.85
L_{acc} ergs-sec^{-1}	1.9×10^{45}	5.4×10^{45}	1.4×10^{46}	7.5×10^{45}	7.8×10^{46}
Derived Quantities					
u_e ergs-sec^{-1}	10.8	6.3	4.7	12.1	4.5
u_B ergs-sec^{-1}	7.0×10^4	1.2×10^5	3.2×10^4	2.7×10^4	5.7×10^4
u_{UV} ergs-sec^{-1}	3.8×10^3	3.0×10^4	3.1×10^4	4.9×10^4	2.3×10^4
R/R_{Sch}	37.	37.	8.1	23.	35.
L_{acc}/L_{Edd}	0.27	0.49	0.18	0.47	1.56
$l = (L_{acc}/R)(\sigma_T/mc^3)$	68.	122.	200.	185.	406.

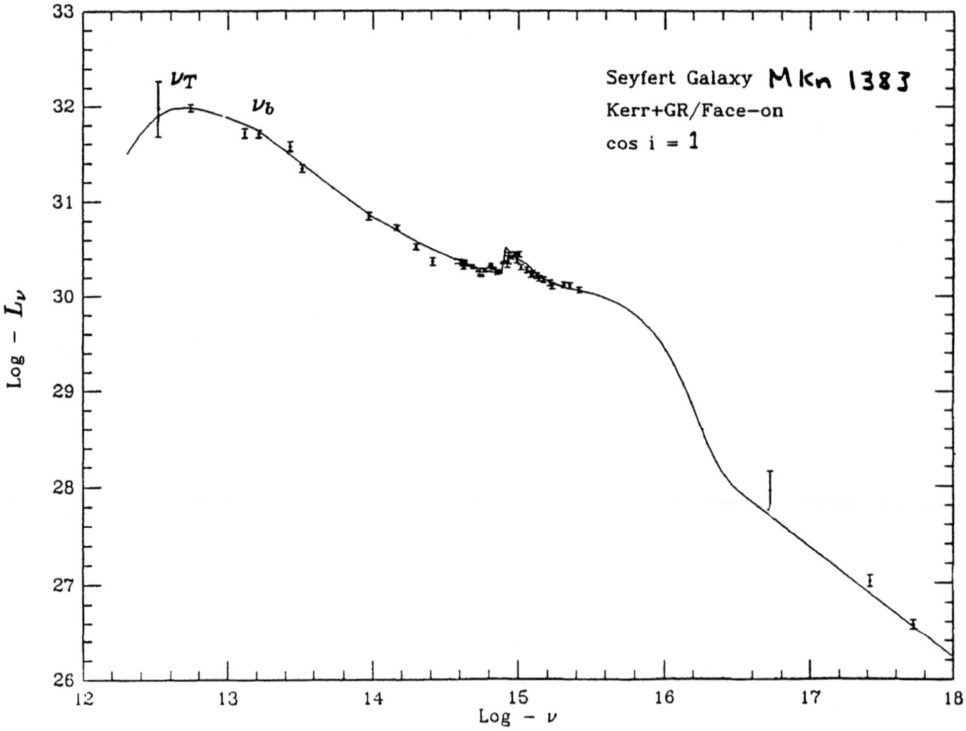

Figure: Fit to observed spectrum of the Seyfert Mkn 1383.

EFFECTS OF SELF-GRAVITY IN AGNs

Isaac Shlosman and Mitchell C. Begelman
Joint Institute for Laboratory Astrophysics
University of Colorado and National Bureau of Standards
Boulder, CO 80309-0440

Accretion onto supermassive black holes (SBHs) is widely believed to be responsible for the phenomena of AGNs. While there has been considerable theoretical work on the inner regions of such accretion flows, little is known about the outer boundaries of AGNs. Except for the broad and narrow emission lines, the observational constraints are weak, and evidence linking the emission line regions to the accretion flow is ambiguous at best. It is not even known whether the accretion flow is fueled by mass loss from a dense cluster of stars surrounding the SBH, or enters the nucleus from the galactic interstellar medium (e.g. Begelman et al. 1984). We adopt the latter view, assuming that most of the incoming fuel forms a thin accretion disk at distances of order $10\text{-}10^3$ parsecs from the SBH. Such a disk must be vertically self-gravitating. We analyze its thermal and dynamical properties and conclude that its energetics is likely to be dominated by backscattered AGN radiation. We give a necessary condition for Jeans instability to develop in a disk, and elaborate on the consequences of this instability (Shlosman and Begelman 1987a,b). The effect of star formation on the disk structure and evolution, as well as on the accretion process onto the SBH is considered.

For an isothermal Keplerian α-disk with effective thickness h, the Toomre criterion for instability (Toomre 1964) can be expressed by the threshold accretion rate

$$\dot{M}_{crit} \simeq 2 \times 10^{-5}\ \alpha\ (T/10\ K)^{3/2}\ M_{\odot}\ yr^{-1}. \qquad (1)$$

For accretion rates $\dot{m} > \dot{M}_{crit}$ the disk will become clumpy on the characteristic scale of the most rapidly growing wavelength ~h, with growth times $\sim t_{orb}\hat{n}^{-1/2}$. Here $\hat{n} \equiv n/n_{crit}$ is the density in the disk in units of the critical density at the onset of instability, $n_{crit} \simeq 2 \times 10^7 (v_{\phi 2}/r_{pc})^2\ cm^{-3}$, $v_{\phi 2} \equiv v_{\phi}/100\ km\ s^{-1}$ is the Keplerian velocity, r_{pc} is the distance in parsecs and t_{orb} is the local orbital time in the plane of the disk. For the Jeans instability not to saturate, the cooling time, t_{cool}, must be less than the local (compressional) heating time. The necessary condition, therefore, for the disk fragmentation which is independent of α implies that (Shlosman and Begelman 1987a):

$$\frac{t_{cool}}{t_{orb}} < \hat{n}^{-1/2} \quad . \tag{2}$$

The energetics of the outer regions of realistic accretion disks at $r_{pc} \sim$ few will be probably dominated not by internal viscous dissipation but by the external sources, particularly by the backscattered hard radiation from the immediate vicinity of the SBH or by the jet above the disk. For disks whose temperature is regulated either by molecular cooling or by dust, the condition (2) can be easily satisfied at the distance of a few parsecs from the SBH.

Noting that the local Jeans mass in the disk is a decreasing function of the disk density, i.e.

$$M_J \simeq 2 \times 10^{-2}\ (T/10\ K)^{3/2}\ v_{\phi 2}^{-1}\ \hat{n}^{-1/2}\ r_{pc}\ M_{\odot} \quad , \tag{3}$$

the entire self-gravitating disk can be divided into three major regions of star formation: the inner disk consisting of self-gravitating molecular clouds slowly contracting on the Kelvin-Helmholtz time scale; the intermediate disk where the low-mass stars are formed and the outer disk with high-mass stars. The limits between these regions depend upon the accretion rate onto the SBH.

The onset of star formation in the disk leads to a gradual conversion of the gas into stars on the time scale of $t_* \sim \varepsilon^{-1} t_{orb} \hat{n}^{-1/2}$, $\varepsilon \simeq 10^{-1}$-$10^{-3}$ being the efficiency of star formation in molecular clouds (e.g. Silk 1985, 1987). In the absence of any replenishment process, the disk would be depleted of the gas on the time scale of few t_* and the SBH would be fed at the rate not exceeding $\dot{M}_{crit}$. The viscous dissipation operates on the time scale of $t_{in} \simeq 10^9 \alpha^{-1} (T/10\ K)^{-1} v_{\phi 2} r_{pc}$ yr. For $\alpha \lesssim 1$ at these radii, this is a very inefficient process to bring in the new material. An alternative process, a global disk instability, will be discussed elsewhere.

A low-mass star has its greatest effect on the surrounding ISM during its T Tauri phase which lasts $t_T \simeq 10^7\ (M_T/M_{\odot})^{-2}$ yr. It is well known that this phase in stellar evolution is characterized by an extensive outflow, $\dot{M}_T \simeq 10^{-8}\ M_{\odot}\ yr^{-1}$, with asymptotic velocity $v_{W2} \simeq v_W/100\ km\ s^{-1} \sim 1$. The T Tauri wind acts as a piston sweeping up the disk gas and producing a cavity with the radius

$$R_W \simeq (\dot{M}_T/4\rho v_W)^{1/2} \simeq 2 \times 10^{-3}\ \dot{M}_T^{1/2} v_{\phi 2}^{-1}\ (T/10\ K)^{-1/2} \hat{n}^{-1/2}\ r_{pc}\quad pc \lesssim h, \tag{4}$$

where ρ is the unshocked density in the disk. Our estimates show that the T Tauri stars cannot substantially alter the energetic balance in the disk.

High mass stars have a three-fold effect on the cool disk producing (i) HII regions (ii) line-driven winds (LDWs) with $v_{LDW} \sim$ few $\times\ 10^3$ km s^{-1} and (iii) supernovae. These compact HII regions with the radius given by (using McKee [1986] estimates):

$$R_{HII} \simeq 66.9(S_{49}/n^2)^{1/3}\mathrm{pc} \simeq 10^{-5} S_{49}^{1/3} v_{\phi 2}^{-4/3} \hat{n}^{-2/3} r_{pc}^{4/3}\ \mathrm{pc} \quad (5)$$

will have only local effect on the disk. Here $S_{49} = S/10^{49}$ is the number of ionizing photons per second, e.g. $S_{49} \simeq 0.04$ for B0 star, $S_{49} \simeq 1$ for O6.5 and $S_{49} \simeq 8$ for O4 star. The cavities created by the LDWs and the supernovae will discharge their internal pressure above the disk creating and sustaining a hot corona with the thickness $\sim r$.

Assuming that an AGN is powered by a molecular and dusty accretion disk at ~10-1000 pc, we conclude:

(1) The disk tends to preserve its low temperature, <1000°K, even when its energetics is dominated by external radiation.

(2) The disk is subject to Jeans instabilities for a moderate accretion rate in excess of $\sim 2 \times 10^{-5}\ \alpha\ (T/10\ K)^{3/2}\ M_\odot\ yr^{-1}$.

(3) A necessary condition for such a disk to fragment is given. The local Jeans mass in the disk is a decreasing function of the density in the disk (i.e. of accretion rate).

(4) The effect of star formation on the disk is considered. Massive stars ($> 2\ M_\odot$) form only in a narrow range of accretion rates $\sim 10^{-4}$-$10^{-5}\ M_\odot\ yr^{-1}$. The effect of stellar winds and supernovae is mostly to eject the hot gas, $\sim 10^6$- 10^7°K, above the disk.

(5) Low mass stars ($< 2\ M_\odot$) have favorable conditions to form in the disk. Without mass supply to the disk, operating on the dynamical time scale, the entire disk will be depleted of gas during 10^6 - 10^7 yr. The result: stellar disks in galactic nuclei consisting of low mass stars. Accretion rates below 10^{-5}- $10^{-6}\ M_\odot\ yr^{-1}$ are predicted.

This work was supported in part by NASA Astrophysical Theory Center Grant NAGW-766 and NSF Grant AST83-51997 (Presidential Young Investigator).

References

Begelman, M. C., Blandford, R. D., and Rees, M. J. 1984, Rev. Mod. Phys. 56, 255.

McKee, C. F. 1986, Ap. Space Sci. 118, 383.

Shlosman, I. and Begelman, M. C. 1987a, Nature, 329, 810.

Shlosman, I. and Begelman, M. C. 1987b, in preparation.

Silk, J. 1985, Ap. J. Lett. 282, L71.

Silk, J. 1987, in Star Forming Regions, IAU Symp. No. 115, eds. M. Peimbert and J. Jugaku (Reidel, Dordrecht) p. 663.

Toomre, A. 1964, Ap. J. 139, 1217.

MASS AND LENGTH SCALE OF BLACK HOLES IN QUASARS AND ACTIVE GALACTIC NUCLEI

Yukio Tomozawa

Randall Laboratory of Physics
University of Michigan, Ann Arbor, Michigan 48109

Abstract

The mass M and the length scale R of black holes are deduced from the experimental data of variability times and luminosities of quasars and active galactic nuclei. A simple power law relation, $M \propto R^{\lambda}$, emerges. As a possible explanation of the variability and of the relationship of M and R, oscillation of black holes is considered.

The large luminosities ($10^{40} \sim 10^{46}$ erg/sec) and relatively small variability times ($10^2 \sim 10^6$ sec) of various kinds of quasars and active galactic nuclei AGN (Seyfert galaxies, x-ray and radio galaxies) suggest massive black holes ($10^2 \sim 10^8 M_{\odot}$) as the source of energy. In this article, a relation between the mass and the length scale of black holes is deduced from experimental data (Wandel and Mushotzky 1986) on the variability time Δt and the x-ray luminosity L_x and a possible model for a black hole dynamics is put forward.

A natural relationship between Δt, M and R is given by

$$\Delta t = \kappa\sqrt{\frac{R^3}{2GM}}, \tag{1}$$

where G is Newton's constant and κ is a constant of order unity. The right hand side of eq. (1) is the time scale for collapse of a mass from a distance R into a heavy mass M. The amount of energy emitted in a variability time, $L_x\Delta t$, can be expressed as a fraction η of the available gravitational energy,

$$L_x\Delta t = \eta\frac{GM^2}{R}, \tag{2}$$

where η is given by

$$\eta \approx \frac{2\Delta M}{M} - \frac{\Delta R}{R}, \tag{3}$$

ΔM and ΔR being the change of M and R in Δt.

From eq. (1) and (2), it follows that

$$\log M = \log\Delta t + \frac{3}{5}\log L_x + \text{const} \tag{4}$$

$$\log R = \log\Delta t + \frac{1}{5}\log L_x + \text{const} \tag{5}$$

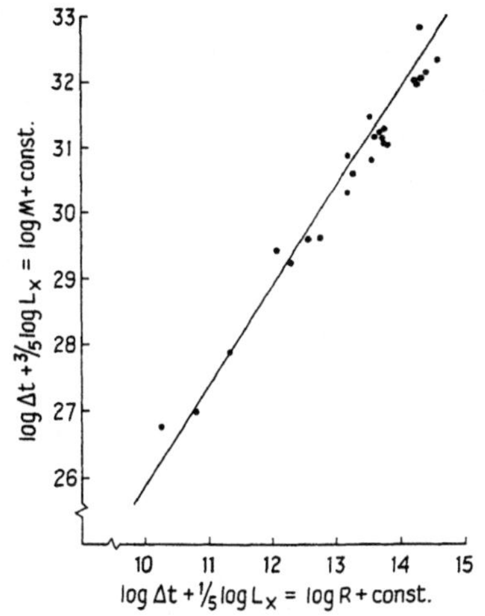

Fig. 1.

Relation between log M and log R based on eq. (4) and (5). The data of Δt and L_x are taken from (Wandel and Mushotzky 1986). The circles correspond to (in the order of M) 3C273, Mrk 509, F9, 3C382, Mrk 205, 4U0241+62, III Zw 2, Mark 841, NGC 526A, NGC 7469, MCG 5-23-16, NGC 2992, NGC 5506, OX 169, NGC 2110, NGC 7582, NGC 4151, NGC 3516, NGC 3227, NGC 4593, Mrk 766, NGC 4051, NGC 3031 (M81), NGC 6814. The straight line is $M \propto R^{1.5}$ (eq. 8), while the χ-square fit yields $M \propto R^{1.37}$.

Fig. 1 is the plot of log M vs. log R computed from eqs. (4) and (5), using the experimental data for L_x, and Δt of Wandel and Mushotzky 1986. This figure suggests the relationship

$$M = \mu R^{\lambda} \tag{6}$$

where μ and λ are constants. A relation such as eq. (6) can be obtained from the following consideration: The energy emitted in $\Delta t, L_x \Delta t$, is proportional to R^2, since it originates from infalling matter on a surface area of radius R. Combining this with eq. (2), we obtain

$$L_x \Delta t = \eta \frac{GM^2}{R} = \xi R^2 \tag{7}$$

Assuming that η and ξ are constants or slowly varying functions, we have the relationship (6) with

$$\lambda = \frac{3}{2} \tag{8}$$

This result is quite compatible with the results of the χ-square fit, $\lambda = 1.37$. In Fig. 1, a straight line $M \propto R^{1.5}$ is drawn for the purpose of comparison with the experimental data points.

The implication of eqs. (6) and (8) is that the mass M and length scale R for black holes in quasars and AGN progress along this line from the lower left to the upper right as time progresses, due to the addition of infalling matter. The next important question is what is the reason for variability, or what is a model that justifies eqs. (1) and (2). Below, we present such a model.

The starting point of our argument is an observation on the behavior of gravitational collapse. The gravitational collapse of a massive object ($\gtrsim 3M_{\odot}$) leads to a black hole. The matter indefinitely collapses into the singularity. However, the consideration of quantum theory changes this scenario completely. Recently, the author (Tomozawa 1985 a,b,c,; 1986 a,b) solved the Einstein equation with quantum corrections (Anderson 1983, 1984; Birrell and Davies 1982) and suggested that the gravitational potential is repulsive at short distances due to quantum effects. As a result, gravitational collapse leads to an oscillation. Such an oscillation may extend spatially from the outside of the Schwarzschild radius to the inside. We call such an oscillation "a black hole oscillation". In the earlier work (Tomozawa 1986 b), the author discussed how a motion of in and out of the horizon can occur without violating causality.

Of course, the possibility of the existence of angular momentum for black holes and the detailed dynamics of the infalling matter would give a complicated structure to the variability. Nevertheless, the variability time should be a reflection of the black hole oscillation.

Finally, a black hole which progresses along the straight line of Fig. 1, has the density distribution, $\rho(r) \sim r^{-\frac{3}{2}}$ so that $m(r) \sim r^{\frac{3}{2}}$, where $\rho(r)$ and $m(r)$ are the distribution of the mass density and the mass inside the radius r when the black hole is in the maximally extended state.

In summary, we have presented a simple regularity between the masses and length scales of black holes in quasars and AGN, $M \propto R^{1.5}$, based on the experimental data on variability time and the x-ray luminosity. Eqs (1), (2), and (7) which are essential in obtaining the above result, can be naturally understood in the model of black hole oscillation which is an outcome of quantum effects on gravity.

REFERENCES

Anderson, P., 1983 Phys. Rev., D28 , 281.

Anderson, P., 1984, Phys. Rev., D29 , 615.

Birrell, N.D. and Davies, P.C.W., 1982, Quantum Fields in Curved Space (Cambridge University Press).

Tomozawa, Y., 1985a, Proc. of the International Symposium on Quantum Field Theory, Positano, Italy (ed. Mancini, F.) p. 241.

Tomozawa, Y., 1985b, The XIX International Conference on Cosmic Rays, La Jolla, 3 , 187.

Tomozawa, Y., 1985c,, The INS International Symposium on Composite Model of Quarks and Leptons (ed. Terazawa, H. and Yasue, M.) p. 386.

Tomozawa, Y., 1986a, "Quantum Corrections to the Gravitational Potential and Gravitational Collapse", The 26th International Astrophysical Colloquium, Liege, p. 137.

Tomozawa, Y., 1986b, "Cosmic Rays, Quantum Effects on Gravity and Gravitational Collapse", Lectures given at the Second Workshop on Fundamental Physics, University of Puerto Rico, Humacao (Ed. E. Esteban) p. 144.

Wandel, A. and Mushotzky, R., 1986, Ap. J., 306 , L61.

X-Ray Spectra of Active Galaxies

Richard F. Mushotzky
Laboratory for High Energy Astrophysics
NASA/Goddard Space Flight Center
Greenbelt MD 20771

Introduction

The x-ray band, "traditionally" 0.1-100 kev, covers as much dynamic range in frequency space as the ultraviolet to mid-infrared. As such, one might expect rather diverse x-ray spectral shapes such as is seen in other regions of the spectrum across such a large energy range. It was therefore surprising that the first x-ray spectral surveys (Rothschild et al. 1983, Mushotzky 1984, Petre et al. 1984) indicated that simple power law models modified by the effects of absorption by cold matter were adequate descriptions of the x-ray spectra of broad line active galaxies (and most particularly Seyfert I galaxies) . As is becoming clearer as results from the EXOSAT and Einstein (IPC) observatories become available, the x-ray spectra of broad line AGNs at low energies, $E<2$ kev, are complex and many (if not most) objects are not well modeled by simple power laws in the 0.1-10 kev band. However, because of the lack of moderate resolution spectroscopy the form of the spectrum of most AGN in the 0.1-2 kev band is still not clear.

I Old Results

For the sake of completeness I will briefly review the "old" picture of the broad band x-ray spectra of line emitting AGN; (In this brief review I shall omit discussion of the x-ray spectra of BL Lacertae objects which show certain distinct differences from that of line emitting AGN; see Urry 1987 for a discussion of their spectra).

As shown by Rothschild et al. (1983) and Mushotzky (1984) using HEAO-1 A-2 and A-4 data, the 2-100 kev spectrum of active galaxies (a sample size of 12 objects in the 2-100 kev and 26 objects in the 2-30 kev band) could be well fit by a power law model $F_\nu=A\nu^{-\alpha}$ with a narrow distribution in the value of α. These authors claimed that the distribution in α was consistent with a Gaussian distribution with $<\alpha>=0.68$ with a variance $\sigma=0.17$. It was not clear from their analysis how much of the variance was due to measurement error and how much was due to intrinsic differences in α. There was no correlation of x-ray spectral index with luminosity or radio-loudness. Similar results were obtained with Einstein MPC data by Halpern (1982) in the 1.5-10 kev band. Almost all the objects used in these samples had an x-ray flux in excess of 2×10^{-11} ergs/cm^2-sec in the 2-10 kev band.

Petre et al. (1984) using the Einstein Solid State Spectrometer (SSS) data showed that the 0.7-4.5 kev spectrum of luminous, log $L(x)> 43.5$, AGN (this sample was similar to that used by Rothschild et al. 1983 and Mushotzky 1984) was consistent with the same distribution. However,

Reichert et al. (1984), also using the SSS data found that the 0.7-4.5 kev spectrum of low luminosity objects was frequently not consistent with the simple power law form modified by the effects of uniform absorption. They modeled this deviation, which was always in the sense of a soft x-ray excess above the best fit power law, by partial covering of the underlying nucleus by a patchy absorber. However, they could not rule out models in which there was an additional unabsorbed soft component.

II Recent Results

A. EXOSAT

Recently K. Pounds and T. Turner (1988) have reviewed the EXOSAT results on active galaxies. Most of the 42 objects in their sample are classical Seyfert I galaxies. Their x-ray flux threshold was a factor of 2 lower than that of HEAO-1. These authors find a distribution in spectral index (both mean and variance) almost identical to that found in the HEAO-1 sample, with $<\alpha>=0.70$ and variance $\sigma=0.174$. They have attempted to remove the additional variance due to measurement error and find that, for those sources whose individual spectra are not distinguishable from the mean, that $\sigma\leq0.12$. For ~2/3 of their objects the simple power fit is acceptable over the full EXOSAT spectral range (0.1-10 kev). The spectral index for this subsample has the same mean and variance as for the objects measured only at higher energies. However, ~1/3 of their objects require an additional soft component. There is no significant correlation between the presence of the soft component and the 2-10 kev x-ray luminosity.

At low x-ray luminosity only two of the objects determined by Pounds and Turner to have soft components are the same ones noted by Reichert et al.(1985) to have soft excesses. It seems clear that some of the EXOSAT results are not consistent with the patchy absorber interpretation discussed by Reichert et al. 1985 and require that the soft x-ray "excess" is indicative of a new spectral component.

Eight of the higher luminosity unabsorbed objects reported to have soft components by Turner and Pounds were observed with the Einstein SSS in the 0.6-4.5 kev band. Only two of them (Fairall 9 and Mkn 509) have x-ray spectral indices, as measured by the SSS, significantly different from α~0.7 in this energy band. This indicates that either the soft components are strongly variable on long, t> 1 year, timescales or that they are very soft and do not contribute much in the 0.6-4.5 kev band. However, there is one object, ESO141-G55, which has a steep index in the SSS data and no reported soft excess from EXOSAT data. This may indicate significant variability in the either in the intensity of the soft component or in the ratio of soft to hard flux.

Timing results on several AGN (NGC 7314 ,Turner 1987; Mkn 335, Warwick, Pounds and Turner; 1988, 3C120, Maraschi et al. 1987) indicate that the soft x-ray components are not simply related to the hard, E> 2 kev component. In particular, in Mkn 335 the time scale for significant variability seems to be shorter at higher energies.

There is not much evidence for variation in the 2-20 kev x-ray spectral index with either time or intensity. Comparison of the published EXOSAT and HEAO-1 data for the ~20 objects that overlap in the two samples shows no significant spectral index variability on an object to object basis. The best data are for Cen-A, the x-ray brightest AGN, in which the source has varied by an order of magnitude with no variation in spectral index in the 2-500 kev band; the limits on $\delta\alpha$ are ± 0.1. However there are reported to be variations in the x-ray spectrum of NGC 4151 (Perola et al. 1986), 3C120 (Halpern 1985), and NGC 7314 (Turner 1987). These variations are all in the sense of the source spectrum steepening as it brightens. It is not clear at present what, if anything, is different about these objects or even if the interpretation of a change in spectral index is a unique interpretation of the data. Reanalysis of EXOSAT observations of NGC 4151 by Yaqoob, Warwick and Pounds (1988) indicate that the best fitting models do not require variation in the power law index.

B. Einstein IPC

Recently the Einstein Imaging Proportional Counter (IPC, bandpass 0.1-4.5 kev, $E/\Delta E \sim 1$) spectra of 33 quasars (Wilkes and Elvis 1987) and ~70 Seyfert I galaxies (Kruper et al. 1988) have been analyzed. These authors show that the distribution of the x-ray spectral indices in this energy range is not well represented by the "canonical" $<\alpha>=0.68$ Gaussian distribution. In particular Wilkes and Elvis find a broad distribution with radio loud quasars tending to have flat spectra in this band with $<\alpha>=0.50$ and with radio quiet objects tending to have $<\alpha>=1.0$, a disjointedness not seen in the higher energy 2-20 kev band (although the fraction of radio loud objects in the 2-20 kev sample is smaller than in the IPC sample). Very few of the objects in the Wilkes and Elvis sample have error bars small enough to rule out $\alpha \sim 0.7$. Wilkes and Elvis attribute the ubiquity of the $<\alpha>=0.68$ spectrum seen in the higher energy data to the appropriate linear superposition of the two components seen most clearly in the radio bright and quiet objects (figure 1). An additional point in the favor of this model is that in some cases (most particularly the radio loud objects) the extension of the "IR power law", whose slope is often close to 1.2, extrapolates well to the soft x-ray spectrum (Wilkes 1987). Wilkes and Elvis also point out that detailed comparison of the x-ray derived absorption measures with the minimum galacitc column along the line of sight implies the existence of an "ultra-soft" component to the quasar spectra. Thus their complete model has 3 components (see figure 1)

The existence of this range of spectral indices in the low energy datais not due just to the higher luminosities of the IPC quasars. The Seyfert I galaxies measured by the IPC (Urry et al. 1986), objects very similar to those measured by HEAO-1 and EXOSAT at higher energies, have an x-ray spectral index distribution tn the 0.1-4 kev band consistent with that of the radio quiet quasars of Wilkes and Elvis (1987).

III Soft X-ray Components in QSO's and Seyfert Is - A Discussion

Comparison of the HEAO-1, EXOSAT and Einstein results shows that there exist a variety of spectral forms at energies below a few kev; the question is which, if any, of the suggested spectral decompositions is correct. Analysis of the low energy EXOSAT CMA data for a small sample of Seyfert I galaxies (Turner et al. 1987 (Mkn 335), Morini et al. 1987 (Mkn 509), Morini et al. 1986 (Fairall-9), Arnaud et al. 1985 (Mkn 841) and HEAO-1 LED data (Singh et al. 1985, Mkn 509 and NGC 2992)) has suggested that the best fit to the 0.1-3 kev spectral data consists of the "canonical"

Seyfert I galaxies (Turner et al. 1987 (Mkn 335), Morini et al. 1987 (Mkn 509), Morini et al. 1986 (Fairall-9), Arnaud et al. 1985 (Mkn 841) and HEAO-1 LED data (Singh et al. 1985, Mkn 509 and NGC 2992)) has suggested that the best fit to the 0.1-3 kev spectral data consists of the "canonical" $<\alpha>=0.68$ spectrum plus a relatively steep low energy component which can be fit by a power law of index $\alpha \sim 2$, or by a exponential model with a "temperature" of less than 0.3 kev. This suggests the model shown in figure 2 in which the $\alpha \sim 1$ IPC spectrum is an "artifact" of the superposition of only two components, the the steep low energy component and the $<\alpha>=0.68$ spectrum. However, it is not clear that the EXOSAT data can rule out the model spectrum suggested by Wilkes and Elvis.

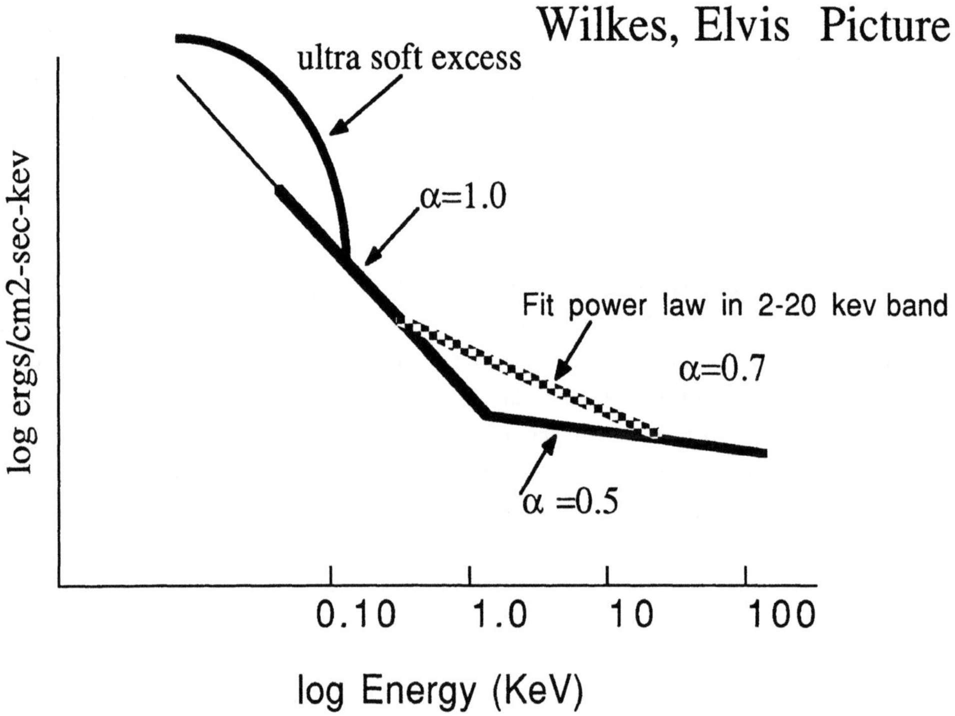

Figure 1 A Schematic of the decomposition of the 0.1-100 kev spectrum of AGN as described by Wilkes and Elvis (1987).The checkered line represents a sum of two components . The flat $\alpha=0.5$ component is stronger in radio loud than radio quiet objects.

Probably neither this simple idea, nor the one suggested by Wilkes and Elvis is "correct". It is more likely that there exist a variety of spectral forms at low energy which have their origin in different physical components. This is already suggested by the EXOSAT timing data for NGC 4151 (Pounds et al. 1987) and NGC 7469 (Barr 1986) in which there appear to exist low energy components which are larger in physical size compared to others that vary. Presumably in these cases the x-ray radiation is due to hot, photoionized gas in the volume around the nucleus (Krolik and Vrtilek 1984) .There are several from several such components (as seems to be the case in NGC 4151). It will require "true", i.e. higher resolution, x-ray spectroscopy combined with detailed time variability analysis over a broad band, to settle the issue.

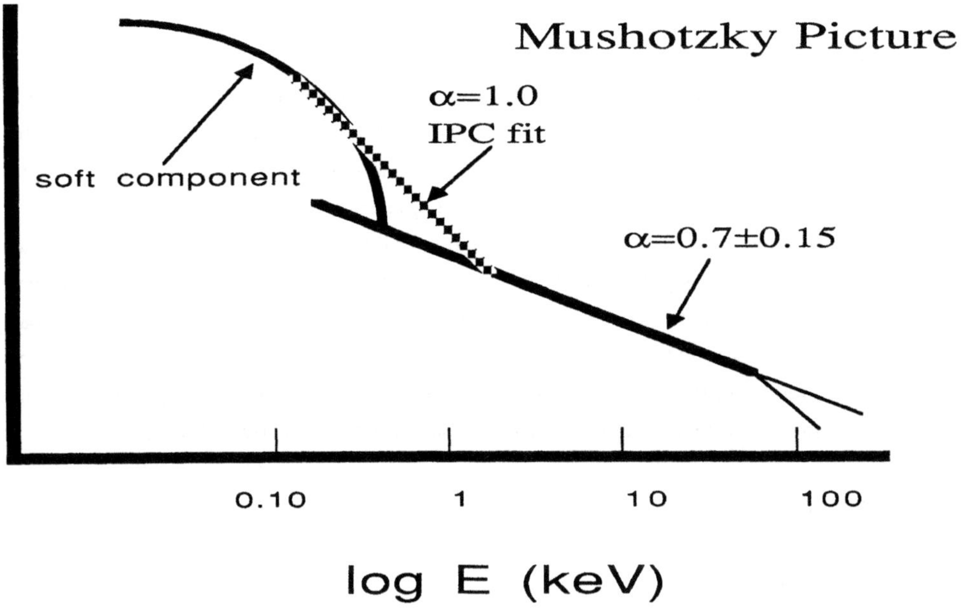

Figure 2 A Schematic of one possible decomposition of the 0.1-100 kev spectrum of AGN. The checkered line represents a sum of two components .

IV "Other types" of AGN

There are several recent results from the Einstein, EXOSAT and Ginga observatories on the x-ray spectrum of Seyfert IIs and LINERS . These objects were too weak for the previous generation of x-ray observatories to obtain sensible spectra, although, it was well known that Seyfert II galaxies, as a class, had lower 2-10 kev x-ray luminosities than Seyfert Is.

A. Seyfert IIs

There now exist high quality x-ray spectrum of two Seyfert II galaxies in the 2-20 kev band and a relatively large number of medium quality IPC spectra in the 0.1-4.5 kev band (Kruper et al. 1988). NGC 1068 (Monier and Halpern 1987, Turner 1988, Ginga team 1988) shows a "complex" x-ray spectrum. The IPC data are well fit by a power law of energy slope α~2.0 with very little, if any, low energy photoelectric absorption. The higher energy 2-20 kev data seem to require a flattening, with the best fit power law having a slope α~0.7 (Turner 1988) . The higher quality Ginga data show that the 2-20 kev spectrum is more complex than a simple power law with a very strong Fe emission feature at 6.4 kev. The Ginga spectrum of Mkn 348, shows a "canonical" α~0.7 power law with a very large column density log N_H~23.1 at/cm^2, similar to that of several narrow emission line galaxies (Mushotzky 1982). Kruper et al. (1988) have analyzed the IPC spectra of a sample of ~ 25 Seyfert II galaxies which have significant detections and find that, on the whole, they are not significantly different than the Seyfert I galaxies in their sample. Thus at this early stage, with the possible exception of NGC 1068, there is not much evidence, from the x-ray band, for a systematic difference between the x-ray spectral properties of Seyfert Is and IIs.

B. LINERs

Recent EXOSAT spectra of two LINERs, M81 (Barr, Mushotzky and Giommi 1988) and Pic-A (Barr, Mushotzky and Reichert 1988), seem to show a qualitatively "different" x-ray spectrum. Barr and co-workers find that the data are best fit in the 0.1-10 kev band not by a power law but by a function which is flat at low energies and steep at high energies such as an exponential spectrum or a broken power law. Because M81 is strongly variable in the EXOSAT data on time scales of less than 1000 seconds it is clear that the x-ray emission in this object is nuclear in character. There is at least one other object that shows a similar spectrum, that of the nucleus of the "non-AGN" M33 (Gotwald et al. 1987). Barr and co-workers speculate that this spectral form may be a characteristic of objects radiating well below their Eddington limit. If such a spectrum is typical of very low luminosity active galaxies they cannot make a strong contribution to the x-ray background at energies above a few kev.

VI Spectral Features due to Iron

The strongest spectral feature seen in the 0.6-10 kev band in AGN is that due to iron, in particular fluorescent emission at 6.4 kev and and the absorption edge of cold Fe at 7.1 kev. The strength of the Fe emission line is proportional to the solid angle occupied by the "cold" reprocessor, $\Omega_{Fe,}$ and the column density of iron (Makishima 1987). For low column densities $N_H < 5x\ 10^{23}$ at/cm^2 and a typical AGN spectrum the equivalent width of the line is,

$$EW \sim 100(N_H / 1x\ 10^{23})(\Omega_{Fe} / 4\pi) A_{Fe}\ ev$$

(where A_{Fe} is the Fe abundance normalized to "solar", $4x10^{-5}$ by number, Halpern 1982).We have assumed that the radiation from the central object is unbeamed. Of course, if the radiation is beamed the Fe reprocessing region may see a higher x-ray flux than our line of sight, producing more Fe radiation than expected in the spherical case. Because present day x-ray proportional counters are basically limited to detecting lines stronger than ~60 ev in weak sources, the mere detection of a 6.4 kev line indicates a large value of $\Omega_{Fe} / 4\pi$ and $N_H / 1x\ 10^{23}$.

If one assumes that the low energy x-ray absorption seen in some AGN is due to the same material responsible for the Fe 7.1 kev absorption feature the abundance of Fe can be determined. The analysis indicate a solar abundance in Cen-A and twice solar in NGC 4151 (see Perola et al. 1986, Makishima 1987, Mushotzky et al. 1977 and Holt et al. 1980) . The fairly strong Fe line seen in these objects indicates that $\Omega_{Fe} \sim 4\pi$ in NGC 4151 and is $> 2\pi$ in Cen-A, quite different from the situation supposed to exist in QSOs if the region responsible for the Fe line in these objects is also the region responsible for the broad optical and UV lines. If there exist reprocessing clouds out of the line of sight one can have Fe 6.4 kev emission with little or no low energy or 7.1 kev absorption. This may be the case in Mkn 509 (Morini et al. 1987). A large statistical sample of Fe emission lines from AGN will allow the determination of the mean coverage of the central source by thick clouds.

In a fashion similar to the "reverberation" analysis presented by Krolik (this symposium) the x-ray data (fig 3) can be used to "map" the region producing the Fe emission feature. The best data so far have been obtained by the EXOSAT observatory for Cen-A (Morini et al. 1988) and NGC 4151 (Perola et al. 1986); the latter is shown in figure 3 .

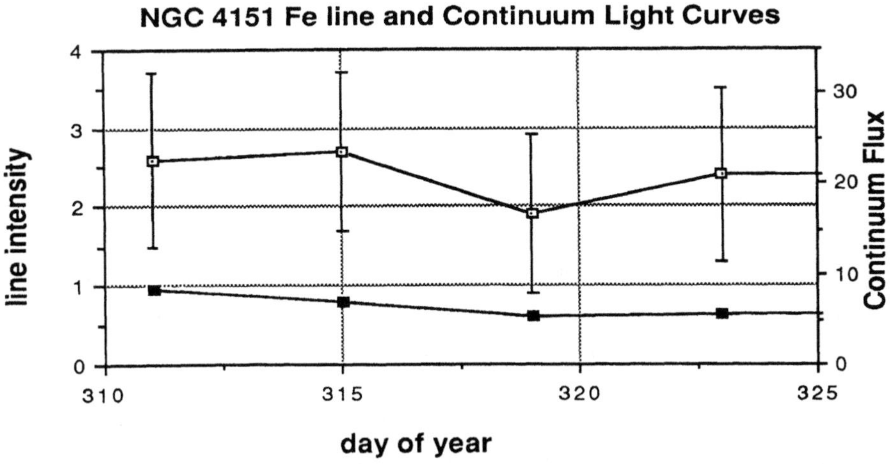

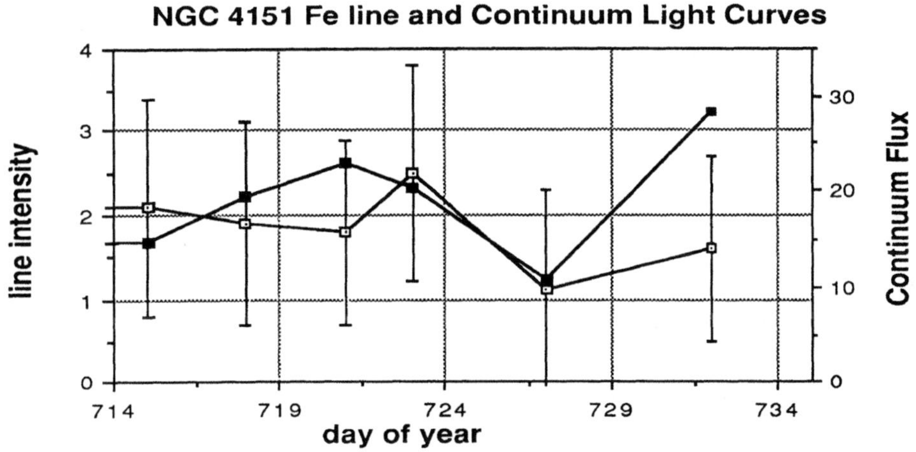

Figure 3: The time series of Fe emission line strength (open squares with error bars, left axis) and continuum strength (closed squares, right axis; the error bars are small) versus time from Perola et al. 1986. Note that the line flux remains approximately constant despite a factor of greater than 3 change in the average continuum flux between the two epochs (I have used a "running" day number, e.g. day number 714 is in the next year).

While the EXOSAT line intensity data are of relatively low signal to noise it is clear that the line flux has remained more or less constant on a timescale of a year while the continuum intensity has changed by a factor of 3. This may indicate that the Fe line emitting region is larger than the Broad Line region in NGC 4151.

Conclusion

It is clear that x-ray spectroscopy has much to tell us about the nature of the active galactic nuclei phenomenon. In particular, with much more sensitivity and better spectral resolution devices available on future missions like BBXRT (a shuttle based x-ray spectrometer to fly in 1989), AXAF (the NASA Great Observatory for x-ray astronomy), XMM (the European Space Agency large multi-mirror x-ray mission scheduled to fly in 1998) and ASTRO-D (the Japanese x-ray spectroscopic mission scheduled to fly in 1993), we can anticipate that future observations will be quite interesting. In particular Fe line x-ray spectroscopy of AGN can be expected to strongly constrain models of the Broad Line Region in AGN. Continuum spectroscopy, in particular the determination of the nature of the ionizing continuum in the soft x-ray band, will be crucial in understanding these objects. I have neglected detailed discussion of the high energy, E>30 kev, continuum because of the paucity of data in this energy range. It is hoped that this band, which may contain the spectral signatures of the physics of the underlying energy generation mechanism, will be explored in more detail by future missions such as XTE (the U.S. X-ray Timing Explorer) and SAX (the Italian X-ray Astronomy saellite).

Acknowledgements. I would to like to express my gratitude to T. Turner and K. Pounds for sending me EXOSAT results prior to publication and to Y. Tanaka and F. Makino for sending Ginga results prior to publication. This review would have been impossible without access to their important recent work. I thank M. Urry for sending me the recent IPC analysis of Seyfert I and II galaxies prior to publication . I would like to thank G. Reichert for extensive discussions and G. Madejski and M. Urry for editorial criticism.

References

Barr, P., Mushotzky, R and Giommi, P. 1988 Ap.J. submitted

Barr, P., Mushotzky, R. and Reichert, G. 1988 in prep

Barr, P. 1986 M.N.R.A.S. 223, 29P

Czerny, B. and Elvis, M. 1987 Ap.J. 321

Edelson, R. and Malkan, M. 1986 Ap.J. 308, 59

Gottwald, M. Pietsch, W. and Hasinger, G. 1987 Astron. and Astrophys. in press

Halpern, J. 1982 Ph.D. Thesis Harvard University, unpublished

Halpern, J. 1985 Ap.J. 290, 130

Holt, S.S. et al. 1980 Ap.J. Letters 241, L13

Krolik, J. and Vrtilek, J. 1984 Ap.J. 279,521

Kruper, J. et al. 1988 in preparation

Makishima, K. 1987 in Physics of Accretion Onto Compact Objects pg 249 eds K. Mason, M. Watson and N. White Springer-Verlag

Monier, R. and Halpern, J. 1987 Ap.J. (Letters) 315 L17

Morini, M. et. al 1986 Ap.J. 307, 486

Morini, M., Lipani, M. and Molteni, D. 1987 Ap.J. 317,145

Morini, M. et al. 1988 in preparation

Mushotzky, R. 1982 Ap.J. 256, 92

Mushotzky, R. 1984 Advances In Space Research 3, no 10-13 pg 157

Perola, G.C. et al. 1986 Ap.J. 306, 508

Petre, R. Mushotzky, R., Krolik, J. and Holt, S.S. 1984 Ap.J. 280, 499

Pounds, K. and Turner, T. 1988 in "X-ray Astronomy with EXOSAT"

Reichert, G, Mushotzky, R, Petre, R. and Holt, S.S. 1985 Ap.J. 296, 69

Rothschild, R., Mushotzky, R. Baity, W. Gruber, D., Mateson, M. and Peterson, L. 1983 Ap.J. 269,423

Turner, T. 1987 M.N.R.A.S. 226, 9P

Turner, T. 1988 PhD. Thesis Leicester University

Urry, M. 1987 in Physics of Accretion Onto Compact Objects pg 357 eds K. Mason, M. Watson and N. White Springer-Verlag

Urry, M., Kruper, J. Canizares, C., Rohan, M. and Oberhard, M. 1986 in Proceedings of the Conference on Variability of Galactic and Extragalactic X-ray Sources ed A. Treves pg 15

Warwick, R., Pounds, K. and Turner, T. 1988 M.N.R.A.S in press

Wilkes, B. 1987 Proceedings of the IAU/COSPAR Symposium "The Physics of Compact Objects" Edited by N. White

Wilkes, B. and Elvis, M. 1987 Ap.J. 323

Yaqoob, T., Warwick, R. and Pounds, K. 1988 in preparation

CLUES TO THE X-RAY EMISSION MECHANISMS IN FLAT RADIO SPECTRUM AGNs

D.M. Worrall
Harvard-Smithsonian Center for Astrophysics, Cambridge, MA, U.S.A.

Abstract

Clues to the X-ray emission mechanisms in flat radio spectrum AGNs arise from comparing radio, optical, and X-ray luminosities for a sample of 80 known-redshift radio-selected AGNs for which the radio luminosity is dominated by flat-spectrum compact-core emission. The X-ray spectral luminosity lies on average above an extrapolation from lower frequencies for all the QSOs, despite a range in optical polarization and activity. Such an X-ray "excess" is often interpreted as support for the presence of a Compton-scattered component. In contrast, for the BL Lac objects, the X-ray spectral luminosity on average lies on the extrapolation from lower frequencies. This is particularly interesting in view of evidence that the X-ray emission of BL Lac objects may be unbeamed while the radio emission is beamed. If the 80 sample sources are assumed to be preferentially beamed towards the observer in radio emission, because of their radio selection, a test supports the X-ray emission also being beamed. However, the BL Lac sub-sample is too small for the test to be applied to it alone.

I. Introduction

Extragalactic sources selected by flat-spectrum compact-core radio emission cover a range of optical properties. One distinction is the optical emission-line strength, the primary characteristic used to separate QSOs from the barren BL Lacertae-type objects. A second distinguishing feature of BL Lac objects, high optical variability and polarization strength (the so-called "blazar" characteristic), has been observed in only a subset of such radio-selected QSOs, although there is limited observational support for the suggestion that in others the blazar activity is merely swamped by the "normal" QSO emission (*e.g.*, Impey 1987). X-ray emission is touted as a diagnostic of the central engine of an AGN, but observational knowledge of the spectral, timing, spatial and polarization properties lags similar knowledge for lower energies. The X-ray data alone contain insufficient clues to the emission mechanism by which, and the environments in which, the X-radiation was produced. However, the connections between emission at different wavelengths can be used to make certain inferences. This paper reports some results of a continuing project on the relationships between BL Lac objects and flat-spectrum, compact-core, radio QSOs of varying degrees of optical polarization and variability.

The data included in this analysis are described by Worrall (1987a), with a few objects discarded for not being radio selected. The resulting sample contains 10 BL Lac objects, 20 highly-polarized QSOs (HPQs), and 50 other compact flat-spectrum radio QSOs (FRS QSOs) which are not classified as highly polarized. They were all observed with the Einstein Observatory Imaging Proportional Counter (IPC), and all but 5 of the FRS QSOs were detected.

II. X-ray relative to Optical Emission

The question to be considered is, where does the X-ray spectral luminosity lie with respect to an extrapolation from lower frequencies? Ideally, this is addressed by a campaign of coordinated multiwavelength observations of the (variable) sources of interest. However, the demands of acquiring the necessary space-experiment data have limited the number of sources for which such observing has been successfully accomplished (see Bregman, Maraschi and Urry (1987) for relevant references). A statistical approach makes a useful alternative. The effects of variability are expected to average out in an ensemble of objects. The concave curvature which is generally observed in the infrared to ultraviolet spectrum of BL Lac objects and HPQs should be taken into account, and Bregman, Maraschi and Urry (1987) then find that for a sample of blazars of mixed radio-, optical-, and X-ray-selection origin, the extrapolation to X-ray energies falls at least a factor of two below the X-ray flux in about 65% of the objects. Here, I wish to compare results for radio-selected BL Lac objects, HPQs and FRS QSOs.

Landau *et al.* (1986) reached an interesting observational conclusion about the radio to ultraviolet curvature of BL Lac objects and HPQs. These authors were able to fit coordinated observations of each of their sources from these classes to a parabolic spectral shape of the form $\log(S/S_m) = [\log(\nu/\nu_m)]^2/2A$, where S_m, ν_m and A are constants for a source. Using this, and assuming a radio spectral luminosity and an optical spectral luminosity for a source, just one more constraint is sufficient to predict an X-ray spectral luminosity to be compared with observations. For this I will use the correlation between bolometric luminosity and half-width of the parabola found by Jones, Rudnick and Landau (1986) (see also Figure 4 of Rudnick 1987). This correlation is in the sense that the higher the bolometric luminosity, the smaller the half-width of the parabola. Figure 1 shows the results for the 80 objects in my sample. Plotted is the predicted 2 keV spectral luminosity in ergs s^{-1} Hz^{-1} (on the abscissa) *vs.* the difference between the logarithm of the observed 2 keV spectral luminosity and that which is predicted from a parabolic fit. Friedman cosmology with $H_o = 50$ km s^{-1} Mpc^{-1} and $q_o = 0$ is assumed.

Several conclusions can drawn from Figure 1.

(1). The distribution is strongly weighted to values greater than zero, implying that on the whole the sources are producing excess emission over an extrapolated optical component. This agrees with Bregman, Maraschi and Urry (1987) (see above).

(2). The HPQs and other FRS QSOs are distributed similarly along the ordinate axis, and show the largest excess. The sources 3C 345 and 3C 446 are among those with a large excess, consistent with conclusions of Bregman (1986) from coordinated multifrequency observations.

(3). The BL Lac objects appear to show a mean roughly consistent with zero.

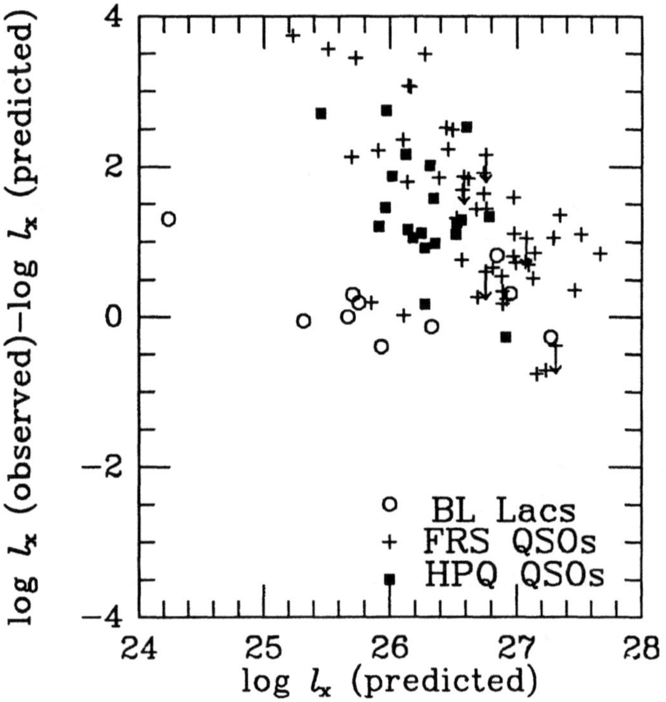

Figure 1:
The observed "excess" 2 keV spectral luminosity *vs.* predicted value. Luminosities are in ergs s^{-1} Hz^{-1} for H_o = 50 km s^{-1} Mpc^{-1}.

Point (3) is particularly interesting in view of evidence that the X-ray emission of BL Lac objects may be unbeamed while the radio emission is beamed. This evidence has been presented by Stocke *et al.* (1985) and Maraschi *et al.* (1986), and comes from the observation that X-ray-discovered BL Lac objects have lower luminosities at frequencies below the X-ray than radio-discovered BL Lac objects — so X-ray selection does not find preferentially beamed radio/optical sources. A BL Lac-object model must therefore reconcile the evidence that the X-ray emission is more isotropic (*i.e.*, presumably of smaller Lorentz factor) than emission at lower frequencies with the observation that the radio-selected objects (*i.e.*, those with jets more pointed towards the observer) show a smooth spectral continuity between radio and X-ray energies. The smooth parabolic fits to data throughout radio to X-ray energies does not automatically suggest that inverse Compton radiation should be an important contributor to the X-ray emission.

The HPQs are similar to radio-selected BL Lac objects in the parabolic curvature of the radio to ultraviolet continuum. However, they appear to be more similar to the FRS QSOs than the radio-selected BL Lac objects in showing an X-ray excess. Such an excess might be due to inverse Compton emission, although more direct evidence is required before this can be concluded.

III. X-ray relative to Radio Emission

Figure 2 shows the correlation of 2 keV spectral luminosity, l_x, with 5 GHz spectral luminosity, l_r, in units of ergs s^{-1} Hz^{-1}, for the 80 sample sources. The twelve known superluminal sources are indicated by open symbols. The figure is based on Figure 1 of Worrall (1987a), who describes the K corrections used. The reduced scatter in Figure 2 (over Figure 1 of Worrall 1987a) results from discarding some sources which were optically or X-ray selected. Note that the correlation is quite tight, particularly given that three different classes of source are included and that variability undoubtedly contributes to the spread. However, it should be pointed out that it is the FRS QSOs and HPQs which define the correlation, since the BL Lac objects cover such a small range of l_r. A fit to the correlation, using the method described by Avni *et al.* (1980), gives a dispersion of the $\log l_x$ values of $\sigma = 0.44 \pm 0.06$ (90% error).

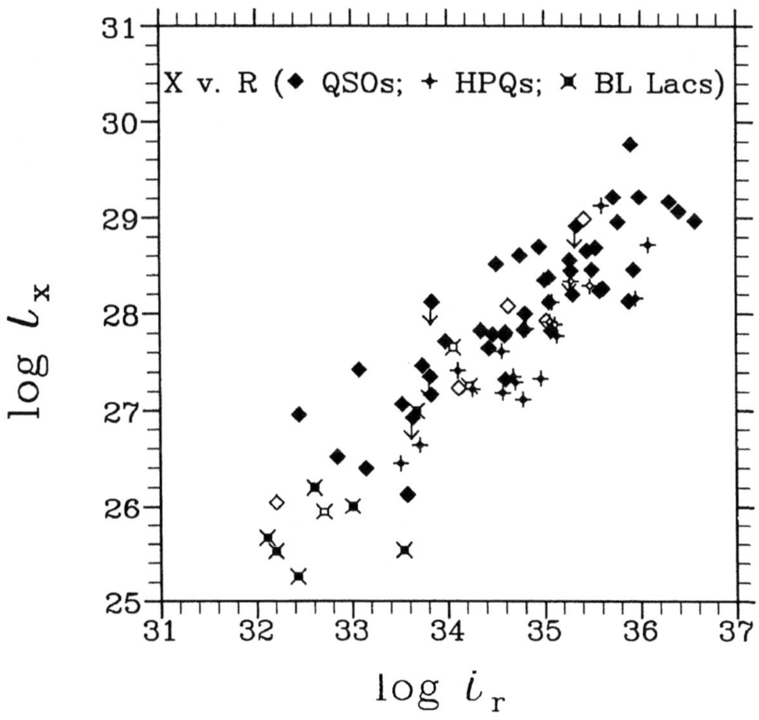

Figure 2:

The correlation of 2 keV and 5 GHz spectral luminosity. Open symbols indicate superluminal sources. Luminosities are in ergs s^{-1} Hz^{-1} for $H_o = 50$ km s^{-1} Mpc^{-1}.

It may be reasonable to assume that the objects are preferentially beamed towards the observer in the radio, because of the radio selection criterion, but what about the X-ray emission? It would be interesting to address this question for the BL Lac sample alone, in view of the evidence discussed above that the X-ray emission may be isotropic. However, the following analysis, based on Figure 2, applies only to the 80 objects as a whole, and the result is primarily defined by the QSOs.

For relativistically boosted emission, luminosity scales as a power of the bulk relativistic Doppler factor, δ. If the radio emission is beamed and the X-ray emission unbeamed, an expected dispersion in the correlation of X-ray and radio luminosity due to the distribution of δ can be calculated. If the predicted dispersion is greater than the observed dispersion of 0.44, we must conclude that the X-ray emission is wholly or partly beamed.

The calculation requires the orientation and Lorentz-factor distributions of the sources to be known, or, in this case, estimated. Figure 3a shows the distribution of δ with angle to the line of sight for values of the Lorentz factor, γ, of 2, 4, 8 and 16. I assume the sources are randomly oriented between some minimum and maximum angle to the line of sight. Reasonable values based on the fact that the sources are radio selected but include both BL Lac objects and normal FRS QSOs are a minimum of zero and a maximum of $> 20^o$. For identifying suitable values of γ, Figure 3b is a plot of transverse velocity *vs.* angle, also for $\gamma = 2, 4, 8$, and 16. The y-axis positions of the 12 superluminal sources in the sample are shown, under the conservative assumption of $H_o = 100$ km s^{-1} Mpc^{-1}. Were $H_o = 50$ km s^{-1} Mpc^{-1}, these would move up by a factor of two. It is evident that $\gamma \gtrsim 8$ is required to accommodate several of these sources.

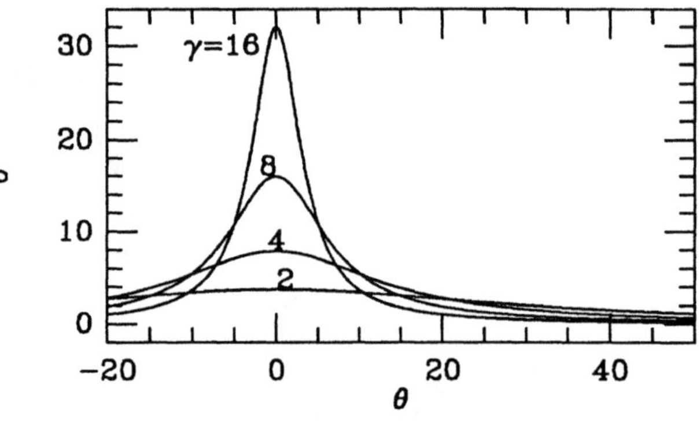

Figure 3a:
Bulk relativistic Doppler factor, δ, as a function of angle to the line of sight (in degrees) for various Lorentz factors.

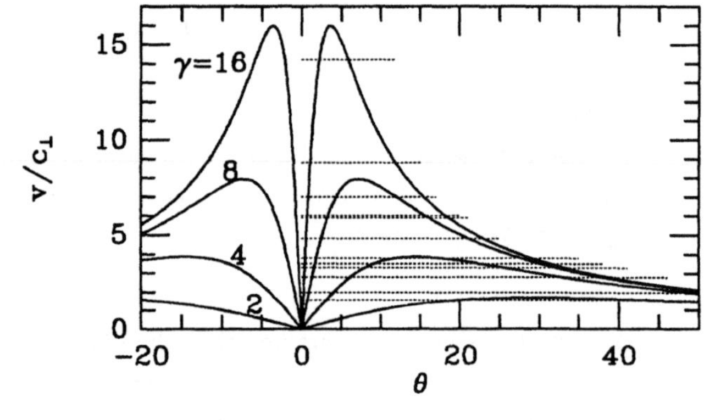

Figure 3b:
Apparent transverse velocity *vs.* angle to the line of sight for various Lorentz factors. Plot is for $H_o = 100$ km s^{-1} Mpc^{-1}. Dotted lines show y-axis values for the 12 superluminal sources.

Figure 4 shows predictions of the dispersion in $\log l_x$ (in the $\log l_x - \log l_r$ correlation) as a function of minimum and maximum angles to the line of sight for the sources. If indeed we take take $\theta_{min} = 0^o$ and $\theta_{max} \gtrsim 20^o$ (see arrows in Figure 4), we would predict dispersions of greater than the observed value of 0.44. The conclusion is that there is a strong suggestion that all or part of the X-ray emission must also be beamed.

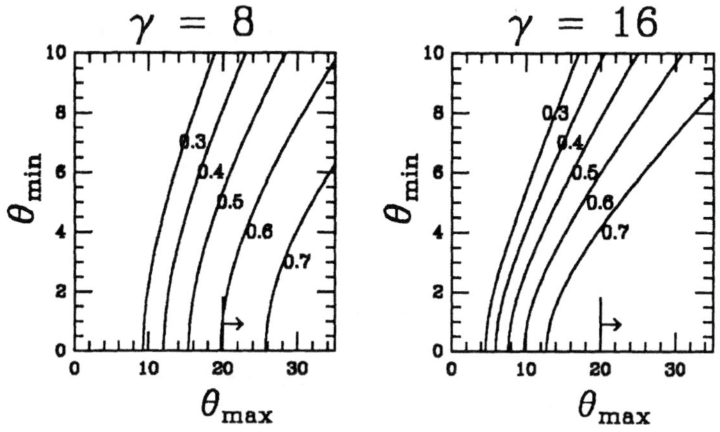

Figure 4:

Contours of constant dispersion (in the $\log l_x - \log l_r$ correlation) predicted for sources uniformly oriented between angles θ_{min} and θ_{max} to the line of sight, for two values of the Lorentz factor. See text concerning arrows.

IV. Discussion

The interesting discovery of Landau *al.* (1986) of the parabolic spectral curvature between radio and ultraviolet energies for the HPQs and BL Lac objects awaits physical interpretation. Meanwhile, it is interesting to wonder whether the observation of a more "humped" spectrum with increasing bolometric luminosity (Jones, Rudnick and Landau 1986) is related to the effect observed in radio-quiet QSOs of a steepening with increasing source luminosity of the effective optical to X-ray spectral index (Avni and Tananbaum 1982,1986; Kriss and Canizares 1985) and infrared to X-ray spectral index (Worrall 1987b; Kriss 1988). Whatever the cause of the parabolic form, there does appear to be a difference between radio-selected BL Lac objects and QSOs in the strength of the X-ray emission — QSOs on average show an excess, whereas the BL Lac objects do not. For the BL Lac objects, the consistency of the level of X-ray emission with an extrapolation from lower frequencies must be understood together with evidence that the X-ray emission is more isotropic than that at lower frequencies (see references above). For the objects with an X-ray excess, however, the evidence presented above shows that, if radio selection effects are dominated by relativistic beaming, the X-ray emission must be beamed also.

More detailed knowledge of the X-ray spectra may help in relating X-ray emission to that at lower frequencies. Already a difference between the average X-ray spectrum of FRS QSOs and BL Lac objects has been found. Wilkes and Elvis (1987) have fit a subset of the FRS QSOs to power-law energy indices grouped around a value of 0.5, and they relate the X-ray and radio emission. In contrast, Madejski (1985) has found that BL Lac objects of mixed selection tend to fit X-ray spectral indices centered around 1.2, but with a large spread, and he concludes that the X-ray emission is primarily synchrotron and related to the optical. A greater pool of observational evidence is required before it will be possible to understand the relationship between the HPQs and BL Lac objects. Their radio to ultraviolet continuum spectra appear similar, and yet the classes seem to show differences at X-ray energies.

Support for this work was provided by NASA contract NAS8-30751 and Smithsonian Institutional funds.

References

Avni, Y. and Tananbaum, H. 1982, *Astrophys. J. (Letters)*, **262**, L17.
Avni, Y. and Tananbaum, H. 1986, *Astrophys. J.*, **305**, 83.
Avni, Y., Soltan, A., Tananbaum, H., and Zamorani, G. 1980, *Astrophys. J.*, **238**, 800.
Bregman, J.N. 1986, in *Continuum Emission in Active Galactic Nuclei*, ed. M.L. Sitko (Tucson: Kitt Peak National Observatory, National Optical Astronomy Observatories), p. 102.
Bregman, J.N., Maraschi, L., and Urry, C.M. 1987, in *"Scientific Accomplishments of the IUE"*, ed. Y. Kondo (Dordrecht: Reidel), p. 685.
Impey, C. 1987, in *"Superluminal Radio Sources"*, eds. J.A. Zensus and T.J. Pearson (Cambridge: Cambridge University Press), p. 233.
Jones, T.W., Rudnick, L., and Landau, R. 1986, in *Continuum Emission in Active Galactic Nuclei*, ed. M.L. Sitko (Tucson: Kitt Peak National Observatory, National Optical Astronomy Observatories), p. 122.
Kriss, G.A. 1988, *Astrophys. J.*, **324**, in press.
Kriss, G.A. and Canizares, C.R. 1985, *Astrophys. J.*, **297**, 177.
Landau, R., Golisch, B., Jones, T.J., Jones, T.W., Pedelty, J., Rudnick, L., Sitko, M.L., Kenney, J., Roellig, T., Salonen, E., Urpo, S., Schmidt, G., Neugebauer, G., Matthews, K., Elias, J.H., Impey, C., Clegg, P., and Harris, S. 1986, *Astrophys. J.*, **308**, 78.
Madejski, G. 1985, Ph.D. Thesis, Harvard University.
Maraschi, L., Ghisellini, G., Tanzi, E.G., and Treves, A. 1986, *Astrophys. J.*, **310**, 325.
Rudnick, L. 1987, in *"Superluminal Radio Sources"*, eds. J.A. Zensus and T.J. Pearson (Cambridge: Cambridge University Press), p. 217.
Stocke, J.T., Liebert, J., Schmidt, G.D., Gioia, I.M., Maccacaro, T., Schild, R.E., Maccagni, D., and Arp, H.C. 1985, *Astrophys. J.*, **298**, 619.
Wilkes, B.J. and Elvis, M. 1987, *Astrophys. J.*, **323**, 243.
Worrall, D.M. 1987a, in *"Superluminal Radio Sources"*, eds. J.A. Zensus and T.J. Pearson (Cambridge: Cambridge University Press), p. 251.
Worrall, D.M. 1987b, *Astrophys. J.*, **318**, 188.

QUESTIONS:

C.M. Urry:

You posed a problem: your correlation of l_x *vs.* l_r suggests if the radio emission is beamed, so is the X-ray; however, the relative $< l_r >$ *vs.* $< l_x >$ suggests X-rays are isotropic. Ghisellini *et al.* (1985) have suggested a model that can solve both problems: a non-conical jet, with fan-like geometry where the X-rays are emitted, and a narrower radio jet. Then both are Doppler boosted but the selection effects are mitigated.

R. Antonucci:

You distinguish BL Lacs and HPQs from flat-spectrum radio quasars in general, but B. and D. Wills, R. Barvainis and I have found high optical polarization in most flat-spectrum radio sources. Previous low-polarization QSOs were highly polarized (and vice versa) in our survey. So the highly optically polarized objects are essentially the same as flat-spectrum sources.

Also, there is another good argument that if the radio cores of radio-selected blazars are beamed, so must their X-rays. The blazars have such strong diffuse (isotropic) radio emission that their misdirected parents must be normal double radio galaxies and quasars.Those objects don't show the strong X-rays, so the X-rays must be misdirected.

M.A. Malkan:

You and Richard Mushotzky both presented evidence that OVV, HPQ, and flat-radio-spectrum quasars have a distinct hard X-ray component with a flat ($-0.5 \leq \alpha_{HX} \leq 0.0; S \propto \nu^{\alpha_{HX}}$) slope. Is this signature related to (or produced in) the jets these objects are believed to have? Do you feel it's a pure coincidence this is the same as the slope of the X-ray background?

D.M. Worrall:

It is an interesting possibility that flat X-ray spectra are associated with the presence of jets, but I don't wish to speculate! M. Schmidt (at this meeting) has estimated that QSOs may produce a few hundred per cent of the X-ray background, but the contribution is dominated by assumed luminous high-redshift progenitors of local Seyfert galaxies. The percentage with bright flat-spectrum radio cores, however, may be relatively small. Incidently, the X-ray background contribution from higher luminosity QSOs is reduced by the effect I mentioned of the effective optical to X-ray spectral index steepening with increasing luminosity. There is no evidence that the radio-loud QSOs behave drastically different in this respect.

W. van Breugel:

If strong flat-spectrum sources are preferentially pointed towards us, and the radio emission is relativistically beamed, it does *not* necessarily follow that (for objects with an X-ray excess) an X-ray/radio correlation means that the X-rays are *also* beamed. If the nuclei are surrounded by dusty tori, as some theorists would propose, then an alternative explanation for the X-ray excess could be that since we are looking down the jet, we would also have less X-ray absorption, and hence there would be an apparent X-ray excess (from X-ray emission produced by Inverse Compton or whatever process).

D.M. Worrall:

That's interesting, and not something I've considered.

X-RAY TIMING OF ACTIVE GALACTIC NUCLEI

C. Megan Urry
Space Telescope Science Institute
3700 San Martin Way, Baltimore, Maryland 21218

I. Introduction

X-ray variability studies can have particularly important consequences for our understanding of active galactic nuclei (AGN), for several reasons. First, X-rays are produced by very energetic particles — such as thermal plasmas at several million degrees or relativistic nonthermal plasmas — which might exist at the heart of the active nucleus. Depending on the details of the X-ray emission (spectrum, variability time scales, etc.), X-rays may tell us about magnetic fields and particle acceleration in jets, or about densities and temperatures at the hot, inner edge of an accretion disk.

Second, AGN tend to be more rapidly variable at higher energies. Shorter time scales are generally associated with smaller emitting regions, possibly close to the central engine. Minimum variability time scales can be used to indicate maximum sizes for the emitting region (under some assumptions), which in turn limits the size of a supermassive compact object (a.k.a. black hole) that provides gravitational potential energy to power the AGN.

Third, and perhaps most importantly, recent X-ray results have changed the way we look at light curves, in a way that has applications to all wavebands. Because of recent results from the European X-ray satellite EXOSAT, X-ray studies of extragalactic objects are becoming increasingly sophisticated, to a degree previously encountered only in studies of *galactic* X-ray sources. The obvious reason is that we can now (with some effort) detect as many photons from a nearby Seyfert galaxy as were detected a decade ago in [shorter] observations of bright X-ray binaries (which are $\lesssim 10^{-7}$ the luminosity of Seyfert galaxies but as much as $\sim 10^3$ times brighter in X-ray flux). This dramatic improvement in the data has lead directly to sophisticated analysis techniques and exciting discoveries, and promises to revolutionize our approach to timing studies of AGN.

Before discussing this new outlook, I begin by summarizing (in §2) some traditional arguments about the implications of rapid variability. Much of this is simple, and certainly naïve, but it forms a good basis for §3, in which the X-ray variability of blazars (mainly BL Lacertae objects) is described. These can vary with large amplitude on very short time scales, and generally do not resemble other AGN in their variability characteristics. So-called "normal", non-blazar AGN (Seyfert galaxies and quasars), for which the recent EXOSAT results were obtained, are discussed in §4. Finally, in §5 I summarize the discussion and reiterate the main point of this review: that the new discoveries in X-ray timing of AGN, whether universally applicable or not, challenge our ingrained naïveté, and suggest that variability studies of extragalactic objects in all wavebands should aspire to the more advanced analysis techniques typically used for bright galactic X-ray sources.

II. The Old Orthodoxy about Timing Studies

The arguments presented in this section depend on the notion that a meaningful variability time scale can in fact be measured. Most definitions of time scale incorporate the magnitude of the change in flux, *viz.* the e-folding time, $t_{var} \equiv (d\ln F/dt)^{-1}$, or the doubling time, $t_{var} \equiv \bar{F}/\dot{F}$. The former assumes that the underlying variation is exponential, and the latter that it is linear. For example, a 10% change in 1 hour would imply a "time scale" of ~ 10 hours which is then presumed to have some physical meaning. This is fine if the 10% change is actually part of an uninterrupted, roughly linear trend. However, as we will argue in § 4, this may not be true in general, and therefore the observed fluctuations in intensity must be large — comparable to the mean intensity — in order for a time scale as defined above to reflect the size of the emitting region. Estimates of a time scale from smaller fluctuations are probably misleading.

The ideas that follow have been reviewed by many others (*e.g.* Fabian 1986a) and are only briefly summarized here. A simple causality argument says that variability cannot be faster than the light-crossing time of the emitting volume. Thus the observed variability time scale gives an upper limit to the size: $R \leq ct_{var}$. If the emitting volume moves at relativistic velocity, there is a special relativistic correction to this estimate, as illustrated in Figure 1.

According to the observer (us), an emitting blob moves with velocity βc at an angle θ to the line of sight. At point A it emits light, and a time Δt_{em} later (as measured in the emitter frame), it emits light from point B. By that time, light emitted at point A has already trav-

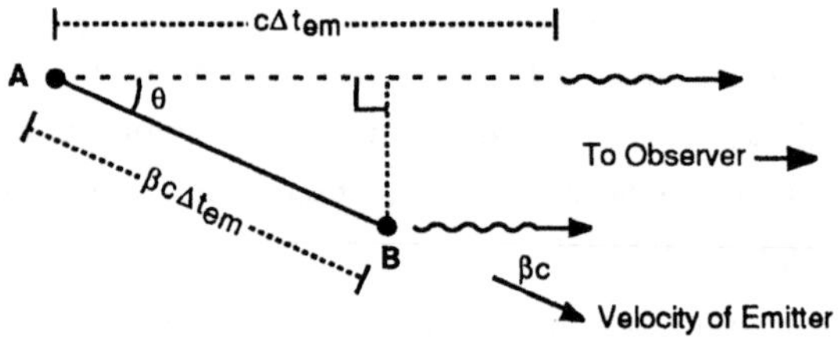

Figure 1. Schematic diagram of an emitting volume moving (relative to us, the observers) at relativistic speed βc from A to B, at an angle θ to the line of sight, in time Δt_{em} in the emitter frame. The distance between a light pulse emitted at A and one emitted at B is smaller than if the emitter were stationary, so the observed time interval between the two pulses is shorter than in the rest frame of the emitter.

eled a distance $c\Delta t_{em}$, so the difference between the path lengths toward the observer is $c\Delta t_{em} - \beta c \Delta t_{em} \cos\theta$. Therefore the observed interval between the two events is $\Delta t_{obs} = \gamma(1-\beta\cos\theta)\Delta t_{em}$. (The factor of $\gamma \equiv [1-\beta^2]^{-1/2}$ enters because of relativistic time dilation.)

This difference between observer and emitter frames is often expressed in terms of the kinematic Doppler factor, defined by $\delta \equiv (\gamma(1-\beta\cos\theta))^{-1}$, so that

$$\Delta t_{obs} = \delta^{-1} \Delta t_{em} . \tag{1}$$

In other words, for relativistic velocities and small angles to the line of sight, the observed time interval is shortened. For exactly the same reason, the observed radiation is blueshifted:

$$\nu_{obs} = \delta \, \nu_{em} . \tag{2}$$

An even more important effect is that photons are beamed in the forward direction (the so-called "lighthouse effect") as a consequence of the Lorentz transformation of solid angle, which is proportional to δ^2. Thus the observed flux density is modified:

$$F_{obs}(\nu_{obs}) = \delta^{3+\alpha} F_{em}(\nu_{em}) , \tag{3}$$

with two powers of δ from the solid angle, one power from the time, and the α from blueshifting a power law with that index ($F_\nu \propto \nu^{-\alpha}$). The observed flux can be orders of magnitude higher than the isotropically emitted flux for small θ and large γ. Figure 2 shows the variation of δ with angle to the line of sight, θ, for several values of the Lorentz factor of the bulk motion: $\gamma=2$ ($\beta \sim 0.866c$), $\gamma=5$ ($\beta \sim 0.980c$), and $\gamma=10$ ($\beta \sim 0.995c$). The function is very sharply peaked, more so for higher values of γ. Arrows indicate the angles at which the Doppler factor is unity for each of the three velocities; Doppler beaming is important when the relativistic bulk motion is nearly along the line-of-sight.

To return to the causality argument, the proper statement in the presence of relativistic motion is,

$$R \leq \delta \, c \, t_{var} , \tag{4}$$

since t_{var} is an observer-frame quantity. Relativistic beaming may be important in blazars, where variability time scales are short and luminosities can be very high, but it is probably not important for most other AGN.

Rapid variability implies a small emitting volume. It is a leap of faith to assume that the small volume is spherical and centered on a massive black hole. (Actually, any compact massive object will do if it has a deep potential well, but for convenience I will simply say black hole from here on.) It might instead be a thin planar region of shocked plasma well away from any central engine, black hole or not (*e.g.* Marscher and Gear 1985). Nevertheless, there is circumstantial evidence that at least makes this leap of faith attractive, such as the correlation of variability time scale with luminosity (Barr and Mushotzky 1986) or the correlation of line widths with luminosity (Wandel and Mushotzky 1986).

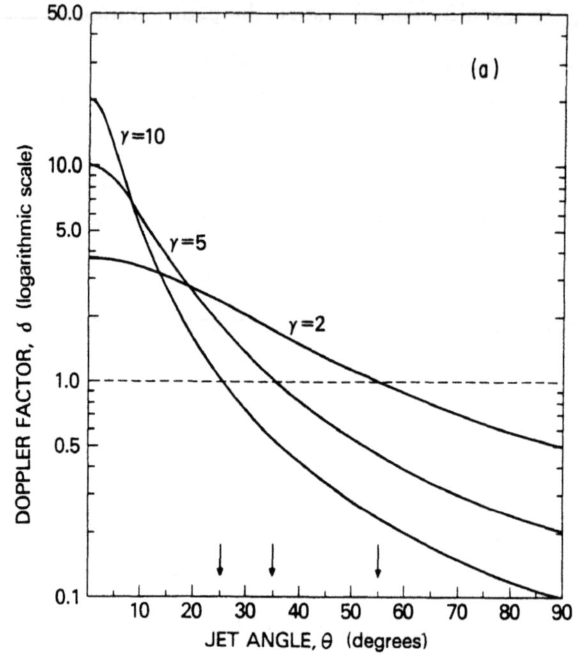

Figure 2. Variation of the kinematic Doppler factor, δ, with angle of the relativistic motion to the line of sight, θ, for three relativistic velocities: $\beta \sim 0.866c$ ($\gamma=2$), $\beta \sim 0.980c$ ($\gamma=5$), and $\beta \sim 0.995c$ ($\gamma=10$). The vertical scale is logarithmic, so δ is a very sharply peaked function of θ. Arrows indicate the angles for which the Doppler factor is unity, for each of the three velocities. Inside this angle, observed time scales are shorter than emitter-frame time scales and the observed flux is strongly enhanced by beaming; outside this angle, the time scales are lengthened and the flux is greatly diminished.

Suppose emission does come from the vicinity of a black hole. Because of the depth of the potential well, high energy emission like X-rays probably arises close to the black hole, say from $R \geq 3R_S$, where $R_S \equiv 2GM/c^2$ is the Schwarzschild radius. Then the absolute minimum variability observable time scale is the light-crossing time,

$$t_{\min} \sim \frac{R_{\min}}{\delta c} \sim \frac{6GM}{\delta c^3} \sim 30\,\delta^{-1} M_6 \text{ s}, \tag{5}$$

where M_6 is the black hole mass in units of $10^6 M_\odot$. Many authors have discussed more physical time scales, such as the dynamical time scale or heating and cooling time scales, but these are always longer, so the limit in Equation 5 is still valid.

If we observe rapid variability in an active galaxy, it can tell us more than the size of the emitting region. For spherical accretion, the emitted luminosity must be less than the Eddington limit, which is derived by balancing the forces on the accreting matter (here assumed to be protons); specifically, the gravitational attraction of the central object versus the radiation pressure of the outflowing photons:

$$L_{Edd} = \frac{4\pi G M_{bh} m_p c}{\sigma_T} \sim 10^{44} M_6 \text{ ergs/s}. \tag{6}$$

Then the largest variation ΔL that should be seen in time Δt is, from Equations 5 and 6,

$$\frac{\Delta L}{\Delta t} \leq \frac{10^{44}\, M_6\, \delta \text{ ergs/s}}{30\, M_6 \text{ s}} \sim 3\times 10^{42}\, \delta \text{ ergs/s}^2 \,. \qquad (7)$$

This represents an extreme case in which matter is converted to energy with 100% efficiency and the time scale reflects the light-crossing time, so it is obviously an upper limit. (A similar limit was derived more rigorously by Cavallo and Rees 1978 and Fabian 1979.)

One of the important physical quantities in AGN is their compactness (Guilbert, Fabian, and Rees 1983), the dimensionless version of which is,

$$l = \frac{L\, \sigma_T}{R m_e c^3} \,. \qquad (8)$$

For Eddington-limited accretion within $3R_S$, the compactness parameter is, from Equations 6 and 8, $l = 2\pi m_p / 3m_e \sim 4000$. If photons have energy near 1 MeV, pairs can be produced, and can dominate the energetics of compact sources. For l(at 1 MeV) $\gtrsim 10^{-3}$, pair production is important; for $l \gtrsim 10$ annihilation is an important cooling process (Svensson 1982 and reviews by Fabian 1986b and by Zdziarski and Lightman 1986).

These quickly sketched considerations show why timing studies are interesting — perhaps they can indicate the size of the black hole, the efficiency of converting matter to radiation, and so on. But these arguments are very crude. The Eddington limit applies only to spherical accretion, whereas AGN may well have cylindrical symmetry. Actual variability is likely to be much slower than the light crossing time. And, the main point of this review, observed time scales are often poorly measured, and may not reflect any characteristic time scale of the AGN.

III. X-ray Variability of Blazars

By definition, blazars tend to be highly variable. Their variability seems to become shorter with wavelength (see review by Bregman, Maraschi, and Urry 1987), and is quite pronounced at X-ray energies. Some examples of well-resolved X-ray variations are: 3C371, with $t \sim$ 10 days (Snyder *et al.* 1982); PKS2155−304, with $t \sim$ 6 hour (illustrated in Figure 3; Urry and Mushotzky 1982, Snyder *et al.* 1980) and $t \sim$ 1 hour (Morini *et al.* 1986); and H0323+022, with $t \sim$ 30 s (Doxsey *et al.* 1983, Feigelson *et al.* 1986).

One can compare an observed value of $\Delta L / \Delta t$ (assuming $\delta = 1$) to the limit in Equation 7. If it is significantly higher than 3×10^{42} ergs/s^2, then either efficiencies are much greater than 1, or $\delta > 1$, or the simple assumptions which went into the derivation of Equation 7 are incorrect. Generally the highest $\Delta L / \Delta t$ come from blazars, although not all blazars exhibit such behavior. This is apparent in Figure 4, adapted from Barr and Mushotzky (1986; see also Bassani, Dean, and Sembay 1983), which shows for a variety of AGN the shortest time scale of a resolved X-ray event versus the mean X-ray luminosity. The BL Lac objects, plotted as open circles, cluster to the right of the diagram (higher luminosity) and

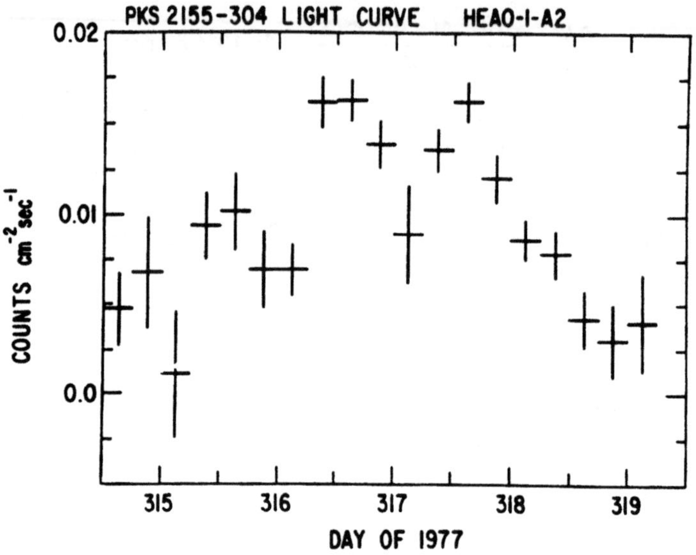

Figure 3. X-ray light curve of the BL Lac object PKS2155−304 (from Urry and Mushotzky 1982). The 2-10 keV intensity goes up and down by a factor of ≥2 in less than a day, for a $\Delta L/\Delta t \sim 5\times10^{41}$ erg/s^2. This implies a very high efficiency of converting accreted matter into X-ray luminosity, $\eta \sim 0.5$.

have a range of time scales that, unlike other AGN, is uncorrelated with luminosity.

Also plotted in Figure 4 are lines corresponding to various efficiencies, $\eta \equiv L/\dot{M}c^2$, of converting accreting matter to X-rays. (Eqn. 7 gives the relation for $\eta = 1$.) Although there are ways to produce values greater than 1 — for example, storage of energy in pairs followed by a global annihilation event — and although "super-efficient" $\Delta L/\Delta t$ flares have been seen in the sun, large implied efficiencies are usually taken to mean that relativistic effects are Doppler boosting the observed flux and shortening the observed time scale. In this connection, it is interesting that blazars have the highest observed values of $\Delta L/\Delta t$ (calculated assuming isotropic emission), since arguments that they are beamed have been put forward on other grounds (*e.g.* Blandford and Rees 1978; Urry and Mushotzky 1982; Worrall *et al.* 1982, 1984a,b,c, 1986). Particularly in a case like the BL Lac object H0323+022, the most rapidly variable extragalactic object known, relativistic effects are almost certainly indicated (Feigelson *et al.* 1986).

One very interesting question concerns periodicities, since these would most likely be associated with orbital phenomena. Most periods reported for BL Lac light curves are either short ($t \lesssim 1$ year) and do not persist over long time scales (*e.g.* Visvanathan and Elliot 1973; Frolich, Goldsmith, and Weistrop 1974; Kiplinger 1974; Carrasco, Dultzin-Hacyan, and Cruz-Gonzalez 1985; Valtaoja *et al.* 1985) or long-term ($t \gtrsim$ several years) with few cycles sampled (Barbieri *et al.* 1985; Webb *et al.* 1988; and Smith *et al.*, this volume). If the underlying process is stochastic flickering, uneven sampling over a short time span can produce spurious periodicities, or at least spurious peaks in the autocorrelation function (*e.g.* Press 1978, and Krolik, this volume).

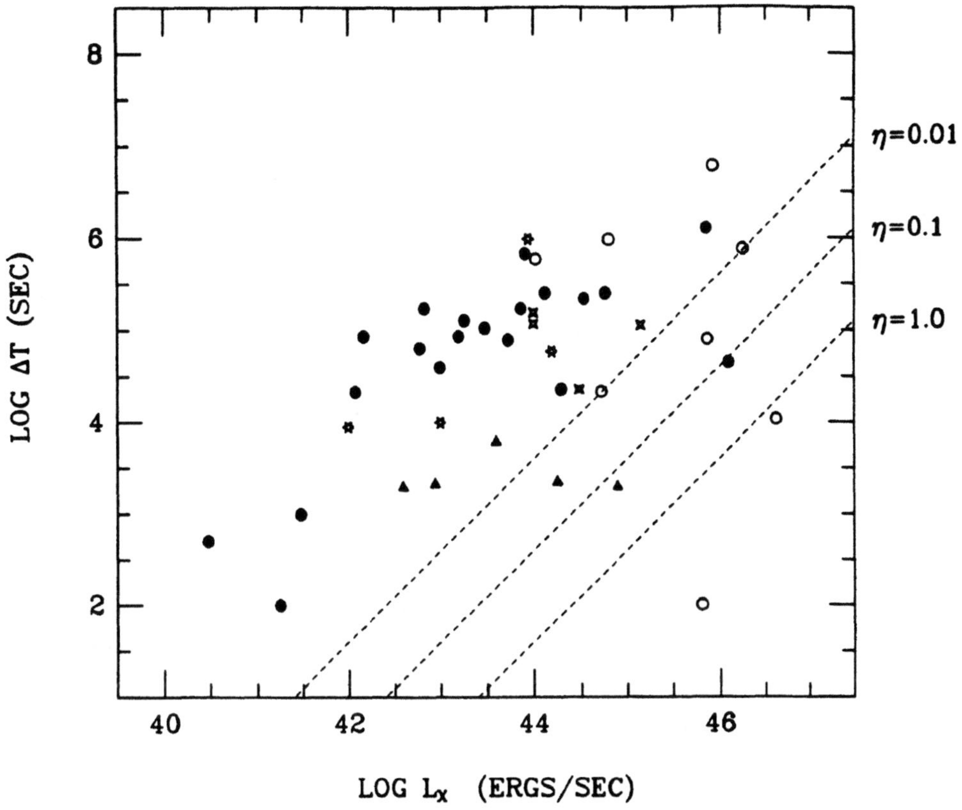

Figure 4. Two-folding variability time scale, t_{var}, versus mean 2-10keV X-ray luminosity, L_x (assuming isotropic emission). Filled circles are Seyfert galaxies and quasars from Barr and Mushotzky (1986); filled triangles are Seyfert galaxies and quasars from Warwick (1986); four-pointed stars are quasars from Zamorani *et al.* (1984); six-pointed stars are Seyfert galaxies from Urry *et al.* (1986); and open circles are BL Lac objects from Barr and Mushotzky (1986). The three dashed lines indicate efficiencies $\eta \equiv L/\dot{M}c^2 = 1.0$, 0.1, and 0.01, assuming all the luminosity emerges as 2-10keV X-rays. Note that $\eta \geq 1$ for some BL Lac objects, and Δt is uncorrelated with ΔL, while for the non-blazar AGN, there is a roughly linear correlation at efficiencies much less than 1.

Important variability studies of blazars remain to be done. In the X-ray, few high quality light curves are available. Also, the spectral evolution during an intensity change is a particularly important diagnostic of the emission process, and this requires still better data. Long term, high quality optical light curves do exist for a handful of blazars; see especially the 20-year records compiled by the Rosemary Hill Observatory (Smith *et al.*, this volume). These show a variety of behaviors, from slow, multi-year trends to more rapid flickering, and it is not clear whether the objects are different from one another, or are simply in different states over the brief span of the observations.

IV. X-ray Variability of Non-Blazar AGN

(a) Observational Picture

The first large X-ray variability survey of active galaxies, using HEAO1-A2 pointed observations of 38 AGN (Tennant and Mushotzky 1983), suggested that short time scale variability ($t_{var} \lesssim 10^4$ s) was rare. Only NGC6814 varied rapidly, with $t_{var} \sim 100$ s (Tennant *et al.* 1981). On the other hand, Tennant and Mushotzky found that many AGN varied by factors of ~ 2 over longer time scales, typically six-months or one-year.

Similar results were obtained from surveys of Seyfert galaxies (Urry *et al.* 1986) and quasars (Zamorani *et al.* 1984) with the *Einstein Observatory* Imaging Proportional Counter (IPC). For example, in a sample of more than 50 Seyfert galaxies detected with the IPC, only three varied on time scales less than a day (and none with $t_{var} < 10^4$ s), while 3 of 6 in the same sample (in 6 of 9 repeated observations) varied over six months or a year (Urry *et al.* 1986).

However, EXOSAT observations of AGN appear to tell a very different story: X-ray variability with $t_{var} \lesssim 10^4$ s seems relatively common (see review by Warwick 1986). The EXOSAT-observed targets are similar to or identical to the HEAO1-A2 objects, so intrinsic differences between the samples cannot explain the apparent discrepancy. What is the origin of this contradiction, and what does it indicate about the X-ray temporal characteristics of AGN?

The answer may be provided by EXOSAT "long looks" (*i.e.* observations lasting $\gtrsim 2\times 10^5$ s) of two low-luminosity Seyfert galaxies, NGC4051 (Lawrence *et al.* 1987) and NGC5506 (McHardy and Czerny 1987). EXOSAT sensitivity was somewhat better than the HEAO1-A2 sensitivity, but most importantly, the length of the uninterrupted data stream was unprecedented for extragalactic objects. In brief, both galaxies evidenced variability *on all time scales sampled* — each had a power density spectrum (PDS) that resembled $1/f$-noise (equal power per logarithmic bandwidth in frequency). If this PDS is characteristic of most AGN, then the detection of significant short time scale variability would be a function of signal to noise and of observation time. It would be rare in short or low signal-to-noise observations, like those of HEAO1-A2 and the *Einstein Observatory*, but more common in long observations ($t_{obs} \gtrsim 10^4$ s) like those of EXOSAT. In order to show this, the following section describes $1/f$-noise and its implications for variability studies.

(b) Flicker Noise, or $1/f$-Noise

Flicker noise is a characteristic of many physical systems. These can be completely unrelated to one another — indeed they may share only the single characteristic that their power density spectra are inversely proportional to frequency. Press (1978) gives a very nice review of flicker noise, describing both its mathematical properties and its occurrence in nature. Table 1 lists some of the diverse circumstances under which flicker noise is found (taken from the Press review and references therein). The last entry, stock prices on the New York Stock Exchange, can be either reassuring (the October 19 crash was just a statistical fluctuation?) or frightening (fluctuations of arbitrary amplitude are possible!), depending on your temperament. In any case, here we summarize the characteristics of flicker noise as they pertain to AGN.

Table 1. Examples of 1/*f*-Noise in Nature†

- Electric current in vacuum tubes and other current-carrying devices (at low frequencies)
- Sunspot number
- Underwater ocean currents
- Flow rate of the Nile over 2000 years
- Loudness and/or pitch of commercial radio broadcasts (including classical music, rock music, and talk shows)
- Flow of sand in an hourglass
- Electrical potential across nerve membranes
- Stock market prices.

† Taken from Press (1978).

A power density spectrum is essentially the square of the Fourier amplitude as a function of frequency, whether spatial, or temporal, or whatever. For the purposes of this paper, we are interested in an amplitude (the X-ray intensity) as a function of time, so "frequency" means temporal frequency. Suppose the amplitude at any time is independent of the amplitude at any other time. Then the intensities in this hypothetical light curve are essentially laid down by a random number generator. This is known as "white noise", and its PDS is a constant; in other words, there is equal power per Hertz at all frequencies. The instantaneous value of the intensity is undefined since adjacent points are unrelated, but over a long enough time, the mean *is* defined because a random number generator samples a normal distribution between fixed limits.

On the other hand, consider the integral of this process, where the amplitude is a random displacement from the adjacent point rather than from the mean. This is just the random walk. Here the instantaneous value is well-defined — it is related to the preceding value — but the mean is not. The PDS of the random walk goes as $1/f^2$ (the Fourier transform brings in one power of frequency because of integrating over white noise, and then it is squared). The process is called "red noise" because the power is significantly higher at low than at high frequencies. In fact the power diverges at the low frequency limit (the random walker can travel arbitrarily far from his starting point).

Flicker noise, or $1/f$-noise as it is eponymously called, is halfway between white noise and red noise, with [unfortunately] some of the worst characteristics of each. By definition, the PDS goes as $1/f$, and so the power diverges, albeit logarithmically, at both low and high frequencies. Therefore the mean is defined only after observations of infinite length. The change in amplitude from one point to the next depends on the sampling, and could be quite different as the interval is changed. For a light curve, this point-to-point change, $\dot{L} = \Delta L / \Delta t$, is related physically to the efficiency of converting accreting mass to light (see § 2). Then for pure $1/f$ processes, an estimate of the efficiency from observed "events" is meaningless. Also, the time scale of such events, $\bar{L}/\dot{L}$, is undefined, since neither $\bar{L}$ nor $\dot{L}$ are well-defined. In fact, the observed time scale is a function of the signal to noise, since this dictates the highest sampling frequency, and of the length of the observation, since this determines the mean intensity (and the lowest sampling frequency; Lawrence *et al.* 1987, McHardy and Czerny 1987).

One final note on the mathematical expression of flicker noise. In the past few years, fractals have been a popular topic in physics. X-ray light curves can be analyzed in terms of fractals as well (McHardy and Czerny 1987). Consider the well-worn but simple example of a craggy coast-line. The length of the coastline depends on the ruler by which it is measured. If the ruler is 100 miles long, one averages over many inlets and bays, and the coastline is relatively short, but as the measurement scale becomes finer, the measured length of the coastline increases. This variation in measured length with measuring scale is called the fractal dimension.

An X-ray light curve is a one-dimensional version of the coastline, and its fractal dimension, $D \equiv -d\log L(\Delta t)/d\log\Delta t$, depends on the "length" (the sum of the distances $|\Delta L|$ between points separated by a fixed length Δt) as a function of the "ruler size" (the interval Δt). For $1/f$-noise, this fractal length will be a power law of the time interval with index $D=0.5$, which is characteristic of one-dimensional turbulent processes. The general conversion from fractal dimension to PDS index (for a power law PDS) is $r=2(D-1)$ (McHardy and Czerny 1987).

In summary, if the X-ray light curves of AGN are dominated by $1/f$-noise over the time scales sampled, then individual flares or events are simply fluctuations that do not by themselves reveal anything about characteristic time scales or length scales in the AGN. Observers generally make note of the shortest apparent time scales, and these would just depend on the signal to noise and on the length of the observation — variations with arbitrarily short time scales could occur if you wait long enough. The eye is sensitive to long-term, low-frequency trends, and apparently observers tend to pick out, by eye, "periodicities" with ~ 3 cycles per light curve (Press 1978). These problems exist for optical light curves and radio light curves as well as for X-ray light curves. In other words, *in order to understand any light curves in any variability study, one must first know something about the character of the underlying variations (e.g. the PDS).*

Incidentally, just because a process is governed by $1/f$-noise does not mean that all is lost. There is information in the light curves that is neglected in calculating the PDS. For example, higher order correlation functions can reveal whether the fluctuations are retarded (a fast rise time and slow decay time) or advanced (a slow rise time and fast decay time), which is obviously very important for physical models of the emission process(es). Energy dependences, and cross-correlations between different energy bands might be very important. Model-fitting is also a useful tool, even though shot models, used successfully for Cygnus X-1 (Sutherland, Weisskopf, and Kahn 1978), do not have a satisfactory physical explanation. Still, if all light curves are dominated by $1/f$-noise over the observed frequency range, we are in a particularly ugly regime, with divergent power at both low- and high-frequency limits. The following section concerns the incidence of $1/f$-noise in cosmic X-ray sources, and the frequency range(s) over which it is seen. These observations raise as many questions as they answer, but in any case there are profound implications for observations of AGN at any wavelength.

(c) 1/f-Noise in Cosmic X-ray Sources

Even if careful studies showed that all AGN have $1/f$-noise over some frequency range, there would still be limits to that frequency range. At the high frequency (short time scale) end, the causality argument discussed in § 2 still holds — the fastest variability expected is the light-crossing time of the emitting region — so the PDS must go to zero. At the low frequency (long time scale) end, the PDS must turn over, since the overall amplitude of AGN variations, monitored for $\gtrsim 10$ years at X-ray and optical wavelengths, does not keep growing with new observations (*e.g.* Kunkel 1967, Webb *et al.* 1988, Mushotzky 1987).

The phenomenon of $1/f$-noise in X-ray light curves is already well-known for galactic sources, and the black hole candidate Cygnus X-1 is one of the best-studied cases. Between ~ 1 Hz and $\sim 10^3$ Hz its PDS goes as $1/f$, but there are breaks at a few milliseconds and at a few seconds (Meekins *et al.* 1984, Nolan *et al.* 1981). The former corresponds roughly to the size of a $10 M_{\odot}$ black hole, while the latter is unexplained.

There are several reasons why Cygnus X-1 is relevant to AGN. (1) Cygnus X-1 is a black hole candidate. Perhaps activity in the vicinity of a black hole could generate flicker noise, even over 7 orders of magnitude in the mass or luminosity. (2) The Cygnus X-1 PDS has been sampled over a wider frequency range than for AGN, and as predicted above, changes in shape (flattening at low frequencies, steepening at high frequencies) have been seen. Scaling the Cyg X-1 time scales by the bolometric luminosity — both are plausibly proportional to the black hole mass — gives, for low luminosity AGN like NGC4051 or NGC5506, ~ 2000 s and 2×10^5 s (Mushotzky 1987). These are just outside the range to which the EXOSAT long-looks were sensitive! To probe the high-frequency end, a factor of a few in sensitivity is needed; for the low frequency end, a total observation time of ~ 10 days is needed (longer if coverage is not continuous). (3) The minimum time scale for Cyg X-1 does seem to correspond to the size of the black hole estimated so crudely in

Equation 5. Then by analogy, the minimum time scale for AGN may reflect the black hole size. (4) Because of its relative nearness, Cygnus X-1 is much brighter than any AGN. In terms of the number of detected photons, an EXOSAT long look corresponds roughly to a 10 minute observation of Cygnus X-1 — imagine how much more could be learned about AGN with a 100-fold increase in sensitivity! The Japanese X-ray satellite Ginga, with several times the EXOSAT area and typical observation times of a few days up to a week, will probably get interesting timing data for a small sample of AGN, but high quality data of a sufficient cross section of active galaxies will probably require a low-background, large-area, high-telemetry-rate experiment like XTE (NASA's X-ray Timing Explorer). Unlike EXOSAT, both these satellites are hampered, at least for timing studies, by low earth orbits that cause frequent gaps in the data.

At this point we do not know whether flicker noise is important in all AGN, or in only a few. For MCG-6-30-15 (Pounds and Turner 1986), NGC4051 (Lawrence *et al.* 1987), and NGC5506 (McHardy and Czerny 1987), a power law with index $r \sim -1$ is a reasonable description of the PDS. (Some may be slightly steeper, $r \sim 1.2$, or have a break; McHardy, private communication). But several other AGN with comparable EXOSAT observations look very different. The Seyfert galaxies NGC4593 (Barr *et al.* 1986) and NGC4151 (Pounds and Turner 1986) have a steeper power-law PDS, with $r \gtrsim 2$. The rapidly variable NGC6814 seems to have much of its power in short-time-scale periodic variations (Branduardi-Raymont and Mittaz 1988). And some higher luminosity AGN, like III Zw 2 have remarkably "flicker-free" light curves, with a few isolated flares superimposed on a slowly varying or constant background (Pounds and Turner 1986). In the original variability study, Tennant and Mushotzky (1983) suggested that low luminosity AGN were more variable than high luminosity AGN. Perhaps the character of variability is fundamentally different, and low luminosity AGN have power density spectra with $r \sim$ 1-2 while high luminosity AGN have smooth light curves with flares superimposed. Incidentally, nothing is really known about the PDS of BL Lac objects. They may show flicker noise or something completely different. Estimating a time scale from the light curve shown in Figure 3 is probably reasonable, however, since the change in intensity is ~ 100%.

(d) How 1/f -Noise Can Account for Incidence of Variability

In § IV*a* an apparent contradiction emerged from comparing the HEAO1-A2 and *Einstein Observatory* variability studies of AGN to the EXOSAT results. This can be explained if the light curves of most AGN are dominated by flicker noise. The HEAO1-A2 pointed observations were not nearly as long as the EXOSAT long looks, although they have comparable signal to noise. (HEAO1-A2 had a smaller effective area but lower background rates.) Crudely, a typical HEAO1-A2 observation was ~ 15% as long, so there would be $\lesssim$15% of the probability of finding a rapid event in an EXOSAT observation. Also, fast or moderate trends might have been washed out by the sparse sampling and coarse binning in HEAO1-A2 compared to EXOSAT. For $1/f$-noise, more power is in the low frequency variations, and HEAO1-A2 observations were too short to be very sensitive to these.

The *Einstein Observatory* IPC had very high signal to noise because of the focusing optics (its effective area was much less than for HEAO1-A2 or EXOSAT but the background was smaller still). The IPC was able to detect a large number of AGN, many never previously observed. However, the observations were very short, usually $\lesssim 3000$ s. Because of the high signal to noise for bright X-ray sources, IPC observations were actually sensitive to a higher frequency range in the PDS than EXOSAT observations. If the PDS of a typical AGN were $1/f$-like up to such frequencies, the rarity of rapid variability in IPC observations (see § IV*a*) could simply reflect the lack of power at high frequencies. Had the IPC observations been longer, and continuous, flicker noise should have been readily apparent.

Figure 5 shows examples of the kind of data we are discussing: (a) HEAO1-A2 pointed observations of nine AGN (from Tennant and Mushotzky 1983); (b) *Einstein Observatory* IPC light curves of two Seyfert galaxies, one variable and one constant (Urry *et al.* 1986); and

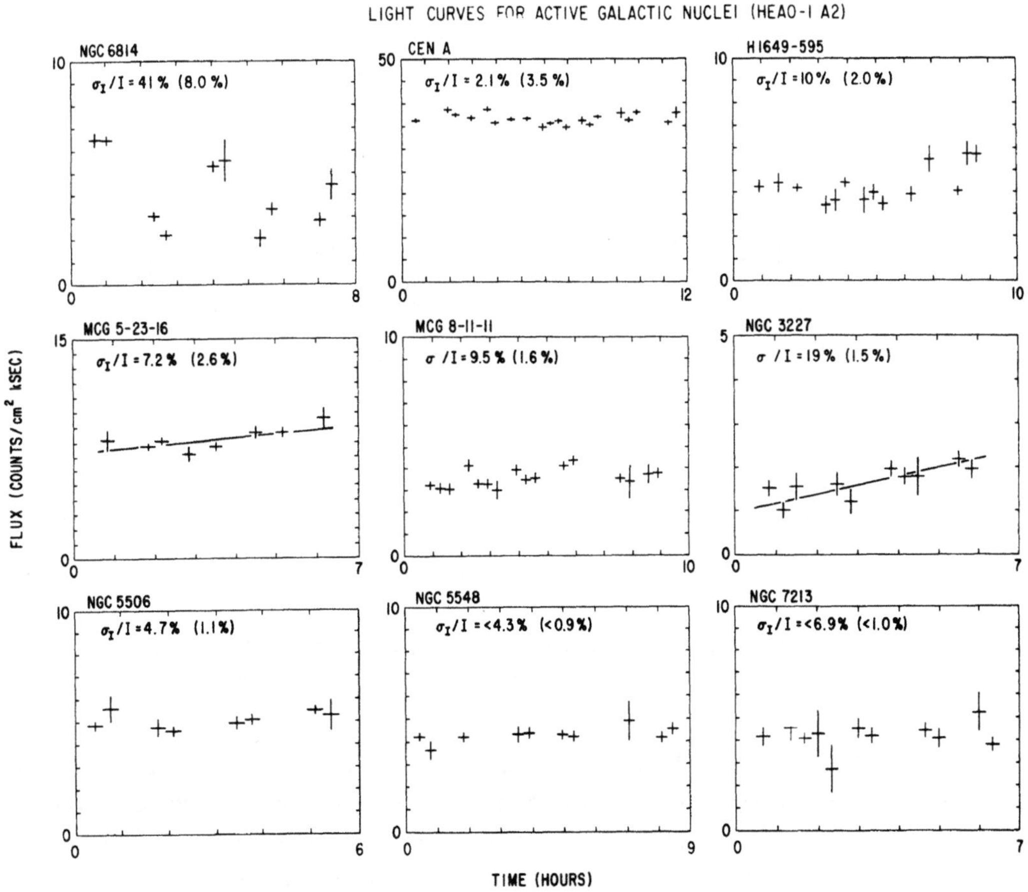

Figure 5a. HEAO1-A2 pointed observations of 9 active galaxies, mostly Seyfert 1s (from Tennant and Mushotzky 1983). Gaps in the data are due to Earth occultation or high background. The rapidly variable NGC6814 is at the upper left; the others are consistent with a constant intensity or a slow linear trend.

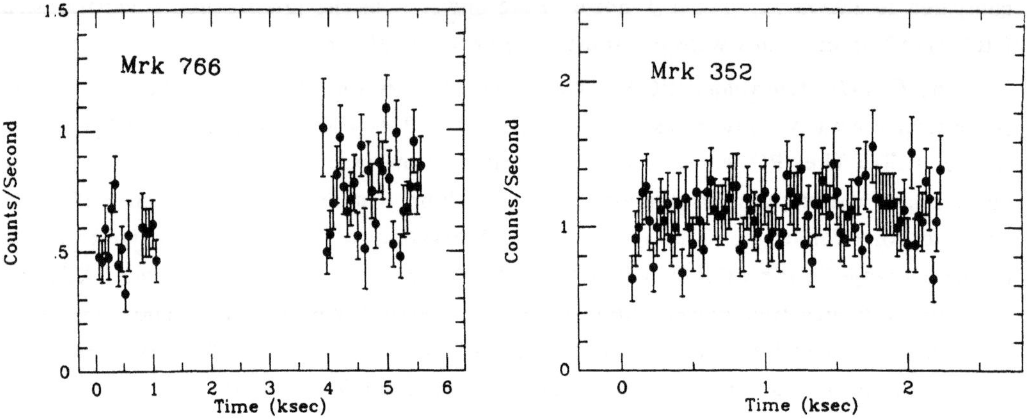

Figure 5b. *Einstein Observatory* IPC light curves of two Seyfert 1 galaxies. *left*: Mrk766, one of the few variable objects in the sample (with $t_{var} \sim 10^4$ s); *right*: Mrk352, a brighter source observed for half the time, which has constant intensity.

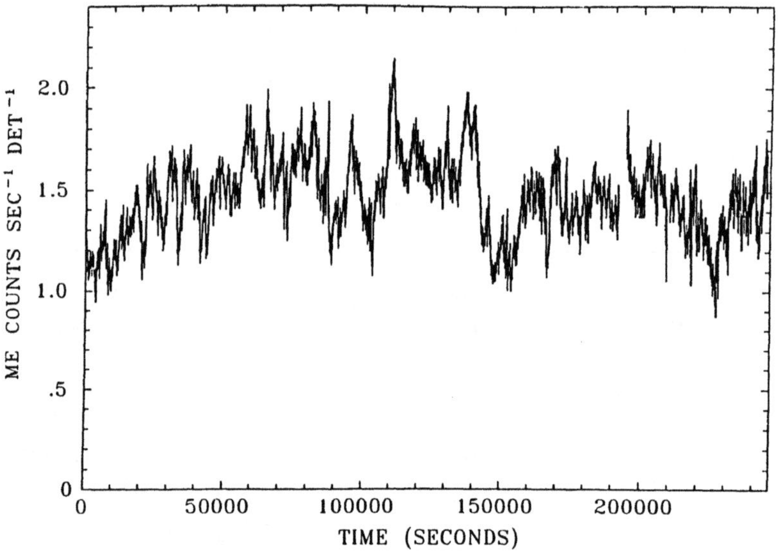

Figure 5c. An EXOSAT light curve of the moderate-luminosity Seyfert 2 galaxy NGC5506, taken from McHardy and Czerny (1987). The bin size is 400 s and the length of the continuous observation is nearly three days.

(c) the EXOSAT light curve of NGC5506 (from McHardy and Czerny 1987). One tick-mark in the time scale of the latter figure is 10,000 seconds, longer than most IPC observations; the bin size is 400 s, and a typical IPC observation would have had ~ 5 such bins, far too few to see variability except for the rare sharp spikes seen in light curve (c). A comparable HEAO1-A2 light curve, had it been as long as curve (c), would have had ~ 80 data points (rather than the ~ 600 shown) and ~ 45 gaps, doubtless obscuring a lot of the fluctuations.

As can be seen from Figure 5b, typical HEAO1-A2 light curves had ~ 10 data points (~ 1200 s bins) with data and ~ 4-5 gaps.

Consider the example of NGC5506, one of the flicker-noise AGN. The HEAO1-A2 light in the lower left corner of Figure 5a is unremarkable. Each bin is ~ 1200 s, with gaps that are nearly twice as long as the data train. The same lack of variability was inferred from a short EXOSAT observation of NGC5506 (McHardy 1983). (Both these observations occurred before the long look of Fig. 5c.) If we imagine sampling light curve (c) with the window functions of the light curves in (a) or (b), it is not hard to believe that the conclusions about rapid variability could be different. The different-looking light curves are actually consistent with one another. If so, Figure 5c indicates the richness of light curves that in less extensive data appear uninteresting. In some ways this is frightening (how much we are missing!) and in other ways it is inspiring (how much there is to learn!).

V. Conclusions

Recent results from the EXOSAT satellite have created new interest in X-ray timing of active galactic nuclei. High signal-to-noise, continuous, long-duration X-ray light curves of a handful of low luminosity AGN show a remarkable presence of variability on all time scales sampled. In particular, some have flicker noise, which is characterized by a power density spectrum inversely proportional to frequency. If this is a common characteristic of AGN, at any wavelength, then interpretation of "individual, resolved events" is problematic. All previous work in the field needs to be re-examined. Earlier failures to detect rapid variability may well be due to details of the PDS, specifically: (1) that there is more power at low frequencies, to which short observations are insensitive, (2) that there is little power at high frequencies, so the detection of rapid variability during short observations will be rare (it also requires high signal to noise), and (3) that sparse sampling can obscure a great deal of activity. New observations of comparable or better quality, of many more AGN, are badly needed. Some may be provided by the Ginga satellite; the rest may come from the XTE mission. Until we understand the character of the observed variations in a light curve, by measuring its PDS and perhaps higher order correlation functions, the interpretation of individual "flares" or "trends" or "periods" is not possible.

Acknowledgements — I would like to thank Dick Miller and Paul Wiita for organizing such a useful meeting as the Georgia State Conference on AGN, and for inviting me to speak. This paper was inspired in large part by a compelling talk on a similar topic given by Andy Lawrence at the Como Meeting on Variability in Galactic and Extragalactic X-Ray Sources, and I thank him, Ian McHardy, Ken Pounds, and Mike Watson for many illuminating discussions on that occasion. I also thank Richard Mushotzky for critical reading of this manuscript.

References

Barbieri, C., Cristiani, S., Omizzolo, S., and Romano, G. 1985, *Astr. Ap.*, **142**, 316.

Barr, P., Clavel, J., Giommi, P., Mushotzky, R.F., and Madejski, G. 1986, in *Variability of Galactic and Extragalactic X-Ray Sources*, ed. A. Treves, (Milan: Assoc. Advanc. Astron.), p. 43.

Barr, P., and Mushotzky, R.F. 1986, *Nature*, **320**, 421.

Bassani, L., Dean, A.J., and Sembay, S. 1983, *Astr. Ap.*, **125**, 52.

Blandford, R.D., and Rees, M.J. 1978, in *Pittsburgh Conference on BL Lac Objects*, ed. A.M. Wolfe, (Pittsburgh: University of Pittsburgh), p. 328.

Branduardi-Raymont, G., and Mittaz, J.P.D. 1988, preprint.

Bregman, J.N., Maraschi, L., and Urry, C.M. 1987, in *Exploring the Universe with the IUE Satellite*, ed. Y. Kondo, (Dordrecht: D. Reidel), p. 685.

Carrasco, L., Dultzin-Hacyan, D., and Cruz-Gonzalez, I. 1985, *Nature*, **314**, 146.

Cavallo, G., and Rees, M.J. 1978, *M.N.R.A.S.*, **183**, 359.

Doxsey, R., Bradt, H., McClintock, J., Petro, L., Remillard, R., Ricker, G., Schwartz, D., and Wood, K. 1983, *Ap. J. (Letters)*, **264**, L43.

Fabian, A.C. 1979, *Proc. Roy. Soc.*, **366**, 449.

Fabian, A. C. 1986a, in *The Physics of Accretion onto Compact Objects*, eds. K.O. Mason, M.G. Watson, and N.E. White, (Heidelberg: Springer-Verlag), p. 229.

Fabian, A.C. 1986b, in *Variability of Galactic and Extragalactic X-Ray Sources*, ed. A. Treves, (Milan: Assoc. Advanc. Astron.), p. 111.

Feigelson, E.D., Bradt, H., McClintock, J., Remillard, R., Urry, C.M., Tapia, S., Geldzahler, B., Johnston, K., Romanishin, W., Wehinger, P.A., Wyckoff, S., Madejski, G., Schwartz, D.A., Thorstensen, J., and Schaefer, B.E. 1986, *Ap. J.*, **302**, 337.

Frohlich, A., Goldsmith, S., and Weistrop, D. 1974, *M.N.R.A.S.*, **168**, 417.

Guilbert, P.W., Fabian, A.C., and Rees, M.J. 1983b, *M.N.R.A.S.*, **205**, 593.

Kiplinger, A.J. 1974, *Ap. J. (Letters)*, **191**, L109.

Kunkel, W.E. 1967, *A. J.*, **72**, 1341.

Lawrence, A., Watson, M.G., Pounds, K.A., and Elvis, M. 1987, *Nature*, **325**, 694.

Marscher, A.P., and Gear, W.K. 1985, *Ap. J.*, **298**, 114.

McHardy, I.M. 1983, in *Nonthermal and Very High Temperature Phenomena in X-ray Astronomy*, eds. C. Perola and M. Salvati, (Rome: Inst. Astr. Univ.), p. 117.

McHardy, I., and Czerny, B. 1987, *Nature*, **325**, 696.

Meekins, J.F., Wood, K.S., Hedler, R.L., Byram, E.T., Yentis, D.J., Chubb, T.A., and Friedman, H. 1984, *Ap. J.*, **278**, 288.

Morini, M., Chiappetti, L., Maccagni, D., Maraschi, L., Molteni, D., Tanzi, E.G., Treves, A., and Wolter, A. 1986a, *Ap. J. (Letters)*, **306**, L71.

Mushotzky, R. F. 1987, *Nature*, **325**, 666.

Nolan, P.L., Gruber, D.E., Matteson, J.L., Doty, J.P., Levine, A.L., Lewin, W.H.G., and Primini, F.G. 1981, *Ap. J.*, **246**, 494.

Pounds, K.A., and Turner, T.J. 1986, in *Variability of Galactic and Extragalactic X-Ray Sources*, ed. A. Treves, (Milan: Assoc. Advanc. Astron.), p. 1.

Press, W. H. 1978, *Comm. Ap.*, **7**, 103.

Snyder, W. A., Davidsen, A. F., Wood, K., Kinzer, R., Smathers, H., Shulman, S., Meekins, J.F., Yentis, D.J., Evans, W.D., Byram, E.T., Chubb, T.A., Friedman, H., and Margon, B. 1980, *Ap. J. (Letters)*, **237**, L11.

Snyder, W.A., Wood, K.S., Yentis, D.J., Meekins, J.F., Smathers, H.W., Byram, E.T., Chubb, T.A., and Friedman, H. 1982, *Ap. J.*, **259**, 38.

Sutherland, P.G., Weisskopf, M.C., and Kahn, S.M. 1978, *Ap. J.*, **219**, 1029.

Svensson, R. 1982, *Ap. J.*, **258**, 335.

Tennant, A.F., Mushotzky, R.F., Boldt, E.A., and Swank, J.H. 1981, *Ap. J.*, **251**, 15.

Tennant, A.F., and Mushotzky, R.F. 1983, *Ap. J.*, **264**, 92.

Urry, C.M., Kruper, J.S., Canizares, C.R., Rohan, M.L., and Oberhardt, M.R. 1986, in *Proceedings of the Conference on Variability in Galactic and Extragalactic X-Ray Sources*, (held in Como, Italy, October 1986), ed. A. Treves, p. 15.

Urry, C.M., and Mushotzky, R.F. 1982, *Ap. J.*, **253**, 38.

Valtaoja, E., Lehto, H., Teerikorpi, P., Korhonen, T., Valtonen, M., Terasranta, H., Salonen, E., Urpo, S., Tiuri, M., Piirola, V., and Saslaw, W.C. 1985, *Nature*, **314**, 148.

Visvanathan, N., and Elliot, J.L. 1973, *Ap. J.*, **179**, 721.

Wandel, A., and Mushotzky, R.F. 1986, *Ap. J. (Letters)*, **306**, L61.

Warwick, R.S. 1986, in *The Physics of Accretion onto Compact Objects*, eds. K.O. Mason, M.G. Watson, and N.E. White, (Heidelberg: Springer-Verlag), p. 195.

Webb, J.R., Smith, A.G., Leacock, R.J., Fitzgibbons, G.L., Gombola, P.P., and Shepherd, D.W. 1987, preprint.

Worrall, D.M., Puschell, J.J., Bruhweiler, F.C., Miller, H.R., Rudy, R.J., Ku, W.H.-M., Aller, M.F., Aller, H.D., Hodge, P.E., Matthews, K., Neugebauer, G., Soifer, B.T., Webb, J.R., Pica, A.J., Pollock, J.T., Smith, A.G., and Leacock, R.J. 1984a, *Ap. J.*, **278**, 521.

Worrall, D.M., Puschell, J.J., Bruhweiler, F.C., Sitko, M.L., Stein, W.A., Aller, M.F., Aller, H.D., Hodge, P.E., Rudy, R.J., Miller, H.R., Wisniewski, W.Z., Cordova, F.A., and Mason, K.O. 1984b, *Ap. J.*, **284**, 512.

Worrall, D.M., Puschell, J.J., Jones, B. Bruhweiler, F.C., Aller, M.F., Aller, H.D., Hodge, P.E., Sitko, M.L., Stein, W.A., Zhang, Y.X., and Ku, W.H.-M. 1982, *Ap. J.*, **261**, 403.

Worrall, D.M., Puschell, J.J., Rodriguez-Espinosa, J.M., Bruhweiler, F.C., Miller, H.R., Aller, M.F., and Aller, H.D. 1984c, *Ap. J.*, **286**, 711.

Worrall, D.M., Rodriguez-Espinosa, J.M., Wisniewski, W.Z., Miller, H.R., Bruhweiler, F.C., Aller, M.F., and Aller, H.D. 1986, *Ap. J.*, **303**, 589.

Zamorani, G., Giommi, P., Maccacaro, T., and Tananbaum, H. 1984, *Ap. J.*, **278**, 28.

Zdziarski, A.A., and Lightman, A.P. 1986, in *Variability of Galactic and Extragalactic X-Ray Sources*, ed. A. Treves, (Milan: Assoc. Advanc. Astron.), p. 121.

QUESTIONS AND COMMENTS

C. Dermer: Is any additional information obtained by the autocorrelation function?

C.M. Urry: The PDS is the Fourier transform of the autocorrelation function and so contains the same information. However, higher order (3- and 4-point) correlations do contain more information and can distinguish among light curves with the same PDS.

J.R. Webb: Professor Buchler at the University of Florida did a fractal analysis of some well sampled OVV light curves and found some structure, but the noise level was high, indicating we need more data.

C.M. Urry: Very interesting. Your data are probably good enough to have useful power density spectra as long as the window function is properly accounted for. Perhaps a discrete autocorrelation function is the answer.

J.R. Webb: Can 1/f noise account for high amplitude flares seen in light curves such as those of 0235+164 and 2345-16?

C.M. Urry: This depends upon the PDS (power at the frequency of the flare) and the length of observations (probability of observing such an event). I would like to see simulated light curves, sampled with real optical-observing-type windows, for a variety of PDS: $1/f$, $1/f^2$, and periodic or aperiodic flares. I wonder how these would resemble your real data, some of which looks like flicker noise to me.

R.F. Mushotzky: There exists at least one object, NGC 4151, which shows repeatable flares and has a PDS which is not 1/f. HEAO-1 long timescale data (days) indicate a flattening of the PDS at long timescales.

C.M. Urry: Right.

X-RAY AND OPTICAL STUDIES OF IRAS SELECTED AGN

Martin J. Ward
Department of Astronomy, FM-20, University of Washington
Seattle, WA 98195, USA

Summary

The samples of galaxies selected by virtue of their warm infrared colors defined from IRAS measurements contain large numbers of previously unknown AGN. X-ray measurements are reported for 7 of these. Their optical properties suggest different percentages of Seyfert 1.9s and narrow line Seyfert 1s than have been found using other selection techniques. The one non-Seyfert x-ray detection is of a galaxy with an HII type optical spectrum. It has an x-ray luminosity several times higher than the starburst galaxy NGC7714.

1. Introduction

Seyfert galaxies have been selected by diverse methodology, including continuum surveys at radio, optical/ultraviolet and x-ray frequencies. In addition, emission line criteria have been used to identify them from objective prism and slit spectroscopic surveys.

A new method of discovering additional AGN and Seyferts apparently missed by the above techniques, makes use of their mid-far infrared properties (12 μ - 100 μ). The IRAS detected about 20,000 galaxies, of which approximately 50% are in existing catalogs. Considerable effort has been devoted to construction of an IRAS selection criteria that will select AGN from this large data base, while excluding *normal* galaxies. These criteria select galaxies with so-called *warm* IRAS colors, meaning that they have IRAS energy distributions that are flatter than observed in *normal* galaxies. We should remember that the term *warm*, should not necessarily be assumed to imply a thermal component. In Seyfert 1s the 12–100 μ continuum *may* be significantly non-thermal.

De Grijp, Miley and Lub (1987) have reported the results of a spectroscopic survey of a sample of *warm* IRAS galaxies. A similar study has been undertaken by Kailey and Lebofsky (1987). Close to 100% of these galaxies exhibit an emission line spectrum, but since the exact criteria adopted by these workers are slightly different, this leads to somewhat different fractions of Seyfert and other types emission line galaxies. Kailey and Lebofsky report that 74% of their galaxies are Seyferts, *after* exclusion of previously known Seyferts, and of these 70% are of type 2 and 30% are of type 1. De Grijp, Miley and Lub claim that amongst the IRAS Point Source Catalog galaxies there will be nearly 1000 AGN. Based on their survey, between a third and a half of which will have been previously cataloged as Seyferts. In view of the size of this new sample it is of obvious interest to compare the properties of these warm IRAS selected Seyferts (henceforth, for convenience called WIRS), with the properties of Seyferts selected by other means. One such comparison involves their x-ray characteristics. Unfortunately, the discovery of WIRS barely overlapped with the lifespan of the *EXOSAT*. As a result very few WIRS were the subject of targeted x-ray observations (Ward *et al.* 1988). It is possible to supplement these in an *a posteriori* fashion, by searching for positional coincidences of the WIRS with the *Einstein* observatory data archive (Green *et al.* 1988).

I shall summarize the results from both of these studies.

2. The Sample

For the reasons given above, the sample is not large. In Table 1 is given a list of WIRS Seyfert 1s observed in x-rays. The term Seyfert 1 is taken generically to include objects with some evidence for a broad line component in their optical emission lines ie. Seyfert 1–1.9s. The table also includes data on the one x-ray detected starburst galaxy. Since this was a previously known object, and raises different issues, I will discuss it separately in section (5). No new Seyfert 2s were detected in x-rays. The x-ray upper limits for the 7 WIRS of type 2 included in the *Einstein* serendipitous sample, are given in Green *et al.*

In Table 1 it should be noted that the x-ray luminosities of the first 4 Seyferts refer to the hard x-ray band (2–10 keV), while the *Einstein* observations cover the lower energy band (0.5–4.5 keV). These differences will be significant when considering the absorption effects of high column densities, and also if a soft x-ray excess is present.

It is necessary to emphasize that many previously known Seyferts have both warm IRAS colors, and have been observed in x-rays. But the sample here discussed fullfils the following criteria. Either, they were x-ray targeted observations of Seyferts that had been selected on the basis of their warm IRAS colors, or, they are warm IRAS galaxies that were not targeted, but were serendipitously covered by *Einstein* observations. Clearly, those in the latter group could eventually be included in the *Einstein* medium x-ray survey. But that survey also includes many AGN that are not detected at all by IRAS, or do not have warm IRAS colors. Thus the sample in Table 1 forms a specific sub-set of WIRS. Obviously there is an important distinction between a complete sample of WIRS, and one comprised only of those Seyferts that were previously unknown (within the criteria above). Below I am referring specifically to the latter group.

3. X-ray Properties

The only Seyfert with an x-ray upper limit (IRAS05218) is anomalous, as discussed by Ward *et al.* Relative to a large sample of hard x-ray detected AGN, it is about 10 times underluminous in x-rays with respect to both its H α and IRAS infrared luminosities. Whether this is simply an extreme case of x-ray variablity, or a rare example of an x-ray quiet Seyfert

1, must average unreddened ratiointrinsic The other three hard x-ray detected Seyferts have x-ray, H α and IRAS luminosities typical of known Seyfert 1s. Ward *et al.* have used a large sample of hard x ray detected Seyferts, including these three IRAS selected examples, to study the obscuration of the broad line region. They find a good correlation between log H α / H β and log Lx / H β. This is interpreted as being due to reddening of the BLR. The form of the relation defines both the extinction law intrinsic to the AGN (between the wavelengths of H α and H β), and the average unreddened ratio of H α to H β in the BLR - which is found to be 3.5.

The x-ray properties of three *Einstein* observed WIRS are diverse, reflecting the different types of Seyfert 1s represented (see below). For example, the x-ray luminosity of IRAS14434 is low for a Seyfert 1, but this may be underestimated due to the presence of a high column density implied by the very steep broad line Balmer decrement. By analogy, the hard x-ray detected Seyfert 1.9 IRAS18325 was not detected in the low energy x-ray band of *EXOSAT*, suggesting a high column density. Therefore, IRAS14434 may be considerably brighter at hard x-ray energies. The other two WIRS detected by *Einstein* have x-ray luminosities typical for Seyfert 1s.

With the exception of IRAS05218, the WIRS sample is not dramatically different from other samples observed in x-rays. We detected all three WIRS of type 1 in the serendipitous *Einstein* sample, but no new Seyfert 2s (4 observed) were detected. This is consistent with the known fact that Seyfert 2s are 100-1000 times less luminous in x-rays than Seyfert 1s. It follows that in the sensitivity range of the *Einstein* observations ($1 - 3 \times 10^{-13}$ ergs cm^{-2} s^{-1} for a 2.6 sigma upper limit), we should expect a large percentage of x-ray detections for WIRS of type 1 which are closer than redshift about 0.05.

4. Optical Properties

Optical spectroscopy is available for all of the galaxies listed in Table 1. Figure 1 shows the optical spectra of two of these. A statistical investigation of the emission line properties of a large sample of WIRS is underway (De Grijp 1988), but some comment on the much smaller x-ray sample is justified.

The Balmer decrements of the six x-ray detected objects are generally higher than typically observed in the previously cataloged Seyfert 1s. Indeed, for those listed as type 1.9, the broad component of H β is absent, and the decrement quoted refers to the integrated profile ie. broad plus narrow component of H α and H β. Good limits to the broad line decrement alone, would require better signal/noise and higher resolution data. Another interesting fact is the incidence of Seyferts 1s with *narrow* permitted lines that are only slightly broader than the forbidden line profiles - the so-called narrow line Seyfert 1s (NLS1). Of the detections, only IRAS0412 has both the typically broad permitted lines of Seyfert 1s, and a clearly discernable broad component to the H β line. But in this case also, the Balmer decrement is steeper than the average for Seyfert 1s. It is unfortunate that Seyfert 1.9s and NLS1 are both prone to misclassification as Seyfert 2s, based on low resolution spectra of only moderate signal/noise.

Such a problem would modify the 70% to 30% division of Seyfert 2s and Seyfert 1s mentioned above. This underlines the importance of careful classification of WIRS if we are to interpret the statistical results correctly.

The incidence of Seyfert 1.9s in the WIRS sample could be related to the presence of dust. Osterbrock and De Robertis (1985) have reported that the Seyfert 2s selected by the warm IRAS color criteria, are more heavily reddened than the average Seyfert 2. This phenomena may also explain the large numbers of Seyfert 1.9s, although the dust heating mechanism *may* differ. Further evidence is found in the polarization properties of WIRS. It has been noted that WIRS have a higher level of optical polarization than found in previous samples (Brindle *et al.* 1988).

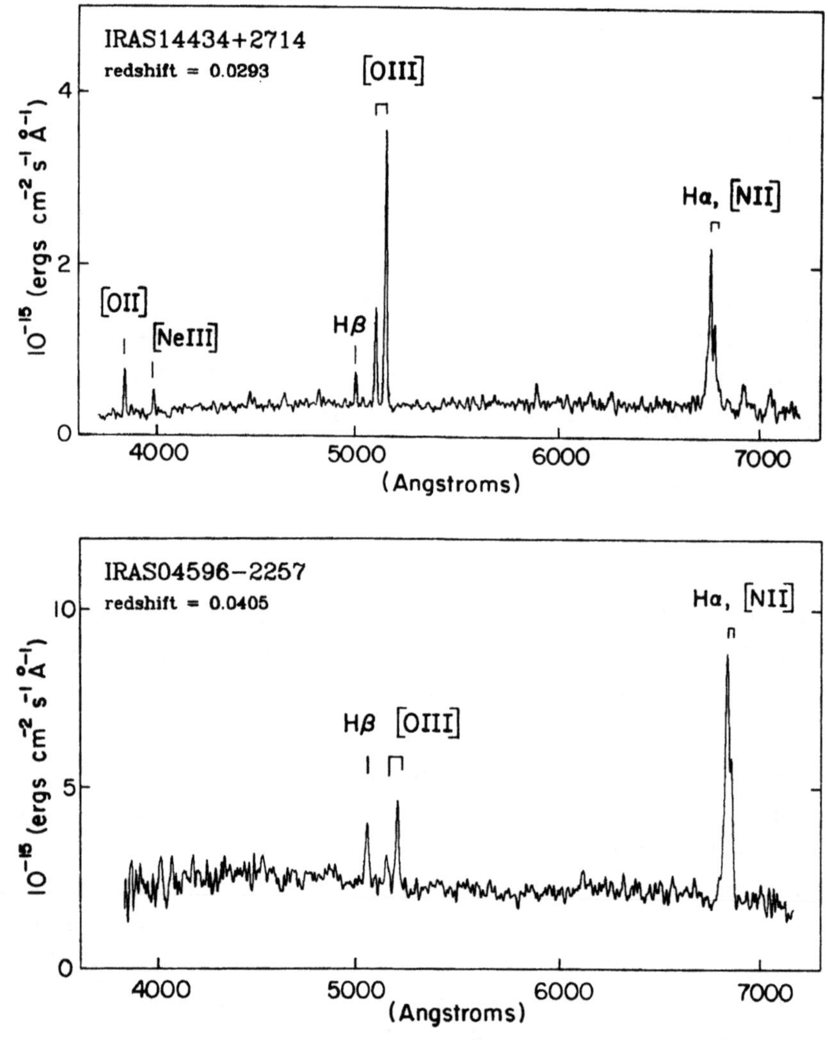

Figure 1: Optical spectra of two WIRS observed by the *Einstein* observatory

It is less obvious why there should be a high incidence of NLS1 amongst the WIRS sample. The most extreme case of a NLS1 in our x-ray sample is IRAS04596. This Seyfert has an $H\beta$ equivalent width of only 20 Å , although based on a careful examination of the $H\beta$ line and the x-ray luminosity, this is definitely a Seyfert 1. The $H\beta$ equivalent width is smaller than all except one of the NLS1 studied by Osterbrock and Pogge (1985). It is interesting, although given the numbers not compelling, that of the 5 out of 9 narrow NLS1 studied by Osterbrock and Pogge and detected by IRAS, *all* have warm IRAS colors. Yet, as Lawrence (1988) has pointed out, only about a third of a complete hard x-ray sample, and a somewhat lower fraction of Seyferts in the CFA redshift survey, have warm IRAS colors. The exact percentage of all Seyfert 1s that are NLS1 is not known, but the work of Osterbrock and Pogge indicates that it is very much less than 50%. This suggests that such objects are to be found preferentially amongst the WIRS sample. We can use the sample in Table 1 to test this assertion. Of these, we find that 43% (3/7) are NLS1.

Thus, while admitting that the statistics need considerable improvement, the incidence of *previously known* NLS1 that also have warm IRAS colors, and the fraction of the *new* WIRS that are NLS1, both suggest some link between warm IRAS colors and the presence of atypically low velocities in the BLR. Perhaps there may be large numbers of NLS1 that have been missed using the other selection techniques, but which are more fairly represented in the IRAS sample. Objects such as the NLS1 IRAS04596 *are* found in other surveys. For example the spectrum of MKN334 is remarkably similar to IRAS04596 (Veron and Veron 1986). Interestingly, the 60 and 100 μ fluxes of MKN334 are nearly equal, indicating a warm far infrared component. The important question to answer is, do the IRAS selection criteria offer an efficient means for discovering such objects? If this is a selection effect we might also expect to discover a higher percentage of NLS1 amongst the AGN in the *Einstein* medium survey. Unfortunately, no detailed spectroscopic results are yet available. What intrinsic property might connect a warm infrared continuum and narrow lines in Seyfert 1s? One possibility is the inclination of a nuclear disc. However, there are several difficulties with this suggestion. Osterbrock and Pogge have shown that NLS1 are not preferentially face-on galaxies, as predicted in a simple model with a BLR rotational velocity component in the plane of the host galaxy. This objection could be circumvented if the compact inner disc and the plane of the galaxy are not aligned. There remains the question of whether this disc, assumed responsible for the warm IRAS component, is optically thick. A quite different model could explain the observations, if the warm infrared component were non-thermal and we invoked some beaming mechanism.

These speculations should shortly be tested by detailed statistical analysis of the optical spectra for a large WIRS sample (De Grijp 1988).

5. X-rays from Starburst and HII Galaxies

Depending on the exact criteria adopted, samples of galaxies with warm IRAS colors will contain from 30–50% non-Seyfert emission line galaxies. Discusion of the sub-classes such as, starburst, HII galaxy and LINER is beyond the scope of this paper. I will briefy mention the only such non-Seyfert detection amongst the *Einstein* serendipitous sources.

The galaxy is NGC838 (MKN1022), and it has an HII type optical spectrum (Keel *et al.* 1985). It is interesting because it has a high x-ray luminosity for a non-Seyfert galaxy, and is several times more x-ray luminous than NGC7714, the prototype starburst galaxy (Weedman *et al.* 1981). Further, its x-ray luminosity may be underestimated since the Balmer decrement suggests an A_v of 2.5 mags, which could translate into a gas column density leading to low energy x-ray absorption. The origin of the x-rays from such objects is the subject of debate. The most favored mechanisms being supernova remnants and massive pop I binary systems (Fabbiano 1986; Ward 1988). Some key parameters that will determine the x-ray properties are; the IMF, age and metalicity of the starforming regions. An interesting question is, how x-ray luminous can such objects be, without having to resort to explanations involving obscured Seyfert nuclei ? At present the record is held by NGC1614(MKN617), which exhibits the optical and infrared signatures of a powerful starburst, and which has an x-ray luminosity of 5×10^{41} ergs s^{-1} (Kriss 1982). A high absorbing column may be present, thus increasing the intrinsic x-ray luminosity.

There has been some consideration of the possible contribution to the low energy diffuse x-ray background from starburst and HII region galaxies (Weedman 1987). The results are particularly sensitive to the luminosity functions of these objects. The accuracy of the estimates also suffer from the small number of x-ray detections. In principle the IRAS selected non-Seyferts should provide a good sample from which to derive the luminosity functions. Existing x-ray data together with some new x-ray observations of a sub-set from the IRAS sample, should enable us to better quantify their contribution to the x-ray background.

Table 1

Name	Redshift	log L_x (ergs s^{-1})*	Hα/Hβ†	Type
IRAS 05218	0.0492	43.02*H*	3.1	S1
IRAS 12496	0.0136	42.93*H*	3.2	NLS1
IRAS 15091	0.0436	44.12*H*	4.6	NLS1
IRAS 18325	0.0196	43.58*H*	19.1	S1.9
IRAS 04124‡	0.0386	43.14*L*	6.4	S1
IRAS 04596	0.041	42.81*L*	5.5	NLS1
IRAS 14434	0.029	41.70*L*	~10	S1.9
IRAS 02071(NGC838)	0.013	41.41*L*	6.6	H II

Notes:

H = hard X-rays (2–10 keV)
L = low energy X-rays (0.5–4.5 keV)
* $H_0 = 50$ km s^{-1} Mpc^{-1}
† Balmer decrement is derived from the integrated broad plus narrow components
‡ X-ray luminosity from Steiner, Grindlay and Maccacaro (1982)

References

Brindle, C., *et al.* 1988 preprint.

De Grijp, M.H.K., Miley, G.K., and Lub, J. 1987, *Astron. and Astrophys. Suppl.*, **70**, 95.

De Grijp, M.H.K. 1988, in preparation.

Fabbiano, G. 1986, *P.A.S.P.*, **98**, 525.

Green *et al.* 1988, *Ap.J.*, submitted.

Kailey, W.F., and Lebofsky, M.J. 1987 preprint.

Keel, W.C., Kennicutt, R.C., Hummel, E., and Van der Hulst, J.M. 1985, *A.J.*, **90**, 708.

Kriss, G.A. 1982, PhD Thesis MIT.

Lawrence, A. 1988, *Proceedings of the 3rd IRAS Conference, pub. Springer-Verlag.*

Osterbrock, D.E., and Pogge, R.W. 1985, *Ap. J.*, **297**, 166.

Steiner, J.E., Grindlay, J.E., and Maccacaro, T. 1982, *Ap. J.*, **259**, 485.

Veron-Cetty, M-P., and Veron, P. 1986, *Astron. and Astrophy. Suppl.*, **65**, 241.

Weedman, D.W., Feldman, F.R., Balzano, V.A., Ramsey, L.W., Sramek, R.A., and Wu, Chi-Chao. 1981, *Ap. J.*, **248**, 105.

Weedman, D.W. 1986, in *Star Formation in Galaxies*, ed. C.J. Lonsdale, NASA Conference Publication (2466) p. 351.

Ward, M.J., Done, C., Fabian, A.C., Tennant, A.F., and Shafer, R.A. 1988, *Ap.J.* in press January 15th.

Ward, M.J. 1988, *M.N.R.A.S*, in press.

Question, R. Mushotsky: Is there differential reddening of the emission lines in velocity space? Could the reduction in apparent line width for the narrow line Seyfert 1s be due to lower signal due to reddening?

Answer, M. Ward: Our data are of insufficient spectral resolution to address this.

Question, R. Goodrich: I have a comment on the previous question. In the case of the narrow line Seyfert 1 IRAS1509, we find that the ratio of Hα to Hβ is not a function of velocity within the line, so reddening is unlikely to be the source of the narrow line profile. I also have a question. For dust transmission in our own Galaxy the percent polarization divided by E(B-V) is less than or equal to 9% . For your three highly polarized IRAS Seyferts, does this relation apply? In other words, is the polarization consistent with dust transmission?

Answer, M. Ward: You are referring to the optical polarizations I quoted from Brindle *et al.* We can only estimate E(B-V) for 2 of these Seyferts. For IRAS1509 we derive 0.6 based on the best fit for the x-ray column density. For IRAS1832, E(B-V) from the Balmer decrement is about 1.0. So, for IRAS1509 the predicted polarization is close to that observed, but for IRAS1832 it is nearly twice as high. This question will best be answered by studies of the wavelength dependence of polarization.

Question, M. Malkan: It would be nice to see the correlation of log $H\alpha/H\beta$ against log $Lx/H\alpha$, instead of log $Lx/H\beta$. Then there would be no doubt that the positive correlation does not arise from correlating two ratios with the same denominator ($H\beta$). Could the increasing contamination in the Balmer lines from the narrow-line region be responsible for this correlation?

Answer, M. Ward: We too were concerned that such a correlation might arise as a mathematical artifact. In our paper Ward *et al.* (1988), we discuss a test that implies that the correlation is in fact real. On your second point, deconvolution of the broad and narrow line components is desirable, but unfortunately not available for all the galaxies in our hard x-ray sample. However, inspection of the location in the correlation plot of those galaxies with appreciable narrow line components suggests that they do not bias the result.

LEAKAGE OF UHE PHOTONS FROM AGNs: PRODUCTION OF X-RAY AND γ-RAY HALOS WITHIN 10-30 KPC

Isaac Shlosman and Marek Sikora
Joint Institute for Laboratory Astrophysics
University of Colorado and National Bureau of Standards
Boulder, CO 80309-0440

Relativistic electrons/positrons, responsible for nonthermal radiation of AGNs, can result from a pair cascade, initiated by injection of ultrarelativistic pairs by ultrarelativistic protons and supported by synchrotron radiation (Sikora et al. 1987, hereafter SKBS). It was shown by several authors that relativistic protons can be effectively produced in collisionless shocks (e.g. Blandford and Eichler 1987). They are cooled due to interactions with cold matter, i.e. in proton-proton collisions, and due to interactions with soft radiation resulting in meson or electron-positron pair production (Blumenthal 1970; Colgate 1983; SKBS). For strong shocks, supposed to operate in the central region of AGN, the most efficient cooling mechanism of untrarelativistic protons is photoproduction of ultrarelativistic neutral pions (π^o). Pions decay into ultra-high energy (UHE) photons ($>10^{14}$ eV), and, due to their absorption by soft radiation of the central source, the ultra-relativistic e^+e^- pairs are created. They are followed by next generations of pairs in the synchrotron-supported pair cascade process. Such a model avoids the difficulties of direct energy transfer from shocks to relativistic electrons.

We raise the possiblity that the UHE photons, in fact, can leak out of the central region, calculate the fraction of escaping photons and discuss their importance for our understanding of the ongoing physical process in the galactic nuclei (Sikora and Shlosman 1987).

The UHE photons are known to be effectively absorbed by the microwave background radiation with a mean free path ~10-30 kpc (see Fig. 1). Electron-positron pairs created due to this absorption will produce synchrotron radiation in the interstellar magnetic fields, i.e. a nonthermal halo around AGNs. The spectral shape of the halo synchrotron radiation and its frequency range will depend upon the spectrum of the UHE radiation leaked out from the central source, as well as on the interstellar magnetic fields. It will extend down to UV-soft X-ray range and, since it is produced in situ, it may affect physical conditions in broad and narrow emission line regions of AGN, as well as in nearby molecular clouds of the host galaxy.

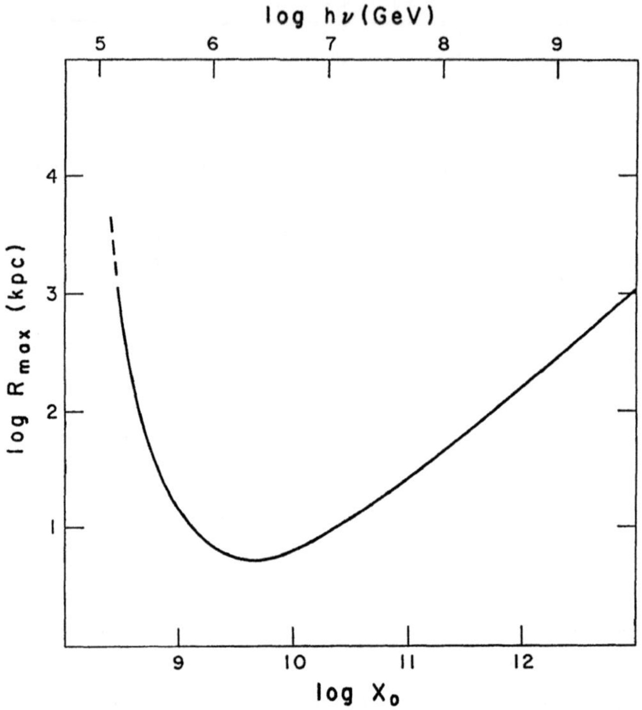

Fig. 1. The mean free path of the UHE photons interacting with the microwave background radiation at present epoch as a function of their energy $x_o \equiv h\nu/m_e c^2$.

Most of the UHE radiation is absorbed on the spot and supports the pair cascade inside the central source of ~10-20 Schwarzschild radii and luminosity L_o. The efficiency of escape, ξ_{esc}, depends on the two parameters -- the mass of the black hole and the compactness of the source, $\ell = \sigma_T L_o / 4\pi R m_e c^3$, approximately as

$$\xi_{esc} \approx 1.5 \times 10^{-2} (\ell/0.1)^{-1/2} (M_{BH}/10^8 M_\odot)^{-1/2} . \tag{1}$$

The upper limit of halo synchrotron spectrum is given by

$$(h\nu/m_e c^2)_{max} \sim 10^2 \ell^{-1/2} (M_{BH}/10^8 M_\odot)^{1/2} (B_{ISM}/1\ \mu G), \tag{2}$$

while the lower limit can be obtained by a comparison of the synchrotron cooling time scale with the age of the UHE source, t_{AGN}, and is

$$(h\nu/m_e c^2)_{min} \sim 10^{-7} (B_{ISM}/1\ \mu G)^{-3} (t_{AGN}/10^7 yr)^{-2} . \tag{3}$$

We show that this nonthermal radiation, when detected, can provide an independent estimate for the mass of the black hole in the AGN. If both the nonthermal luminosity of AGN and the halo luminosity are known, Fig. 2 can be used trivially, as a precalculated map, to recover the compactness parameter of the central source ℓ and the mass M_{BH}. There are, however, technical constraints which must be overcome. First, to measure the halo nonthermal luminosity, it is necessary to have instrumental resolution corresponding to <5 kpc, in order to extract the much more luminous central source. Instrumental angular resolution must be then $\delta \sim (z/0.01)^{-1}$ arc min, where z is the cosmological redshift. Second, the absolute luminosity of halo is quite low, $L_{esc} \sim 10^{41}(M_{BH}/10^8 M_{\odot})^{1/2}$ ergs s^{-1}, and hence a high sensitivity is required to detect the distant AGNs.

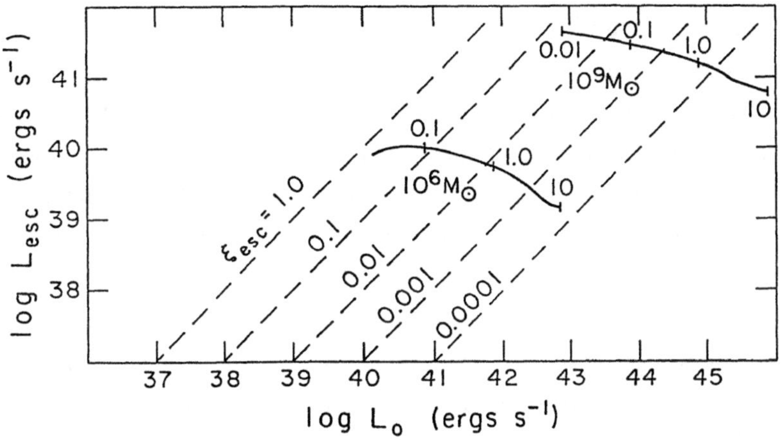

Fig. 2. The total luminosity of synchrotron halo, L_{esc}, versus the nonthermal AGN luminosity L_o for different M_{BH} and ℓ. Also shown are the lines of constant efficiency, ξ_{esc}, for escaping UHE radiation (from Sikora and Shlosman 1987).

Detection of nonthermal halos around AGNs would provide a strong argument in favor of efficient proton acceleration and of proton-photon injection of ultra-relativistic pairs, responsible for the formation of nonthermal continua in AGNs.

The authors acknowledge support under the NASA Astrophysical Theory Center Grant NAGW-766 and the NSF Grant 83-51997.

References

Blandford, R. D. and Eichler, D. 1987, preprint.

Blumenthal, G. R. 1970, Phys. Rev. D 1, 1596.

Colgate, S. A. 1983, 18th Intern. Cosmic Ray Conf., Vol. 2, p. 230.

Sikora, M., Kirk, J. G., Begelman, M. C., and Schneider, P. 1987, Ap. J. Lett. 320, L81 (SKBS).

Sikora, M. and Shlosman, I. 1987, submitted to Ap. J.

RADIATION DYNAMICS AND PAIR CREATION IN AGN ACCRETION SHOCKS

John M. Blondin[1]
NASA/GSFC, Laboratory for High Energy Astrophysics
Code 665, Greenbelt, MD 20771

Arieh Königl
Dept. of Astronomy and Astrophysics, University of Chicago
5640 S. Ellis Ave. Chicago, IL 60637

If the high-energy emission from active galactic nuclei is generated within a small enough region, the optical depth to photon-photon collisions will exceed unity and the energy emitted in gamma-rays will be converted into e^+e^- pairs. When this pair production can maintain a positron density, n_+, greater than the background ion density, n_i, then the Thomson optical depth of the emission region will be enhanced by a factor of $(1+z)$, where $z = 2n_+/n_i$. Provided that the pairs remain well coupled to the accreting matter, the mean mass per Thomson scatterer will be reduced to $\bar{m} \approx m_p/(1+z)$, where m_p is the mass of the proton. This, in turn, reduces the *effective* Eddington luminosity by the factor $(1+z)$, which suggests that the steady-state source luminosity might be limited to a value substantially smaller than the nominal Eddington luminosity (e.g., Lightman, Zdziarski, and Rees 1987). The enhanced optical depth of the accretion flow will also increase the effect of photon trapping, which could further reduce the luminosity reaching the observer. These arguments, however, are based on an equilibrium pair density calculated without considering the dynamical effect of the radiation on the accretion flow, which will be important when the luminosity approaches the effective Eddington limit. We have investigated the dynamical role of the radiation and pairs in compact sources with the help of a simple model of a spherical shock that is mediated by Fermi-accelerated relativistic protons and radiation, and in which e^+e^- pairs are produced through photon-photon collisions. We found that the equilibrium pair density decreases as the luminosity approaches the effective Eddington luminosity due to a feedback effect on pair production. On the basis of these results, we conclude that compact sources are not expected to be Eddington limited to $L << L_E$ on account of e^+e^- pair production.

In order to determine how pair-creation affects the limiting luminosity, we considered a model in which the bulk of the accretion energy is liberated in a spherical shock at a distance r_{sh} from the black hole (cf. Kazanas and Ellison, 1986). We have assumed that the shock is mediated only by relativistic protons (produced by Fermi acceleration in the shock) and by radiation. A fraction, β (of order 1/2), of the energy input into relativistic protons is assumed

[1]NAS/NRC Resident Research Associate

to produce a thermal population of e^+e^- pairs with an efficiency χ. The local pair density is determined by the pair conservation equation:

$$1/r^2\, d/dr(r^2 n_+ v) + 3/8\, \sigma n_+ n_- = \chi (dq/dr)/(2m_e c^2),$$

where dq/dr is the local (pair) energy injection rate in erg/s/cm^3, and σ is the Thomson cross section. The first term represents the advection of the pairs with the accretion flow, and can be ignored in the limit of large optical depth, $\tau >> v/c$. An upper limit for dq/dr is found by assuming that all of the liberated accretion energy (a fraction $1\text{-}f$ of the total accretion energy, where f is the ratio of infall to free-fall velocity) is available for pair production at r_{sh}, so that $dq/dr \leq 0.5(1\text{-}f)^2 GM\dot{M}/(4/3\pi r_{sh}^4)$. In this limit the Eddington reduction factor is given by

$$1+z \leq \{1 + 3\times 10^4\ \chi\, f^2\, (1\text{-}f)^2 / (x_{sh}\dot{m})\}^{1/2}$$

where $x_{sh} = r_{sh}c^2/2GM$, $\dot{m} = \dot{M}c^2/L_E$, and f is of order 1/2. Using a shock radius of $x_{sh} = 5$, $f = 0.5$, $\dot{m} = 1$, and $\chi \approx 0.1$ (e.g., Svensson, 1987), we find a typical pair ratio of $1+z \approx 6$. The local pair number density can thus exceed the ion density by a significant factor, although it is never large enough for the mass density of pairs to become important in the accretion flow (i.e., $\rho_e << \rho_i$).

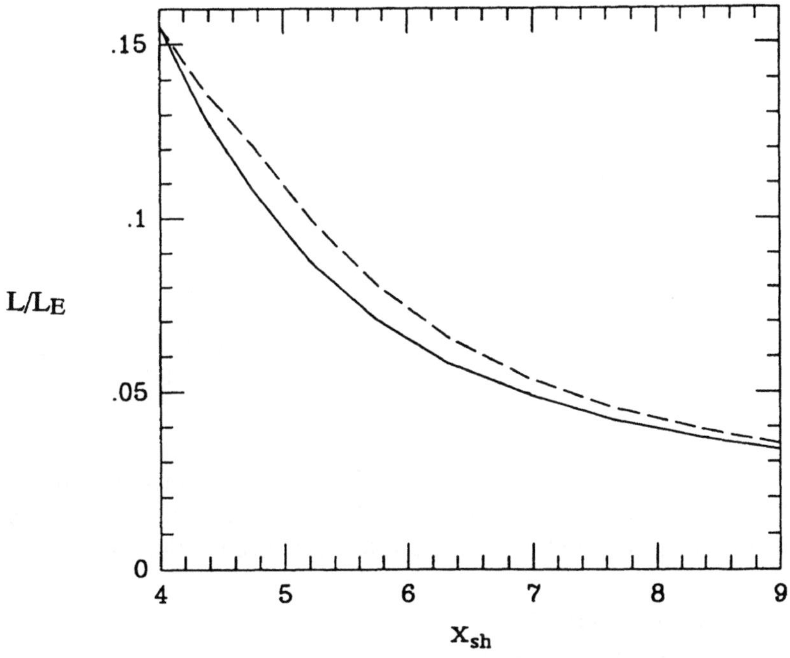

Fig. 1. Variation of the escaping luminosity with the shock radius. The solid curve represents the full solution (including the effects of pairs). The dashed curve represents the corresponding solution without pair creation (but including the dynamical effect of radiation). The normalized mass accretion rate used both here and in figure 2 is $\dot{m} = 1.5$.

A more complete description of the model that we have employed in this investigation is given elsewhere (Blondin and Königl, 1987). The principal results of this model are presented in Figures 1 and 2. One somewhat unexpected result is that radiation trapping by pairs is not a very strong effect in the shock. A plot of the escaping luminosity versus shock radius for a fixed mass accretion rate (Fig. 1) demonstrates that the decrease in the escaping luminosity as a result of the inclusion of pairs is comparatively small. This reflects the feedback of the increase in the radiation-pressure support on z. As the region of dissipation (i.e., r_{sh}) approaches the effective trapping radius $(1+z)r_{tr}$ (where $r_{tr} \approx 2GM/c^2(\dot{m}/2)$; e.g., Blondin 1986), the luminosity approaches the effective Eddington luminosity $L_E/(1+z)$, and radiation-pressure effects become increasingly important. However, as more work is done by the radiation in slowing down the inflow, less accretion energy is available for pair production, and the pair density decreases (see Fig. 2). As a result, the local effective Eddington luminosity increases with the escaping luminosity until $z \approx 0$ and the limiting luminosity is again just L_E.

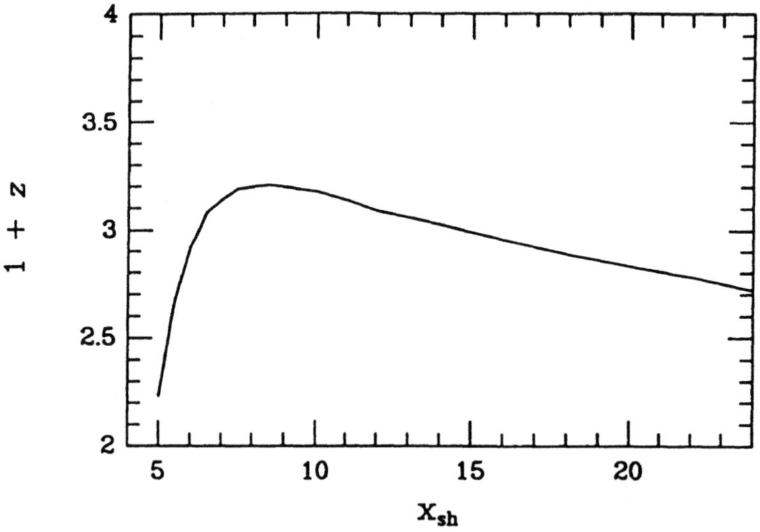

Fig. 2. The maximum value of the pair to ion ratio, $1 + z$, as a function of the shock radius. At large radii $(1 + z)$ decreases roughly as $x_{sh}^{-1/2}$. The rapid decrease in the pair density at small radii is a manifestation of the feedback effect discussed in the text.

Blondin, J. M., 1986, *Ap.J.*, **308**, 755.

Blondin, J. M., and Königl, A. 1987, *Ap.J.*, **323**, 800.

Kazanas, D., and Ellison, D. C. 1986, *Ap.J.*, **304**, 178.

Lightman, A. P., Zdziarski, A. A., and Rees, M. J. 1987, *Ap.J. (Letters)*, **315**, L113.

Svensson, R. 1987, *M.N.R.A.S.*, **227**, 403.

THERMAL COMPTONIZATION IN COMPACT SOURCES AND THE COSMIC X-RAY BACKGROUND

Andrzej A. Zdziarski
Space Telescope Science Institute, Homewood Campus, Baltimore, MD 21218*,
and California Institute of Technology, Pasadena, CA 91125

M. N. R. A. S., in press; Received 30 July 1987; Accepted 20 January 1988

Summary

We model thermal Comptonization and e^+e^- pair production in high viscosity black hole accretion flows. We assume that magnetic fields are in equipartition with the energy density of the flow and that cyclotron radiation is the major source of seed soft photons. The electrons are assumed to receive a constant fraction of the gravitational energy. We find that the resulting spectra have a constant X-ray energy spectral index, $\alpha_x \simeq 0$, for accretion rates $0.5 \gtrsim \dot{m} \gtrsim 20$. Here $\dot{m} \equiv \dot{M}c^2/L_{\rm E}$, and $L_{\rm E}$ is the Eddington luminosity. The spectra are cut off above ~ 150 keV. Such spectra closely resemble that of the residual cosmic X-ray background (after subtraction of the contributions of the known classes of extragalactic sources) if the background originated at a redshift $z \sim 4$–5. Thus, the residual cosmic background can be explained by an early population of active galaxies emitting thermal Comptonized spectra, as originally suggested by Leiter and Boldt (1982).

---

* Present address.

Radio Sources: Small Scale Structure

Marshall H. Cohen
California Institute of Technology

In this review I concentrate on three features of the small-scale structure of radio sources: their morphology and classification, the superluminal motion, and some statistics. Fortunately, a number of excellent review articles have appeared recently which bear on these and related topics, so my task is rather easy. I especially recommend the volumes *Superluminal Radio Sources* [1] and *The Impact of VLBI on Astrophysics and Geophysics* [2], both of which contain informative reviews as well as numerous short articles.

The study of radio sources is still split by instrumental possibilities, with the VLA making beautiful detailed maps at resolutions down to 0.5 arcsec or so, and VLBI making cruder (but improving) maps between about 0.1 and 100 milliarcseconds (mas). Information gained with the latter technique is used in this article, although the larger structure occasionally must be discussed also. The linear scales discussed here range from about 0.02 pc on nearby galaxies, to 1–100 pc on quasars. Throughout this article, $H_o = 100\ h$ km sec^{-1} Mpc^{-1}, and $q_o = 0.5$.

Morphology and Classification

The first step towards a classification scheme for compact radio sources is now being made by Pearson and Readhead [3]. They defined a complete flux-limited sample of 65 objects, chosen at 6-cm wavelength, and identified ten classes of compact sources, based on morphology (both small-scale and large-scale) and spectrum. For this review, however, I compress them into six categories.

1. Asymmetric (25/65)

Most objects are asymmetric on small scales. Pearson and Readhead subdivide them according to F_c, the fraction of the total flux density at 6-cm wavelength which is in the compact core. This is not the VLA core, but rather is the highest surface-brightness component in the VLBI model, with typical size < 1 mas. All Asymmetric objects have a compact flat-spectrum core and at least one secondary component, and the spectrum (including all outer components) must be flat ($\alpha > -0.5$). The secondary, when it has been measured, has a steep spectrum at short centimeter wavelengths. In many cases there are multiple secondary components, always forming a line on one side of the core. Superluminal motion of these components is common. The flux variability, and other indicators of nuclear activity, increase with F_c, and for $F_c > 0.5$ (17/25) the identifications are exclusively with quasars and BL Lac objects. Four objects have $F_c \geq 0.9$, but none is totally unresolved at a wavelength of 6 cm, with angular resolution 1 mas.

Asymmetric sources with the highest values of F_c merely show the core plus a weak close secondary. Those with intermediate values of F_c, such as the variable quasars 3C 273 and 3C 345, show the flat spectrum core plus a number of secondary components. The secondary components usually define a narrow curved jet, as in figure 1. 3C 345 has a projected curvature of at least 60° within 2 mas (8 pc), and the outer jet continues to curve [4]. In 3C 120 the jet is continuous and curves smoothly through more than 90° before it disappears into an outer lobe 100 kpc away [5]. In some lobe-dominated objects (see below) the jet is nearly straight from the core to the outer hot spot.

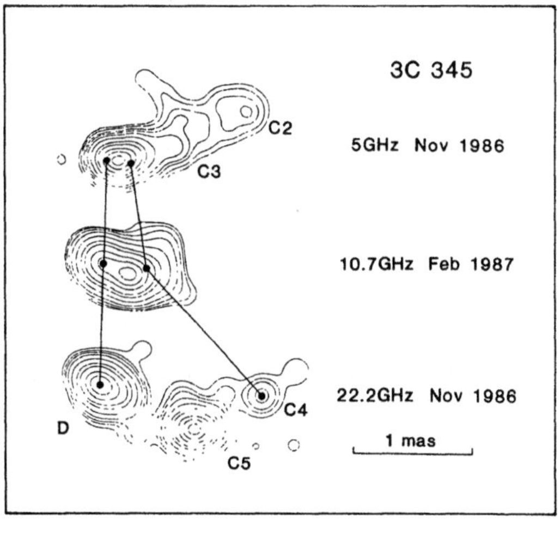

Figure 1. Maps of 3C 345 near 1987.0. The locations of the centroids of components D and C4 are shown as dots. The scale shown in the lower right is appropriate for 22.2 GHz. The bottom contours are 0.5% (5 GHz), 1.5% (10.7 GHz), and 1.0% (22.2 GHz). Taken from Zensus, *et al.*, in preparation.

2a. Compact steep-spectrum double (CSD) (3/65)

These objects are characterized by having two widely-spaced components with similar spectra, which peak at a few GHz and fall-off at high frequencies with $\alpha < -0.5$. They have little or no variability, little polarization, no internal proper motions, and no discernable structure on the arc-second scale. The Pearson-Readhead survey contains only three of these objects, but another half-dozen are known [6]. Some are identified with distant galaxies; others are in blank fields. Hodges and Mutel [6] and Mutel and Phillips [7] suggest that they are progenitors of extended double sources; they are very young and we see the working surfaces where two jets are boring their way through the interstellar medium. Attempts have been made to detect central cores between the two components, but none has been successful. Figure 2 shows CTD 93, a prototypical compact steep-spectrum double.

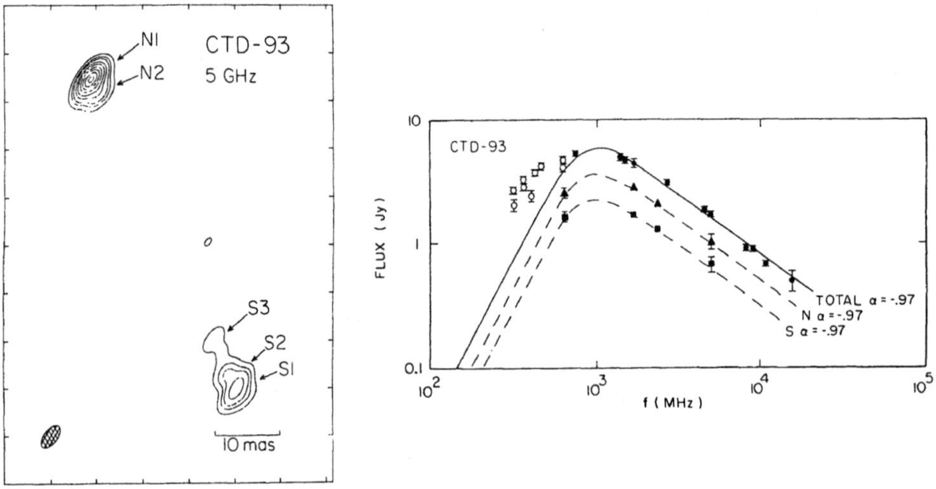

Figure 2. A compact steep-spectrum double source, CTD 93. *Left*: Hybrid map at 5 GHz. ***Right***: Component spectra and total spectrum; taken from Mutel, Hodges, and Phillips [8].

2b. Compact flat-spectrum double (CFD) (4/65)

These have a compact double morphology with a flat high-frequency spectrum ($\alpha > -0.5$). They include the interesting object 4C 39.25, which displays asymmetry, but is not in Class 1 because neither component has the flat spectrum and degree of compactness required of the core in Class 1 [9]. The components have simple synchrotron spectra with peaks at different frequencies. A new component has appeared in 4C 39.25, moving superluminally between the two older components [10]. 3C 395 may also be a member of this class; and, like 4C 39.25, it has spawned a new superluminal component between older stationary components [11].

3. Irregular flat-spectrum objects (2/65)

These two objects are identified with galaxies. The brightness distributions are complex and contain an internal core-jet structure. One of them has subluminal motion (3C 84 [12]).

4. Steep-spectrum compact objects (7/65)

These have a straight steep spectrum, with most of the large-scale emission confined to 15 kpc. The small-scale structure is contorted, sometimes with a jet which apparently bends through 180° or more. It is thought that this extreme distortion is due to interaction of a jet with the ISM of the host galaxy. One of these objects also contains a superluminal core-jet structure (3C 216 [3]).

5. Lobe-dominated objects (19/65)

Only two of these objects (3C 179, 3C 390.3) contain nuclei which were strong enough to be studied by Pearson and Readhead. In these cases the compact components are asymmetric and are like Class 1 objects in most respects, including the display of superluminal motion. The jets of lobe-dominated objects, however, tend to be straight. Figure 3 shows maps of 4C 34.47 which is not in the survey, but is particularly interesting because it is the largest known quasar. It contains a superluminal nuclear component and a one-sided straight jet. As another example, Biretta and Reid have made an excellent map, shown in this volume, of the nearby low-luminosity radio source in M 87. This shows a limb-brightened straight jet containing a sub-luminally moving component.

6. Another five objects have not been observed, or are unclassifiable.

The distinctions among the classes are not sharp, except for the steep-spectrum compact doubles. The others all can have core-jet structures with moving components; and their apparent complexity, and value of F_c, may be partly at least an orientation effect (see below). In any event, variability of the core means that the classification of the objects might be epoch-dependent. There presumably also is a redshift-dependence in the categories, since they are based on observations at a fixed wavelength, while the ratio of steep-spectrum to flat-spectrum emission depends on the emitted wavelength.

Pearson and Readhead emphasize that their classification scheme is provisional and will need to be tested by physical theories and with further data. However, their data are the best we have at the moment. Another, smaller, flux-limited survey is being carried out by Witzel and his colleagues [13, 14]. This survey has 13 objects, which are being monitored intensively. There is some overlap, so that the two surveys together contain 70 objects. Sixteen of these 70 objects now have both a measured redshift (z) and a measured internal proper motion (μ). Progress here is being made rapidly, and the number of sources with both μ and z is increasing steadily.

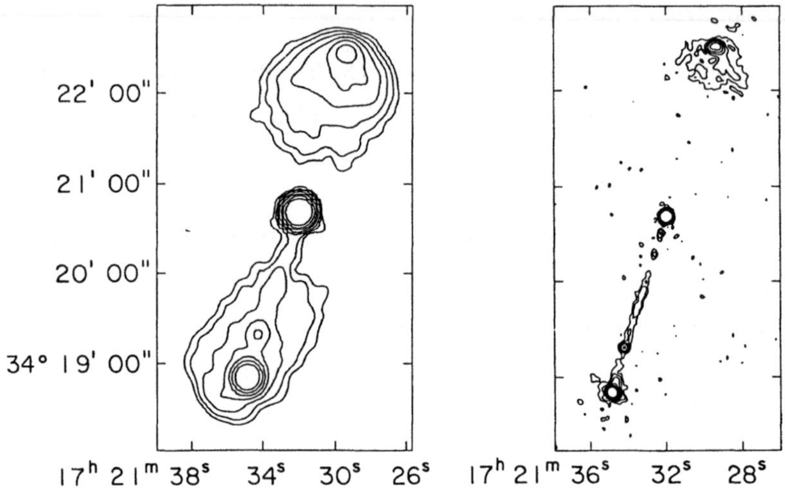

Figure 3. A lobe-dominated source, 4C 34.47, the largest known quasar. The figure shows maps made with the VLA at 1.4 GHz (*left*) and 5 GHz (*right*); taken from Barthel and van Breugel (in preparation). The one-sided jet is typical of powerful double sources. At 5 GHz the nucleus contains a straight asymmetric VLBI source, of overall length about 6 mas, and pointing about 4° from the large jet. The VLBI components are separating superluminally [33].

Survey work is valuable, but is also time-consuming. This is particularly true in VLBI, where the difficulties of observing and analysis are extreme, and the time delays are particularly long and frustrating. The participants in these surveys are to be commended for their foresight and perseverance.

Morphology of Superluminal Motion

The core of 3C 345 is known to be stationary on the sky [15], and we assume that other sources similarly have stationary cores. We therefore have a one-sided phenomenon, and define superluminal motion as occuring when the apparent transverse velocity of a component, relative to the core, is greater than the velocity of light: $\beta_\perp > 1$. (This velocity must be calculated in the coordinate system of the quasar.) About two dozen sources now have a reasonable certification as containing superluminal components [16, 17]. We illustrate their behavior with a brief summary of the properties of 3C 345, the best-studied superluminal source.

Biretta, Moore, and Cohen [18] summarized observations on 3C 345 for the years 1979–1984, and maps through 1986 are being published by Zensus, *et al.* [in preparation]. The structure in late 1986 is displayed in figure 1. At 22 GHz we see the stationary core, D, and two close components,

C5 and C4. These both are moving to the northwest; they first became noticeable as separate components in 1984 and 1981, respectively. D appears elongated; this may be an artifact but it also may herald another new component. At lower frequencies these components are blended together, and components C3 and C2 become prominent. C3 and C2 were apparent in 1979 [18] but lower-resolution maps made in 1977 merely showed them as a northwest extension of the core [19]. A more distant component, C1, has lower surface brightness [20, 21]. It is likely that higher dynamic-range maps would show yet more components, and perhaps the jet could be traced clear to the outer jet, as it is in 3C 120.

The centroids of the components of 3C 345 move steadily away from D. Until 1984 C2 and C3 were moving at different position angles, but during 1984-86 C3 moved north, and now appears to be following the track of C2. C5 may be following the track that C4 occupied four years earlier (Zensus, *et al.*, in preparation).

The major features of the superluminal effect in 3C 345 are

1. New components appear every few years. Their first detectable motion is not radial. C5 first appeared in approximately the same location as C4, and may be following the track that C4 was on.

2. The components do not move ballistically; i.e., individually on radial lines from the core. There may be a fixed wavy track which they follow.

3. The components move with significantly different velocities. C4 accelerated in 1981. C2 has been faster than C3.

4. The core has a flat spectrum from 5 to 22 GHz. After the moving components are well-separated from the core, they have a steep spectrum. Variations in the flux density from the core may be associated with blending of the core and new components. The new components have flux density which is comparable with the core (within a factor 2) when young, but decay with a time constant of perhaps three to five years, at 10 GHz.

5. The separations between the core and the outer components depend on wavelength, with the lower frequency (5 GHz) components being closer to the core by a few tenths of a mas. This can be attributed to frequency-dependent optical depths; the sign and magnitude of the shifts are in good agreement with the Königl theory of a smooth expanding jet [22].

6. The components expand as they move, and crudely define a cone with a projected half-angle 26° [18].

7. Polarization maps of 3C 345 are becoming available [21]. The polarization state of the components changes with time. At one epoch the core was unpolarized, with an exceptionally low upper limit to the fractional polarization.

Some further characteristics of the superluminal effect, seen in other sources, are

8. 4C 39.25 and 3C 395 might both be classified as compact flat-spectrum doubles [10, 11]. In both a new third component appeared at one end of the double, and began moving superluminally towards the other end. The moving component in 3C 395, however, apparently has slowed down or stopped in the middle [34]. The one for 4C 39.25 has steadily continued, and when (if) it reaches the other end the "collision" may shed some light on the physics of these sources.

9. BL Lac has had several components which move along the same track. The moving components appear to stop and then fade away, at a fixed distance from the core [23].

10. The 1981 outburst in 3C 454.3 appeared to produce components which stopped at a fixed distance; but at the same time there were one or two components farther out, which kept up a steady superluminal motion [24].

Further observations will no doubt reveal new features of superluminal motion. Those we already have, however, are more than enough to be a puzzle. In the next section we present some statistics of superluminal motions, and some comments.

Statistics

1. Superluminal Motion

Cohen *et al.* [25] have compiled proper motion observations for 32 objects. In figure 4a we show the corresponding apparent transverse speeds $\beta_\perp$, which are inversely proportional to the Hubble constant $H_o = 100\, h$ km sec^{-1} Mpc^{-1}. We have omitted the two CSD sources in the list of Cohen *et al.*, on the ground that they may represent a physically different population. Many objects have more than one moving component, and in figure 4a we show only the fastest one. The line through the figure represents a proper motion $\mu = 0.05$ mas yr^{-1}, which is a rough observational limit because only about a decade has elapsed since this work started. Few points are expected below the line. Figure 4a shows no clear trend of $\beta_\perp$ with z, so it is legitimate to form the histogram of $\beta_\perp h$, shown in figure 4b.

The two subluminal sources in figure 4a are in nearby galaxies, M 87 and NGC 1275. The four objects with upper limits include two distant quasars (0153+744 and 2134+004) and two galaxies (NGC 6251 and 2021+614). NGC 6251 is a large lobe-dominated radio source whereas the other three can probably all be classified as compact flat-spectrum doubles. Other CFD sources, however,

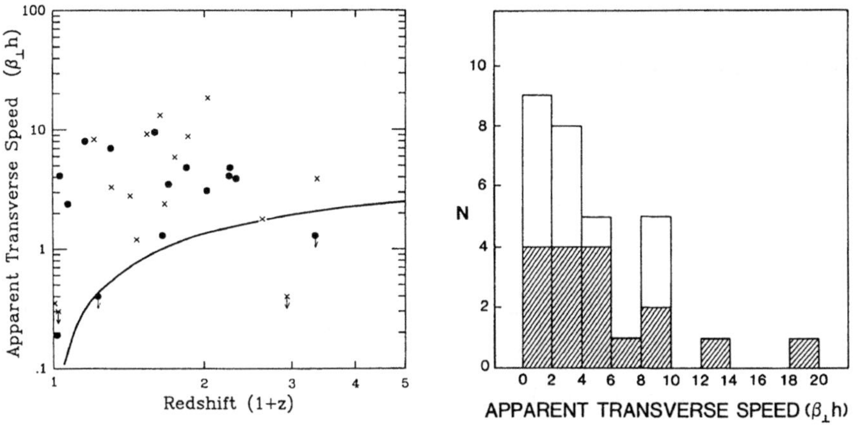

(a) (b)

Figure 4(a). Apparent transverse speed vs. redshift for 30 objects, computed from the compilation in [25]. The line represents a proper motion $\mu = 0.05$ mas yr^{-1}, which is about the observational limit. The solid dots represent observations with more reliability than the crosses. (b). Histogram of $\beta_\perp h$, from figure 4a. Hatched areas represent the crosses.

contain moving components.

The proper motion is measured for some pattern of brightness which moves at angle θ to the line of sight. (We assume that the moving pattern occupies only a small solid angle, as viewed from the quasar.) For each $\beta_\perp$ there is a maximum allowed θ, $\theta_m = 2\cot^{-1}\beta_\perp$, where γ is the Lorentz factor of the pattern. The median for the points in figure 4b is $\beta_\perp h = 3.7$, and the corresponding angle is $\theta_m = 30°$ if $h = 1$, and about half that if $h = 0.5$. Thus, the trajectories represented in figure 4 are not oriented at random. In figure 4b we see a peak at $\beta_\perp h \approx 9$, which corresponds to a peak at angles below $12h°$. Furthermore, since the highest value of $\beta_\perp$ occurs when $\csc\theta = \gamma$, the peak in figure 4b shows that the distribution of γ has a concentration near $\gamma = 9h^{-1}$. The two high points (3C 395 and CTA 102), if confirmed, show that the distribution of γ extends to $20h^{-1}$.

2. Relativistic beaming

The trajectories in figure 4 are aimed close to the line of sight unless $h > 2$, and there evidently is a powerful selection effect which controlled their selection for study. Cohen *et al.*, [25] argue that this is not due to observers systematically ignoring slow sources, but rather that it is due to Doppler boosting, which strengthens sources moving at small θ. Relativistic motion with Doppler boosting is the only viable explanation yet suggested for biasing the selection to sources with small values of θ. Scattering or absorption in a torus (Miller, this volume) is too weak at radio wavelengths.

The phenomena which provide the evidence for relativistic beaming are

1. Calculation of Inverse-Compton X-rays from the observed radio size and spectrum (Worrall, this volume).
2. Selective depolarization in radio lobes (Laing [26]; reported by Bridle, this volume).
3. Rapid variability.
4. High internal proper motions.

Points 1 and 2 are independent of distance; 3 and 4 require cosmological distances with $H_o \lesssim 500$ km sec^{-1} Mpc^{-1}. Point 4 shows merely that there is a relativistic pattern, but points 1, 2, and 3 show that the emitting material itself is in relativistic motion. The common simple model that explains this is a narrow relativistic jet. The radio core is near the apex where the optical depth is unity, and the moving patterns outline shock waves or entrained matter.

Doppler boosting can be very strong. It increases flux density by a factor δ^x where δ is the Doppler (blueshift) factor, and x can be as large as $(3-\alpha)$. In general, however, it will be less than this [27]. Strong boosting, combined with a steep luminosity function, would mean that the sources we see are mainly those which are intrinsically weak but are aimed nearly at us and thus appear strong. These objects have a high probability of being superluminal, with $\beta_\perp$ near γ. The peak at $\beta_\perp \approx 9\,h^{-1}$ in figure 4b is therefore indicative of strong boosting, and of a distribution of Lorentz factors which extends to $\gamma \approx 9\,h^{-1}$.

Strong boosting has a number of implications.

a) In superluminal sources, the core and the new components must have roughly the same boost factor, else their similarity in flux density would be too coincidental. (This is an old point, made originally by Readhead *et al.* [28] and by Blandford and Königl [29].) This similarity has by now been seen in so many cases that I suspect that the Lorentz factors are about the same in the core and the new components, and the exponent x is also about the same. This argument holds also for the new stationary components in 3C 454.3 [24]. There is of course a selection effect which works against our finding weak new components; but note that no new component more than a factor 2 stronger than the core has yet been seen. The acceleration of C4 of 3C 345 does not fit into this picture.

b) Sources oriented near the plane of the sky will be very weak. If the exponent $x = 2$, then a source like 3C 345 but with $\theta = 90°$ might be 10^4 times weaker than 3C 345. At an observed frequency $\nu_{\rm obs}$ of 5 GHz, its flux density would be 1 mJy or less, much too weak to be followed with VLBI. The outer, low-surface brightness lobes are unaffected by beaming, so this object might be a normal double with a very weak core.

If $\nu_{\rm obs}$ is 5 GHz for 3C 345, then the emitted frequency for the relativistic components will be $\nu_{\rm em} \approx 500$ MHz. There is a factor 100 between $\nu_{\rm em}$ for the superluminal sources and $\nu_{\rm em}$ for their counterparts at 90°, and comparisons of different classes of sources must take this into account.

c) The ratio R of core to outer lobe flux density will depend strongly on viewing angle, if the core is strongly beamed and the outer lobes are not. The apparent transverse velocity of superluminal components also depends on viewing angle; thus R and $\beta_\perp$ should be correlated. This has been studied by Hough and Readhead [30] and by Browne [31]. The results are in rough agreement with the idea that the superluminal components have γ = 5–10, for both the pattern and the fluid.

d) Some compact sources which show no evidence for relativistic motion have an enormous luminosity if they radiate isotropically, far more than the strong superluminal sources which are anisotropic. The CFD source 0153+744, with several stationary components [3, 32], has $L_\nu \approx 3 \times 10^{34}$ erg sec^{-1} Hz^{-1} at $\nu_{\rm em} = 15$ GHz if isotropic, whereas the active sources 3C 279, 3C 345, and 3C 454.3 are weaker by two or three orders of magnitude because they are anisotropic. Instead of believing that 0153+744 is intrinsically much more powerful and that the near coincidence of flux densities is an accident, it seems better to believe that they are all similar and are all Doppler boosted by about the same factor. This suggests that the stationary components in 0153+744 and similar sources either have relativistic fluid running through them or in fact are in high-speed motion. We might imagine various possibilities: standing shock waves, a bent or helical pattern which has several tangents towards us, and a jet without a visible core but with two moving components. The last seems unlikely because multiple components seen so far (e.g., in 3C 345) have detectably different velocities.

e) In some scenarios, e.g., [35], the jet is wide, not narrow, and we see that particular piece which happens to be favorably oriented towards us. It is entirely possible that the bright superluminal components are the tail of the beast and not the beast itself, and that we are only investigating a minor side issue.

It is difficult to test the beaming theory because of the powerful orientation bias that presumably exists in flux-limited samples. To evade this problem, Hough and Readhead [36], Zensus and Porcas [37], and Hooimeyer *et al.* [38] defined samples of quasars based on the strength of the outer radio lobes, which (everyone believes) are not relativistic. The central components of these double sources are weak, and only a half-dozen from the combined lists have yet been studied at more than one epoch. Most of them turn out to have superluminal components, but the apparent velocities are generally low and the results to date probably are consistent with the standard beaming theory. A possible problem here is that if the optical continuum radiation from quasars is beamed, then quasars as a class are preferentially aligned towards us, and these three radio samples (all quasars) are not randomly oriented [39]. Nevertheless, these studies remain our best hope for a statistically clean study of beaming.

The relativistic beaming theory explains naturally (if heuristically) so many features of the morphology, spectrum, and variations in compact sources that it surely will play a substantial role in the ultimate elucidation of these objects. However, there is still an "uncomfortable" feature, the overall size [33]. If the superluminal sources really are narrow beams at small θ, then their deprojected sizes are at or above the envelope of linear size as a function of z, for extragalactic sources generally. This would mean that the superluminal sources are not simply normal double sources which happen to be end-on; on the other hand, nearly all superluminal sources have outer lobes and the only available parent population is the normal double sources. Figure 3 illustrates the problem. 4C 34.47 is a normal double source, and is the largest known quasar. It also contains a superluminal compact component. It is surprising that the largest known quasar is aimed close to our line of sight. On the other hand, one of the quasars must be the largest, and maybe all quasars are aimed close to the line of sight. Many superluminal sources are curved, and a misalignment of 30° or so between the core and outer lobes eliminates the size problem; but this is unlikely in 4C 34.47 because it is so straight. Another possibility is to assume that there is a correlation between core Lorentz factor and overall size. Browne [31] gives a detailed discussion of the size problem, and concludes that such a correlation may be its most likely resolution.

I am grateful to Drs. T. J. Pearson and J. A. Zensus for useful comments on the manuscript. This work has been supported by NSF grant AST85-09822.

References

1. *Superluminal Radio Sources*, eds. Zensus, J. A. and Pearson, T. J., (Cambridge University Press, Cambridge, 1987).
2. *The Impact of VLBI on Astrophysics and Geophysics, IAU Symposium No. 129*, eds. Reid, M. J. and Moran, J. M. (Reidel, Dordrecht, 1988).
3. Pearson, T. J. and Readhead, A. C. S., *Ap. J.*, in press (1988).
4. Browne, I. W. A., *et al.*, *Nature*, **299**, 788 (1982).
5. Walker, R. C., Benson, J. M., and Unwin, S. C., *Ap. J.*, **316**, 546 (1987).
6. Hodges, M. W. and Mutel, R. L., *Superluminal Radio Sources*, op. cit., 168.
7. Mutel, R. L. and Phillips, R. B., *The Impact of VLBI on Astrophysics and Geophysics*, op. cit.
8. Mutel, R. L., Hodges, M. W., and Phillips, R. B., *Ap. J.*, **290**, 86 (1985).
9. Marscher, A. P., *et al.*, *Ap. J. (Letters)*, **319**, L69 (1987).
10. Shaffer, D. B. *et al.*, *Ap. J. (Letters)*, **314**, L1 (1987).
11. Simon, R. S. *et al.*, *Superluminal Radio Sources*, op. cit., 72.
12. Backer, D. C., *Superluminal Radio Sources*, op. cit., 76.
13. Eckart, A., *et al.*, *Astron. Astrophys.*, **168**, 17 (1986).
14. Eckart, A., *et al.*, *Astron. Astrophys. Suppl.*, **67**, 121 (1987).
15. Bartel, N., *et al.*, *Nature*, **319**, 733 (1986).

16. Zensus, J. A. and Pearson, T. J., *The Impact of VLBI on Astrophysics and Geophysics*, op cit.
17. Porcas, R. W., *Superluminal Radio Sources*, op. cit., 12.
18. Biretta, J. A., Moore, R. L., and Cohen, M. H., *Ap. J.*, **308**, 93 (1986).
19. Readhead, A. C. S., *et al.*, *Ap. J.*, **231**, 299 (1979).
20. Cohen, M. H., *et al.*, *Ap. J.*, **272**, 383 (1983).
21. Wardle, J. F. C., *et al.*, *Ap. J. (Letters)*, **304**, L1 (1986).
22. Königl, A., *Ap. J.*, **243**, 700 (1981).
23. Mutel, R. L. and Phillips, R. B., *Superluminal Radio Sources*, op. cit., 60.
24. Pauliny-Toth, I. I. K., *et al.*, *Nature*, **328**, 778 (1987).
25. Cohen, M. H., *et al.*, *Ap. J.*, in press (1988).
26. Laing, R., *Nature*, in press (1987).
27. Lind, K. R. and Blandford, R. D., *Ap. J.*, **295**, 358 (1985).
28. Readhead, A. C. S., *et al.*, *Nature*, **276**, 768 (1978).
29. Blandford, R. D., and Königl, A., *Ap. J.*, **232**, 34 (1979).
30. Hough, D. H., and Readhead, A. C. S., *Superluminal Radio Sources*, op. cit., 114.
31. Browne, I. W. A., *Superluminal Radio Sources*, op. cit., 129.
32. Witzel, A., *Superluminal Radio Sources*, op cit., 83.
33. Barthel, P. D., *Superluminal Radio Sources*, op. cit., 148.
34. Simon, R. S., Johnston, K. J., and Spencer, J. H., *The Impact of VLBI on Astrophysics and Geophysics*, op. cit.
35. Blandford, R. D., *Superluminal Radio Sources*, op. cit., 310.
36. Hough, D. H., and Readhead, A. C. S., *Superluminal Radio Sources*, op. cit, 114.
37. Zensus, J. A., and Porcas, R. W., *Superluminal Radio Sources*, op. cit., 126.
38. Hooimeyer, J. R., Barthel, P. D., and Schilizzi, R. T., *The Impact of VLBI on Astrophysics and Geophysics*, op. cit.
39. Scheuer, P. A. G., *Superluminal Radio Sources*, op. cit., 104.

Discussion

Begelman : What is the current evidence on the question of whether these one-sided jets have opening angles which are narrower or wider than their Doppler beaming cones?

Cohen : If the opening angles were wider than the beaming angles we would usually be looking inside the jet and the projected directions of motion might be widely spread around the overall large-scale axis. In fact the alignments are usually good, which suggests that the Doppler beaming cone is wider. However, if the flow is smooth, then wide-cone scenarios are possible (e.g., [35]) and even are attractive in reducing the overall size problem.

Urry : The problem posed by 4C 34.47 is that the large lobe separations ($\geq$ 540 kpc) with the superluminal core (implying a small angle to the line-of-sight) imply that there should be double sources with non-aligned cores that are really huge. Could they be so large (following Ian Browne's suggestion that large sources are associated with high-Lorentz factor jets) that they constitute the blank-field sources?

Cohen : I don't know what the typical blank-field source looks like. It might well be difficult to find an identification for a source consisting of two very widely separated lobes, and no obvious jets or central core.

THE INTERNAL CONDITIONS OF PARSEC–SCALE RELATIVISTIC JETS

P. A. Hughes, H. D. Aller and M. F. Aller
Department of Astronomy, University of Michigan,
Ann Arbor, MI 48109–1090

Introduction

Many compact radio sources display time variable total and polarized fluxes at cm–wavelengths, and are sometimes resolved by VLBI into one or more subcomponents that appear to separate from a stationary core with a speed greater than c. Maps of some well studied sources show a direct link between a compact core and large–scale radio structure (e.g. 3C 120, Walker (1984)), and it is commonly assumed that both the time variable flux and superluminal motion are manifestions of parsec–scale relativistic jets that supply extended components with energy.

We have constructed models of collimated, relativistic flows of synchrotron emitting plasma, with one or more shocks propagating parallel to the flow axis, and integrated the transfer equations for the Stokes parameters to determine the total and polarized fluxes from such jets as a function of time and frequency, and the total and polarized flux profiles along the jet. Comparison of the model values with data from the University of Michigan monitoring program at 4.8, 8.0 and 14.5 GHz, for certain well–defined outbursts, and with published VLBI maps, shows that such models are capable of explaining both observed variability and superluminal motion. These models also provide information on the parameters of the quiescent (unshocked) flow, and so enable us to explore the relationship between the jet and its environment.

Shock Models of Flux Variability

We consider an adiabatic flow of synchrotron emitting plasma, with radius $r \propto d^p$, where d is the axial coordinate, and the flow is taken to start at d_l with radius r_l. The magnetic field, $B \propto r^{-q}$, is represented by two components: a mean field parallel to the flow axis, which contains never more than 5% of the field energy density; and a turbulent component of field which dominates the source emission, and is responsible for the low percentage polarization in the quiescent state. The passage of a shock compresses those field components perpendicular to the source axis, resulting in an effective mean field that produces a high percentage polarization, if viewed in the plane of compression (Hughes, Aller and Aller (1985)). The electron distribution is taken to be a power–law in energy, $N(E)dE = N_0 E^{-\delta} dE, \quad E > E_0$, where N_0 and E_0 are functions of position that reflect the adiabatic energy loss of an element moving parallel to the jet axis.

In our code, shocks are represented by discontinuities in flow velocity and density, with the values on the downstream side of the shock being derived from the quiescent state values using the extreme relativistic jump conditions of Königl (1980). The particle spectrum downstream was derived from that upstream by assuming that the flow suffers only adiabatic compression. We have investigated the possibility that an additional (first order Fermi) acceleration occurs at the shocks, but find that, in general, this leads to too much emission from the shocked component.

We assume that the portion of the jet under study is in the 'adiabatic phase' discussed by

Marscher and Gear (1985) for 3C 273 – inverse Compton radiation is unimportant, and the particles lose negligible energy by synchrotron radiation. Furthermore, we assume that the *extent* of the shocked flow, and the Lorentz factor of the bulk flow do not change appreciably during a cm–wavelength outburst. The validity of these approximations are tested and confirmed in our modelling.

In this model, the quiescent emission is the spatially integrated flux from the jet with no shocks in evidence, whilst the outbursts are associated with the passage of shocks through the $\tau = 1$ surface of the jet. The flux rises as the shocked material becomes optically thin, and falls as this component propagates down the jet, experiencing adiabatic energy losses. Observable relativistic effects (superluminal motion etc.) are a maximum for observers on the critical cone of the flow (at an angle $1/\gamma_{flow}$ to the flow axis) and such an observer sees radiation that, in the source frame, was emitted perpendicular to the direction of motion. An observer on the critical cone of the shocked flow thus sees radiation emitted parallel to the plane of the shock, and hence radiation that exibits the maximum possible polarization as a consequence of the associated compression. We assume that the observer does lie near this critical cone, and we integrate the transfer equations for the Stokes parameters (I, Q, U, V) for many lines–of–sight through the flow in order to predict the total and polarized fluxes as a function of time and frequency. We account for the polarized emission, polarized absorption, rotation of the plane of linear polarization, and conversion of circular to linear polarization, following the discussion of Jones and O'Dell (1977).

Figure 1 shows the application of this model to two well–defined bursts in BL Lac, the lower panel being the total flux and the upper panel the polarized flux. The early phase of the 1982 burst in total flux is heavily 'contaminated' with the residual flux from earlier activity; and continuing

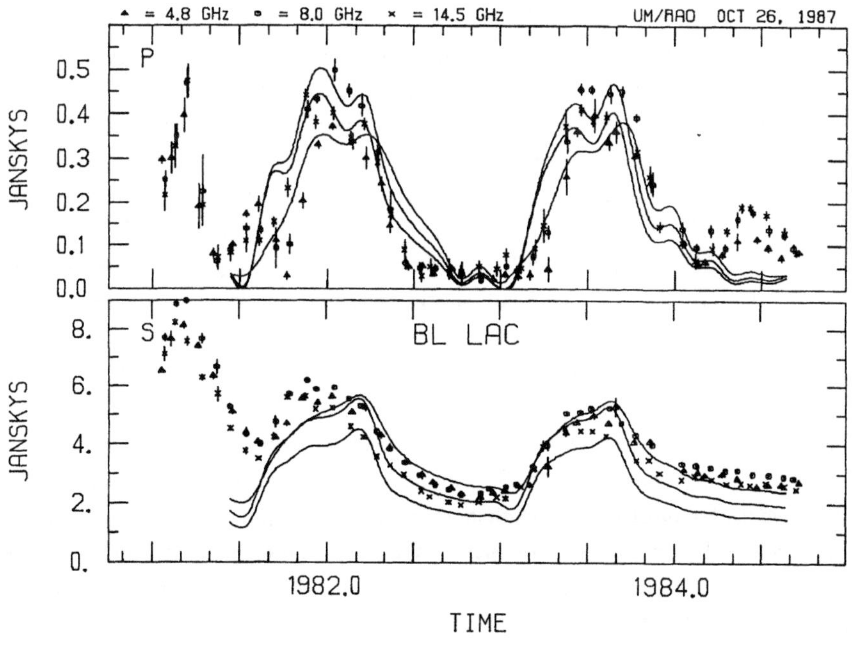

Figure 1. Monthly averaged data for BL Lac; The curves are from the model described in the text.

activity in a third, weak event inhibits the decline of the second burst (as can be seen clearly from an examination of the polarized flux behavior during 1984). Furthermore, it should be born in mind that at any time, only a narrow region of a jet composed of randomly oriented cells is contributing significantly to the observed flux. Thus there is a stochastic character to the model (and, we believe, to the source) that prevents a detailed fit to the polarized flux. Allowing for these caveats, we consider the model to be a very satisfactory fit to the data.

The model fit was accomplished by first combining certain functions of characteristic magnetic field strength, characteristic particle density, source dimensions, Doppler factor etc., into two parameters that scale time and total flux. These, and the model parameters discussed below, were adjusted to optimize the fit of the model to the data. The parameters corresponding to this best fit model are: $p \approx 1$ (i.e. $r \propto d^{1.0}$ – a conical jet, suggesting free expansion); $q \approx 0.75$ (i.e. $B \propto r^{-0.75}$ – field 'regeneration'); $\delta \approx 2$ (i.e. a relatively flat electron spectrum); $\gamma_{flow} \approx 4$; $\gamma_{shock} \approx 1.35$ (i.e. a weak shock – this Lorentz factor for the upstream flow speed seen by the shock corresponds to Mach number $\mathcal{M} \approx 1.28$ and a compression of unit length to 0.67); $\gamma_0 \approx 100.0$ ($\gamma_0 \equiv E_0$); $\gamma_{crit} \approx 1000.0$ and $\nu_{crit} \approx 5 \times 10^8$ Hz (γ_{crit} is a measure of the energy of the particles responsible for emission near the peak in the frequency spectrum, and ν_{crit} gives a measure of the magnetic field in which these particles are radiating – see Jones and O'Dell (1977)). Where appropriate, the above values apply at some arbitrary fiducial point along the flow that is close to the peak in the flux profile of the jet at 4.8 GHz.

The State of the Quiescent Flow

A study of the polarization properties of jets, or of the influence of shocks on the total or polarized flux, requires the construction of numerical models. The total flux from a quiescent jet, on the other hand, is tractable analytically, and we have investigated the state of the quiescent flow by deriving an expression for the total flux from such a source, using the values derived from the shock model discussed above to constrain the parameters involved.

We use the cutoff energy and slope of the electron energy spectrum, together with a Doppler factor corresponding to the quiescent flow Lorentz factor, from the above model. We adopt a distance to BL Lac of 210 Mpc ($H_0 = 100$km/sec/Mpc) and characteristic values of 1pc and 5pc for the axial and radial coordinates of the jet. (The latter values represent a best estimate from published maps (Mutel and Phillips (1984)) together with an estimate from knowing the relative size of shocked region and jet in the numerical models, and the time–scale of the variations.) In an equipartition model (CASE A), we adjust the number density of particles (at some fiducial point) and magnetic field strength, to produce the observed quiescent flux at 8.0 GHz (≈ 2.0 Jy), assuming that there is equipartition between electrons and magnetic field. Of course, field and particle energy density fall as different powers of the jet radius; so if there is equipartition at one point, there cannot be equipartition at any other point. We therefore implement the equipartition constraint *at that point in the jet that contributes most of the radiation at 8 GHz.* The equipartition model ignores the fact that from the model value estimates of the parameters γ_{crit} and ν_{crit} we can make an independent estimate of the source magnetic field. In the model that we call CASE B we adopt this field value (which is 2×10^{-4}G) and adjust the particle density alone, in order to match the observed 8 GHz flux. The values that we derive are summarized in the following table:

CASE A (Equipartition)	CASE B (Variability Model)	QUASAR BLR
$B = 0.07$ G	$B = 2 \times 10^{-4}$ G	
$N = 0.2$ cm^{-3}	$N = 1600$ cm^{-3}	$N = 10^{5,\,6}$cm^{-3}
$P = 3 \times 10^{-4}$ dyn cm^{-2}	$P = 0.5$ dyn cm^{-2}	$P = 10^{-2,\,-3}$ dyn cm^{-2}
$\dot{E} = 2 \times 10^{45}$ erg s^{-1}	$\dot{E} = 5 \times 10^{47}$ erg s^{-1}	$L_{bol} \geq 10^{48}$ erg s^{-1}

B is the magnetic field strength, N the particle number density (i.e. $\int N_0 E^{-\delta} dE$), P is the total pressure, and $\dot{E}$ is the energy flux through the jet. The values given for reference in the third column (from Mathews and Capriotti (1985)) are estimates for the gas in which the broad line clouds are supposedly imbedded, and L_{bol} is representative of BL Lac and OVV object bolometric luminosities. The values apply, where appropriate, to a fiducial point near to the peak 8 GHz flux point in the jet.

It can be seen that the equipartition (CASE A) jet is very light, and thus potentially unstable to the Kelvin–Helmholtz mode (e.g. Birkinshaw 1984), is overconfined, and thus unlikely to be conical, and deposits very little energy in the ambient medium, compared with the total energy output of typical highly luminous objects. The jet with parameters derived from our model fits to the outbursts (CASE B), on the other hand, although still very light, is underconfined, and thus likely to be freely expanding and hence conical; it has the ability to deposit an energy within the nuclear region of its parent galaxy that is a significant fraction of the bolometric luminosity. We do not here propose any mechanism whereby the jet energy may be tapped, but note that it does have the potential to significantly influence its environment, and energize nuclear processes.

Conclusions

We can understand cm–wavelength variability in terms of shocks in relativistic jets.

Using the results of numerical models of such flows, we can study the quiescent state, and find that a consistent picture emerges: we find that the quiescent flow of BL Lac must depart significantly from energy equipartition, and appears to be a light, high pressure, freely expanding flow that has the potential to dump a significant amount of energy into its surroundings. The fact that it is underdense with respect to its surroundings, suggests that instabilities may play a role in generating the turbulent magnetic field which is inferred from the polarization data. This is consistent with the inferred very weak magnetic field that would be inadequate to stabilize the jet.

This work was supported in part by grant AST–8501093 from the National Science Foundation.

References

Birkinshaw, M., 1984. In *VLBI and Compact Radio Sources*, eds. Fanti, R., Kellermann, K. and Setti, G. (Reidel, Dordrecht, Holland), p. 229.

Hughes, P. A., Aller, H. D. and Aller M. F., 1985. *Astrophys. J.*, **298**, 301.

Jones, T. W. and O'Dell, S. L., 1977. *Astrophys. J.*, **215**, 236.

Königl, A., 1980. *Phys. Fluids*, **23**, 1083.

Marscher, A. P. and Gear, W. K., 1985. *Astrophys. J.*, **298**, 114.

Mathews, W. G. and Capriotti, E. R., 1985. In *Astrophysics of Active galaxies and Quasi–Stellar Objects*, ed. Miller, J. S. (University Science Books), p. 185.

Mutel, R. L. and Phillips, R. B., 1984. In *VLBI and Compact Radio Sources*, eds. Fanti, R., Kellermann, K. and Setti, G. (Reidel, Dordrecht, Holland), p. 117.

Walker, R. C., 1984. In *Physics of Energy Transport in Extragalactic Radio sources*, eds. Bridle, A. H. and Eilek, J. A. (NRAO Workshop No. 9), p. 20.

QUESTIONS:

S. P. Reynolds:

Are your jets adiabatic? If so, your derived change in magnetic field along the jet should be reflected in changes in the flow velocity and/or electron energy density, to conserve energy.

P. A. Hughes:

You are quite right – the jet does not conserve energy; and requiring it to do so would certainly impose an additional constraint on the parameters. On the other hand, it is not clear to me that this is the correct thing to do; perhaps the jet does not conserve energy – the field regeneration might be associated with the entrainment of matter.

C. M. Urry:

Have you calculated the integrated spectrum for either CASE A or CASE B?

P. A. Hughes:

I have not computed the spectrum at shorter wavelengths, although this is very easy to do. I have computed fluxes for comparison with some 408 MHz data; considering that this was done 'blind' (in other words, without tweaking parameters to improve the fit in the light of the longer wavelength data), and considering that the model spans a factor of about 30,000 in opacity between 14.5 GHz and 408 MHz, the fit was very encouraging.

M. J. Valtonen:

The outbursts of BL Lac which you used appear to be the same for which I showed spectra yesterday. We obtained an optically thin spectral index (α, $I_\nu \propto \nu^{+\alpha}$) of $\alpha \approx -0.2$. What value of the spectral index do you get from your models?

P. A. Hughes:

In this model the optically thin spectral index would be $\alpha \approx -0.5$. However, the spectrum of a jet is very flat near turnover, and it would be necessary to go to wavelengths much shorter than ours to see the optically thin spectrum.

B. M. Peterson:

Observationally, is centimeter–wavelength variability correlated with variability at any other wavelength?

M. F. Aller:

In a few sources there is good evidence for some correlated events at radio and optical wavelengths. Matching events in very active sources is difficult because of the many outbursts that occur in both wavebands. High and low radio frequency variability shows some evidence of correlation; a recent study by Padrielli *et al* (*Astron. Astrophys. Suppl. Series*, **67**, 63, 1987) found well–established broadband variability in 4 out of 51 objects studied.

THE NUCLEAR JET IN M87

J. A. Biretta and M. J. Reid
Harvard - Smithsonian Center for Astrophysics

The nearest extragalactic jet source in the northern hemisphere is M87 (3C274, NGC4486), the giant elliptical galaxy near the center of the Virgo Cluster. This galaxy contains a 20" jet which has been studied in detail at radio, optical, and x-ray frequencies. Radio observations with 0.12" resolution show a 0.2" nuclear jet, and a complex structure with shock-like features in the 20" jet (Biretta, Owen, Hardee 1983). Because of its proximity M87 provides one of our best opportunities to study a jet as it emerges from the nucleus of a galaxy.

We have made new VLBI observations of the nuclear jet at 1.66 GHz using 18 antennas of the "World Array." The data were global fringe fitted, and self calibrated to remove both antenna and baseline amplitude and phase errors. The resulting map (Figure 1) has a resolution of 5 milli-arcseconds (mas) and a dynamic range of 2300:1.

The combination of high resolution (0.5 pc at 20 Mpc) and high dynamic range permit us, for the first time, to resolve a nuclear jet across its width. The map reveals a very complex structure. At a distance of ≥ 50 mas from the core, the jet appears limb-brightened, particularly along the northern limb. This suggests an interaction of the jet with the external medium. We also find side to side oscillation of the brightness centroid about the position angle of the 20" jet. This oscillation could be due in part to brightness variations in a limb-brightened jet.

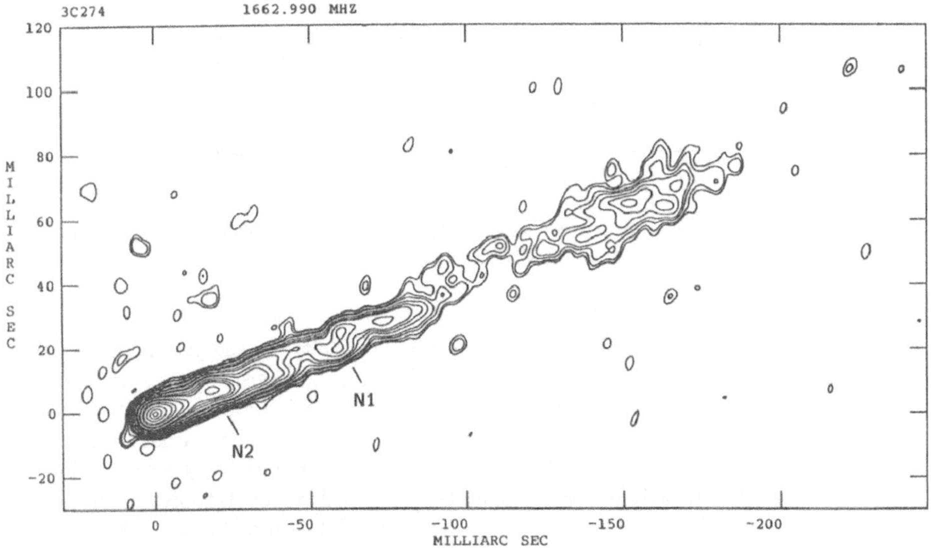

Figure 1. Map of M87 core for epoch 1984.26. Contours are at ± .15, .2, .3, .4, .6, .8, 1, 1.5, 2, 3, 4, 6, 8, 10, 15, 20, 30, 40, 60, 80, and 95 percent of the peak intensity. The CLEAN restoring beam is a 5 mas FWHM Gaussian function, and north is at the top.

The width of the jet seems to broaden until about 50 mas from the core, where the morphology changes. This change occurs at the point where the limb brightening becomes dominant. When one limb of the jet dominates the tranverse brightness distribution smaller values for the half-intensity width would be expected. The data between 4 and 50 mas from the core give an opening angle of $6.5° \pm 0.3°$ (measured to the FWHM points). This is perhaps the smallest linear scale on which an opening angle has been measured, and shows that most of the jet collimation occurs within $< 1/2$ pc of the central engine. A similar measurement on the 20 arcsecond jet (de Vaucouleurs and Nieto 1979, Biretta, Owen, and Hardee 1983) obtains 4.6°, and hence some collimation must occur on scales between 50 mas and a few arcseconds. Strongly magnetized jets are thought to do this (Clark, Norman, and Burns 1987).

The nuclear jet also shows evidence for sub-luminal motion. Comparison of the 1984 data with observations from 1980 and 1982 (Schmitt and Reid 1985) shows a prominent feature, labeled N2 in Figure 2, moving outward at 1.1±0.3 mas/year. This corresponds to a speed of 0.34±0.09c for a distance of 20 Mpc. Motion at sub-luminal speeds has been detected in the nucleus of only one other radio source, 3C84, which shows speeds of 0.33c to 0.49c (Marr, *et al.* 1987). Both of these objects are low luminosity radio galaxies. However, we note that sub-luminal motion cannot be ubiquitous in this class of objects, since 3C120 has a similar luminosity and is super-luminal (Walker, *et al.* 1984).

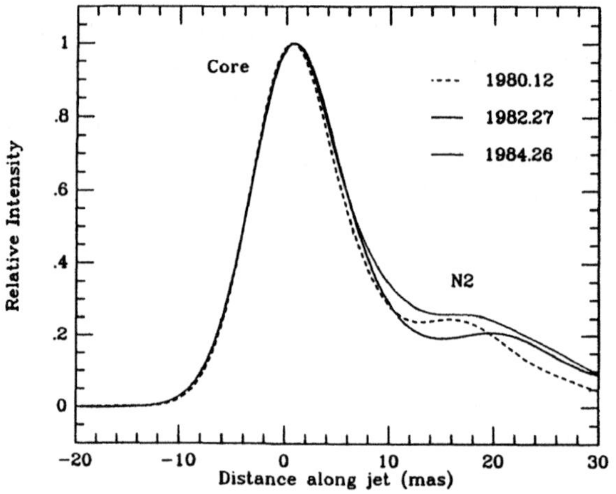

Figure 2. Intensity profiles taken along P.A. −70° for all three epochs. Peak intensities have been normalized.

The observed proper motion, β_{obs}, and jet to counter-jet brightness ratio, R, can be used to constrain the geometry and intrinsic (or bulk flow) velocity of the jet in the context of a relativistic beaming model. To do this we assume (1) that the observed feature and jet fluid move with the same speed, and (2) that there is a counter jet with an intrinsic brightness equal to that of the observed jet. Fig. 3 shows the resulting constraints. The proper motion detection gives the constraint $\beta_{obs} = 0.34 = \beta sin(\theta)/[1 - \beta cos(\theta)]$ where θ is the angle to the line of sight, and β is the jet velocity. In this model the counter-jet is hidden by Doppler beaming, and the observations give a ratio of jet to counter-jet brightness R$= [(1+\beta cos(\theta))/(1-\beta cos(\theta))]^{2.5} \geq 200$ which is the shaded region in Fig. 3. The intersection of the two constraints (shown by the heavy line) requires that $\theta \leq 5.2°$.

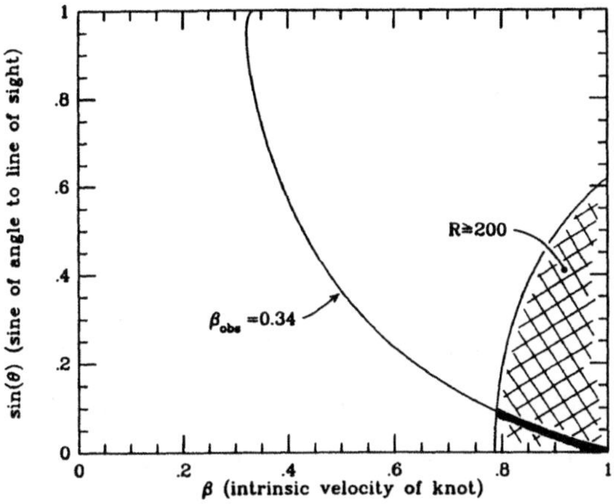

Figure 3. Allowed combinations of $sin(\theta)$ and β for the observed motion $\beta_{obs} = 0.34$ and jet/counter-jet ratio R$\geq$ 200. The heavy line shows the region allowed by both constraints together.

The very small angle to the line-of-sight required in the simple beaming model seems very unlikely for several reasons: (a) The solid angle for $\theta \leq 5.2$ is very small. (b) The deprojected opening angle for the 20 arcsec jet would be less than 0.3°, which is much smaller any other radio jet of similar luminosity. And (c), knot A exhibits a sharp linear feature across the jet, which would be difficult to obtain for small θ. Presumeably this linear feature is a plane seen nearly edge-on, and is very roughly normal to the jet axis. If the jet is non-relativistic, or if the pattern of knot A is stationary (e.g. a standing shock), it would be impossible to see a linear feature for small θ. (Though it would still be possible to see a linear feature if the entire knot A moves relativistically, since aberations can rotate the apparent plane by $Arccos(\beta)$.)

The above arguements together strongly suggest that either the jet fluid has a much higher velocity than that of the observed feature N2 (Lind and Blandford 1985), or that the entire jet is non-relativistic and the counter-jet is intrinsically weak or absent.

Future high dynamic range observations at 1.66 and 5.0 GHz should permit velocity measurement for many of the features seen in the 1984 map, and help decide between these possibilities. They will also show whether the side-to-side oscillation pattern moves or is fixed relative to the core, and should reveal details of the limb-brightened structure.

References

Biretta, J., Owen, F., Hardee, P.: *Ap. J. Lett.* **274**, L27 (1983).
Clark, Norman, and Burns: in preparation (1988).
deVaucouleurs, G. and Nieto, J-L.: *Ap. J.* **231**, 364 (1979).
Lind, K., and Blandford, R.: *Ap. J.* **295**, 358 (1985).
Marr, J. *et al.* : in *IAU Symposium No. 129*, in press (1988).
Schmitt, J., and Reid, M. : *Ap. J.* **289**, 120 (1985).
Walker, C. *et al.* : in *IAU Symposium No. 110*, p.121 (1984).

European VLBI Network Observations of Seyfert Nuclei

A. Pedlar[1], B. Harrison[1], S.W. Unger[2], D.A.Graham[3], E.Preuss[3], D.J.Saikia[1,4], G.J.Yates[1]

[1] University of Manchester, Nuffield Radio Astronomy Laboratories, Jodrell Bank, Macclesfield, Cheshire. SK11 9DL, England.

[2] Royal Greenwich Observatory, Herstmonceux, Hailsham, East Sussex, England.

[3] Max Planck Institut für Radioastromonie, Auf dem Hügel 69, D5300, Bonn 1, FDR.

[4] Tata Institute of Fundamental Research, Post Box 1234, Bangalore 560012, India.

1 Introduction

Detailed observations at a number of wavebands, especially at optical and radio frequencies, have helped immensely to clarify our understanding of the Seyfert nucleus and examine critically many of the models which have been suggested in recent years. Because of the relative proximity of these galaxies, it has been possible to make detailed observations with high spatial resolution at radio frequencies using the VLA, MERLIN and the European VLBI Network (EVN). The large collecting area of the EVN and use of the MK III VLBI recording system result in sensitivities of about 5 mJy in 10 minutes.

In this paper we summarize the work done so far and report the results of more recent observations. We have so far observed seven Seyfert galaxies with the EVN, four of which, namely NGC 2273, 4151, 5929 and 7674 have flux densities greater than 10 mJy. Of these we have mapped the latter three. The sources we did not detect were NGC 3227, 4051 and 4388.

2 NGC 4151

The Seyfert galaxy NGC 4151, with a redshift of 990 km sec^{-1} is perhaps one of the best studied examples of a nearby active galactic nucleus. The EVN observations of this source made on 1984 November 26 at 18cm using the Effelsberg, Jodrell Bank Mk II, Onsala and three telescopes of the Westerbork array have been described in detail by Harrison *et al.* (1986). The VLBI map of NGC 4151 (Figure 1, lower panel) with an angular resolution of 20 milliarcsec shows a collinear triple source with an angular extent of 450 milliarcsec along PA$\simeq 83^{o}$. This axis agrees well with the MERLIN (Figure 1, upper panel) and the VLA maps (Booler *et al.* 1982; Wilson & Ulvestad 1982), as well as the optical continuum polarization PA (Schmidt & Miller 1980). The EVN map confirms that collimated ejection is taking place in NGC 4151. It is not certain whether the eastern or central component is associated with the nucleus of the galaxy. However, the good alignment of the eastern component with the extended emission line region suggests that this component may be the nucleus (Harrison 1988).

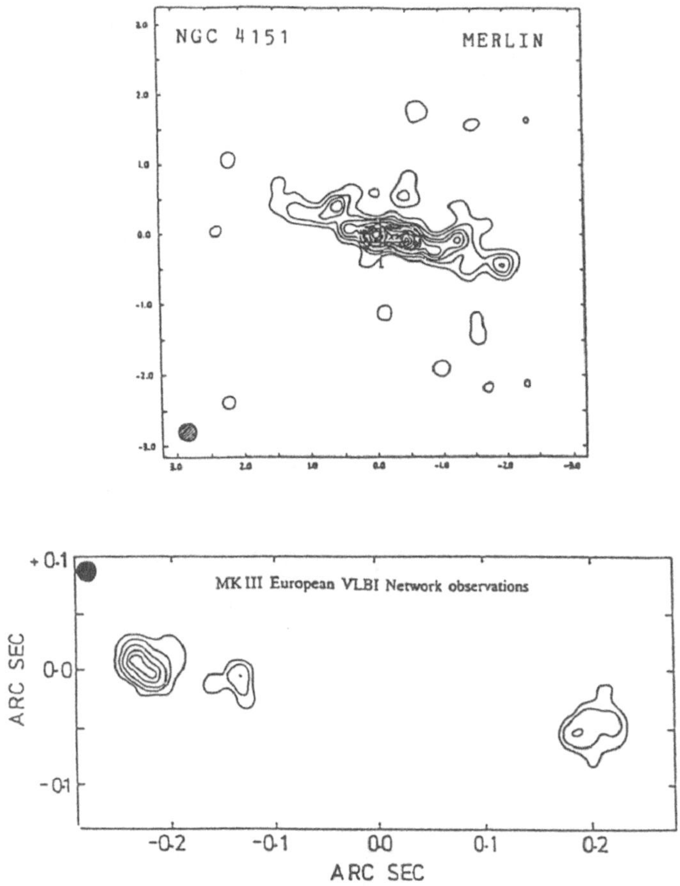

Figure 1: MERLIN map of NGC 4151 at 18cm with an angular resolution of 0.25 arcsec, reproduced from Booler *et al.* (1982) is shown in the upper panel. The lower one shows the EVN 18cm map made with an angular resolution of 20 milliarcsec

3 NGC 5929

NGC 5929 is a Seyfert 2 galaxy at a redshift of 2467 km sec^{-1}. Together with NGC 5930 which is separated from NGC 5929 by 0.5 arcmin along PA 46°, this pair is called Arp 90. Our VLA 2-cm map shows a weak central component in addition to the two outer lobes which are separated by ~1.3 arcsec along PA ~60° (Ulvestad & Wilson 1984). The central component in NGC 5929 has also been detected by Keel & Wilson (private communication).

Short track EVN observations made during the 1985 April observing session at 18cm have been reported by Whittle *et al.* (1986). There was enough data to perform a simple model fit. The visibility data suggests a simple double structure, and the decrease in correlated flux density on the longer baselines implies that the components in the double structure are resolved. Full track EVN observations have since been made, and we are currently reducing the data.

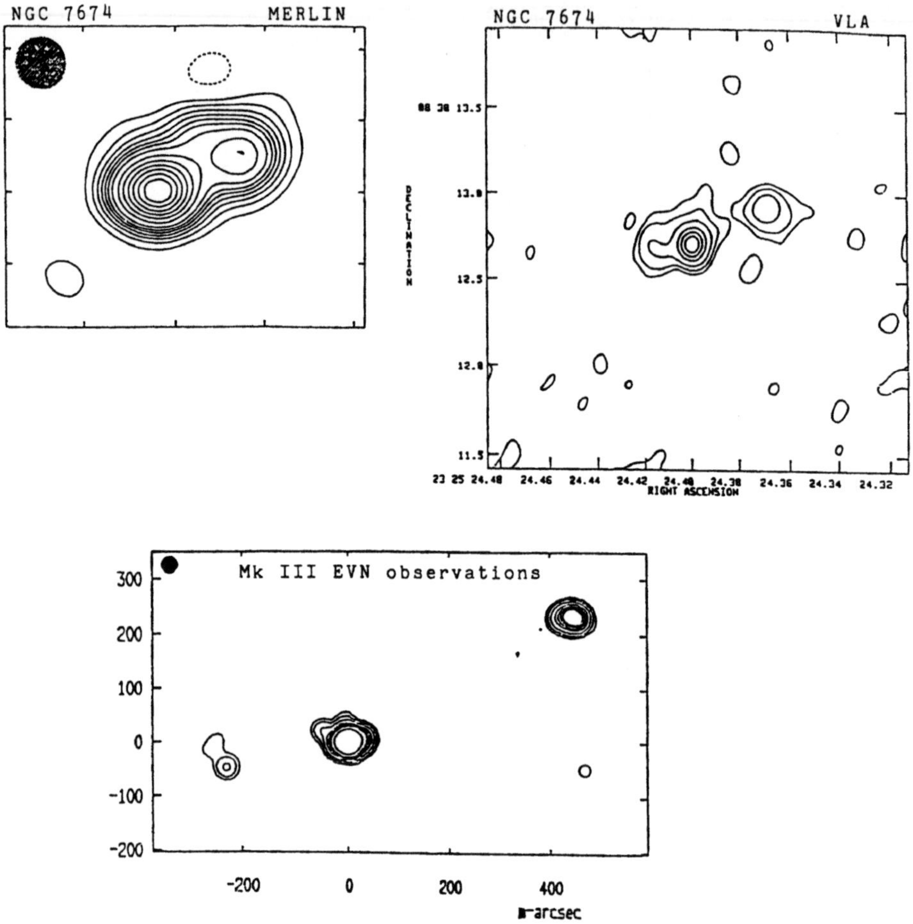

Figure 2: MERLIN map of NGC 7674 at 18cm with a resolution of 0.3 arcsec and the VLA map at 2cm with a resolution of 0.11 arcsec are shown in the upper panel. The EVN map at 18cm has been restored with a beam of 30 milliarcsec.

4 NGC 7674

NGC 7674 is an Sb galaxy containing a Seyfert 2 nucleus and has a redshift of 8698 km sec^{-1}. This source was observed on 1985 April 22 between 0630 and 1400 UT using the Effelsberg, Jodrell Bank Lovell, Onsala and three of the telescopes of the Westerbork array. The EVN map restored with a beam of 30 milliarcsec is shown in Figure 2 (lower panel). A MERLIN map of the source at 18cm with an angular resolution of 0.3 arcsec, and a VLA map restored with a 0.11 arcsec beam are shown in the upper panel of Figure 2. The total flux density detected in the EVN map is about 70 mJy compared to about 160 mJy for the MERLIN map, indicating that significant flux is missing in the VLBI map due to the lack of low-order spacings. The structure seen in the VLA map is consistent with the EVN map suggesting that the easternmost VLBI component, although weak is perhaps real. From the existing data, it is not clear whether the central or easternmost component of the non-collinear triple is associated with the nucleus of the galaxy. A full account of these observations will be described elsewhere (Unger *et al.*, in preparation).

5 NGC 2273, 3227, 4051, 4388

Short tracks at 18 cm using four stations of the EVN, namely Effelsberg, Jodrell Bank Lovell, Onsala and three telescopes of the Westerbork array, were observed in 1985 during the same observing session as that of NGC 7674. NGC 2273 showed signs of structure but there was insufficient data to make a map or model-fit. For NGC 3227, 4051 and 4388 no significant flux density was detected above the noise level; the upper limit on the shortest spacings being about 10 mJy.

6 Concluding remarks

One of the interesting results to emerge from the EVN observations is that the internal pressures in the radio components are often orders of magnitude higher than the external pressures of the narrow line region. This suggests that the radio components are not in static pressure equilibrium but are perhaps traversing outwards (*cf.* Pedlar *et al.* 1985, Wilson 1981). The detection of double and triple radio structure providès us with the exciting possibility of detecting any motions, changes in structure or flux densities. We also plan to make EVN observations of a larger sample of Seyferts which have already been mapped with MERLIN and/or the VLA.

Acknowledgments We thank the editors of Monthly Notices for permission to reproduce the maps shown in Figure 1.

References.

Booler, R.V., Pedlar, A. and Davies, R.D., 1982. *Mon. Not. R. astr. Soc.*, **199**, 229.

Harrison, B., 1988. *Ph.D. Thesis, University Of Manchester.*

Harrison, B., Pedlar, A., Unger, S.W, Burgess, P., Graham, D.A and Preuss, E., 1986. *Mon. Not. R. astr. Soc.*, **218**, 775.

Pedlar, A., Dyson, J.E. and Unger, S.W., 1985. *Mon. Not. R. astr. Soc.*, **214**, 463.

Schmidt, G.D. and Miller, J.S., 1980. *Astrophys. J.* **240**, 759.

Ulvestad, J.S. and Wilson, A.S., 1984. *Astrophys, J.*, **285**, 439.

Whittle, D.M., Haniff, C.A., Ward, M.J., Meurs, E.J.A., Pedlar, A., Unger, S.W., Axon, D.J. and Harrison B., 1986. *Mon. Not. R. astr. Soc.*, **222**, 189.

Wilson, A.S., 1981. *Proc 2nd ESO/ESA Workshop*, 18-19 February 1981. p. 125, ESO, Geneva.

Wilson, A.S. and Ulvestad, J.S., 1982. *Astrophys. J.*, **263**, 576.

VLBI OBSERVATIONS OF THE RADIO LOBE SPIRAL GALAXY, NGC3079

Judith Irwin and E. R. Seaquist
University of Toronto, Toronto, Canada

I. Introduction

NGC3079 is an edge-on spiral galaxy, disturbed and dusty in optical appearance, and displaying a wide variety of nuclear activity. Its most remarkable distinguishing feature is its radio continuum emission (de Bruyn, 1977) which shows two Kpc-scale, nonthermal radio lobes extending along the minor axis. Duric *et al.* (1983) have shown that the lobes resolve into loops with curved bridges connecting them to a compact, bright nucleus. Several suggestions have been advanced as to the origin of the lobes, (eg. see Hummel, van Gorkom and Kotanyi, 1983), usually involving either a starburst or 'prime mover', as there is evidence for both star formation and a CNRS in this galaxy. We have made VLBI observations of the nucleus of NGC3079 to help distinguish between these scenarios, as star formation might result in randomly oriented sources over an extended region, whereas a massive central object may produce aligned emission within a smaller area.

II. Observations and Data Reduction

Observations of NGC3079 were carried out in June, 1986, with the U.S.A. Mark III, VLBI network at λ6cm. Five stations were used, including Bonn and the VLA in phased array mode, with a total of 72 minutes of on-source time within a three hour period. The data were correlated on the Mark III processor at Haystack Observatory and fringes were successfully detected on all baselines. The UV data were then edited, calibrated and then iteratively modelled and self-calibrated (Cornwell and Wilkinson, 1981) using CalTech VLBI software. A good fit to the UV data required that three components of elliptical gaussian brightness distribution contribute to the observed emission. The parameters of the final, best-fit model with 1σ errors are listed in Table 1. Row 1 identifies each component, row 2 lists the flux density, row 3 gives the angular separation of the component from component A which has been placed at the origin, row 4 lists the position angle of the component from the origin, row 5 gives the major axis of the fitted gaussian at half maximum, row 6 lists the axial ratio and row 7 gives the position angle of the component's major axis. The data were also fourier transformed and 'cleaned', with the lowest, believable signal in the map plane corresponding to a beam averaged brightness temperature of 2×10^7 K, implying that all of the detected emission is nonthermal in origin.

III. Discussion and Conclusions

As the energies of the three nuclear components of NGC3079 (assuming a 'minimum energy' criterion) are within the acceptable range for supernovae, it is instructive to compare these sources to the radio supernovae (RSN) in the prototypical starburst galaxy, M82. At a radio power of 270 Cas A's, components A and B would be among the most luminous RSN's yet detected. Only 43.3+591, the brightest RSN in M82, is comparable in luminosity, though its angular size, if placed at the distance of NGC3079, would be ≈ 10 mas. On energetic arguments alone, then, these sources could be ultraluminous radio supernovae.

However, the centers of the three components are roughly aligned at a position angle of $\approx 305°$ within a 1 parsec region in the nucleus of the galaxy. If the lifetimes of the components are comparable to those of similarly bright RSN (cf. SN1979c, Weiler *et al.* , 1986), the supernova rate in NGC3079 would be at least $0.5\,\mathrm{yr}^{-1}\,\mathrm{pc}^{-3}$. In contrast, the SN rate in M82 is 0.1 $\mathrm{SN\,yr}^{-1}$ over a region, 600 pc in linear extent (Kronberg and Wilkinson, 1975) with 43.3+591 offset from the nucleus by 160 pc (Bartel *et al.* , 1987). Moreover, Young *et al.* (1987) find that the star formation rate in NGC3079 is *not* exceptionally high. Also, the similarity in flux between components A and B along with the alignment of all three sources suggest that they are causally connected.

These arguments therefore favour a compact central object and, possibly, bipolar outflow, rather than a starburst as the origin of the activity in the nucleus of NGC3079. If this interpretation is correct, NGC3079 may be the first good example of "classical" double lobed radio activity in a *spiral* galaxy, analogous to that observed in the powerful, radio bright ellipticals.

Table 1

MODEL PARAMETERS

	A	B	C
Flux (mJy)	12.3 ± 2.7	12.1 ± 4.4	19.1 ± 17.1
R (mas)	—	20.2 ± 0.2	3.2 ± 2.5
θ (°)	—	302.6 ± 0.2	309.1 ± 26.1
D (mas)	1.1 ± 1.2	2.1 ± 0.9	11.4 ± 6.2
AR	0.87 ± 0.95	0.46 ± 0.20	0.22 ± 0.13
ϕ (°)	116.9 ± 143.6	95.4 ± 31.0	162.4 ± 8.8

ACKNOWLEDGEMENTS

We gratefully acknowledge the assistance of Bob Phillips and Raymond Rusk with the data acquisition and reductions.

REFERENCES

Bartel, N., Ratner, M., Shapiro, I., Rogers, A. 1987, *I. A. U. Symp.*, **121**, 521.

Cornwell, T. J. and Wilkinson, P. N. 1981, *M. N. R. A. S.*, , **196**, 1067.

de Bruyn, A. G. 1977, *Astr. Ap.*, , **58**, 221.

Duric, N., Seaquist, E. R., Crane, P. C., Bignell, R. C., and Davis, L. E. 1983, *Ap. J.*, , **273**, L11.

Hummel, E., van Gorkom, J. H., and Kotanyi, C. G. 1983, *Ap. J.*, , **267**, L5.

Kronberg, P. P. and Wilkinson, P. N. 1975, *Ap. J.*, , **200**, 430.

Weiler, K. W., Sramek, R. A., Panagia, N., van der Hulst, J. M., and Salvati, M. 1986, *Ap. J.*, , **301**, 790.

Young, J., Claussen, M. and Scoville, N. Z. 1987, *Preprint.*

Compact steep spectrum radio sources

D. J. Saikia
University of Manchester, Nuffield Radio Astronomy Laboratories,
Jodrell Bank, Macclesfield, Cheshire SK11 9DL, England
and
Tata Institute of Fundamental Research,
Post Box 1234, Bangalore 560 012, India

1 Introduction

Extragalactic radio sources exhibit a rich variety of structures on all scales ranging from less than a parsec to several megaparsecs. Classification of these structures involved studying them with the minimum of assumptions and placing them in groups according to their observed properties. It is relevant to bear in mind that there is bound to be a continuous gradation in these properties. Before the advent of imaging with high angular resolution and sensitivity, sources were very broadly classified into the extended, resolved sources which have a steep radio spectrum ($\alpha \geq 0.5$ where $S \propto \nu^{-\alpha}$) and the compact or unresolved sources which usually have a flat spectrum. More recent observations of flat spectrum sources often show steep-spectrum extended emission in addition to the dominant cores. Considerable evidence now suggests that the bulk of these sources are perhaps normal double-lobed sources seen almost end on. Their core flux densities have been Doppler boosted by relativistic beaming.

Many examples of compact sources which have a steep high-frequency spectrum were known in the 1970s. We, however, took notice of them as a possible class or group of sources when observations of two different high-frequency ($\nu > 1$ GHz) selected samples by the Westerbork and Cambridge 5-km telescopes respectively showed that about a third of the steep-spectrum sources were unresolved (Kapahi 1981; Peacock & Wall 1982, hereinafter referred to as PW), not quite consistent with the prevailing folklore. These sources, which have since been christened compact steep spectrum sources or CSSs, are the subject of this review. Since it is more meaningful to consider a physical rather than an angular dimension for the source, we define a CSS to be one with an overall projected linear size of less than a few tens of kpcs, *i.e.* of subgalactic dimensions, and a steep-spectrum ($\alpha \geq 0.5$) over at least a decade in frequency. The primary interest in studying these sources is the hope that they might provide us with interesting clues for understanding the interaction of the radio plasma with the external interstellar medium, and possibly the evolution of individual radio sources with age.

CSSs occur in a variety of active galaxies, including quasars, radio galaxies and Seyferts. They may occur in isolation or in the presence of more extended emission (Bridle & Fomalont 1978) as for example in the radio galaxies 3C 236 (Willis, Strom & Wilson 1974; Barthel *et al.* 1984) and 3C 293 (Bridle, Fomalont & Cornwell 1981). It should, however, also be noted that a number

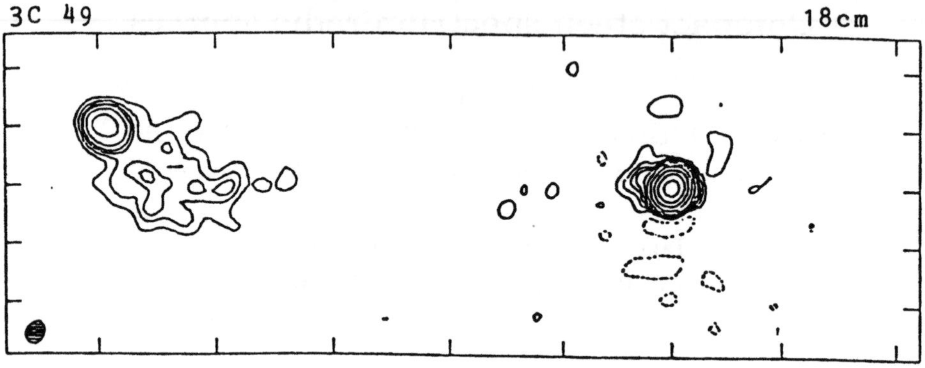

Figure 1: EVN+MERLIN map of the galaxy 3C49 reproduced from Fanti & Fanti (1987). The tick marks in right ascension are separated by 200 milliarcsec.

of cores with prominent jets near them appeared to have a steep spectrum because at the lower resolution the core flux density was significantly contaminated by the jet flux density (*cf.* Saikia, Kulkarni & Porcas 1986). In this review we shall concentrate largely on the more powerful CSSs which have been selected from strong radio source surveys.

2 Structures

A number of authors have made systematic studies of different samples of CSSs with subarcsec resolution both with the VLA and MERLIN. These include VLA observations of sources selected from the 3C and PW samples (*e.g.* van Breugel, Miley & Heckman 1984; Pearson, Perley & Readhead 1985) and VLA and MERLIN observations of a sample of 3C and PW sources (Spencer *et al.* 1988, in preparation), and a sample selected from the S4 survey (Saikia *et al.* 1988, in preparation). Information at even higher resolution has been obtained from VLBI observations (*e.g.* Fanti *et al.* 1985, 1986; Pearson & Readhead 1988; Phillips & Mutel 1982; Hodges, Mutel & Phillips 1984; Wilkinson *et al.* 1984a,b, 1986).

On VLBI as well as subarcsec scales, the observed structures of the resolved sources range from simple doubles to core-jets and more complex morphologies. Figures 1 and 2 illustrate some of the structures seen on VLBI scales while Figures 3 and 4 show some of the MERLIN and VLA maps. In Figure 3 we show the MERLIN maps of the well-known CSS 3C286 made without (left) and with (right) one of the Cambridge telescopes being part of the array. The increased angular resolution and sensitivity using Cambridge has revealed further details of the structure and a new component to the east. In Figure 4 we present VLA maps of some of the sources from the S4 survey, illustrating the observed range in structures.

From their VLBI observations, Wilkinson *et al.* (1984b) and Fanti *et al.* (1985) noted that galaxies appear to be simple doubles while quasars have a core-jet or more complex morphology. This difference in structure between galaxies and quasars is also consistent with the tendency for the compact VLBI-scale doubles, first identified by Phillips & Mutel (1982), to be associated with

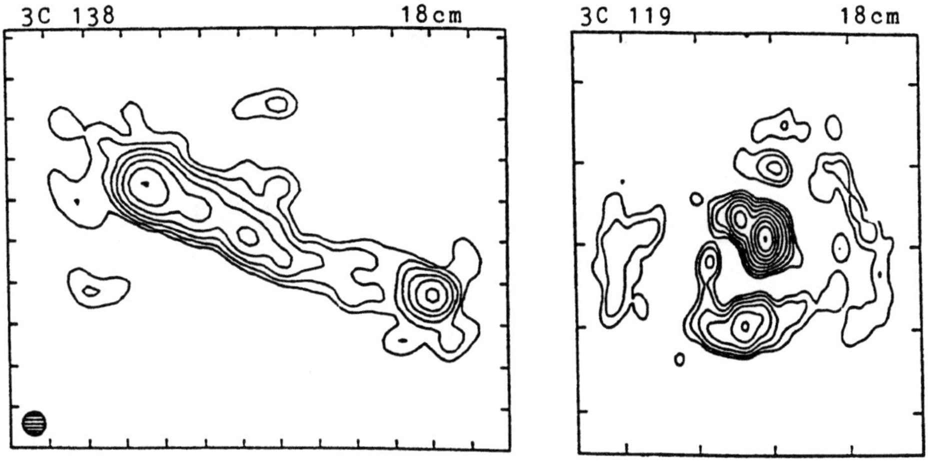

Figure 2: EVN maps of the quasar 3C138 and the probable quasar 3C119 at 18cm, reproduced from Fanti & Fanti (1987). The westernmost component of 3C138 is likely to contain the core. The tick marks are separated by 50 milliarcsec in both maps.

galaxies or empty fields. These compact doubles or CDs have more than 90 per cent of their flux density in two nearly equal components with similar spectra which are separated by 10 to about 200 milliarcsec, with no evidence of arcsec scale structure to a dynamic range of about 3000:1 (Hodges & Mutel 1987). They have low polarization ($\leq$ 2 per cent) and are at best weakly variable ($\leq$ 10 per cent). Flux density measurements of this group of sources at about a mm would be interesting to investigate either whether a core component shows up at these wavelengths or whether there is any evidence of spectral steepening. It is interesting to note at this point that Rudnick & Jones (1982) who made multi-frequency VLA observations of 40 sources with $\alpha < 0.5$ noted that source properties were related to their spectral shapes. They found that unlike the ones with complex spectra, the straight-spectrum sources were not significantly variable and also exhibited evidence of depolarization. These properties, as we shall also see later, are reminiscent of CSSs.

There have also been suggestions that the possible difference in structure between galaxies and quasars seen on VLBI scales extends to arcsec scale structures as well, with the galaxies being simple doubles while the quasars have more complex structures (*e.g.* Wilkinson *et al.* 1984b; Spencer *et al.* 1988, in preparation). It is important to verify such a trend using a reasonably large sample where the galaxies and quasars are matched in redshift in view of a possible dependence of radio source structure on cosmic epoch (Barthel & Miley 1988). Also, it is relevant to bear in mind that in the *enlarged unified schemes* where galaxies are statistically inclined at larger angles to the line of sight than quasars, the effects of projection and relativistic beaming of the core flux density will be more important for quasars than radio galaxies.

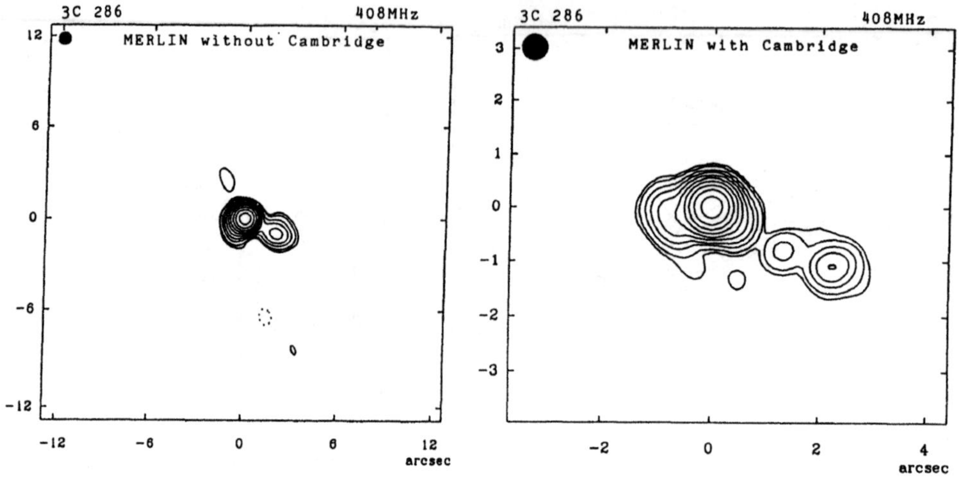

Figure 3: MERLIN maps of the quasar 3C 286 at 408 MHz without and with one of the Cambridge telescopes being part of the array. The angular resolutions are about 0.9 and 0.5 arcsec respectively.

3 Possible scenarios

In this section we shall discuss briefly some of the principal scenarios which have been suggested to explain the structures of this class of sources. These include confinement by a dense external environment, luminous double-lobed sources at an early stage of evolution, and projection effects.

3.1 Dense external environment

There is increasing evidence that the external medium can influence the shapes and sizes of radio sources. Radio and optical studies of relatively nearby galaxies by van Breugel and collaborators have shown that the interaction of jets and lobes with the external medium may deflect and disrupt the jets, ionize the surrounding gas (Miley *et al.* 1981; van Breugel *et al.* 1985b) and perhaps even induce star formation (Brodie, Bowyer & McCarthy 1985; Graham & Price 1981; van Breugel *et al.* 1985a). Evidence for depolarization of radio emission by ionised gas has been seen in a number of low-luminosity sources such as 3C305 (Heckman *et al.* 1982), 4C26.42 (van Breugel, Heckman & Miley 1984), 3C277.3 (van Breugel *et al.* 1985b) and 4C29.30 (van Breugel *et al.* 1986). In 3C277.3 entrained gas appears to be transported over many kiloparsecs and deposited in the lobes. van Breugel *et al.* have also suggested that the depolarization is not internal, but is possibly due to ionized clumpy gas in the outer layers of the source. A similar relationship between depolarization and line-emitting gas has also been seen in the high-luminosity radio galaxy 3C171 (Heckman *et al.* 1984). Correlation of depolarization gradients with structural bends in the quasar 3C2 at a redshift of 1.037, has been interpreted to be due to collisions with clouds of ambient gas (Saikia *et al.* 1987). For the low-luminosity radio galaxy 3C272.1 (M84), Laing & Bridle (1987) find an organised rotation measure (RM) pattern across the source. They suggest a magnetoionic medium which is associated with the galaxy but is in front of and not mixed with the radio emitting plasma. They find evidence of interaction of this medium with the outflow in the two radio jets.

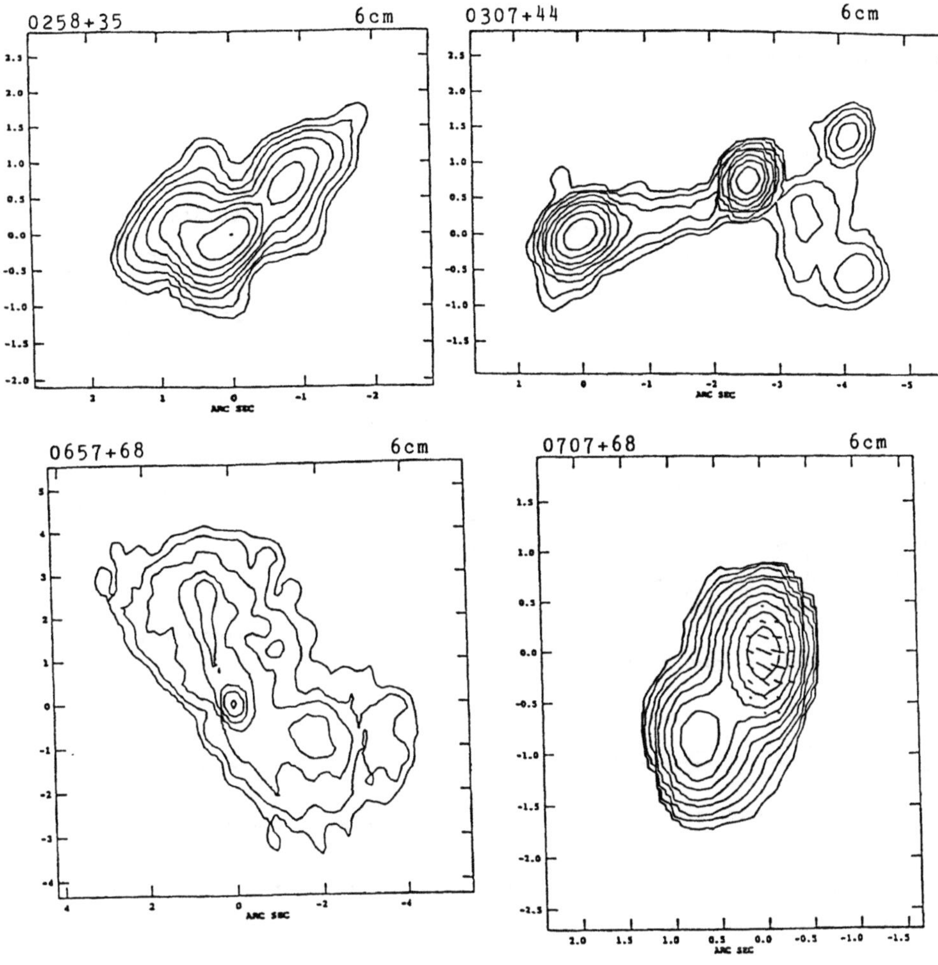

Figure 4: VLA maps of a few compact sources selected from the S4 survey.

Evidence for interaction of the radio plasma with an external medium also comes from correlation of line luminosity with extended radio luminosity (*e.g.* Caganoff *et al.* 1988; Rawlings 1988), reminiscent of such correlations seen in Seyfert galaxies (de Bruyn & Wilson 1978; Wilson & Willis 1980; Whittle *et al.* 1988). In the Seyfert galaxy, NGC 1068, which is a CSS, there is some evidence of depolarization by the line-emitting gas (Wilson & Ulvestad 1983).

For the more distant CSSs, detailed radio-optical studies are difficult and most of the work reported so far have been on radio data alone. For the CSS quasar 3C380, which has a complex structure, Wilkinson *et al.* (1984a) suggested that the morphology could be due to interaction with a dense interstellar medium. From a multifrequency polarization study of this source using both MERLIN and VLA data, Flatters(1987) found no evidence of a high RM arising from such a medium. Saikia, Swarup & Kodali (1985), however, found the RM of a small sample of CSSs to be higher than for larger radio galaxies and quasars as well as flat-spectrum cores. More recently Kato *et al.* (1987) observed about 100 sources suspected of having large RMs and found four, namely 3C119, 147, 218 and 295, to have RMs > 1000 rad m^{-2}. The first two, and possibly 3C295 too, are CSSs. van Breugel has reported more examples of CSSs with high RM in this conference.

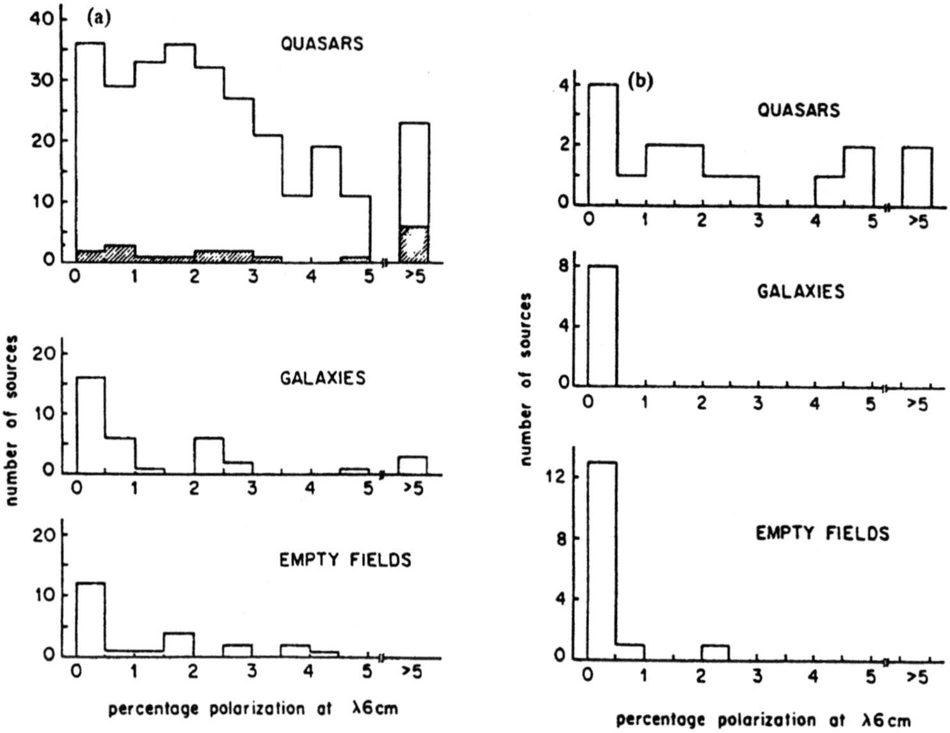

Figure 5: Distributions of core polarization at 6cm for both flat spectrum cores (left) and CSSs (right) identified with quasars, galaxies and empty fields. The hatched area denotes BL Lac objects.

Using a large sample of sources observed with the VLA A-array (Perley 1982), Saikia *et al.* (1985, 1987) investigated some of the polarization properties of cores, and found that the cores in quasars tend to be more strongly polarized at both 6 and 20cm that those in galaxies and empty fields. This appears to be true for both CSSs and flat-spectrum cores or FSCs (Figure 5). It is, however, important to investigate this result further using samples of galaxies and quasars which are well matched in their redshift distributions. In their study, Saikia, Singal & Cornwell (1987) found that the CSSs in quasars showed significant evidence of depolarisation between 6 and 20 cm, while the sensitivity of the observations was insufficent to examine a similar trend for galaxies. van Breugel *et al.* (1984) also reported evidence of depolarization in CSSs but their analysis may have been affected by not considering the galaxies and quasars separately (Saikia *et al.* 1985).

3.2 Tidal interactions and mergers

Optical observations of radio galaxies and quasars have increasingly stressed the importance of tidal interactions and mergers in understanding extragalactic radio sources (Hutchings *et al.* 1984;

Heckman *et al.* 1984, 1986; Stockton & MacKenty 1987). If one of the companions is gas-rich we may have a situation similar to that discussed in the earlier section. However, in addition to this, the galaxies may exhibit orbital motion giving rise to interesting radio structures (*cf.* Blandford & Icke 1978), or the axis of the nuclear engine of the parent galaxy may be disturbed resulting in precession of the ejection axis. Gower *et al.* (1982) have highlighted the rich diversity of observed structures which arise from precessing relativistic jet models, even when they are of the simple ballistic form.

3.3 Young luminous double-lobed sources

Hodges & Mutel (1987) have discussed recently an evolutionary scheme in which the compact VLBI-scale doubles evolve into CSSs with sizes of a few tens of kiloparsecs. The latter in turn evolve into the luminous double-lobed (Fanaroff-Riley class II or FRII) radio sources which can be as large as several hundred kiloparsecs. Their plots of lobe diameter, radio luminosity and surface brightness at 1.4 GHz as a function of lobe separation for a sample of CDs, CSSs and FRII sources, appear to follow broadly the evolutionary tracks predicted by Carvalho (1985), and are consistent with the above scenario. A plot of the turn-over frequency as a function of lobe separation is also consistent with this possibility. It is quite conceivable that a significant fraction of CSSs, especially the ones with well-defined double-lobed structure, are FRII sources at an early stage of their evolution.

3.4 Projection effects

For sources with relativistic jets which are seen almost end-on, the core flux densities may be Doppler boosted so that the overall spectrum of the source is flat. But for sources at somewhat larger angles, especially for the ones with slow jets or a temporarily inactive nucleus, the overall spectrum may appear steeper than 0.5. As discussed earlier, distortions in structure for such sources may appear amplified by projection effects. A possible example of such a CSS is 3C309.1, which has been studied in some detail by Wilkinson *et al.* (1986). However, only a small fraction of CSSs are likely to be close to our line of sight. The majority of CSSs are intrinsically small (Fanti & Fanti 1987).

4 Concluding remarks

The nature of compact steep-spectrum radio sources is perhaps best summarised in the words of Fanti *et al.* (1985) who noted that they consist of "objects of different nature, which for different reasons are compact with a steep spectrum". A fraction of them, at least, provides us with an opportunity of studying how the radio emitting plasma and the external environment interact and influence one another. Some CSSs may be luminous double-lobed sources seen at an early stage of evolution while others appear small because of the inclination of their radio axes towards the line of sight.

The usefulness of optical and radio studies, with polarization information, has been demonstrated clearly by van Breugel and collaborators. Deep narrow-band imaging and spectroscopic observations combined with radio polarimetry for a large sample of CSSs would provide us with useful information on the environments of these sources and their interaction with the radio plasma. For the smaller sources, polarization observations with the enhanced MERLIN at 18 and 6cm combined with VLA data at higher frequencies should provide us with interesting constraints on the properties of thermal material in the central regions. The discovery of an extremely strong correlation of polarization asymmetry with jet sidedness for high-luminosity sources (Laing 1988; Garrington *et al.* 1988), as well as the anti-correlation of depolarization between 0.6 and 1.4 GHz with distance from the optical galaxy for galaxies with prominent bridges (Strom & Jägers 1988) all indicate ionised material associated with the parent galaxies. In addition, the arrival of VLBI polarization mapping will enable us to study at least some of the stronger sources such as 3C286 and 380 with the highest possible resolutions, and perhaps help clarify why the two well-known CSSs 3C286 and 138, which are standard polarization calibrators in the VLA are almost unpolarized in the optical (Liller 1969; Moore & Stockman 1984).

Acknowledgments

It is a pleasure to thank Ian Browne, Alan Pedlar and Peter Wilkinson for their comments on the manuscript. The maps of 3C 286 were made in collaboration with Bill Junor and I thank him for his consent to present them here. I also thank Carla and Roberto Fanti for permission to use their maps shown in Figures 1 and 2, the editors of Monthly Notices for permission to reproduce Figure 5, and Dick Miller and Paul Wiita for organising a good conference.

References

Barthel, P.D. & Miley, G.K. 1988. *Nature*, in press.

Barthel, P.D., Schilizzi, R.T., Miley, G.K., Jägers, W.J. & Strom, R.G. 1985. *Astr. Astrophys.*, **148**, 243.

Blandford, R.D. & Icke, V. 1978. *Mon. Not. R. astr. Soc*, **185**, 527.

Bridle, A.H. & Fomalont, E.B. 1978. *Astr. J.*, **83**, 704.

Bridle, A.H., Fomalont, E.B. & Cornwell, T.J. 1981. *Astr. J.*, **86**, 1294.

Brodie, J.P., Bowyer, S. & McCarthy, P. 1985. *Astrophys. J. Lett.*, **293**, L59.

Caganoff, S., Ekers, R.D., Bicknell, G.V. & Carter, D. 1988. In *Active Galactic Nuclei*, eds. Miller, H.R. & Wiita, P.J., Springer-Verlag.

Carvalho, J.C. 1985. *Mon. Not. R. astr. Soc.*, **215**, 463.

de Bruyn, A.G. & Wilson, A.S. 1978. *Astr. Astrophys.*, **64**, 433.

Fanti, C. & Fanti, R. 1987. in *Superluminal Radio Sources*, ed Zensus, J.A. & Pearson, T.J., Cambridge University Press, Cambridge, p. 174.

Fanti, C., Fanti, R., Parma, P., Schilizzi, R.T. & van Breugel, W.J.M. 1985. *Astr. Astrophys.*, **143**, 292.

Fanti, C., Fanti, R., Schilizzi, R.T., Spencer, R.E. & van Breugel, W.J.M. 1986. *Astr. Astrophys*, **170**, 10.

Flatters, C. 1987. *Nature*, **326**, 683.

Garrington, S.T., Leahy, J.P., Conway, R.G. & Laing, R.A. 1988. *Nature*, **331**, 147.

Gower, A.C., Gregory, P.C., Hutchings, J.B. & Unruh, W.G. 1982. *Astrophys. J.*, **262**, 478.

Hutchings, J.B., Crampton, D., Campbell, B., Duncan, D. & Glendenning, B. 1984. *Astrophys. J. Suppl.*, **55**, 319.

Graham, J.A. & Price, R.M. 1981. *Astrophys. J.*, **247**, 813.

Heckman, T.M., Bothun, G.D., Balick, B. & Smith, E.P. 1984. *Astr. J.*, **89**, 958.

Heckman, T.M., Miley, G.K., Balick, B., van Breugel, W.J.M. & Butcher, H.R. 1982. *Astrophys. J.*, **262**, 529.

Heckman, T.M., Smith, E.P., Baum, S.A., van Breugel, W.J.M., Miley, G.K., Illingworth, G.D., Bothun, G.D. & Balick, B. 1986. *Astrophys. J.*, **311**, 526.

Heckman, T.M., van Breugel, W.J.M. & Miley, G.K. 1984. *Astrophys. J.*, **286**, 509.

Hodges, M.W., & Mutel, R.L. 1987. In *Superluminal Radio Sources*, ed Zensus, J.A. & Pearson, T.J., Cambridge University Press, p.168.

Hodges, M.W., Mutel, R.L. & Phillips, R.B. 1984. *Astr. J.*, **89**, 1327.

Kapahi, V.K. 1981. *Astr. Astrophys. Suppl.*, **43**, 381.

Kato, T., Tabara, H., Inoue, M. & Aizu, K. 1987. *Nature*, **329**, 223.

Laing, R.A. 1988. *Nature*, **331**, 149.

Laing, R.A. & Bridle, A.H. 1987. *Mon. Not. R. astr. Soc.*, **228**, 557.

Liller, W. 1969. *Astrophys. J.*, **155**, 1113.

Miley, G.K., Heckman, T.M., Butcher, H.R. & van Breugel, W.J.M. 1981. *Astrophys. J. Lett.*, **247**, L5.

Moore, R.L. & Stockman, H.S. 1984. *Astrophys. J.*, **279**, 465.

Peacock, J.A. & Wall, J.V. 1982. *Mon. Not. R. astr. Soc.*, **198**, 843.

Pearson, T.J., Perley, R.A. & Readhead, A.C.S. 1985. *Astr. J.*, **90**, 738.

Pearson, T.J. & Readhead, A.C.S. 1988. *Astrophys. J.*, in press.

Perley, R.A. 1982. *Astr. J.*, **87**, 859.

Phillips, R.B. & Mutel, R.L. 1982. *Astr. Astrophys.*, **106**, 21.

Rawlings, S.G. 1988. *Ph. D. thesis*, University of Cambridge.

Rudnick, L. & Jones, T.W. 1982. *Astrophys. J.*, **255**, 39.

Saikia, D.J., Kulkarni, V.K. & Porcas, R.W. 1986. *Mon. Not. R. astr. Soc.*, **219**, 719.

Saikia, D.J., Salter, C.J. & Muxlow, T.W.B. 1987. *Mon. Not. R. astr. Soc.*, **224**, 911.

Saikia, D.J., Singal, A.K. & Cornwell, T.J. 1987. *Mon. Not. R. astr. Soc.*, **224**, 379.

Saikia, D.J., Swarup, G. & Kodali, P.D. 1985. *Mon. Not. R. astr. Soc.*, **216**, 385.

Stockton, A. & MacKenty, J.W. 1987. *Astrophys. J.*, **316**, 584.

Strom, R.G. & Jägers, W.J. 1988. *Astr. Astrophys.*, in press.

van Breugel, W., Filippenko, A.V., Heckman, T. & Miley, G. 1985a. *Astrophys. J.*, **293**, 83.

van Breugel, W., Heckman, T. & Miley, G. 1984. *Astrophys. J.*, **276**, 79.

van Breugel, W.J.M., Heckman, T.M., Miley, G.K. & Filippenko, A.V. 1986. *Astrophys. J.*, **311**, 58.

van Breugel, W., Miley, G. & Heckman, T. 1984. *Astr. J.*, **89**, 5.

van Breugel, W., Miley, G., Heckman, T., Butcher, H. & Bridle, A. 1985b. *Astrophys. J.*, **290**, 496.

Whittle, D.M., Pedlar, A., Meurs, E.J.A., Unger, S.W., Axon, D.J., & Ward, M.J. 1988. *Astrophys. J.*, in press.

Wilkinson, P.N., Booth, R.S., Cornwell, T.J. & Clark, R.R. 1984a. *Nature*, **308**, 619.

Wilkinson, P.N., Kus, A.J., Pearson, T.J., Readhead, A.C.S. & Cornwell, T.J. 1986. In *Quasars:IAU Symp. 119*, eds Swarup, G. & Kapahi, V.K., Reidel, Dordrecht, p.165.

Wilkinson, P.N., Spencer, R.E., Readhead, A.C.S., Pearson, T.J. & Simon, R.S. 1984b. In *VLBI and compact radio sources:IAU Symp. 110*, eds Fanti, R., Kellermann, K. & Setti, G., Reidel, Dordrecht, p.25.

Willis, A.G., Strom, R.G. & Wilson, A.S. 1974. *Nature*, **250**, 625.

Wilson, A.S. & Willis, A.G. 1980. *Astrophys. J.*, **240**, 429.

Wilson, A.S. & Ulvestad, J.S. 1983. *Astrophys. J.*, **275**, 8.

THE CM-WAVELENGTH FLUXES AND LINEAR POLARIZATIONS OF BL LAC OBJECTS

M.F. Aller, H.D. Aller, and P.A. Hughes
Department of Astronomy
University of Michigan
Ann Arbor, Michigan 48109-1090

We present results from an observational program initiated in 1979 with the University of Michigan 26-meter paraboloid to study the cm-wavelength properties of BL Lacertae-type objects; the program includes all BL Lac sources (49) which were known to be stronger than 0.2 Jy at 8.0 GHz in 1979 and which are in the declination range -31° to +80°. The goal of this program is to characterize the spectra and variability with time of the total flux density and polarization in this class of active galactic nuclei at short centimeter wavelengths. Observations were made at three-month intervals at 4.8, 8.0 and 14.5 GHz, and objects found to exhibit variability were incorporated into our intensive monitoring program. The number of observations per source ranges from about 75 to over 1000.

We find that there is a wide range in the degree of variability and in the time scales of the variations in BL Lac objects from very active (e.g. 0235+164 and OT 081) to low amplitude, slow variations (e.g. Mrk 501, Ap Lib). Figure 1 shows a histogram of a variability index, (Smax-Smin)/(Smax+Smin), at 14.5 GHz; this ratio

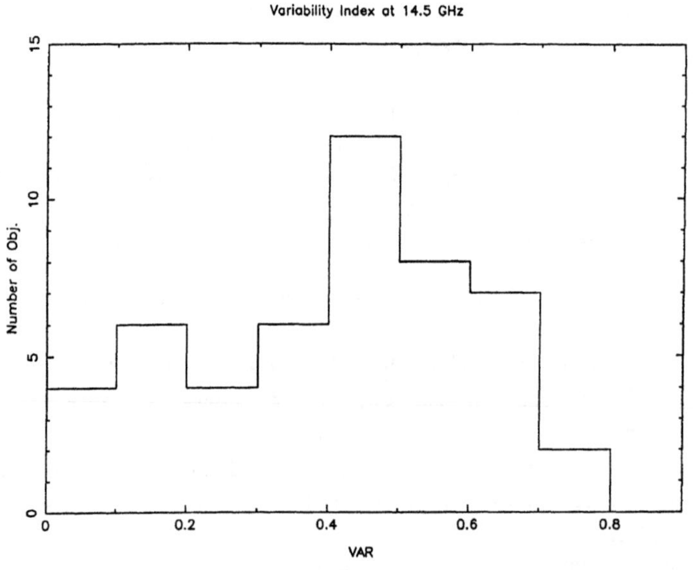

gives a measure of the peak-to-peak flux density variations over the eight-year observing period. At 14.5 GHz over 90% of the sources showed significant variations. The majority of the objects were found to be variable at our other two frequencies although the fractions showing large-amplitude flux density variations were lower.

For most objects the time-averaged spectral indicies between 4.8 and 14.5 GHz (full line in Figure 2) and between 8.0 and 14.5 GHz cluster near zero while the 4.8-8.0 GHz indicies (dashed line in Figure 2) are skewed towards positive values.

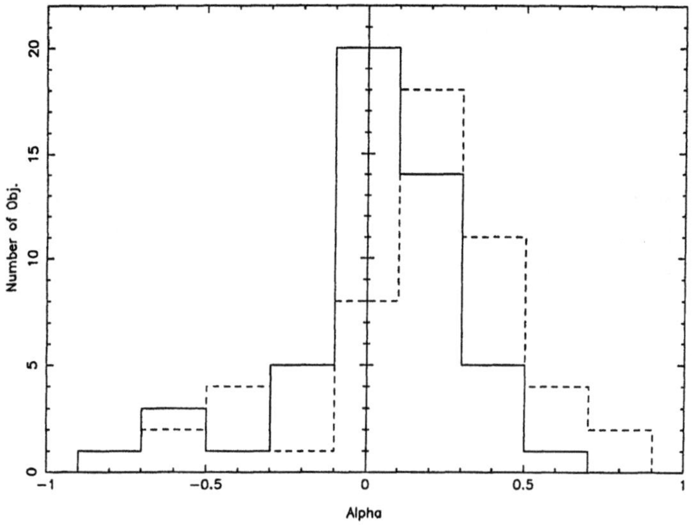

We believe that this difference is caused by self-absorption; the emitting regions are commonly transparent above 8.0 GHz while opacity effects become increasingly important at lower frequencies.

In Figure 3 we compare the time-averaged polarizations at 8 GHz of the BL Lac

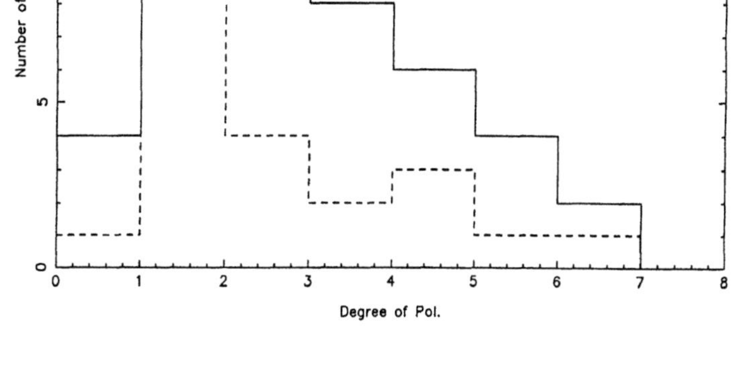

objects (full line) with the values we have obtained for the QSO's in the Pearson-Readhead (1984) flux-limited sample (dashed line). The polarization values have been corrected for bias (Wardle and Kronberg 1974) when the signal-to-noise ratio is less than 10:1, and cases where the signal-to-noise ratio is less than 3:1 have not been included in the histogram. We find that: a) the time-averaged degrees of polarization are nearly the same for the two groups of objects (2.4% for QSO's versus 2.9% for BL Lacs; but b) a larger fraction of the BL Lac objects have relatively large degrees of polarization.

This research was supported in part by grants AST-8301234 and AST-8501093 from the National Science Foundation.

REFERENCES

Pearson, T.J. and Readhead, A.C.S. 1984, in VLBI and Compact Radio Sources, eds. R. Fanti, K. Kellermann and G. Setti (D. Reidel: Dordrecht), p. 15.

Wardle, J.F.C. and Kronberg, P.P. 1974, Ap. J., 194, 249.

LARGE SCALE RADIO STRUCTURES

Alan H. Bridle
National Radio Astronomy Observatory
Charlottesville, Virginia 22903-2475, U.S.A.

"Our choicest plans have fallen through, our airiest castles tumbled over,
because of lines we neatly drew, and later neatly stumbled over."
(Piet Hein, *Grooks, Vol.1*)

1. INTRODUCTION

The radio sources associated with AGNs were once classified either as "compact, flat spectrum" objects, or as "extended, steep spectrum" objects, *e.g.*, [1]. "Compact" sources were studied by VLBI and "extended" sources by connected-element interferometry (often by orthogonal sets of observers). The most significant remnant of this distinction may be that meeting organizers still arrange separate discussions of each class of source, so that radio properties get two reviews rather than only one! It now seems likely that all AGN radio sources with powers above $\sim 10^{22}h^{-2}$ W.Hz^{-1} at 1.4 GHz (for $H_0 = 100h$ km.s^{-1}.Mpc^{-1}) will turn out to have both compact flat spectrum "cores" and extended steeper-spectrum "jets", "plumes" or "lobes" when observed with enough sensitivity and dynamic range at the right resolution. The terms "compact" and "extended" are therefore being replaced by "core-dominated" and "lobe-dominated". Both classes of radio feature are believed to be leakage from a machine whose power plant is at the center of the AGN, whose transmission is the jets (the cores being their optically thick bases), and whose exhaust is the radio plumes and lobes. The physical question raised by the old classification has become: "what determines the *relative prominence* of the radio cores, jets and lobes associated with different AGNs?" Many believe that, at least in powerful sources, orientation to the line of sight is an important part of the answer. The current status of radio structure classification therefore resembles that of the spectroscopic typing of Seyfert nuclei (discussed at this meeting by Don Osterbrock, Joe Miller and Bob Goodrich). The "lines we neatly drew" fifteen or more years ago are being stumbled over today, and we are trying to decide how much of what we see is governed by the viewing angle, and how much by intrinsic properties.

With this in mind, I will review four main topics, interpreting the "large" scale in my title to mean "kiloparsec and larger" scale. First, what properties of the large scale radio sources are generic, *i.e.*, shared by most examples of their general type? Second, how are these properties described by recent models of radio source dynamics? Third, are there observations that may require significant re-thinking

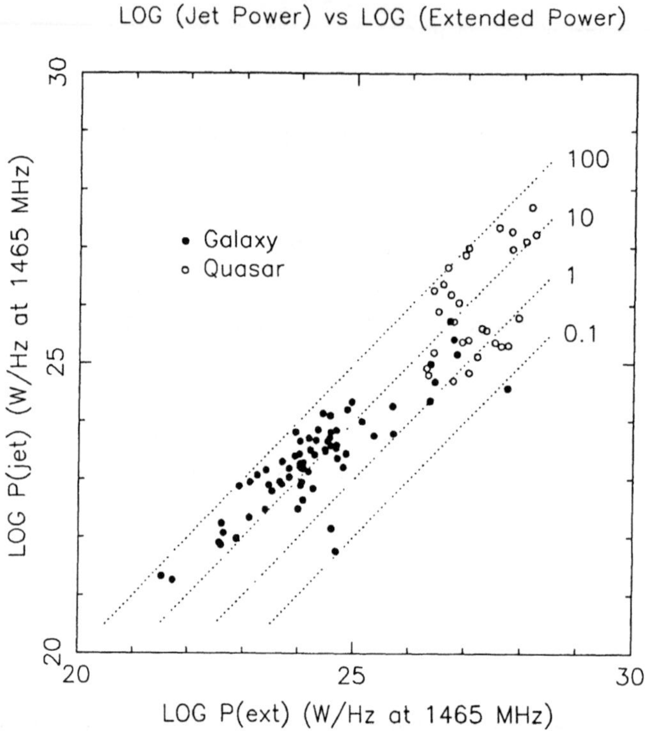

Figure 1. Logarithmic plot of jet power against total extended power at 1465 MHz for 98 sources, distinguishing galaxies and quasars. The dotted lines show where the jets contain 100, 10, 1 and 0.1% of the total extended power respectively. Points above the 100% line would be unphysical; points below the 0.1% line can be obtained only from data of exceptionally high dynamic range.

of these models? Finally, I will discuss some new observations that may steer us towards understanding the main structural asymmetry of powerful sources—that their lobes are usually two-sided while the jets that supposedly "feed" them usually appear to be one-sided.

2. GENERIC PROPERTIES OF LARGE-SCALE RADIO SOURCES

Most properties of radio-loud galaxies change with their extended (*i.e.*, jet plus lobes) radio power P_{ext} within about a decade of $10^{24.5}h^{-2}$ W.Hz^{-1} at 1.4 GHz.

Below this power, most of the large scale structures are edge-*darkened* [2] with ill-defined outer boundaries. The steepest radio spectra (a rough tracer of the most radiatively-aged particles) are in the *outer* extended regions. The sources have *prominent* jets that often contribute more than 10% of the extended power (see Figure 1). These jets are mostly *two*-sided and *rapidly*-widening. The straight jets are dominated by $\mathbf{B}_{\|}$ magnetic field components near the core but by $\mathbf{B}_{\perp}$ further out; the bent jets sometimes remain $\mathbf{B}_{\|}$-dominated on the outside of the bend [3]. The jets turn into broad plumes or trails, in which $\mathbf{B}$ tends to follow the intensity ridges (sometimes with superposed vortex-like structure). The jet-plume transition is not always well-defined, but is sometimes associated with a *sudden* widening and brightening of the radio structure. The parent galaxies of nearby low-power sources are typically "big, round, pink and friendly", *i.e.*, large, not very flattened, slightly redder-than-normal elliptical galaxies with extended outer envelopes, often in regions of high galaxy density [4,5]. Their nuclear emission lines are generally weak.

Above this power, most radio galaxies have large scale structures that are edge-*brightened* [2] with well-defined outer boundaries, though a few edge-darkened structures occur up to $P_{ext}^{1.4} \approx 10^{26}h^{-2}$ W.Hz^{-1}. The steepest spectra are in the *inner* extended regions. The jets and cores are *less prominent* than in the weaker radio galaxies, rarely having >10% of the total extended radio power (if indeed they are detected at all). The detected jets are usually *one*-sided, *i.e.*, they have side-to-side brightness asymmetries >4:1 [3]. They widen *slowly*, and are $\mathbf{B}_{\parallel}$-dominated (except near bright knots, where $\mathbf{B}$ is sometimes perpendicular to the steepest brightness gradient [3,8]. They usually end *near* bright emission, or, in the most powerful sources, near compact *hot spots*. The magnetic fields in the lobes are usually circumferential around the outer lobe "caps", and parallel to the major axis in the inner lobe "bridges". The parent galaxies tend to be "disturbed, blue and lonely", *i.e.*, bluer-than-normal ellipticals that prefer regions of low galaxy density but for which deep broad-band optical images often show disturbed morphologies [4,5]. They often have strong emission lines.

Several notable phenomena occur *near* the transition regime of $P_{ext}^{1.4} \approx 10^{24.5}h^{-2}$ W.Hz^{-1}. The local radio luminosity function changes its slope significantly near this power. In a poster at this meeting, Saul Caganoff and colleagues show that the nuclear (Hα+ [NII]) emission line strength is positively correlated with the extended radio power above, but not below, this regime.

The radio properties of quasars generally resemble those of galaxies with similar extended radio powers, except that their radio cores and jets are often more prominent, and the visible jet tends to end at or near the most compact "hot spot" in either lobe [6]. Figure 1 shows how jet prominence varies with extended power for 98 sources whose integrated jet powers and total extended powers are both known, distinguishing the sources by their optical identifications. The decrease in prominence of the radio galaxy jets above $P^{1.4} \approx 10^{24}h^{-2}$ W.Hz^{-1} is not likely to be a *radio* selection effect — the jet detection rates in well-observed complete samples [7,8] confirm the deficiency of powerful radio galaxies with prominent jets. It might however be an optical selection effect, if core and jet prominence are correlated, as the "radio galaxies" and "quasars" clearly differ in the prominence of their optical cores (nuclei).

The re-appearance of *prominent* jets in some quasars (but not, so far, in radio galaxies) with $P_{ext}^{1.4} > 10^{26.3}h^{-2}$ W.Hz^{-1} (Figure 1) is, however, also related to their extended radio morphologies. Figure 2 shows this explicitly by distinguishing the sources according to the type of extended emission that remains *after subtracting the jet(s)*. Many powerful radio quasars with extended edge-brightened double lobes — FR class II [2] — have relatively weak radio jets. But some quasar extended structures are complex — the lobes are neither clearly edge-brightened nor clearly bilaterally symmetric. In others, the known extended structure is entirely one-sided. Figure 2 distinguishes these "non-classical" morphologies and shows that *the jets are often more prominent in quasars with complex or one-sided extended structures*. These prominent jets are all "one-sided". Section 6 returns to this in detail.

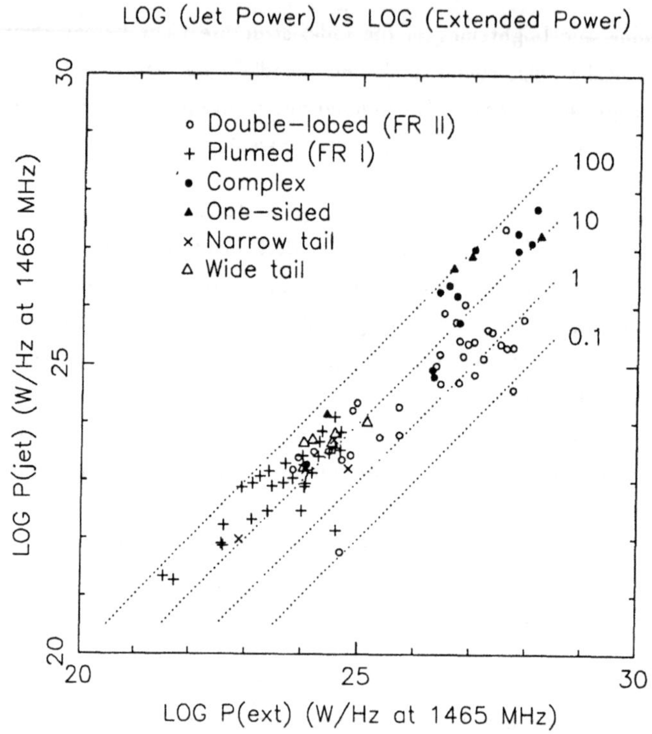

Figure 2. Logarithmic plot of jet power against total extended power at 1465 MHz for 95 sources, distinguishing different extended morphologies. The basic Fanaroff-Riley [2] classes I (plumed) and II (double-lobed) are distinguished (a) from the two classes of severely bent "tailed" sources and (b) from the sources whose extended structures, after subtracting the jet(s), are either "complex" (not clearly edge brightened or bilaterally symmetric) or entirely one-sided.

3. OUTFLOW DYNAMICS

Dynamical models of radio sources are expected to explain the generic structural features outlined above. In recent years, important insights have come from comparing the radio data a) with laboratory observations of supersonic and transonic jet propagation in ambient media, and b) with numerical modeling of supersonic and hypersonic jet propagation in media with finite pressure gradients. It appears that the two principal types of large scale radio structure may be related to two principal jet propagation regimes in the domain of jet Mach number ($M_J = v_J/c_J$) and density contrast $\eta = \rho_J/\rho_{ext}$ [8]. The radio structures will then be governed by how jets evolve from different starting conditions in the (M_J, η) domain as they propagate down pressure gradients in galactic haloes and the IGM.

An outline of jet dynamics at powers below $P_{ext}^{1.4} \sim 10^{24.5}h^{-2}$ W.Hz^{-1} has been given by Geoff Bicknell [9,10,11]. Many features of such low-power sources can be understood if their jets are *transonic* and have well-developed turbulent boundary layers. Such jets interact with ambient gas primarily by ingesting (entraining) it. They spread rapidly (in the laboratory, about as rapidly as the fastest-spreading radio jets). Rapid spreading decreases $\mathbf{B}_\perp$ more slowly than it does $\mathbf{B}_\parallel$. Jet propagation in this regime is an ongoing competition between pressure gradients and buoyancy (which try to accelerate the jet), and entrainment (which decelerates it and increases η). This competition can explain why individual jets spread at non-constant rates [7,8,11]. Adiabatic deceleration can help to keep the jets "lit up", by longitudinally compressing the particle population and $\mathbf{B}_{\perp}$ [8,9]. Even if such jets reach kiloparsec scales at low density contrast ($\eta \ll 1$), η increases towards unity as they propagate further. Eventually they become subsonic plumes and trails at the mercy of buoyancy effects and pressure gradients in the surrounding medium. The relativistic particles are swept to the outer parts of the sources, where the radio spectra are steepened by synchrotron and inverse Compton losses. The spectacular U–shaped "narrow angle tails" are low-power sources; they can be modeled as low-momentum jets of this type, bent by ram pressure as the parent galaxy moves through an ambient medium.

Models relevant to the dynamics at powers above $P_{ext}^{1.4} \sim 10^{24.5}$ h^{-2} W.Hz^{-1} have been given by Tony Williams and colleagues [12,13,14], and by Mike Norman and colleagues [15,16,17]. The large scale structures of high-power sources can be explained if their jets are initially hypersonic ($M_J \geq 10$) and light ($\eta \leq 10^{-2}$), so that they interact with surrounding gas more through shocks than by entrainment. Such jets remain supersonic until they reach a shock-dominated interface that travels slowly into the ambient medium. The region near this interface is radio-loud because the shocks compress the field components that are parallel to them, and may both compress and reaccelerate the relativistic particle population. (Recent models of the interface *e.g.*, [13,17], refine the view of the "beam working surface" sketched by Blandford and Rees in [18]). If $\eta \ll 1$, the outward velocity of the interface is much less than the jet velocity, so the flow first decelerates, then deflects back around the jets. Light hypersonic jets therefore propagate through extensive shock-heated backflow cocoons, not through undisturbed IGM. (Parts of the jet may thus be significantly overpressured relative to the undisturbed IGM, especially near internal X-shocks associated with the saturation of reflection-mode instabilities on the jet-cocoon interface [15,16]. This may help to explain why the minimum pressures p_{min} inferred from synchrotron properties of jet knots sometimes exceed p_{IGM} estimated from low-resolution X-ray data, even though the jets seem not to expand freely.) In this picture, the relatively clear outer boundary of the radio lobes traces the contact discontinuity between the backflow and shocked IGM. The relativistic particles in the outflow from the jet are left behind as the interface moves outward, and may also be actively swept back towards the galaxy by backflow. The extended emission should thus have a steeper high-frequency spectrum towards the center of the source, as is indeed observed [19].

Williams *et al.* showed how some further common features of the powerful sources can be understood if large-scale flow indeed continues *beyond* the well-collimated radio jets. If a jet deviates from strict axi-symmetry (either due to wobbling of the "central engine" or to the growth of instabilities in the jet-cocoon boundary layer) it will enter the interface region across an *oblique* shock, rather than across a perpendicular "Mach disk". By integrating the pressure through numerical simulations of the "splatter" flows beyond such oblique initial shocks, Williams and Gull [13] produced model radio brightness distributions that strongly resemble the multiple off-axis hot spots and edge-brightened *lobes* that are often seen in powerful sources. The spectral indices and magnetic field structures of multiple hot spots reinforce this picture [6]. The L-, X- and C-shaped distortions of the inner lobes ("bridges") of many sources can also be accounted for by deflections of the backflows as they try to flow "uphill" in misaligned pressure gradients in the atmosphere of the parent galaxy [14].

Plausible dynamics have therefore been suggested for the two main types of large scale radio sources, though the origin of some features remains obscure. For example, it is not clear what shapes the "wide angle tails" [19] — large, bent sources near the transition power whose structures are less severely distorted than the "narrow tails", but whose parent galaxies are dominant cluster members that should not move rapidly or far through the ambient gas. Are these sources a sign that an important dynamical factor has still to be identified?

Despite the promise of these dynamical descriptions, two major issues must be addressed before we can be sure that they are on the right track. The first is to explain why the critical transition in the (M_J, η) domain usually occurs near $P^{1.4}_{ext} = 10^{24.5} h^{-2}$ W.Hz^{-1} now (and then to predict how it varies with cosmic epoch). Paul Wiita's papers at this meeting suggest a direction in which the explanation may lie. The second is to reassess whether synchrotron radiation traces flow properties as directly as modelers usually assume when they relate jet dynamics to the "facts" of Section 2. Section 4 reviews why this reassessment is needed.

4. IS WHAT WE THINK WE SEE WHAT'S REALLY GOING ON?

Synchrotron radiation is far from an ideal flow visualization tool. To relate radio observables to the supposed flows we should compute how four Stokes parameters vary with frequency while relativistic and thermal particle densities, magnetic fields, particle energy spectra and pitch angle distributions are processed through shocks and turbulence. In practice, we can do neither the computations nor the physics. We know little about the composition or temperature of the jets. Are the "positive ions" mainly protons, or positrons? How and where are relativistic particles accelerated, and with what efficiency? Are hydrodynamic flow models adequate, or do the jets carry net currents and magnetic fields that influence their dynamics? How are the magnetic fields modified by velocity shear, or by reconnection? Different models make different guesses, with fragile, or no, constraints from the data. This is unlikely to improve soon.

The data are, however, sending us a few signals that we should not ignore. Are the kiloparsec-scale "radio jets" indeed synchrotron emission from a volume filled by outflowing material? When these jets were first transversely resolved, their apparent center-brightening (at resolutions of a few HPBW per jet FWHM) was used as evidence that the emission comes from a filled volume. This view was reinforced by the discovery that the wider jets were $\mathbf{B}_\perp$-dominated — in contrast to what was expected in boundary layers or in backflow cocoons, where the fields are likely to be stretched into $\mathbf{B}_\parallel$ configurations. But the spectacular VLA images of the M87 jet, which Phil Hardee describes here, say we should go back over this ground again. The filamentary, edge-brightened structure and the dark "central thread" force us to ask anew: "what parameters of the putative flow control the synchrotron emissivity?" If most of the radio flux resolves into filaments, the process that governs the filamentation controls how we diagnose the underlying flow — are we seeing shocks, or helical instabilities in the boundary layer, or individual strands in a magnetic "flux rope"? Does the process that produces the fine structure modify the vector parameters that affect the synchrotron emissivity (the pitch angle distribution of the particles and the field geometry), and not just the scalar densities and field strengths? Similar problems are raised by the larger-scale filaments and rings detected in the lobes of Cygnus A [21] and Hercules A [22]. Until we understand what creates such structures we must be skeptical of how well the envelope of the synchrotron emission diagnoses the overall shape of a *flow*. Might the M87 data be interpreted as a dark "jet" in a bright "cocoon"?

A related question is whether the apparent "overpressures" in some jets are evidence that $\mathbf{J}\times\mathbf{B}$ forces are needed to confine them. The pure hydrodynamic (HD) models suggest that light, hypersonic jets can cocoon themselves in overpressured backflow for significant fractions of their length. But should the synchrotron radiation from the long lines of sight through the backflow then be expected to overwhelm that from the jet at modest resolution, contradicting observation? The answer depends on how uniquely the synchrotron *emissivity* diagnoses the *total pressure* — are there ways that "radio-loud" jets can have "radio-soft" backflows at similar total pressures? We need to understand this before judging whether the wide-open (*i.e.*, rich but poorly constrained) parameter spaces of MHD must be added to those of HD when making models for the large-scale structures.

We should also ask if the centimeter-decimeter chauvinism of most data from today's imaging radio telescopes (VLA, MERLIN, WSRT) biases our view of large-scale radio structures. Might jet sidedness, cocoon sizes and shapes, or the relative prominence of jets and cocoons, vary with wavelength? Our radio view is occasionally checked from the short-wave side by optical images of jets, such as those shown at this meeting by Bill Keel. But the meter- and decameter-wavelength side is mostly *terra incognita*, despite its formal inclusion in the standard equipartition calculations! Section 5 reports a jet system in 3C288 whose sidedness varies with wavelength, as a cautionary tale.

Finally, if jets in powerful sources have high Mach numbers and pressures but low densities, their velocities are high. In modeling Cygnus A, Williams [12] estimates $M_J \sim 10$, $\eta = 10^{-4}$ to 10^{-3} and $v_J \rightarrow c$. How far do bulk relativistic motions extend beyond the parsec-scale phenomena reviewed at this meeting by Marshall Cohen? May relativistic boosting, light travel time, and aberration effects distort our view of larger-scale sources significantly? For example, is jet one-sidedness in the powerful sources (a) a relativistic (Doppler boost) effect, (b) an asymmetry in the synchrotron emissivities of two otherwise symmetric flows, or (c) an indicator of intrinsic one-sidedness in the energy outflow from the AGN? Are there intrinsically "radio-dark" flows in two-lobed sources with one-sided jets (or with no detected jets), or are jets that we *don't* see being "hidden" by beaming their radio emission away from us? There is no evidence for, and substantial evidence against, bulk relativistic motion in the two-sided jets of low-power sources, but the flow velocity in high-power jets is poorly constrained [7,8]. An important trend in high-power sources, as Peter Scheuer has emphasized [23,24], is that the one-sided parsec-scale jets always point towards the bases of the one-sided kiloparsec-scale jets when both are seen in the same source. The cause of the one-sidedness is therefore likely to be the same on both scales. If we are convinced that the Doppler boost produces the parsec-scale brightness asymmetry, we must ask if it produces the kiloparsec-scale one. Conversely, if we are convinced that the kiloparsec-scale asymmetry is intrinsic, we must question the relativistic-jet interpretation of the parsec-scale phenomena. (Unless, of course, the jet designer is perverse enough to make *intrinsically one-sided relativistic jets*, in which case we can only detect her perversion statistically [25]). The rest of this review discusses new data that bear on the interpretation of large-scale jet sidedness in powerful sources.

5. "TALES FROM THE DARK SIDE", OR — COUNTERJETS IN STRONG SOURCES

We need to know *how* one-sided the kiloparsec-scale jets in powerful sources are, as the probability of observing a given brightness asymmetry due to the Doppler boost in a randomly oriented sample decreases as the asymmetry increases. It is therefore crucial to detect counterjets in these sources.

Several examples of counterjet emission have now been documented in radio galaxies with $P_{ext} > 10^{26}h^{-2}$ W.Hz^{-1} at 1.4 GHz. I will distinguish those in which the main jet appears not to continue all the way to the bright parts of the lobes from those that appear to be more continuous.

In 3C 219 ($P^{1.4}_{ext} = 10^{26.44}h^{-2}$ W.Hz^{-1} [26]) and 3C 288 ($P^{1.4}_{ext} = 10^{26.36}h^{-2}$ W.Hz^{-1} — Bridle, Byrd, Fomalont and Valtonen, in preparation) the brighter jets disappear well before they reach the edge-brightened parts of the lobes. In both sources, the counterjets ($P^{1.4}_{cj} = 10^{23.22}$ and $10^{23.68}$ h^{-2} W.Hz^{-1} respectively) have bright tips that are closer to the core than the tips of the main jets. In 3C288, the spectral index of the main jet between 4.9 and 15 GHz is 0.72±0.04, while that of the counterjet is 1.30±0.17. These are all asymmetries that are *expected* in "born-again" relativistic jets that have restarted after a period of inactivity. In this picture, the tips of the counterjets are where an unfavorable Doppler factor that hides their inner sections can be removed, or decreased, at shocks. The brightest part of a Doppler-hidden counterjet opposite a "born-again" main jet should therefore be furthest from the core. As the counterjet must (in this picture) be on the far side of the source, it is seen at an earlier time, so its bright tip must appear to be closer to the core than the tip of the main jet. The counterjet is also red-shifted, whereas the main jet is blue-shifted. Any steepening of the radio spectrum with frequency in the rest frames of the jets must translate to the counterjet having a steeper spectrum than the jet in the observer's frame. The asymmetries of the jet/counterjet pairs in 3C219 and 3C288 are thus *all* of the kind expected for intermittent relativistic jets. The brightness maxima of the counterjets are, however, anticorrelated in position with those of the main jets; this could be taken as evidence for "flip-flop" behavior, as an alternative to the relativistic asymmetry.

In three other powerful radio galaxies, the main jet is detected most of the way into the lobe; the counterjet is either similarly extended, or it is hard to tell because both jets are confused with lobe emission. The sources are 3C 341 ($P^{1.4}_{ext} = 10^{26.78}h^{-2}$ W.Hz^{-1}), 3C 438 ($P^{1.4}_{ext} = 10^{26.84}h^{-2}$ W.Hz^{-1}) and Cygnus A ($P^{1.4}_{ext} = 10^{27.74}h^{-2}$ W.Hz^{-1}). The counterjets in these sources ([7] and private communications from R. A. Laing and R. A. Perley) have $P^{1.4}_{cj} \approx 10^{24}$ to $10^{25}h^{-2}$ W.Hz^{-1}. 3C438 is remarkable for the exceptional symmetry of its jet and counterjet, given its high power.

I am also, with many collaborators, systematically looking for counterjets in 3CR quasars with angular sizes >10 arcsec, using 8-hr integrations on the VLA at 4.9 GHz. We have yet to find a clear, continuous, counterjet extending all the way from a core to the lobe, but there are "bits and pieces" of possible counterjets in three or four of the dozen sources observed so far. The 1.4-GHz extended powers of these sources range from $10^{26.9}$ to $10^{28.2}h^{-2}$ W.Hz^{-1}, and the counterjet candidates again have $P^{1.4}_{cj}$ from 10^{24} to $10^{25}h^{-2}$ W.Hz^{-1}. A counterjet of similar power has also been seen in the quasar 1928+73 ($P^{1.4}_{ext} = 10^{26.32}h^{-2}$ W.Hz^{-1}), which has superluminal motions on parsec scales [27].

Several conclusions can be drawn from these first glimpses of counterjets in strong sources. First, there *is* detectable emission on the counterjet side of some powerful sources in images made with dynamic ranges of which the VLA, MERLIN and the WSRT are all capable with modern calibration techniques. Intensive searches for counterjets with these instruments may therefore find enough examples for the statistics of jet/counterjet symmetries to become useful constraints on the jet-asymmetry problem. Second, integrated jet to counterjet ratios (in the sources where counterjets have been detected) range from around two to several hundreds. Such ratios can be accounted for by Doppler boosting without requiring Lorentz factors as high as those inferred for the superluminal parsec-scale jets, or demanding improbable orientations. Third, the jet/counterjet ratio may depend on frequency, as in 3C288. Whether or not this effect is a hallmark of a relativistic jet, it may lead to observational selection problems, and some counterjet hunting should be done at low frequencies to allow for this possibility. (To get enough angular resolution, this may require the extended MERLIN array or a combination of the VLA and VLBA). Fourth, the detected *counterjets* in the more powerful sources generally have $P_{cj}^{1.4}$ from 10^{24} to $10^{25}h^{-2}$ W.Hz^{-1}, and are thus more powerful than the *symmetric* jets in the weak radio galaxies. Even if Doppler boosting and dimming produces the apparent side-to-side asymmetries of the jets in these powerful sources, *the underlying flows must emit more radio power than those in weaker sources* (as we see directly in 3C438).

6. BRIGHT JETS IN POWERFUL SOURCES

I will separate two interesting phenomena that fall under this heading, although they may later prove to be related. The first was discovered by Peter Barthel, Colin Lonsdale, George Miley and Richard Schilizzi [28,29,30] in a sample of 80 steep-spectrum radio sources identified with quasars at $z>1.5$. The source structures in this sample are more bent than those in sources of similar powers at lower redshifts, and the prominence of their jets (the jet-to-lobe intensity ratio) correlates positively with the jet curvature but not with not with the core-to-lobe ratio [30]. They interpret this as evidence for epoch dependence of the environment at high redshifts, *e.g.* that sources at $z>1.5$ must force their way into a denser ambient medium than that around the local sources. An alternative (suggested to me by Jim Condon) is that the radio sample may be biased, despite the steep-spectrum filter, if the quasars must have *both* an unusually luminous optical continuum and prominent emission lines for the sources to have been optically identified. This could introduce an *orientational bias* if the optical continuum or lines escape anisotropically (this need not require that the light is *relativistically beamed* — obscuration by an accretion disk or torus would do). It could also introduce a *bias towards immature sources* that are still establishing their double lobed structures, if optical cores are brighter while the sources are young. Whatever the physical reason for the Barthel *et al.* effect, I wish to minimize it in what follows, so I restrict the analysis below to sources with $z<1.5$.

The second effect is present in the sample of all 27 sources in the redshift range $0.15<z<1.5$ with $P_{ext}^{1.4} >10^{26}h^{-2}$ W.Hz^{-1}, for which integrated jet and lobe flux densities are available. (These luminosity and redshift limits are chosen to give a sample in which largest linear size is not *strongly* correlated with redshift or with source power, but which contains enough sources to be interesting). Figure 3 plots the ratio between the integrated flux densities of the *kiloparsec scale* jet emission and the lobe emission against the largest linear size of the source. The symbols distinguish the structure type of the emission that remains after subtracting the jet and the core, as in Figure 2. The sources

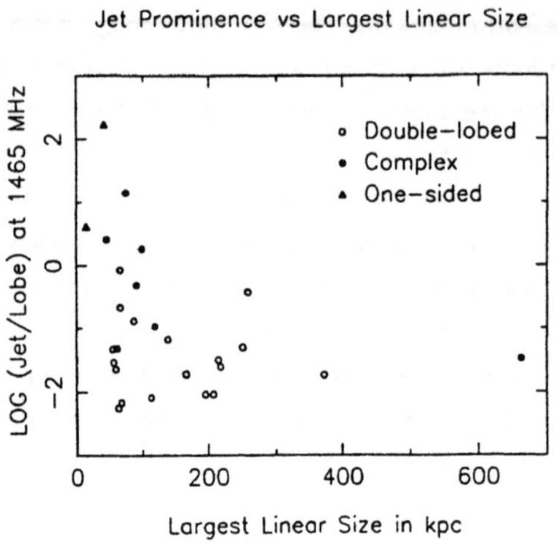

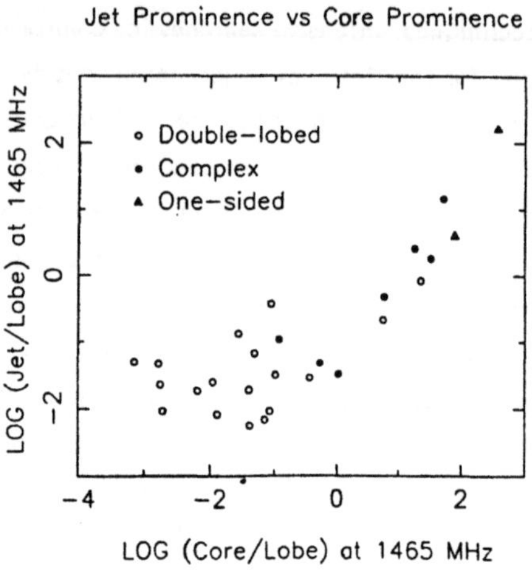

Figure 3 (left). Plot of the logarithm of the ratio of the jet flux density to the lobe flux density against largest linear size, for 27 sources with $P_{ext}^{1.4} > 10^{26}h^{-2}$ W.Hz^{-1} and $0.15<z<1.5$, distinguishing different extended morphologies.

Figure 4 (right). Logarithmic plot of the jet-to-lobe flux density ratio against the core-to-lobe flux density ratio, for the same 27 sources.

with $S_{jet}/S_{lobe} > 1$ all have "complex" or "one-sided" extended structures, and apparently small linear sizes. There are (at least) two ways to interpret this. If apparently small linear size and apparently complex morphology *occurring together* in powerful sources are taken to indicate orientation away from the plane of the sky, Figure 3 suggests that some *kiloparsec-scale* jets at $z<1.5$ are Doppler boosted, by analogy with the arguments for Doppler boosting of the *cores* in double radio sources **[31,32]**. A viable alternative is that some intrinsically small (young, smothered?) sources have intrinsically complex structures *and* intrinsically bright jets. Such sources might be regarded as "local" examples of the Barthel *et al.* high-redshift effect, if this is intrinsic.

Colin Lonsdale **[30]** has emphasized that jet prominence and core prominence are uncorrelated for the 25 jetted sources in his high-redshift sample. Figure 4 plots the jet-to-lobe ratio against the core-to-lobe ratio for the 27 sources in Figure 3, for comparison with his Figure 2(a). Such plots are prone to systematic error, as the data at high prominence are obtained by dividing the jet and the core flux densities by the same, weak, lobe flux density. Even so, it is clear that the most prominent jets in the lower-redshift sample are accompanied by prominent cores. This difference between the two samples should be checked when more data are available, as it may signify that the factors linking jet and core prominence also change with redshift. I urge observers to record integrated jet and lobe flux densities when publishing their data, so that larger samples can be compiled for such studies.

One should not conclude from Figures 3 and 4 that core prominence, jet prominence and linear size are *un*correlated for all classical double-lobed sources with $P_{ext}^{1.4} > 10^{26}h^{-2}$ W.Hz^{-1} and $0.15<z<1.5$. Many such sources have no detected jets and are thus not shown in these Figures, which must be only an *upper envelope* to the full distributions. We cannot judge whether jet prominence is correlated with largest linear size or core prominence at *low* levels of jet prominence until most of the jets have been detected in a complete sample of such sources. (Upper limits cannot be given reliably for undetected jets, as one does not know *a priori* which areas of sky the jets may occupy.)

To distinguish the Doppler boosting interpretation of Figures 3 and 4 from the alternative that there is a class of small sources whose cores and jets are intrinsically prominent, we need diagnostics for source and jet orientations that are independent of apparent linear size and morphology. Section 7 discusses a recently-discovered candidate for this diagnostic role.

7. FARADAY SCREENS, DEPOLARIZATION AND JET SIDEDNESS

Evidence is rapidly accumulating that many of the large-scale radio structures are surrounded by magnetoionic media whose Faraday depths are significant at centimeter wavelengths. In most cases, the media appear to be outside the radio sources, but associated with the parent optical object. I will not discuss the physical conditions in, or the origin of, the material here. I will instead concentrate on the evidence for, and the consequences of, its finite Faraday depth.

a) Evidence from rotation measure data

There is now evidence for extremely large Faraday rotation measures (RM $>$ 1000 rad.m^{-2}) in front of several extragalactic sources. The best documented is Cygnus A **[33]**, where the RM ranges from -4000 rad.m^{-2} to $+3000$ rad.m^{-2}, with ∇RM reaching 400 rad.m^{-2}.kpc^{-1} (if the screen is at the distance of the radio source). As there is no internal depolarization of either the lobes or the jet, most of the Faraday rotating material must be *outside* the source (in a sheath or in the surrounding IGM). If the screen was mixed with the radio source, differential rotation between the front and the back sides would have depolarized the emerging radiation so that we could not have detected the very signal from which the rotation measure was inferred! RMs $>$1000 rad.m^{-2} have since been detected in several other extended sources: M87 (Frazer Owen, private communication), 3C295 and 3C218 **[34]**. It will be important to learn whether these exceptionally thick Faraday screens correlate with the presence of cluster X-ray sources, of cooling flows, or (for high-redshift quasars) of the broad systemic-velocity Lyα absorption discussed at this meeting by Ray Weymann.

Possibly more significant here however is the evidence that Faraday screens with more modest rotation measures (10's to 100's of rad.m^{-2}) may be widespread. Screens with RMs in this range have been imaged in several radio galaxies **[35]**. In several cases their symmetries and ∇RM amplitudes show that they are associated with the source's parent galaxy, rather than with foreground gas in our galaxy. The RM data also show directly that the Faraday depths across these screens do not vary smoothly, but have substructure on 1 to 10-kiloparsec scales in which ∇RM can be $\sim$ 10 to 20 rad.m^{-2}.kpc^{-1}. In most cases, there is again little evidence for *internal* depolarization of the radio emission, so the screens are generally *outside* the radio emitting volumes, in sheaths or in larger-scale ambient gas. The RM substructure will however cause differential Faraday rotation across the synthesized beams of imaging radio telescopes at decimeter wavelengths, producing low-frequency depolarization whether or not the lobes and jets contain thermal material. It is therefore not surprising that evidence for Faraday screens around radio galaxies is also accumulating from low-frequency depolarization data.

b) Evidence from depolarization data

Systematic depolarization gradients have been detected in several samples of radio galaxies observed at 0.6 and 1.4 GHz with the WSRT [36,37]. The degree of linear polarization $p^{0.6}$ is systematically less than $p^{1.4}$ in the *inner* parts of radio galaxies, *i.e.*, the ratio $p^{0.6}/p^{1.4}$ increases with distance from the galactic nuclei over scales that are typically from $50h^{-1}$ to $100h^{-1}$ kpc.

These results, and the direct RM imaging discussed above, together suggest that there are *structured* Faraday screens on $\sim$ 100 kpc scales around *many* radio galaxies. If the screens are not thin sheaths, but surround the sources on scales greater than or comparable to those of the radio lobes, depolarization data can distinguish the *front* side of the source from the *back* side. The front is seen along a shorter path through the screen, so it depolarizes less than the back at a given frequency.

c) Depolarization asymmetries

It has been known for several years that depolarization asymmetries are common in large-scale double radio sources [38]. It has only recently become clear, however, that in powerful sources these asymmetries are strongly correlated with the brightness asymmetries of the radio jets. Robert Laing and co-workers [39,40] have determined the depolarization asymmetries between 1.4 and 4.9 GHz of two samples of powerful extended sources with one-sided radio jets. They find that in about 33 of 35 cases, the jetted side of the source is less depolarized than the unjetted side at 1.4 GHz. Of the other two cases, one can be ascribed to the unjetted side having depolarized at a higher frequency, the other to observational uncertainties.

If the depolarizing screens are symmetrically distributed around the sources, *the one-sided jets are systematically on the side facing us* in the cases studied by Laing *et al.*. This is good news for the Doppler-boost model of the jet brightness asymmetry, but I expect that the effect will also be explained *post hoc* in terms of an intrinsic linkage between jet brightness and the screen parameters. While we wait for this to happen, the uncertainty about the geometry and scale of the screen can be addressed directly by imaging ∇RM across sources that show the depolarization asymmetries. Do the RM distributions show the edge effects expected of thin sheaths around the lobes or jets, or do they suggest larger-scale media? The depolarization of low-power sources with *symmetric* jets will also be relevant. If these sources also depolarize asymmetrically, it will be harder to believe that depolarization and jet brightness are intrinsically anti-correlated in sources with asymmetric jets.

8. SUMMATION

It seems likely that hydrodynamic models for the outflows in AGN radio sources will be able to explain many generic properties of the large-scale structures (Sections 2 and 3). The Mach numbers and density contrasts of the flows, plus the ambient pressure gradients, appear to be the key variables.

There is evidence that one-sided kiloparsec-scale jets are more prominent in the powerful sources at $z<1.5$ that have apparently small sizes *and* complex morphologies (Section 6). The one-sided jets in double sources may also be systematically on the side of the source that faces us (Section 7). These results, the gradually increasing number of counterjet detections in powerful sources, and the detailed asymmetries of some jet/counterjet pairs (Section 5), are all consistent with some kiloparsec-scale outflows in powerful sources being *mildly relativistic, e.g.* $v_J \sim 0.8c$. This hypothesis connects a wide range of properties of the large scale structures with the parsec-scale phenomena described at this meeting by Marshall Cohen, and explains the strong correlation between kiloparsec-scale and parsec-scale brightness asymmetries (Section 4). It thus offers a *coherent* explanation of a range of phenomena in powerful sources that otherwise must be attributed to a "conspiracy" of seemingly unrelated effects — some involving small, complex sources; some involving intrinsically one-sided (and intermittent) jets; and others involving jet-related asymmetries in Faraday screens.

The major difficulties of the hypothesis are (a) that the symptoms may be seen more often than they should be in a randomly oriented sample, and (b) that "de-projecting" some sources by the required angles implies that they have larger physical sizes than some feel comfortable with. These difficulties could evaporate if there is an orientational bias in present "complete samples" of *optically identified* radio sources (or if relativistic jet velocity fields are as complex as those in present hydrodynamic models). The discovery that some Seyfert nuclei may look different from different directions emphasizes the issue of whether there is orientational bias in the radio source identifications. Could samples of "radio quasars" be biased more towards the line of sight than samples of the (more often jetless) powerful radio galaxies [7]?

Although we do not yet have decisive evidence in favor of either the relativistic-jet or intrinsic-asymmetry pictures, dynamical models for the large-scale structures should consider the possibility that mildly relativistic bulk motions extend all the way to the lobes in the powerful sources. Modelers should also attempt to describe the detailed linkages between their flow variables and the synchrotron emissivity (Section 4). We also need diagnostics for whether "disappearing" jets like those in 3C219 and 3C288 (Section 5) are indeed being "born again", or whether something else reduces the synchrotron emissivity for most, but not all, of their lengths. *Direct* evidence of variability in AGNs on time scales from hours to decades is discussed in detail at this meeting. Could features of the large-scale structures of sources such as 3C219 and 3C288 tell us about variability on time scales of tens of thousands of years, if properly interpreted? Numerical models of *intermittent* jets may help us to do this.

Ultimately, our models must explain why so many large-scale features of radio sources change qualitatively near a local 1.4-GHz extended luminosity of $P_{ext} = 10^{24.5}h^{-2}$ W.Hz^{-1} (Section 2). There are so many correlations between large-scale radio morphology and radio power that we must ask for *quantitative* interpretations of them in terms of the interactions between the outflows from AGNs and their environment. Our goal should be to find predictive relationships between source power, jet variables such as initial Mach number, density and collimation, and environmental variables such as density and pressure gradients (possibly functioning both as an "exhaust system" and as a "fuel tank" for the central engine). The new computer-intensive tools of radio imaging and of numerical jet modeling have suggested plausible directions to go in, but most of the road is still ahead.

I thank Geoff Bicknell, Jean Eilek, Steve Gull, Robert Laing, Mike Norman, Chris O'Dea, Frazer Owen, Rick Perley and John Wardle for illuminating discussions of the relationships between radio data and the dynamics of large-scale sources, and Jim Condon for valuable advice on selection effects and luminosity functions. I am also grateful to Jim Condon and Rick Perley for critically reading a draft of this paper. The NRAO is operated by Associated Universities, Inc. under contract with the National Science Foundation.

REFERENCES

1. Kellermann, K. I., in *Galactic and Extragalactic Radio Astronomy*, eds. G. L. Verschuur and K. I. Kellermann, Springer-Verlag (1974), p. 220.
2. Fanaroff, B. L. and Riley, J. M., *Mon. Not. Roy. Astron. Soc.* **167**, 31P (1974).
3. Bridle, A. H., *Astron. J.* **89**, 979 (1984).
4. Smith, E. P. and Heckman, T. M., in *Radio Continuum Processes in Clusters of Galaxies*, eds. C. P. O'Dea and J. M. Uson, NRAO (1986), p. 305.
5. Heckman, T. M., Smith, E. P., Baum. S. A., van Breugel, W. J. M., Miley, G.K., Illingworth, G. D., Bothun, G. D. and Balick, B., *Astrophys. J.* **311**. 526 (1986).
6. Laing, R. A., in *Physics of Energy Transport in Extragalactic Radio Sources*, eds. A. H. Bridle and J. A. Eilek, NRAO (1984), p.128.
7. Bridle, A. H. and Perley, R. A., *Ann. Rev. Astron. Astrophys.* **22**, 319 (1984).
8. Bridle, A. H., *Can. J. Phys.* **64**, 353 (1986).
9. Bicknell, G. V., *Astrophys. J.* **286**, 68 (1984).
10. Bicknell, G. V., *Astrophys. J.* **300**, 591 (1986).
11. Bicknell, G. V., *Astrophys. J.* **305**, 109 (1986).
12. Williams, A. G., *The Fluid Dynamics of Radio Sources*, Ph. D. thesis, Cambridge University (1985).
13. Williams, A. G. and Gull, S. F., *Nature* **313**, 34 (1985).
14. Leahy, J. P. and Williams, A. G., *Mon. Not. Roy. Astron. Soc.* **210**, 929 (1984).
15. Norman, M. L. and Winkler, K.-H., *Supersonic Jets*, Los Alamos Science No. 12, Los Alamos National Laboratory (1985), p. 38.
16. Smarr, L. L., Norman, M. L. and Winkler, K.-H., *Physica* **12D**, 83 (1984).
17. Smith, M. D., Norman, M. L., Winkler, K.-H. and Smarr, L., *Mon. Not. Roy. Astron. Soc.* **214**, 67 (1985).
18. Blandford, R. D. and Rees, M. J., *Mon. Not. Roy. Astron. Soc.* **169**, 395 (1974).
19. Alexander, P. and Leahy, J. P., *Mon. Not. Roy. Astron. Soc.* **225**, 1 (1987).
20. Burns, J. O., *Can. J. Phys.* **64**, 373 (1986).
21. Perley, R. A., Dreher, J. W. and Cowan, J. J., *Astrophys. J. Letts.* **285**, L35 (1984).
22. Dreher, J. W. and Feigelson, E. D., *Nature* **308**, 43 (1984).
23. Scheuer, P. A. G., in *VLBI and Compact Radio Sources*, Proc. I.A.U. Symp. No. 110, eds. R. Fanti, K. Kellermann and G. Setti, Reidel (1984), p. 197.
24. Scheuer, P. A. G., in *Astrophysical Jets and their Engines*, ed. W. Kundt, Reidel (1987), p. 129.
25. Bridle, A. H., in *Physics of Energy Transport in Extragalactic Radio Sources*, eds. A. H. Bridle and J. A. Eilek, NRAO (1984), p.135.
26. Bridle, A. H., Perley, R. A. and Henriksen, R. N., *Astron. J.* **92**, 534 (1986).
27. Johnston, K. J., Simon, R. S., Eckart, A., Biermann, P., Schalinski, C., Witzel, A. and Strom, R. G., *Astrophys. J. Letts.* **313**, L85 (1987).
28. Barthel, P. D., *Radio Structure in Quasars*, Ph. D. thesis, Leiden University (1984).
29. Miley, G.K., in *Observational Cosmology*, Proc. I.A.U. Symp. No. 124, eds. A. Hewitt, G. Burbidge and L.-Z. Fang, Reidel (1987), p. 267.
30. Lonsdale, C. J., *Can. J. Phys.* **64**, 445 (1986).
31. Hine, R. G. and Scheuer, P. A. G., *Mon. Not. Roy. Astron. Soc.* **193**, 285 (1980).
32. Kapahi, V. K. and Saikia, D. J., *J. Astrophys. Astron.* **3**, 465 (1982).
33. Dreher, J. W., Carilli, C. L. and Perley, R. A., *Astrophys. J.* **316**, 611 (1987).
34. Kato, T., Tabara, H., Inoue, M. and Aizu, K., *Nature* **329**, 223 (1987).
35. Laing, R. A. and Bridle, A. H., *Mon. Not. Roy. Astron. Soc.* **228**, 557 (1987).
36. Jägers, W. J., *The Polarization of Radio Galaxies*, Ph.D. thesis, Leiden University (1986).
37. Strom, R. G. and Jägers, W. J., *Astron. and Astrophys.*, in press (1987).
38. Conway, R. G. and Strom, R. G., *Astron. and Astrophys.* **146**, 392 (1985).
39. Laing, R. A., *Nature*, in press (1987).
40. Garrington, S. T., Leahy, J. P., Conway, R. G. and Laing, R. A., *Mon. Not. Roy. Astron. Soc*, in press (1987).

DISCUSSION

Marshall Cohen: You are convincing us that the powerful kiloparsec-scale jets are two-sided and relativistic. What then of 3C273, which remains one-sided no matter how hard we look?

Alan Bridle: I can think of two possibilities. The first would be that we have still not looked hard enough, despite the Herculean (Perleyan?) efforts made so far — it's been hard to find convincing *lobe* emission on *either* side of this source. If this is right, the "other side" may appear in an extremely high dynamic range image (possibly at a low frequency!). We might then conclude that the kiloparsec scale jet in this source has a higher-than-average Lorentz factor. The second possibility is that this is a "born-again" relativistic jet at such a small angle to the line of sight that the emission from the far side has not reached us yet. In that case the "news from the other side" will appear if we wait long enough. The counterjet tip could be too far out and too resolved to be detected by VLB, but still be too close to the core to be detected by the VLA.

Wil van Breugel: Do the brighter, or one-sided, jets always point to the lobe with the larger total flux density?

Alan Bridle: I have not looked at that recently. But there is no correlation between jet sidedness and the maximum distance from the core to the edge of the lobe, in the sample of jets that I used here. As several groups have found that the lobe with the larger total flux density is usually closer to the core, I would be a bit surprised if there is a strong effect of this kind. But I agree that it should be looked at again, as there are over 200 jetted sources known now.

Oved Dahari: Are curved jets found to be more luminous on average?

Alan Bridle: In the weak U-shaped "head-tail" sources, the jet with the greater curvature is often the brighter one. In the powerful sources, the only correlation that anyone has published is the one between jet prominence and jet curvature at $z>1.5$ [30]. I have not yet looked at that correlation in my own sample.

Oved Dahari: In the picture where relativistic jets slow down, do counter-jets appear further from the core?

Alan Bridle: In a source with "born-again" relativistic jets, the counterjets should slow down, or even splash back, at their tips. The observer sees an asymmetry because the counterjet tip is on the far side, so that it is seen at an earlier stage of development and thus *closer* to the core than the main jet tip. The slowdown also helps us to see the far tip, by reducing an unfavorable Doppler factor. If jets in a steady flow slow down, their *lengths* are defined by the history of how the interface has moved relative to the cocoon wall, but the counterjets may then become *more visible* in the outer parts of the source. If jets slow down between parsec and kiloparsec scales, we might find sources with counterjets in VLA images but not in VLBI images of equal dynamic range.

TRIPLE RADIO STRUCTURE IN THE "DOUBLE-NUCLEUS" GALAXY MARKARIAN 266

J. M. Mazzarella
Department of Astronomy, University of Michigan, Ann Arbor, MI 48109-1090

R. A. Gaume
Naval Research Laboratory, Code 4134 G, Washington, DC 20375-5000

H. D. Aller
Department of Astronomy, University of Michigan, Ann Arbor, MI 48109-1090

1. INTRODUCTION

Markarian 266 (= NGC 5256 = IZw67) appears optically as a peculiar system containing two compact nuclei separated by 10″. Petrosian *et al.* (1980) showed that both nuclei exhibit optical Seyfert 2 properties, while Kollatschny and Fricke (1984) re-classified the NE nucleus as a Liner based on IUE data. We present mutli-frequency radio synthesis observations of Mrk 266 which reveal the presence of three separate non-thermal continuum sources at a resolution of HPBW = 2″. Two of the radio components coincide with the optical nuclei, while the third is located *between* the nuclei. Higher resolution VLA data (HPBW = 0″.3) reveal complex sub-structure in the central and SW (Seyfert 2) radio components. Radio synthesis data at 20 cm, 6 cm, and 2 cm are used to produce spectral indices for the three components, and possible interpretations of these results are discussed.

2. OPTICAL IMAGING

Four CCD images of Mrk 266 were obtained using the 1.3-m telescope of the McGraw-Hill Observatory equipped with an RCA CCD camera. B, V, and R images were calibrated using observations of de-focused standard stars, and reduced in the standard fashion. In order to detect faint extended emission in the outer halo of Mrk 266, all four images were registered and co-added. A contour diagram of the result is displayed in Figure 1. The optical morphology, including the diffuse outer halo, asymmetrical spiral or tidal features, and overall disturbed appearance, suggest that this is an ongoing collision/merger between two previously separate galaxies. This interpretation of the double optical nuclei and peculiar optical morphology is also supported by the presence of strong far infrared emission as measured by IRAS, and a large value of $L_{60\mu}/L_{100\mu} = 0.61$, which is common only among strongly interacting galaxies (Young *et al.* 1986).

3. RADIO SYNTHESIS OBSERVATIONS

Mrk 266 was observed at the VLA (National Radio Astronomy Observatory) in the A-, C/D-, and C- configurations during observing runs in August, February, and January 1987, respectively. Figure 1 displays contours of four co-added CCD images superposed on a grey-scale representation

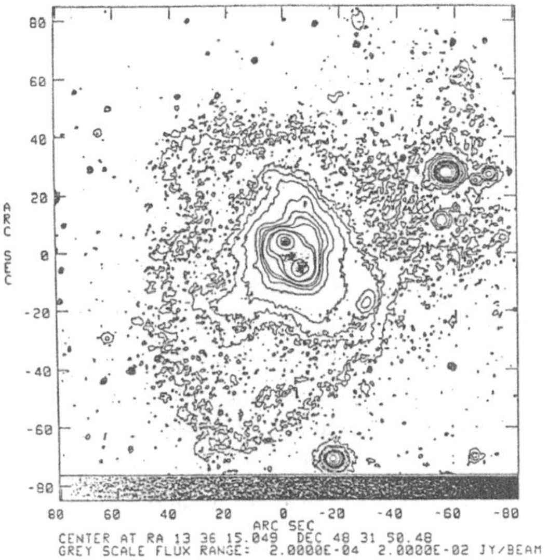

Fig. 1– Optical contour diagram of four co-added CCD images (B, B, V, and R; 2400 sec total integration) superposed on a grey-scale representation of the 20 cm synthesis image. The spatial scale in all contour diagrams presented here is 540 pc to 1 arcsec, assuming $H_o = 75$ km$\cdot$ s^{-1}$\cdot$ Mpc^{-1}.

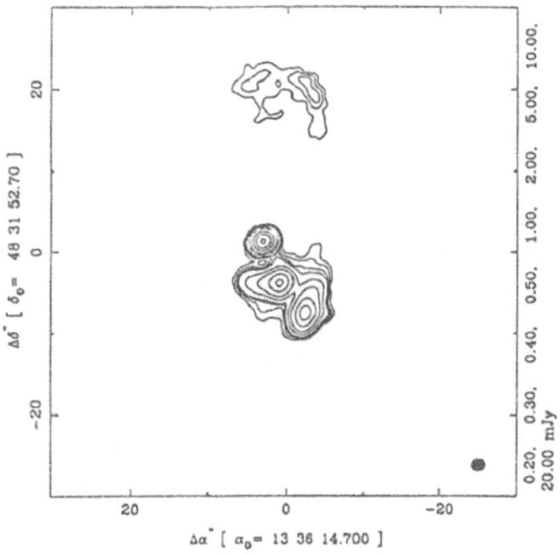

Fig. 2– Contour diagram of A-config. VLA 20 cm data. Sidelobes were removed with the CLEAN algorithm, and the image was mapped with natural weighting. The half-power-beam-width (HPBW) of 1″.8 x 1″.5 is represented by the ellipse in the lower right hand corner. The r.m.s. noise in the map is 0.05 mJy, and the contour levels are displayed starting at 4 times the r.m.s. noise.

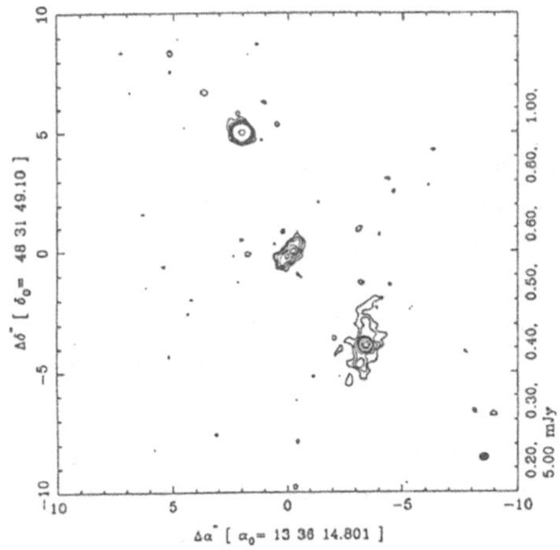

Fig. 3– Contour diagram of C- and A-config. VLA 6 cm data. Sidelobes were removed with the CLEAN algorithm, and the image was mapped with uniform weighting to obtain the highest possible resolution (HPBM = 0″.3 x 0″.4). The r.m.s. noise in the map is 0.07 mJy, and the contour levels are displayed starting at 3 times the r.m.s. noise.

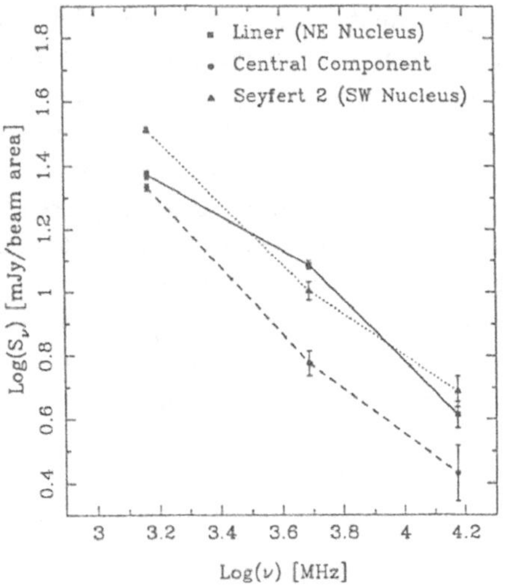

Fig. 4– Radio spectra computed from 20 cm, 6 cm, and 2 cm VLA synthesis data convolved to the same HPBW of 1″.8 x 1″.5. The spectral indices α, where $S_\nu \sim \nu^{-\alpha}$, are as follows. Liner: $\alpha_{20,6} = 0.54$, $\alpha_{6,2} = 0.97$; Central component: $\alpha_{20,6} = 1.06$, $\alpha_{6,2} = 0.71$; Seyfert 2: $\alpha_{20,6} = 0.97$, $\alpha_{6,2} = 0.65$.

of a 20 cm continuum image. The beam width at 20 cm of HPBW = 1"8 x 1"5 provides very nearly the same resolution as the optical image (1"5 seeing). Note the faint optical halo that extends approximately 110"x 75"(59 kpc x 41 kpc, assuming H_o = 75 km· s^{-1}· Mpc^{-1}), as well as the unexpected strong radio source located between the active nuclei. There is also weak 20 cm emission associated with the northern edge of an optical tidal feature. The morphology of the 20 cm emission is more clearly visible in the contour diagram in Figure 2. The central radio component has no obvious optical counterpart in the broad-band images.

Additional insight into the nature of this complex system is revealed by high-resolution 6 cm observations (HPBW = 0"3 x 0"4) presented in Figure 3. The SW source coincident with the optical Seyfert 2 nucleus presents a "linear" morphology indicative of the double/triple continuum sources found in many Seyfert nuclei observed with the VLA (e.g., Ulvestad and Wilson 1984, and references therein). This morphology suggests that the SW Seyfert component is transporting energy over the scale of ~1"(540 pc) along an axis at position angle (PA) -10°. The source between the optical nuclei presents a double structure (separation = 0"4 = 200 pc) aligned along PA ≈ -45°. Figure 4 is a plot of the 20 cm/6 cm/2 cm spectra of the three components.

4. DISCUSSION

We discuss four possible interpretations for the radio components visible in the 2"–resolution synthesis data. i) All three components have steep, non-thermal spectral indices that indicate the presence of synchrotron radiation (Figure 4). These data alone are highly suggestive of an optically obscured central "engine" which may be transporting energy into the surrouding medium, thus producing the radio continuum emission and accompanying optical emission-line regions associated with the AGNs. Triple radio structure in the double Seyfert 2 nuclei of NGC 5929 has been interpreted in this fashion (Keel 1985). ii) Alternatively, the optical morphology of this object and its warm infrared emission as measured by IRAS indicate that two radio continuum sources associated with classical Seyfert-type nuclei are in a stage of violent interaction, perhaps merging. Strong shocking would be expected to occur near the center of gravity of the collision, producing a burst of star formation and associated supernova remnants. The lack of an obvious optical couterpart to the central radio component (a giant H II region) requires strong optical obscuration in this region. Also, the presence of a very steep spectral index suggests that another process may be responsible for much of the radio emission in the central component. iii) Collisions of interstellar clouds induced by the merger can accelerate electrons and produce synchrotron emission in the region between the nuclei where shocking may be most intense. The current data suggest this is the most likely hypothesis. iv) We consider the possibility that the central source is a background object to be highly unlikely. Detailed analyses and discussion of these data will be published elsewhere.

We are grateful to P. Hughes for helpful discussions. We also acknowledge that the central radio component in Mrk 266 has been detected in 20 cm VLA data obtained by J. B. Hutchings.

REFERENCES

Keel, W. C. 1985, *Nature*, **318**, 43.
Kollatschny, W., and Fricke, K. J. 1984, *Astr. Ap.*, **135**, 171.
Petrosian, A. R., Saakian, K. A., and Khachikian, E'. E. 1980, *Astrofizika*, **16**, 621.
Ulvestad, J. S., and Wilson, A. S. 1984, *Ap. J*, **285**, 439.
Young, J. S. *et al.* 1986, *Ap. J. (Letters)*, **311**, L17.

VLA OBSERVATION OF THE M87 JET AND SMALL SCALE JET STRUCTURES

Philip E. Hardee
Department of Physics and Astronomy, University of Alabama
Tuscaloosa, AL 35487
Frazer N. Owen
and
Timothy J. Cornwell
National Radio Astronomy Observatory*
Socorro, NM 87801

OBSERVATIONS

New radio observations of the M87 jet have been made with the VLA* in the A array at 2cm which yields a resolution of 0.″1. At a distance of 17 Mpc, 0.″1 corresponds to ≈ 8 pc. Using new observing and reduction techniques, a dynamic range of about 50,000 to 1 has been achieved. The new images show a complex pattern of filamentary and other fine scale structure in the jet. In general, the minimum pressures are 3 to 30 times higher than the pressures in the external medium estimated from x-ray profiles (Lea et al. 1982; Schreier et al. 1982; Stewart et al. 1984). The intensity maxima in the jet initially refered to as knots are now at least partially resolved and show complex structure. These features are G, D, E, F, I, A, B and C and are located at angular separations from the nucleus of about 1.″3, 3.″5, 6.″0, 8.″5, 11″, 12″, 14″ and 17″, respectively.

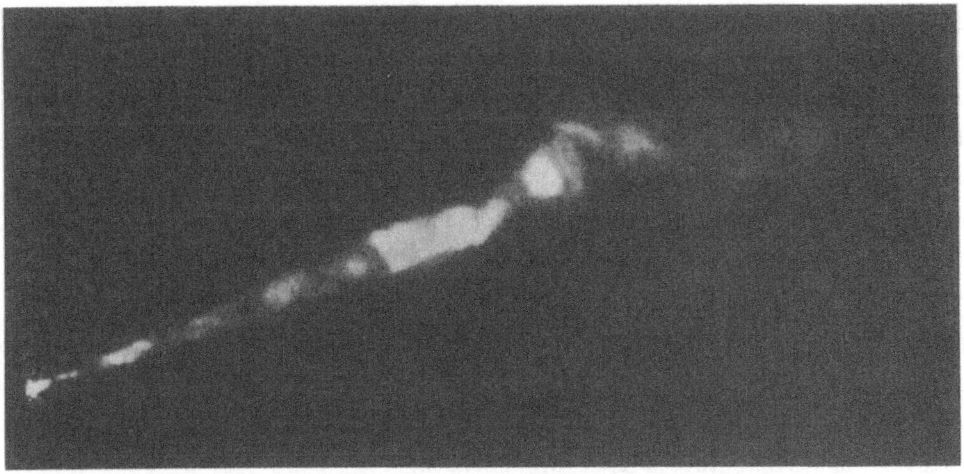

The M87 jet at 2 cm with 0.″1 resolution observed by the VLA*
*The National Radio Astronomy Observatory is operated by Associated Universities, Inc., under contract with the National Science Foundation.

G consists of a single bright sinusoidally oscillating filament which can be traced back to the nuclear jet, which at 0.1" resolution points along a line coincident with the northern boundary of the jet. The jet expands relatively uniformly to A with an opening angle of ≈ 6.5°. This is identical to the opening angle of the VLBI jet (Biretta and Reid 1987). At A the jet stops expanding and contracts slightly. Downstream of A the jet clearly bends south to B and then north to C. Beyond C the jet then begins a southwards bend which is coincident with the outer edge of the northern radio lobe.

The jet is limb brightened and intensity profiles across the jet are flat topped or edge brightened. Extreme limb brightening can be seen in the image between E and F. The images and profiles also show the presence of one or more bright filaments. The filaments appear to wrap around the jet in a helical pattern. Out to F the helical pitch is ≈ 30°. VLBI images of the innermost 0.1" show similar structure (Biretta and Reid 1987). From F through I the filaments appear to merge into sheets. In general there are only small intensity minima between filaments in the sheets. The exception is a dark diagonal line across the jet between F and I. The helical pitch in F and I is ≈ 45°. A filament is prominent in the A B region (overexposed in this image) where it appears to wrap relatively tightly around the jet. Close examination of other images shows that the tight wrapping begins upstream between I and A. The wrapping of filaments in the A B region implies a helical pitch of about 85° which decreases slightly away from A towards B. The apparent emission ring downstream of C visible in the image may be a tightly wrapped filament viewed in projection.

Not visible in this image is a dark line that runs down the center of the jet between A and C. This is caused by a central dip in the intensity of about 15%. A similar dip can be found in profiles between D and F. In general, the intensity profiles can be fit by a combination of a surface layer of thickness $a/R < 0.3$ with high average emissivity containing filaments of higher than average emissivity, an interior region with emissivity < 0.25 of the emissivity in the surface layer and a central core with $r_{core}/R < 0.3$, in which the emissivity must be $<< 0.1$ of the surface emissivity.

Typically the polarization along the jet's centerline is 20%. The polarization increases at the jet's edge and the jet is edge brightened in polarized flux as well as in total intensity. At the jet's edge the electric vectors are normal to the jet's surface. The magnetic field in the A B complex runs along the filamentary structure.

INTERPRETATION

The morphology suggests that emission from the jet is dominated by surface interaction and the emission comes primarily from a surface layer in which the filaments are embedded. The low emissivities permitted at the jet's core suggest that energetic particles in the jet's core do not experience significant radiative losses. This suggests that only relatively weak magnetic fields, a field-aligned distribution of relativistic electrons or cold electrons exist in the jet's core. Highly energetic electrons must either be generated in situ within the boundary layer or be deposited in the layer from the jet's interior. In situ acceleration by wave turbulence (Eilek 1982) confined to the jet's boundary (Hardee 1983) or other boundary phenomenon is somewhat more likely because it is difficult for relativistic electrons to cross field lines. Particle acceleration is likely to occur in a magnetized current-carrying shear layer in which both current and flow instabilities play an important role.

If the jet were a filled cylinder in equilibrium with the external medium, the nearly constant opening angle of the jet out to feature A and slowly varying radius from A to C require peculiar conditions inconsistent with x-ray derived pressure models. This alone is not a sufficiently strong arguement against global thermal confinement because the jet probably consists of a core and surface boundary layer which need not be in static pressure balance. However, a magnetically confined jet can expand with a nearly constant opening angle determined entirely by the initial conditions (Hardee 1985) and the overpressure in the filamentary emission regions suggests magnetic confinement.

The morphology of a filament twisted helically around a jet can be used to estimate the orientation of the jet. If the wavelength is short, as is the case for helical structures in A B C, then a wrapped filament will appear to loop back on itself if the jet is rotated sufficiently far out of the plane of the sky. In fact the structure from A to B could consist of a single filament that loops back on itself about 4 times between A and B and the emission ring at C can be interpreted similarly. The appearance of these loops when combined with the observed wavelength of twist suggest that the jet is about 25^0 out of the plane of the sky.

The linearized theory of fluid jet stability predicts that jets will be sinusoidally or helically distorted (Hardee 1987). Based on the observed sinusoidal oscillation in the plane of the sky a minimum Mach number for the jet is $M_j \equiv u/c_{s,j} \approx 1.5$. For a magnetized jet replace the sound speed by the magnetosonic speed (Ferrari et al. 1981; Hardee and Norman 1988). The theory also predicts that the angle of helical patterns relative to the flow direction will be $\phi \approx 35^0$ if the jet is near the plane of the sky. At least approximately this is what is observed along the inner jet. Numerical simulation of a slab jet containing a small magnetic field shows that filamentary structures predicted by the linearized theory develop and then merge into large scale sideways oscillation at some distance down a jet (Norman and Hardee 1987) and this is similar to what is seen along the inner portion of the jet.

The tightly wrapped filaments in A B and C cannot be produced by Kelvin-Helmholtz instability of a supermagnetosonic shear because they are too tightly wrapped. However, the observations point to a picture in which the jet contains a helically twisted field within a boundary layer with a high probability that jet currents are also contained within the boundary layer. In this case, the filamentary structure from A through C can be produced in a thick submagnetosonic shear layer by a current driven rippling or tearing mode (Furth et al. 1963), the pitch of filaments will reflect the initial field pitch and the structure will be dominated by magnetic effects, as is suggested by the high minimum pressures.

The brightness of the filaments in the M87 jet raises the possibility that a jet with one filament wrapped helically around a jet's surface on a jet pointed close to the line of sight would present the appearance of a strongly bent jet even if the underlying jet were straight. Interpretation of some bent VLBI jets as being the result of bending of the entire jet may need to be reassessed. Finally, relativistic motion of a discontinuity down a jet combined with local brightening of a wrapped filament at the point of contact can lead to interesting superluminal screen effects.

P.E. Hardee acknowledges support by the National Science Foundation through grant AST85-16921 and EPSCoR grant RII 8610669.

REFERENCES

Biretta, J.A. and Reid, M. 1987, abstract AGN meeting at Georgia State.
Eilek, J.A. 1982, *Ap.J.*, 254, p472.
Ferrari, A., Trussoni, E. and Zaninetti, L. 1981, *M.N.R.A.S.*, 196, p1051.
Furth, H.P., Killeen, J. and Rosenbluth, M.N. 1963, *Phys. Fluids*, 6, p459.
Hardee, P.E. 1983, *Ap.J.*, 269, p94.
Hardee, P.E. 1985, IAU Symposium #107, ed. M. Kundu & G. Holman, (Dordrecht: Reidel).
Hardee, P.E. 1987, *Ap.J.*, 313, p607.
Hardee, P.E. and Norman, M.L. 1988, in preparation.
Lea, S.M., Mushotsky, R. and Holt, S.S. 1982, *Ap.J.*, 262, p24.
Norman, M.L. and Hardee, P.E. 1987, preprint.
Schreier, E.J., Gorenstein, P. and Feigelson, E.D., 1982, *Ap.J.*, 261, p42.
Stewart, G.C., Canizares, C.R., Fabian, A.C. and Nulsen, P.E.J. 1984, *Ap.J.*, 278, p536.

DISCUSSION

Burns: How do you interpret the origin of the filaments? Are they magnetic flux ropes?

Hardee: The filaments are created by Kelvin-Helmholtz or by a rippling or tearing mode instability depending on the structure of the shear layer between the jet and the external medium. They might be thought of as self-pinched magnetic flux ropes.

Biretta: The filaments caused by reflection modes in the 2-D numerical simulation would be planes in a 3-D jet. These planes would appear as filaments only if seen precisely edge-on, which seems unlikely. Please comment.

Hardee: In three dimensions Kelvin-Helmholtz fundamental fluting modes have a very rapid decrease in perturbation amplitude away from the jet's surface and would appear as filaments wrapping the jet. Current driven rippling or tearing modes would be confined to the shear layer at the jet's surface and so would appear as filaments wrapping the jet.

Sulkanen: Models developed by R. Lovelace, J. Wang and myself of current carring force free jets produce magnetic fields that are purely axial along the centerline of the jet and then become completely toroidal at the outer surface of the jet. Maybe somehow flux tubes associated with the field structure are preferentially lit near the jet surface?

Hardee: The most likely explanation is that in situ particle acceleration is occuring in self-pinched flux tubes in a surface shear layer. This could be a natural consequence of turbulent or resistive effects with energy supplied by the flow or magnetic field. Confinement of in situ acceleration and emission to these flux tubes would explain the similarity in structure in the radio and in the optical.

Benford: As the jet toroidal field stacks up (perhaps from jet slowing) the kink instability condition is met, $B_\theta > B_z 2\pi R/\lambda_j$. So it is unsurprising that the jet kinks. But how could pedestrian hydrodynamics explain this?

Hardee: The kinking of a purely hydrodynamic jet is the result of the Kelvin-Helmholtz instability of the helical mode with $\lambda_j/R \approx 5\gamma M_j/(1+\gamma\eta^{1/2})$.

Cohen: The filament topology suggests that the jet is 25^0 or so from the plane of the sky, but the Biretta-Reid observations of the VLBI jet suggest an alignment close to the line of sight unless there is a drastic difference between the pattern and fluid velocities. Can you comment on this?

Hardee: In fact their observations place no restrictions on the alignment unless their velocity estimate is assumed needed to explain the lack of a counterjet. It is interesting to note that patterns produced by Kelvin-

Helmholtz instabilities associated with a jet that is much less dense than the external medium can move much slower than the jet's flow speed and there can be a drastic difference between the pattern and fluid velocities. The relationship is $v_p \approx [\gamma\eta^{1/2}/(1 + \gamma\eta^{1/2})]v_f$ where $\eta \equiv \rho_j/\rho_{ex}$. However, note that Doppler boosting would be associated with emission regions moving with the pattern speed and not the fluid velocity.

Gaskell: What predictions does your 25^0 angle (65^0 from the line of sight) make for the M87 counterjet? Or putting it another way, what β does the intensity ratio limit correspond to?

Hardee: The counterjet must be down in intensity by a factor of about 100. Using the simplest assumptions $R \approx [(1 + \beta \cos \theta)/(1 - \beta \cos \theta)]^{2.5} < 10$ and we should detect a counterjet unless it is intrinsically fainter by at least an order of magnitude. Perhaps there is no counterjet.

Begelman: A second possible way to confine a jet at a pressure higher than that of its surroundings is to have it embedded in a cocoon which has not yet come into pressure balance with the ambient medium. The requirements for this are a low jet density and sufficiently low Mach number. Are there reasons to rule out this confinement mechanism in M87, in favor of magnetic confinement?

Hardee: Certainly the requirements of low jet density and Mach number can be met. It is unclear as to whether the incredibly constant opening angle from about 10 mas out to 12 arcsec can be produced in such a model embedded in a cooling accretion flow. The advantage to magnetic confinement is that it is more or less independent of the surrounding conditions.

EXTENDED OPTICAL LINE EMISSION IN POWERFUL RADIO GALAXIES

Wil van Breugel and Patrick J. McCarthy
Astronomy Department, UC Berkeley, CA 94720, USA

ABSTRACT

Optical and radio observations of nearby and distant powerful radio galaxies with extended optical emission-line regions are discussed. Comparisons of optical emission-line images with radio continuum maps, supplemented with detailed optical spectroscopy, show that the interaction of radio jets and lobes with ambient gas can have dramatic consequences for both the radio source and its surrounding medium. We will discuss evidence for the occurence of large scale star formation and collimated ionizing radiation along the axes of radio galaxies. We also present preliminary results indicative of large rotation measures in two distant radio galaxies.

I. Star Formation Induced by Radio Galaxies

Evidence for star formation along radio jets is known to exist in the nearby radio galaxies Centaurus A (eg. Graham and Price 1981) and Minkowski's Object (M.O., van Breugel *et al.* 1985), and has been suggested to possibly occur along the jets in 3C 120 (Wlerick *et al.* 1985) and Fornax A (Graham 1987). In this section we briefly discuss further observations of Minkowski's Object, and report on the recent discovery (McCarthy *et al.* 1987*a*) that star formation may commonly occur along the axes of high redshift radio galaxies.

A. Minkowski's Object

Minkowski's Object (M.O.) is a small irregular galaxy or extra-galactic HII-region located at the end of a radio jet (PKS 0123-016A), which emanates from the elliptical galaxy NGC 541 in the cluster of galaxies A 194. Comparison of detailed radio and optical images strongly suggest that M.O. and the radio jet are interacting (Fig. 1). These observations suggest therefore, that the radio jet has entered an extensive region of relatively cool gas, triggering the large scale collapse of clouds and subsequent star formation.

The origin of this gas is unclear and could be attributed to 1) a pre-existent extra-galactic HII region, 2) the ISM of NGC 541 or 3) cool(-ing) intracluster gas. Together with Dr. J. van Gorkom we have searched for cold gas clouds (HI), in the vicinity of M.O. using the VLA in the D-array. HI was detected only from M.O. in two velocity channels near its systemic velocity. The centroids of the HI emission were located slightly downstream from the bright optical ridge of M.O. and its fainter counter part at the opposite side of the jet (see Fig. 1). We derive a total neutral hydrogen mass M_{HI} of 3.9×10^8 M_o, which is fairly typical of small irregular galaxies, and estimate that the HI column density is $\sim 4.0 \times 10^{21}$ atoms, slightly above the threshold for massive star formation in HII regions on scales of ~ 0.5 kpc (Skillman 1987).

Together these radio and optical data suggest that star formation in M.O. is probably continuing in the downstream direction. Unfortunately our HI observations still do not give an unambiguous clue as to the origin of the gas. Since no extended, other HI was detected in the galactic bridge, nor anywhere else in the cluster, none of the above mentioned alternatives for the origin of the gas can definitely be ruled out.

B. Distant Radio Galaxies

Optical and IR photometry of distant ($z \gtrsim 1.0$) radio galaxies suggests that they are undergoing episodes of large scale star formation (Lilly and Longair 1984; Spinrad and Djorgovski 1987; Djorgovski 1988). Broad-band images show that at large redshifts many galaxies are elongated and exhibit multimodal structure (Djorgovski *et al.* 1987; Le Fèvre *et al.* 1987). The major axes of these galaxies are generally aligned with those of their associated radio sources (McCarthy *et al.* 1987*a*; Chambers *et al.* 1987), suggesting a fundamental relationship between their morphologies. Histograms showing the correlation for distant 3CR and (primarily) 4C radio galaxies are shown in Figure 2*a*, *b*

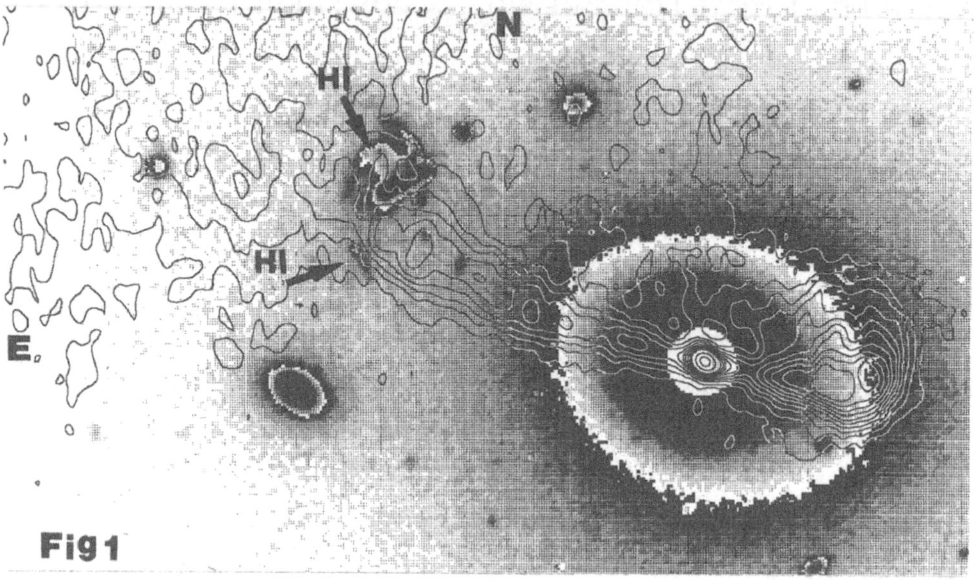

Figure 1: Overlay of a 21-cm radio continuum map ($\sim 3''$ resolution) on an optical narrow-band image of Minkowski's Object that includes the $H\alpha$ emission-line (from van Breugel *et al.*, 1985). Neutral hydrogen (HI) observations resulted in the two detections as indicated by arrows.

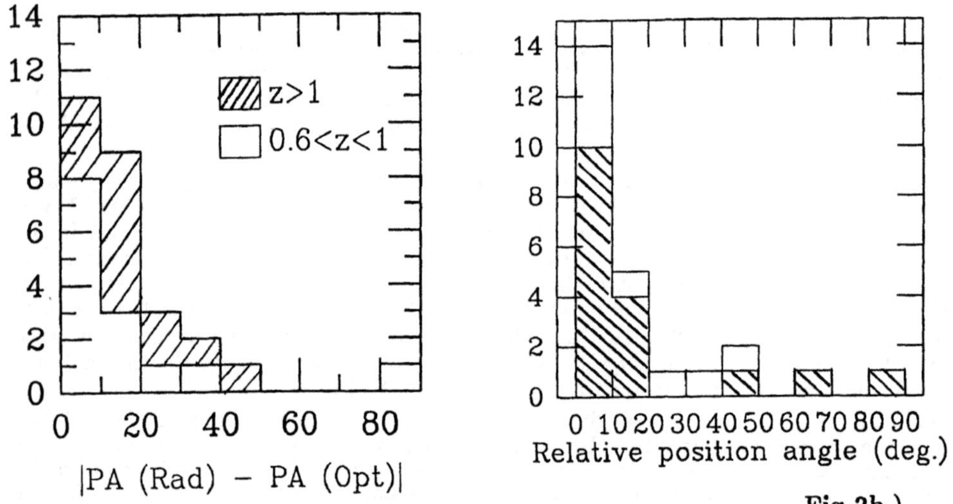

Fig 2a.)

Fig 2b.)

Figure 2: Histogram of the difference between the radio and optical position angles for distant 3CR radio galaxies (from McCarthy *et al.*, 1987*b*, left) and for faint, steep spectrum 4C radio galaxies (from Chambers *et al.* 1987).

If one accepts the evidence for large scale starformation in these distant galaxies, then the most straight forward interpretation of the radio/optical alignments is that enhanced starformation has been triggered by outflow (jets, winds) associated with the creation of the radio sources. The larger fraction of elongated galaxies at high redshifts could then mean that their interstellar media are denser when compared to nearby objects. This might be expected since these galaxies are young and therefore may still contain large amounts of cool, starforming clouds. Such a scenario implies that an active nucleus develops very early on in the formation of the parent galaxy, perhaps even before the bulk of the star formation is completed. A possible example of such a "proto-galaxy" may be 3C 326.1 (McCarthy *et al.* 1987*a*). Theoretical models of galaxy formation along these lines have been proposed by Silk and Norman (1981) and by Silk and Szalay (1987).

II. Collimated Ionizing Radiation From AGN's in Radio Galaxies

There is increasing evidence, much of it discussed at this Conference, that different viewing angles towards AGN's might, at least in part, be responsible for the variety in active galaxies. In particular it has been proposed that partial blockage of AGN radiation by accretion disks/tori might result in collimated (broad-) line and continuum emission (Krolik and Begelman 1987). This continuum might range from radio to X-ray wavelengths, and some of it might be relativistically beamed. In cases where AGN's also exhibit bulk outflow such as jets and winds, it is likely that the collimated radiation escapes in the same direction (perpendicular to the accretion disks) and along a path that may have been partially cleared of opaque material by the outflowing plasma.

Collimated AGN radiation might help explain the Seyfert I-II dichotomy (Osterbrock 1988; Miller 1988), the observed anti-correlation between X-ray luminosity and absorbing column density in Seyfert II's (Mushotzky 1988), the presence of high excitation line emission along radio source axes in Seyfert II's (Unger *et al.* 1987; Miller 1988) and in radio loud, steep-spectrum Quasars (e.g. Boroson *et al.* 1985; Stockton and MacKenty 1987). By lack of credible alternatives it has also tentatively been proposed as a possible ionization mechanism in some radio galaxies (eg. Fosbury 1985; van Breugel *et al.* 1986). Here we present new evidence which would seem to support the presence of such collimated ionizing radiation in powerful radio galaxies.

A. Photoionization

As part of our optical and radio investigations of powerful (3CR) radio galaxies and their environment (with Dr. H. Spinrad and collaborators), we have found that a large fraction exhibit extended optical line emission along their radio axes. In a few cases we find extended emission from highly ionized species (e.g. [NeV], [NeIII], CIV) at distances of up to 100 kpc from the nucleus. For the high redshift 3CR galaxies, most of the standard diagnostic emission lines have been redshifted out of the visible spectrum. Line ratio plots involving the near UV lines [NeIII]λ3869/[OII]λ3727 and [NeV]λ3426/[OII]λ3727 provide a sensitive diagnostic over the redshift range of 0 to 1.4. For the 3CR galaxies with measured nuclear and extranuclear [NeIII] and [NeV] emission, the data agree best with power-law photoionization models ($\alpha \sim -1$, Log U= $-1 \sim -3$) , with lower ionization parameters for the off-nuclear versus nuclear emission lines (Fig. 3). The luminosties

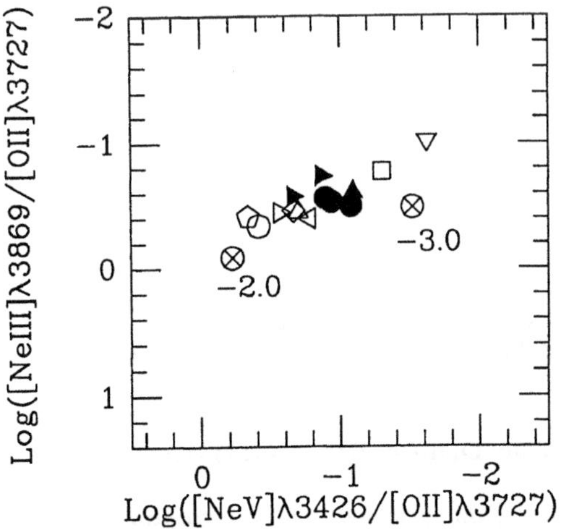

Figure 3: A line-ratio diagram, after Baldwin *et al* (1981), employing forbidden lines of [OII], [NeIII], and [NeV]. The open symbols refer to the nuclei of 3CR galaxies with $z \gtrsim 0.5$, the filled symbols refer to the off-nuclear emission for the same galaxies. The crossed circles represent the result of photo-ionization calculations with a powerlaw index of -1, the numbers shown next to the symbols are the logarithm of the ionization parameter U.

of the extranuclear emission line regions are usually $\sim 0.5 - 3$ times that of the nuclear emission lines ($L_{[OII]} \sim 10^{42} - 10^{43}$). Since the observed *local* continua are far too weak to provide the ionizing photons, ionizing radiation from the nucleus is required. The large luminosities of the off-nuclear regions then requires that they see as much to a few times more ionizing photons than the gas in the classical narrow line region. Further evidence that the ionizing radiation is emitted anisotropically may be inferred from the spatial distribution of the gas.

B. Emission-Line Gas Beyond Radio Galaxy Hotspots

The extended line emission in powerful radio galaxies is generally found within $\sim 15°$ of the radio axes (McCarthy *et al.* 1987*a*), implying a preffered direction for either the source of the gas or the source of ionization. [Note: there are cases where some of the ionized gas is located at large angles to the radio sources (Baum *et al.* 1988). In at least one well studied case, PKS 0349-278, the ionization in these directions is distincly lower however (Danziger *et al.* 1984)]. Recently, we have found several cases, one of which is illustrated in Figure 4, where extended optical line emission is found along the radio axis well beyond the radio hotspots. Since the emission- line diagnostics favor photoionization, the presence of ionized gas at large distances beyond the radio hotspots ($\sim$ 50 kpc in the example shown in Fig. 4) would suggest the presence of an anisotropic photoionization source. Considering the circumstantial evidence for collimated ionizing radiation in many other AGN types a similar explanation for the ionization of the gas along the axes of powerful radio galaxies is very attractive.

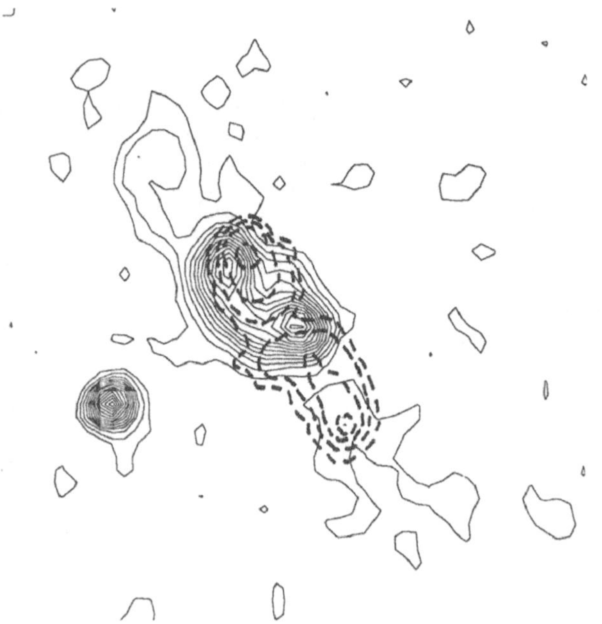

Figure 4: Contour map of the optical line emission ([OII], continuum subtracted) associated with 3C 435A as observed through a narrowband filter at Lick observatory. Overlaid (dashed lines) is a 21 cm map obtained with the VLA (1.2″ resolution).

III. Large Rotation Measures in Two Distant Radio Galaxies

A further result of our radio and optical studies is our discovery of large rotation measures associated with two of the most distant radio galaxies, 3C 326.1 (z = 1.825) and 3C 256 (z = 1.819). Both sources were observed at 2 cm and 6 cm with the VLA in the A-configuration. Upon inspection of the polarization maps surpisingly large rotations were observed between the polarization position angles in these two wavelength bands. Although the resolution in the 2 cm and 6 cm maps differ (0.15″ and 0.4″ respectively), a preliminary comparison between the polarization position angles is still possible since most of the polarized emission appears to be unresolved.

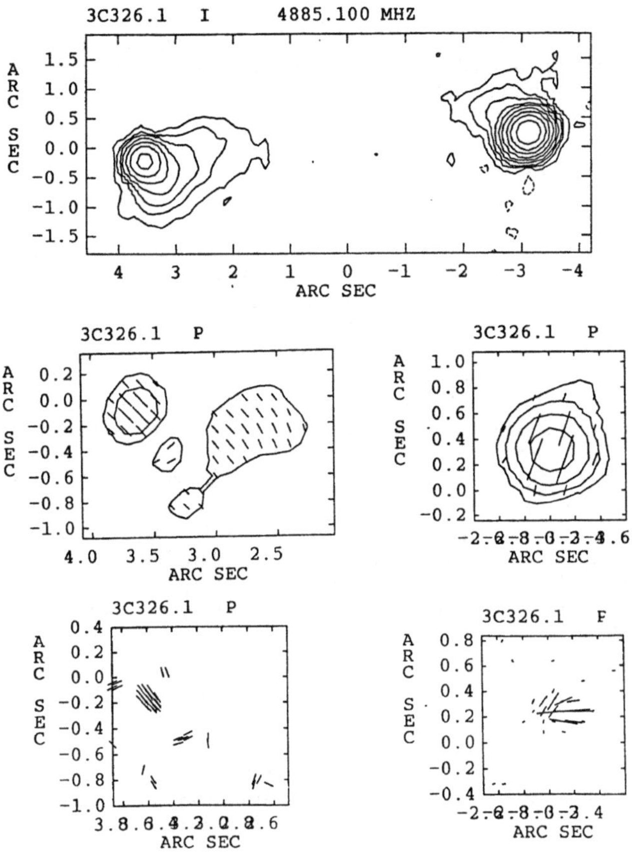

Figure 5: Contour maps of 3C 326.1 as observed with the VLA in the A-configuration. a) 6 cm total intensity, b) 6 cm polarized intensity + position angles (E and W hotspots), c) 2 cm polarized intensity + position angles (E and W hotspots). The 6 cm and 2 cm maps have resolutions of 0.4″ and 0.15″ respectively.

In 3C 326.1 (Fig. 5), the *minimum observed*, absolute RM values are ~ 302 rad m^{-2} for the W hotspot, and ~ 0 rad m^{-2} for the E hotspot. In deriving the RM value for the W hotspot we integrated the 2 cm Q and U polarization maps over a $0.4'' \times 0.4''$ region since some fine structure can be seen in the hotspot (Fig. 5c). The E hotspot actually consists of three independent regions, all of which have the same (minimum) RM value within the measurement uncertainties.

In 3C 256 (not shown), polarized emission was detected at both wavelengths only from the NW hotspot. Again the polarized emission was slightly resolved at 2 cm, though the polarization angles did not vary, and using procedures similar to those described above for 3C 326.1 we derive a minimum observed, absolute RM of 329 rad m^{-2}.

To derive the intrinsic RM (RM_{int}) a $(1+z)^2$ redshift correction must be applied. This results in *minimum* RM_{int} values of 2500 rad m^{-2} for the N and NW hotspots in 3C 326.1 and 3C 256 respectively. Similar high RM_{int} values have recently been found in Cygus A (Dreher *et al.* 1987), 3C 119, 3C 147, 3C 218 and 3C 295 (Kato *et al.* 1987). Very large RM values have also been reported for the central radio emission in M87 (Owen 1987).

Various explanations of these large (RM_{int}) values are possible, including Galactic foreground rotation, local magneto-ionic gas - either in a cluster or a screen surrounding the source - or Faraday rotation internal to the source. A discussion of these various possibilities as applied to Cygnus A has been given by Dreher *et al.* (1987). These authors concluded that strong ($2 - 10\mu$ G), large scale magnetic fields threading the cluster gas surrounding Cynus A would provide the most plausible explanation for the large RM values.

Optical narrow-band (Lyα) observations of 3C 326.1 (McCarthy *et al.* 1987*b*) have shown that the W hotspot coincides, to within the astrometric uncertainties, with the brightest emission-line gas. A similar radio/optical comparison for 3C 256 shows that its NW hotspot also appears embedded in optical emission-line gas. The two small steep spectrum sources (quasars) 3C 119 and 3C 147 are also likely to be surrounded by their Narrow Line Regions, and 3C 218, 3C 295 and M87 all exhibit extended line emission associated with their radio sources as well (Baum *et al.* 1988). No convincing evidence has been reported sofar which indicates that Cynus A is also embedded in such an emission-line cloud, but in this case the observations are hampered by the Galactic extinction of this low latitude object.

Detailed observations of relatively nearby radio galaxies (e.g. van Breugel 1986) have shown that usually the brightest emission-line gas is located along the boundaries of bright radio components, suggesting that these sources are surrounded by a sheath of dense ionized gas. Thus we believe that perhaps a better explanantion for the large intrinsic RM values in the above mentioned sources might be Faraday rotation of the polarized radio emission in such sheaths ('Faraday Screens'). The emission-line gas in these sheaths is probably dense (in nearby radio galaxies emission-line gas densities of several 100 cm^{-3} have been measured) and very clumpy. It is likely with the finite resolution available in the radio images that this clumpy gas (beam-) depolarizes the embedded radio emission. The large RM_{int} values however might be caused by the pervasive. more tenuous intercloud gas. In young, starforming galaxies such 3C 326.1 and 3C 256 the ISM gas densities are probably fairly large. Assuming, for illustrative purposes, $n_e \sim 0.1$ cm^{-3} ($\sim$ 10 times that in our own Galaxy) and a screen thickness of $\sim$ 1 kpc we find that magnetic field strengths of several tens of μ G would be required. We can only speculate whether this is realistic. Relatively large magnetic fields in the ISM of massive proto-galaxies might perhaps be generated through dynamo action (Pudritz and Silk 1987), and further enhancement could occur as a result of the propagation of the radio source through this medium.

We are currently undertaking multifrequency observations of several distant radio galaxies with matched array configurations which would allow us to better constrain these large RM values and to correlate these polarization properties with our optical data.

We wish to thank the excellent support by the observing staffs at Lick Observatory, Kitt Peak National Observatory and the VLA. The work described here was supported in part through NSF grants AST 84-16177, AST 85-13416 and from the California Space Institute.

REFERENCES

Baldwin, J. A., Phillips, M. M., and Terlevich, R. 1981, *Pub. A. S. P.*, **95**, 5.

Baum, S. A., Heckman, T. M., Bridle, A. H., van Breugel, W. J. M., and Miley, G. K. 1988, *Astron. and Astroph. Suppl.* in press.

Boroson, T. A., Persson, S. E., and Oke, J. B. 1985, *Ap. J.*, **293**, 120.

Chambers, K. C, Miley, G. K, and van Breugel, W. J. M. 1987, *Nature* **329** 604.

Danziger, I. J., Fosbury, R. A. E., Goss, W. M., Bland, J., and Boksenberg, A. 1984, *M.N.R.A.S*, **208**, 589.

Djorgovski, S., Spinrad, H., Pedelty, J., Rudnick, L., and Stockton, A. 1987a, *A. J.* **91**, 1267.

Djorgovski, S. 1988, in *Starbursts and Galaxy Evolution*, Th. Montmerle (ed.), Paris: Editions Frontieres, in press.

Dreher, J. W., Carilli, C. L., and Perley, R. A., 1987, *Ap. J.*, **316**, 611.

Fosbury, R. A. E. 1985, in *Structure and Evolution of Active Galactic Nuclei*, Giuricin *et al.* (editors) pg. 297, Dordrecht: Reidel.

Graham, J. A., and Price, R. M. 1981, *Ap. J.*, **247**, 813.

Graham, J. A. 1987, *BAAS Abstract, 171th AAS meeting*, in press.

Kato, T., Tabara, H., Inoue, M., and Aizu, K. 1987, *Nature*, **329**, 223.

Krolik, J. H., and Begelman, M. C. 1987, preprint.

Le Fèvre, O., Hammer, F., Nottale, L., and Mathez, G. 1987, *Nature*, **326**, 268.

Lilly, S., and Longair, M. 1984, *M. N. R. A. S.*, **211**, 833.

McCarthy, P., van Breugel, W., Spinrad, H., and Djorgovski, S. 1987*a*, *Ap. J. Lett.*, **321**, L29.

McCarthy, P., Spinrad, H., Djorgovski, S., Strauss, M. A., van Breugel, W. J. M., and Liebert, J. 1987*b*, *Ap. J.*, **319**, L39.

Miller, J. S. 1988, *These Proceedings.*

Mushotzky, R. F. 1988, *These Proceedings.*

Osterbrock, D. E. 1988, *These Proceedings.*

Owen, F. N. 1987 (private communication).
Silk, J., and Norman, C. 1981, *Ap. J.*, **247**,59.
Silk, J., and Szalay, A. 1987 *Ap. J.*, in press.
Skillman, E. D. 1987, ***Star Formation in Galaxies***, NASA Conference Publication 2466, pg. 263, ed. C. J. Lonsdale Persson
Spinrad, H., and Djorgovski, S. 1987, ***Observational Cosmology***, proceedings of IAU Symposium 124, ed. G. Burbidge. Dordrect: Reidel, in Press.
Stockton, A., and MacKenty, J. W. 1987, *Ap. J.*, **316**, 584.
Unger, S. W., Pedlar, A., Axon, D. J., Whittle, M., Meurs, E. J. A., and Ward, M. J. 1987, *M.N.R.A.S.*, **228**, 671.
van Breugel, W. J. M., Filippenko, A. V., Heckman, T. M., and Miley, G. K. 1985, Astrophys. J. **293**, 83.
van Breugel, W. J. M., Heckman, T. M., Miley, G. K., and Filippenko, A. V. 1986, *Ap. J.* **311** 58.
van Breugel, W. J. M. 1986, in *Jets from Stars and Galaxies*, eds. R. N. Henriksen and T. W. Jones; *Can. J. Phys.* **64**, 392.
Wlérick, G. *et al.* (1985), In *IAU Symposium 119 on Quasars*, edited by G. Swarup and V. K. Kapahi, p. 129.

DISCUSSION

Jack Burns: The implied star formation for the 3CR galaxies with major axes aligned between the radio and optical must be orders of magnitude more than in Minkowski's Object. Are you not asking too much from even a powerful jet?

Wil van Breugel: From a direct comparison of the number of ionizing photons we estimate that in 3C326.1 the starburst is $\sim$ 50 times stronger than in the center of M 82, and $\sim$ 500 times stronger than in M.O. The star formation in these distant galaxies occurs over a much larger area however. In our scenario the radio source merely acts as a trigger, and as such its power is largely irrelevant (although it must of course be able to propagate through the - relatively dense - medium). What does count is the amount of cool gas which is ready to collapse and forms stars, and the cross-section swept up by the source. Cool gas may be abundant in these young galaxies, while the lobes can have cross-sections of many kpc's.

Alan Bridle: In relation to Jack Burns' question, how much more powerful than Cen A are the 3CR sources within which you have found the optical major axis alignment that you attribute to stimulated starfromation?

Wil van Breugel: The distant 3CR sources are all in the same (radio) power range as Cynus A, i.e. 4 - 5 orders more powerful than Cen A and M.O.

D. Kazanas: Is there enough ionizing flux to produce the observed line emission in 3C 441 and 3C 435?

Wil van Breugel: For these two sources we do not yet have that information. In other cases, like 4C 29.30, the observed nuclear optical continuum is insufficient by almost a factor of 10. Hence either the ionization occurs in situ, by some as of yet unknown process, or the ionization is due to collimated radiation from the nucleus which is partially hidden from view.

Mark Whittle: I have a comment to the question by Kazanas. In the case of Seyferts it is often the case that the extended line emission requires a central ionizing luminosity a few times that observed along the line of sight. It would seem that the extended gas in the direction of the radio structure sees a much more luminous source than we do. This, of course, is just what Miller, Antonucci and Goodrich conclude from their polarization studies.

Mark Whittle: I would like to comment that the evidence you find for collimation of ionizing radiation, we also find in Seyferts. High excitation gas with low internal velocity dispersion is found well beyond the radio kpc scale structure. The extended emission is not seen in a direction perpendicular to the radio axis. As for your radio galaxies, we interpret this in terms of a source of ionizing radiation which is collimated in a direction parallel to the collimation of the radio plasma (Unger *et al.* 1987).

THE INTERACTION OF RADIO JETS WITH THE NARROW LINE REGION IN SEYFERT GALAXIES

M. Whittle (University of Virginia)
A. Pedlar (University of Manchester)
E.J.A. Meurs (European Southern Observatory)
S.W. Unger (Royal Greenwich Observatory)
D.J. Axon (University of Manchester)
M.J. Ward (University of Washington)

ABSTRACT

Long-slit spectroscopic observations of Seyfert galaxies reveal sub-structure in the [OIII] λ5007 emission-line profile. The sub-structure is spatially resolved and located close to the position of the near-nuclear radio lobes. The velocity of the sub-structure emission is quite different from galactic rotation and almost certainly involves radial outflow. The emission probably originates either in the post-shock region of the radio lobe or from entrained gas in the jet itself. It is not yet clear why some Seyferts have conspicuous [OIII] profile sub-structure while others do not. However, it seems likely that the detailed association of radio and line-emitting gas is at least partly responsible for the well known correlations between 'global' radio and emission-line properties.

BACKGROUND

It has been known for some time that in Seyfert galaxies, the forbidden (narrow) emission line region and steep spectrum radio emitting region are closely coupled. The evidence for this comes from several relations between 'global' emission line and radio properties. (a) The radio and line-emitting regions are approximately cospatial, having similar size and morphology (~few kpc, often elongated; Ulvestad, Wilson and Sramek 1981; Wilson and Heckman 1985). (b) The total [OIII] λ5007 luminosity correlates well with the steep-spectrum radio luminosity (de Bruyn and Wilson 1978; Whittle 1985). (c) The [OIII] line width (FWHM) also correlates with the steep spectrum radio luminosity (Wilson and Willis, 1981; Whittle 1985). (d) Seyferts with resolved double or triple radio morphology usually have square topped [OIII] profiles while those with resolved 'non-linear' or unresolved sources tend to have peaky [OIII] profiles (Whittle 1985).

Do these relations between 'global' line-emitting and radio-emitting properties have their root at a 'local' level ? Radio observations of many Seyferts have revealed linear double or triple lobed morphology (e.g. Ulvestad and Wilson 1984a, b). It is important to ascertain whether the emission-line properties also share this double structure — either in spatial distribution or velocity field. Such double emission-line structure might then be interpreted in terms of direct interaction between radio- and line-emitting gas in the vacinity of individual radio lobes.

Observations of a related class of objects, Steep Spectrum Core (SSC) radio galaxies, have revealed evidence for strong interaction between radio-emitting jets and line-emitting gas (e.g. Heckman *et al.* 1982). In the present project, we also aim to study radio/emission-line interactions but in the somewhat different context of the nuclear regions of Seyfert galaxies. The angular scales are small ($\lesssim$ few arecsec) and it is necessary to make high spatial resolution observations in good seeing to trace the emission-line properties across the radio emitting region.

OBSERVATIONS

Long-slit spectra were obtained along and perpendicular to the radio axis using the Image Photon Counting System on the Isaac Newton 2.5 meter Telescope. The spectral resolution was 45 km s^{-1} (FWHM) with a slit-width and spatial pixel size of 0.63 arcsec. Observations were obtained in the Hβ λ4861 to [OIII] λ5007 region in 1.5 – 2.0 arcsec seeing. Ten Seyferts were selected from the samples described by Ulvestad and Wilson (1984a, b). The objects are : Mkn 3, Mkn 6, Mkn 34, Mkn 78, Mkn 270, Mkn 573, NGC 2273, NGC 5929, MCG 8-11-11. All these have linear double or triple radio structure in which the radio 'lobes' have separations > 0.75 arcsec.

From our sample of ten, 5 show clear evidence of [OIII] components related to radio lobes (see below), 2 show this to a lesser degree, 1 has profile sub-structure which is spatially unresolved, and 2 show no profile sub-structure. A control sample of 8 Seyfert galaxies with radio structure classified as unresolved or slightly resolved has also been observed. Of these none show evidence of spatially extended [OIII] profile sub-structure.

RESULTS

The principal results from this study are summarised below while fuller descriptions are given in Whittle *et al.* (1986, 1988).

—**(1)** Most objects show profile sub-structure (humps, shoulders) in the [OIII] λ5007 emission line which are spatially coincident with the radio lobes (see Figures 1 and 2).

—**(2)** This profile sub-structure has a velocity which is unrelated to (and usually greater than) the velocity field due to normal galactic rotation (see Figure 3).

Together, these imply the line-emitting gas in the NLR is present in two distinct forms : (a) "Ambient gas" — generally distributed and following galactic rotation, and (b) "Component gas" — associated specifically with the radio lobes and probably undergoing radial outflow.

—**(3)** The relative strength of the [OIII] component emission varies greatly from object to object and does not seem to depend in any simple way on, for example, radio lobe power, total [OIII] luminosity, [OIII] component velocity or host galaxy property. It is not yet clear why some Seyferts have conspicuous [OIII] component emission while others do not.

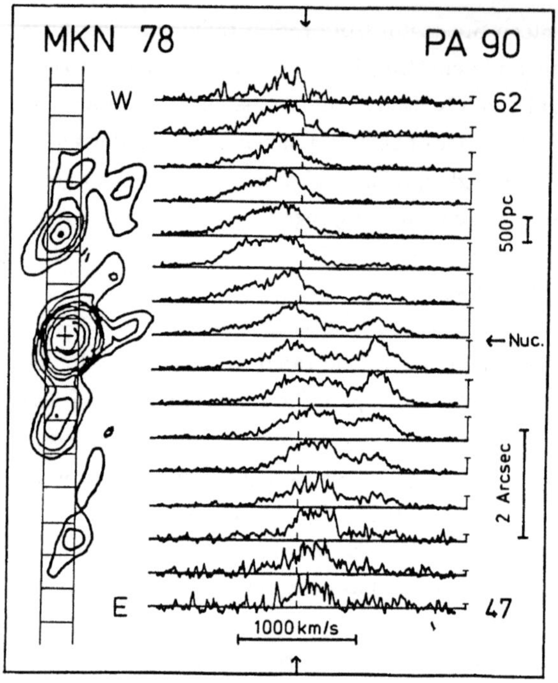

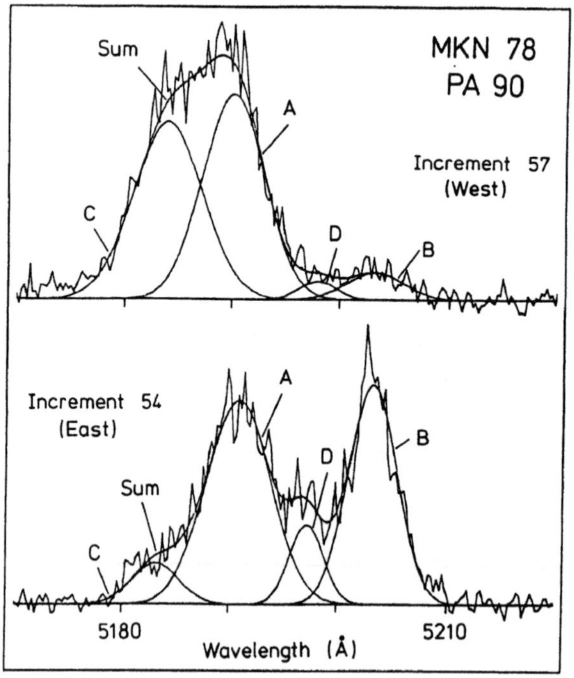

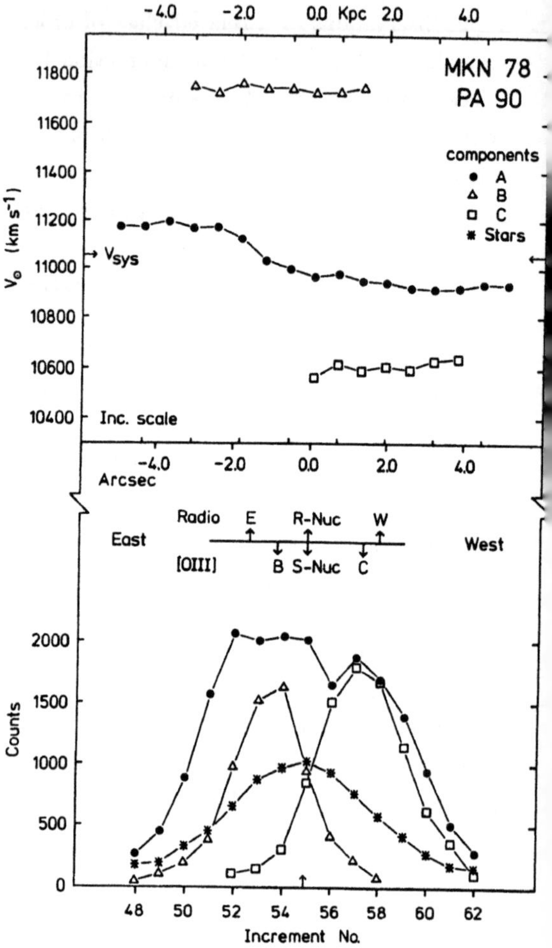

Figure 1 (upper left) : [OIII] λ5007 profiles and radio structure along PA 90 in Mkn 78. The radio map and [OIII] profiles have the same spatial scale and nuclear position. Each spectrum is autoscaled to highlight the profile changes along the slit.

Figure 2 (lower left) : Gaussian component fits to increments 54 and 57 from Figure 1. These illustrate the change in relative dominance of the components east and west of the nucleus.

Figure 3 (above) : Velocity (upper panal) and intensity (lower panal) of emission line components in Mkn 78 derived from the Gaussian profile fits shown in Figure 2.

—**(4)** The profile sub-structure is also apparent in Hβ at a strength which implies high excitation (i.e. [OIII]λ5007/Hβ ≃ 10–15).

—**(5)** There may be a tendency for faster moving components to have greater internal line width.

—**(6)** For the three objects with clearest profile sub-structure (NGC 5929, Mkn 78, Mkn 34) there is a tendency for [OIII] components to lie closer to the nucleus than the radio lobes.

INTERPRETATION

The most natural explanation for the profile sub-structure in Seyfert galaxies involves the interaction of radio jets with the near-nuclear interstellar medium. The radio properties of Seyferts are similar, in many respects, to those of the classical radio galaxies — but smaller and less powerful. It seems likely, therefore, that collimated ejection occurs in Seyferts and the radio lobe structure is caused by the interaction of a jet with the ambient medium at a working surface and/or bow shock. In this scenario, the line-emission giving rise to the profile sub-structure results from gas which is influenced directly by the jet or by the jet-head region. Two possibilities are :

(a) Narrow line region material is entrained in the jet itself and accelerated.
(b) The advancing lobe bow shock compresses and heats interstellar clouds which cool behind the shock, to become relatively dense line-emitting NLR clouds.

In both pictures, the source of ionization for the line-emitting gas is photoionisation by the central UV source. The apparent displacement between the radio lobe and line-emitting components can be explained as the jet-lobe separation in case (a) and a post-shock cooling length in case (b).

REFERENCES

de Bruyn, A.G., and Wilson, A.S., 1978, *Astr. Ap.*, **64**, 433.

Heckman, T.M., Miley, G.K., Balick, B., van Breugel, W.J.M., and Butcher, H.R., 1982, *Ap. J.*, **262**, 529.

Ulvestad, J.S., Wilson, A.S., and Sramek, R.A., 1981, *Ap. J.*, **247**, 419.

Ulvestad, J.S., and Wilson, A.S., 1984a, *Ap. J.*, **278**, 544.

—— 1984b, *Ap. J.*, **285**, 439.

Whittle, M., 1985, *M.N.R.A.S.*, **213**, 33.

Whittle, M., Haniff, C.A., Ward, M.J., Meurs, E.J.A., Pedlar, A., Unger, S.W., Axon, D.J., and Harrison, B.A., 1986, *M.N.R.A.S.*, **222**, 189.

Whittle, M., Pedlar, A., Meurs, E.J.A., Unger, S.W., Axon, D.J., and Ward, M.J., 1988, *Ap. J.*, **326**, in press (March 1st).

Wilson, A.S. and Heckman, T.M., 1985, in *Proc. 1984 Santa Cruz Workshop on Active Galaxies*, ed. J. Miller (Mill Valley, CA : University Science Books), p39.

Wilson, A.S., and Willis, A.G., 1980, *Ap. J.*, **240**, 429.

OPTICAL CONTINUUM SHAPES OF EXTRAGALACTIC RADIO JETS

William C. Keel
Sterrewacht Leiden, Leiden, The Netherlands
and
Department of Physics and Astronomy, University of Alabama
Tuscaloosa, AL 35486

Multicolor images (and in a few cases spectra) have been used to examine the continua of radio jets previously detected in the optical, and to search for additional cases. Six jets (including one knot in NGC 6251) show continuum emission. In all cases except a portion of the 3C 273 jet, the spectra are well fit by a simple synchrotron model, in which the electron energy spectrum is sharply truncated at some value γ_{max}; this corresponds to an emitted critical frequency ν_c of 3×10^{14} to 2×10^{15} Hz. Limits on unseen jets are in accord with similar behavior for them; the exponential form of a synchrotron cutoff implies that an object with ν_c less than 10^{14} Hz will be optically unobservable.

The M87 images were partially seeing-deconvolved, to allow clean separation of the various knots, and a close comparison with the high-resolution radio observations. Individual knots are well represented by the truncated synchrotron model, with significant differences in ν_c from knot to knot. An overall drop in ν_c going outwards is present, interrupted by a higher value in knot A. The optical and radio structures of each knot match at the 30 pc level, suggesting a dominant role for magnetic-field structure in shaping the observed emission.

A full report of this work will appear in the Astrophysical Journal Supplement Series.

OBSERVATIONS OF INTEGRATED JET LIGHT

CCD images were obtained (from KPNO, CTIO, ESO, and La Palma) of 13 radio galaxies and 2 QSOs. The galaxy light was modelled and subtracted using routines based on the GRASP package (Davis et al. 1985). Detected features are the jets in M87 (see below), PKS0521-36, 3C 66B, Coma A = 3C277.3, the strongest knot in NGC 6251, and 3C 273, and a "hot spot" in 3C 303 (see Kronberg et al. 1977). Significant nondetections include 3C 31 (reported, along with 3C 66B, by Butcher, van Breugel, and Miley 1980), and the jet of Cen A upon emerging from behind the dust lane. In all radio-galaxy jets, the integrated spectra are fit very well by a single-component sychrotron model, with electron energies abruptly truncated at energies typically $\gamma = 10^4$ (see figure). In contrast, the hot spots in 3C 303 and Pictor A (Röser and Meisenheimer 1987) show continuous power-law spectra from the radio to at least the B band, and may therefore be in a different regime of particle acceleration.

These results suggest that particle acceleration in jets is sharply bounded in the high-energy limit, either by particle energy losses or due to structure in the accelerating regions. In cases where X-rays are seen, a different population of particles, or reprocessing of radiation from this population, must be responsible. If this behavior is typical, observations of radio jets hitherto undetected at higher frequencies might be fruitfully conducted with arrays in the 1-5 micron regime.

The M87 Jet

BVRI images were obtained in 1.0" (FWHM) seeing, with sufficient sampling (5 pixels/FWHM) that a moderate level of image restoration was possible. A version of CLEAN was used, modified to allow for the different noise characteristics of direct images as compared to interferometer maps. An effective resolution of about 0.5 arsec was reached, sufficient to separate various knots for spectral-index measurement and to compare with a 6 cm map at the same resolution.

Each knot shows a truncated power-law synchrotron spectrum, just as seen in the integrated light of more distant jets. The critical frequency is highest in the inner knots and A, and lowest in B and C (see Hardee, Owen and Cornwell, these proceedings). There is at least a statistical link between location in the source and local particle-acceleration properties.

The structures at 0.45 micron (B) and at 6 cm match exactly at 0.5 arsec resolution, so any differences must be at a level 0.3 aresec (30 pc) or less. The synchrotron lifetimes of radio-emitting particles are much longer when translated to streaming lengths, implying that the jet structure is not due to streaming from a few sites of particle injection, but to a tangled magnetic field and widely distributed particle accelerators (essentially as argued by Shklovskii 1984). Very high resolution data may show differential streaming between electrons of various energies, and allow estimates of field-structure scales.

I am pleased to thank NOAO, ESO, and La Palma for observing time. The M87 6 cm map was provided by Frazer Owen. Conference attendance was supported by EPSCOR grant RII-8610669.

REFERENCES

Butcher, H., van Breugel, W. J. M., and Miley, G. K. 1980, Ap. J. 235,749.
Davis, L., Cawson, M., Davies, R. L., and Illingworth, G. 1965, Astron. J. 90, 169.
Kronberg, P. P., Burbidge, E. M., Smith, H. E. and Strom, R. G. 1977, Ap. J. 218,8.
Röser, H.-J., and Meisenheimer, K. 1987, Ap. J. 314,70.
Shklovskii, I. S. 1984, Astron. Zh. 61,833.

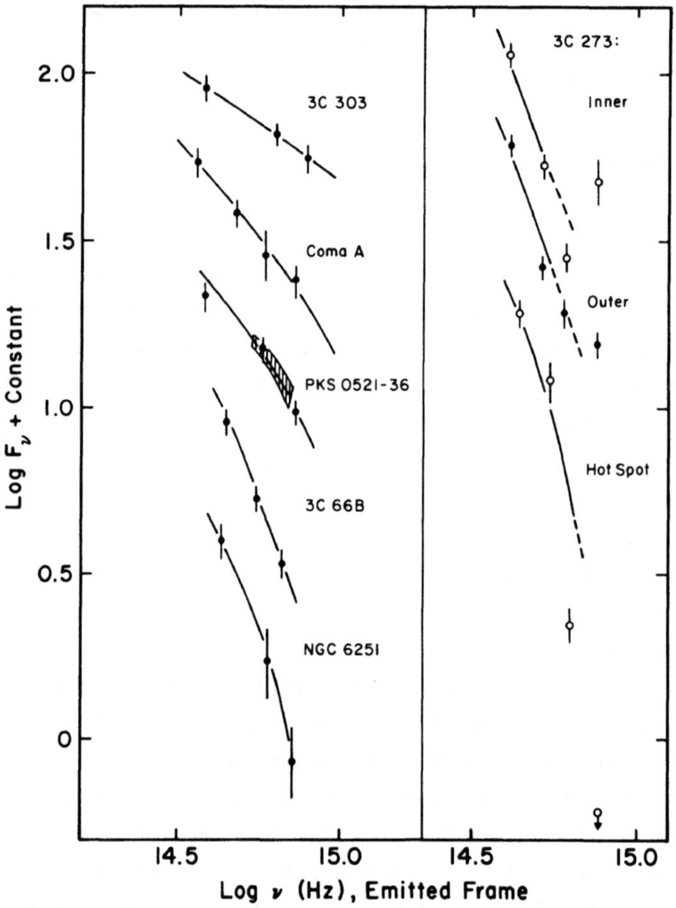

Integrated spectral shapes of jets and hot spots, with truncated synchrotron models shown for comparison.

THE INTERACTION OF RELATIVISTIC JETS IN AGNs WITH THE AMBIENT RADIATION FIELD

Fulvio Melia[1,2] and Arieh Königl[1]

[1]*Dept. of Astronomy and Astrophysics, University of Chicago, Chicago, IL 60637*

[2]*Dept. of Physics and Astronomy, Northwestern University, Evanston, IL 60208*

I. INTRODUCTION

In the "unified" picture of AGNs (interpreted as accreting supermassive black holes), at least three components contribute to the overall spectrum: (i) the quasi-isotropic nonthermal emission, which presumably originates close to the central object, (ii) the thermal emission from an accretion disk, and (iii) the beamed component produced by jets. It is generally believed that the latter component dominates only in the fraction of AGNs whose jets are directed close to the line of sight (blazars; i.e., BL Lac objects, optically violently variable quasars, and highly polarized QSOs).

The radio-through-γ-ray emission from relativistic jets is commonly attributed to incoherent synchrotron self-Compton radiation by the jet plasma (e.g., Königl 1981). Recently, Begelman and Sikora (1987) proposed that a jet accelerated to relativistic velocities close to the black hole may Comptonize the ambient nonthermal component, and thereby produce a beam of highly polarized radiation that could contribute significantly to the IR–X-ray continuum of blazars. They also suggested that the upscattered thermal component could give rise to the soft X-ray excess observed in certain AGNs.

In this paper, we consider the interaction of the jet with the ambient radiation field, using a somewhat different scenario than that adopted by Begelman and Sikora. We focus on the *thermal* radiation component, which we assume can account for the observed soft X-ray excess even without any relativistic boosting (e.g., Czerny and Elvis 1987). Futhermore, we consider much larger initial jet Lorentz factors, such as may arise in certain jet acceleration schemes (e.g., Michel 1987). We suppose that the much lower terminal Lorentz factors that are inferred from the apparent expansion speeds of "superluminal" radio sources are due to the Compton drag exerted on the jet by the disk radiation field (Phinney 1987). In the context of this scenario, we have calculated the detailed dynamical interaction of the jet with the thermal radiation from a model disk and the self-consistent Compton spectrum produced by this interaction. Our results have interesting implications to the dynamical properties of relativistic jets and to the interpretation of the X-ray spectra of BL Lac objects.

II. CALCULATIONAL PROCEDURE

An attractive explanation for the strong UV/soft X-ray excess in certain quasars (e.g., PG1211 +143; Elvis *et al.* 1985) is the release of gravitational energy through an accretion disk with characteristic (maximum) temperature $T_{max} \sim 10^5$ K, with opacity, inclination and reflection effects taken into account (Czerny and Elvis 1987; Madau 1987). In order to bring out the dependence of our results on the essential features of the accretion disk, we have modeled its emission as a simple sum of black bodies. The accretion disk is unlikely to remain thin at the super-Eddington luminosities inferred for these sources. In particular, the disk may develop a funnel, within which reflection effects increase the radiation energy density along the jet in comparison with the net flux (Sikora 1981). In our simple generalization from a thin disk, we "fold" the disk with an angle θ_f to the normal at the inner radius R_{in} and then "fold" it back into a flat disk at R_f. This configuration is therefore completely specified by the parameters R_{in}, R_f, θ_f, and T_{max}.

The motion of particles in a relativistic jet is constrained by the drag due to photons from the hypothesized accretion disk. The jet is rapidly decelerated if the Lorentz factor γ is greater than some critical value, which depends in part on the compactness parameter of the central source (Phinney 1987). In calculating the γ-distribution along the jet, we have ignored the effects of multiple scatterings on the dynamics as these are not energetically important. This is due to the fact that the optical depth for scattering is proportional to $(1 - \vec{k} \cdot \vec{\beta})$, where $\vec{k}$ and $\vec{\beta}$ are the photon direction and electron velocity, respectively. Since the singly scattered photons are highly beamed along the jet, this factor is of order γ^{-2} for higher order scatterings (Begelman and Sikora 1987). Figure 1 shows the results of one such calculation for a model electron-positron jet.

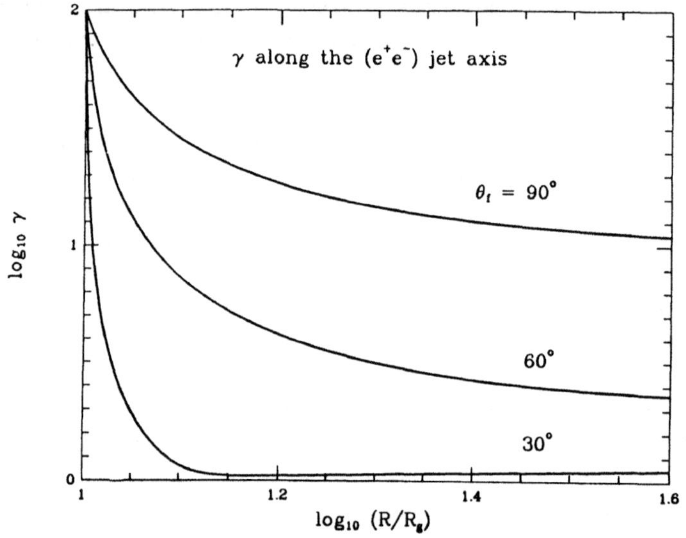

Figure 1. Distribution of γ along the axis of an e^+e^- jet starting with an initial Lorentz factor $\gamma_i = 100$ at a distance $10R_g$ (where R_g is the Schwarzschild radius) from a black hole of mass $M = 10^7\, M_\odot$, for three different funnel opening angles.

The result of a detailed Monte Carlo calculation of the scattered spectrum for an illustrative choice of disk and jet parameters is shown in Figure 2.

III. DISCUSSION

There are several compelling arguments for the existence of a characteristic maximum expansion speed of VLBI knots, corresponding roughly to an asymptotic Lorentz factor $\gamma_\infty \sim 5 - 20$. The numerical calculations reported here show that the deceleration of e^+e^- jets by this mechanism severely limits γ_∞ (to $\lesssim 10$) under a wide range of conditions. This "unified" picture therefore provides a natural explanation for the relatively uniform limit on γ_∞. We have found, however, that the deceleration is sensitive to the inclination angle of the jet with respect to the disk. Thus, a fan-shaped jet with initially uniform γ_i would emerge from the deceleration region with a transverse distribution of Lorentz factors.

This process also produces a significant X-ray component in the spectrum of AGNs whose jets are directed close to the line of sight. The typical X-ray flux density at $1 - 10$ keV is $10^{-2} - 10^{-1}$ of the peak thermal flux density emitted by the accretion disk, and is generally also greater than the beamed self-Compton emission expected from the jet; it could thus account for the hard X-ray component detected at these energies in BL Lac objects such as Mrk 421, PKS 0548-322, and PKS 2155-304 (see, e.g., Urry *et al.* 1982). It is interesting that in this energy range, the spectrum of the hard component is predicted by our model to be correlated with the flux density, in the sense that it rises more steeply in the more luminous sources (which, in this picture, are associated with narrower central funnels). Hence, the sources most likely to be detected should have a hard component with a flat or rising spectral index, which is consistent with the observations. In the context of this model, we would predict that the hard X-ray component should cut off at $\sim$ 100 keV, with the exact cutoff energy depending on the initial jet Lorentz factor and on the opening angle of the funnel. Another feature of the hard X-ray component in BL Lac objects which finds a natural explanation in this scenario is its reported high variability, which we associate with the steep gradients in γ that occur in the emission region. X-ray variability time scales as short as $\sim$ 30 s have been reported in certain BL Lac objects (e.g., Feigelson *et al.* 1986), which is consistent with the Lorentz-shortened light travel time over the jet-deceleration region considered in our model. These results thus further reinforce the interpretation of these sources in terms of beamed relativistic jets.

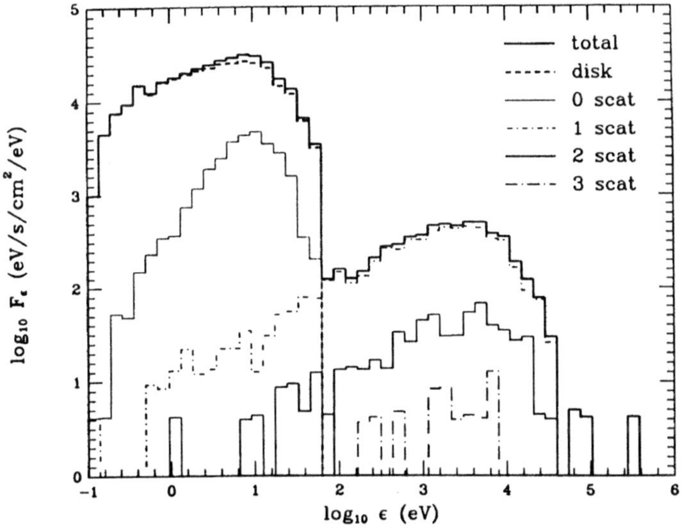

Figure 2. The spectrum arising from the scattering of disk radiation by the particles in a relativistic e^+e^- jet, assuming $\gamma_i = 20,000$, $\theta_f = 70^o$, $R_{in} = 5R_g$, $R_f = 50R_g$, $M = 10^7\,M_\odot$, and $T_{max} = 1 \times 10^5$ K. Also indicated are the contributions to the total spectrum from the disk and the individual multiplicities of scattering within the jet.

REFERENCES

Begelman, M. C. and Sikora, M. 1987, *Ap. J.*, **322**, 650.
Czerny, B. and Elvis, M. 1987, *Ap. J.*, **321**, 305.
Elvis, M., *et al.* 1985, *Ap. J.*, **292**, 357.
Feigelson, E. D., *et al.* 1986, *Ap. J.*, **302**, 337.
Königl, A. 1981, *Ap. J.*, **243**, 700.
Madau, P. 1987, *Ap. J.*, submitted.
Michel, F. C. 1987, *Ap. J.*, **321**, 714.
Phinney, S. 1987, in *Superluminal Radio Sources*, eds. J. A. Zensus and T. J. Pearson, (Cambridge University Press, Cambridge).
Sikora, M. 1981, *MNRAS*, **196**, 257.
Urry, C. M., *et al.* 1982, *Ap. J.*, **261**, 12.

A CIRCUIT ANALOGY FOR ACTIVE GALACTIC NUCLEI

Gregory Benford
Physics Department
University of California, Irvine
Irvine, California 92717

Near galactic center lie several types of ordered, luminous structures visible in the radio bands. Most prominent is the Arc, which has two very different segments: straight nonthermal filaments perpendicular to the galactic plane, and the Arch --thermally emitting filaments above the galactic plane which turn back toward galactic center. These features are weakly echoed at negative galactic latitude, suggesting an overall structure connecting the ends of the straight filaments toward the galactic center. This paper attempts an electrodynamic model for these observed bright elements.

The circuit equation for electrodynamically connected regions is

$$V = IR + \frac{d}{dt}(IL_i),$$

$$V = \int\left(\frac{\vec{v} \times \vec{B}}{c}\right) \cdot d\vec{\ell}, \quad IR = \int \eta\vec{J} \cdot d\vec{\ell}$$

$$\frac{d}{dt}(IL_i) = \frac{1}{c}\int \frac{\partial\vec{A}}{\partial t} \cdot d\vec{\ell}$$

Local conditions contribute to the integrals, and evolution of the circuit depends also on global aspects of the current. We can write

$$R = \frac{4\pi L\nu}{A\omega_p^2} \; ; \; L_i = L\mu \, \ell n\left(\frac{8L}{a}\right) \; ; \; C = NL^*\varepsilon$$

Here the inductance L_i is that of a loop of length L and cross-sectional area $A = \pi a^2$. The capacitance C is that of N current paths of length L* (< L) in parallel, separated by a distance comparable with the radius of the current path. Locally the collision frequency ν governs resistance.

A giant molecular cloud moving opposite to the galactic rotation can induce powerful electric fields by $\vec{v} \times \vec{B}$ motion throughout the partially ionized cloud. The plasma component of the cloud can push the poloidal field $\vec{B}$ ($\sim 10^{-3}$G) out of the way, locally deforming it. Electric fields will drive currents which can become attached to neighboring, straighter field lines, through scattering of the electrons by current-induced turbulence. Once on the ordered field lines, the currents proceed around a large circuit which leads through the galactic plane. Conditions in the plane are unclear, so we cannot judge whether ionization is sufficient there to return the currents through the plane to field lines lying closer to galactic center, and thence back up into the Arch. If not, the currents may proceed to negative galactic latitudes, forming the somewhat weaker filamentary structures there. Presumably the circuit eventually closes through the polarized lobes at the ends of the filaments and thence back toward galactic center, where they connect to the Arch. The ultimate cause of the circuit is the strong milliGauss ordered magnetic field filling the ~ 100 p.c. central volume and arranging systematic current flows.

There will be a steady electron drift in the mean field, $v_D = eE/m\nu$, with the appropriate collision frequency ν set by the level of electrostatic turbulence, E_e, described by $W \equiv \langle E_e^2 \rangle / 4\pi n T_e$.

We shall take a form suggested by extensive simulation and theory, $\nu \cong \omega_p W$, where W includes only $E_e(\omega/k < v_D)$. For steady turbulence, $W \approx 10^{-2}$ to 10^{-5} seems plausible from laboratory experiments.

The resistance R of a pinch of length L and radius is

$$R = \frac{\beta_{-4}^2 W\,(L/100\text{ lt yr})}{n^{1/2}\left(\frac{a}{a_p}\right)^2 T} B_{-3}^2\ 10^9 \text{ ohms}$$

where a_p is the Bennett pinch equilibrium radius for a plasma with temperature T (in eV),

$$a_p = 7 \times 10^8 \text{cm} \left(\frac{W_{-2}\ T^{1/2}}{n\beta_{-4} B_{-3}}\right)$$

This is a very small pinch and thus R is very large for such a constricted channel. With many such pinch paths available, the voltage derived from eq. (1) will doubtless flow through a large number of small pinches in parallel. The driving voltage is

$$V = 3\beta_{-4} B_{-3} \left(\frac{L}{100\text{ lt. yr.}}\right) \sin\phi\ 10^{16} \text{ Volt}$$

The total resistance of N pinches in parallel is $R_t = R/N$ and total current flowing is

$$I_t \equiv \frac{V}{R_t} = \frac{N\sqrt{n}}{\beta_{-4} B_{-3}} \frac{1}{W_{-2}} \left(\frac{a}{ap}\right)^2 T \sin\phi\ 3 \times 10^9 \text{ Amp}$$

while the magnetic field seen near any small pinch, B_e, is still small,

$$B_e = \frac{I_t}{Na_p} \cong 10^{-4}G. \left(\frac{v_2\ (nT)^{1/2}}{B_{-3}\ \beta_{-4}^{2}} \left(\frac{a}{a_p}\right)^2 \cos\phi\right)$$

where v_2 is the cloud velocity in units of 10^2 km/s. This confirms a qualitative expectation that B_e should be a good deal less than B if pinches are stable, as we shall discuss. Note that as B increases B_e decreases, suggesting that B may dominate polarization maps of even a pinched region.

So far we have no idea how many small pinches comprise the ordered regions of illuminated strands. The net energy dissipated in the total circuit is

$$\varepsilon = \frac{V^2}{R_T} \cong (\sin\phi\ a/a_p)^2 \left(\frac{L}{100\ \text{lt. yr.}}\right) \frac{Nn^{1/2}}{W_{-2}}\ 10^{33}\ \text{erg/s}$$

The total observed thermal infrared luminosity ~ 5 x 10^{40}erg/s comes principally from a large molecular cloud in which the inferred plasma density is 10^3 - 10^4 cm^{-3}. To account for this wholly as Ohmic dissipation requires

$$n^{1/2} \left(\frac{L}{100\ \text{lt. yr}}\right) (a/a_p)^2\ (\sin\phi)^2\ \frac{N}{W_{-2}} \sim 5 \times 10^7.$$

Nonthermal (synchrotron) observed emission is about 4 x 10^{33}erg/s, quite easily explained by a few pinches. Ohmic losses should occur where currents choose the highest available conductivity, which scales with $n^{1/2}$. If Ohmic processes do account for the luminosity in the large molecular clouds, then we should expect that luminosity to follow the visible density ridges. Data of Yusef-Zadeh, et al, (Nature, 310 557, 1984) seem to bear this out.

Our principal inferences from the model, which may be checked by observation, are:

(a) luminosity from Ohmic heating should follow current paths, which in molecular clouds should be regions of higher density;

(b) straight, synchrotron filaments may twist about each other from current coupling, and locally may form typical helical force-free geometries;

(c) flares due to changes in resistivity can last months among the intense, straight filaments;

(d) magnetic fields should be polarized along filaments, though some helical confining fields may be observable in the outer regions of a pinch.

The electrodynamic picture can account for the observed total thermal luminosity from regions near molecular clouds with a large number N of small pinches. These can organize into the larger structures observed, perhaps locally forming force-free equilibria. In this view the straight filaments correspond to magnetically ordered pinches, while the Arch is the region where $\vec{v} \times \vec{B}$ induced electric fields provide the dynamo of the circuit. How the circuit closes is not clear, but observed filamentary structure below the galactic plane suggests that ordered current flow may pass through the plane and return to the molecular clouds through a region nearer the galactic center.

How the circuit closes is not clear, but observed filamentary structure below the galactic plane suggests that ordered current flow may pass through the plane and return to the molecular clouds through a region nearer the galactic center. The few isolated "threads" are current paths, perhaps connected to other moving clouds. They could have been pulled away from the filaments by induction of passing conducting clouds, yet remain electrodynamically coupled to the governing potential.

Active Galactic Nuclei may experience similar process, aiding accretion within the outer 10-100 pc. through such an electrodynamic "catcher's mitt." This would greatly enhance coupling of gravitational energy into IR or synchrotron luminosity, and hasten mass inflow toward the central object.

QUESTIONS AND COMMENTS

P.J. Wiita: What is the evidence for the milligauss field in the Galactic Center?

G. Benford: Yusef-Zadeh and Morris (*Ap. J.*, **322**, 721; 1987) give two arguments. First, equipartition with the relativistic particles yields 2×10^{-4} G. Second, and more stringent, the bulging of apparently field-attached filaments suggests that the magnetic pressure may be somewhat larger than the ram pressure of a nearby molecular cloud. This gives 10^{-3} G. Such a strong field is compatible with the helical structure around the linear filaments, perhaps arising from a nearly force-free flow, which has magnetic pressure exceeding thermal pressures.

A.E. Wehrle: Would you please outline what a Bennett pinch is and how it forms?

G. Benford: When a current I_z flows in a plasma, it can pinch into a cylinder. Within it, thermal pressure nkT is confined by $B_\theta^2/8\pi$ locally. This gives a hot core with density falling algebraically with r, while B_θ peaks at the outer perimeter of the current. Pinch equilibrium obeys

$$I^2(\mathrm{Amp}) = 2Nk(T_e+T_i)c^2 \ (\mathrm{cgs}) = 3.2 \times 10^{-4} N \ (T_e+T_i) \ (\mathrm{Amp})^2$$

with N the electron axial density in cm^{-1} and temperatures in MeV. For more discussion, see the text by Krall and Trivelpiece. Relativistic generalizations were discussed by myself and David Book in *Advances in Plasma Physics*, 4, 1971.

I. Schlosman: It seems to me that the conditions you just have mentioned, namely a molecular cloud moving in an external field (not frozen in) can be fulfilled inside giant molecular clouds, where individual clumps move randomly. Why is not the same kind of phenomenon observed under these circumstances?

G. Benford: Such coupling may occur if significantly charged cloudlets do move across an ordered magnetic field inside the large clouds. But notice that if the cloudlets are weakly ionized (perhaps 10^{-7} of the particles) and the $\mathbf{v} \times \mathbf{B}$ force is not strongly correlated through time (the field is not highly ordered), the *net* force will be small. Coupling will still heat these cloudlets, but not decelerate them. This will slowly heat and ionize the cloudlets and may be energetically significant in the evolution of large clouds.

A.H. Bridle: There is another, similar but smaller radio structure in the Sagittarius direction that has been imaged by H. Liszt. Could you make some predictions of how observables such as luminosity of the filaments might scale with size (preferably before you see Liszt's data!)?

G. Benford: I haven't seen this source, but I would predict that luminosity should scale with the strength of the ambient ordered field, the velocity of the motive molecular clouds, and the density. A critical signature would be bright emission in or near the cloud(s) which attaches to well-ordered, nearly straight filaments which follow the magnetic field, possibly crossing the galactic plane.

A MODEL FOR QUASI–ONE-DIMENSIONAL NARROW JETS

T. Koupelis and H. M. Van Horn
Department of Physics and Astronomy and C. E. Kenneth Mees Observatory
University of Rochester, Rochester, N.Y. 14627

Abstract: We develop a quasi–one-dimensional model of a narrow jet by projecting the system of time-independent MHD equations onto the axis of the jet. This is done in analogy with the projection of equations onto the equatorial plane in thin accretion disk models. We use average values for the parameters of two coupled differential equations which relate the "line density" $\lambda(z)$ (or the z-component of the velocity $v_z(z)$) and the radius of the jet. We study the analytic structure of the equations and solve the system using parameters appropriate to some cases of astrophysical interest. We examine the importance of the coupling between rotation and magnetic fields as an acceleration mechanism in a jet and identify the physical consequences of different assumptions. The simplicity of our model, being one-dimensional, enables us to reduce it to one directly comparable with existing jet models.

I. The "Narrow Jet Model"

The purpose of the present paper is to provide a quasi–one-dimensional model of narrow jets. Our goal is to use the simplicity of the model to identify the physical consequences of different assumptions and examine the relative importance of different factors for a wide range of types and parameter values of jets, thus providing a tool useful in interpreting the results of more detailed numerical MHD calculations. We project the system of the ideal MHD equations that describe a jet outflow onto the jet axis assuming φ-symmetry and small opening angles for the jet, so that the parameters of interest can be approximated by their values averaged along the radial distance. A general equation of state can be used, but here we assume for simplicity that $P \propto \rho$ (constant jet temperature). We also assume that at the jet boundary the density (and thus, the pressure) is zero. Note that this assumption is quite different from the conventional assumption of confinement by an external medium. Such "pressure confinement" is of course permitted by our system of equations, but it is unnecessary, and we prefer the simpler assumption of "magnetic confinement." We define a "line density" λ, and the average z-component of the velocity by

$$\lambda(z) = \int_0^{R(z)} \rho(r,z)\, r\, dr \qquad \text{and} \qquad \langle v_z \rangle = \frac{1}{\pi R^2} \int_0^{2\pi} \int_0^{R(z)} v_z(r,z)\, r\, dr\, d\varphi .$$

In the same way we define the averages of the remaining unknown quantities. We define $x \equiv z/z_o$, $R_x \equiv R/R_o$, $f \equiv \langle v_z \rangle / v_o$, $\omega \equiv \langle \Omega \rangle / \Omega_o$, $b_\varphi \equiv \langle B_\varphi \rangle / B_o(\Omega_o R_o / v_o)$, and $\lambda_x \equiv \lambda/\lambda_o$, where $R_x\big|_{x=1} = 1$, $f\big|_{x=1} = 1$, $\omega\big|_{x=1} = 1$, $\lambda_x\big|_{x=1} = 1$, $b_\varphi\big|_{x=1} = \gamma$, and $\lambda_o \equiv \rho_o R_o^2/2$ are the values (in dimensionless form) of the radius, velocity, angular velocity, projected linear density, and weighted φ-component of the magnetic field at the origin of the jet ($x = 1$).

We also use the following definitions (note that for Keplerian orbits $\mu \equiv \varepsilon$): $B_o^2/4\pi\rho_o v_o^2 \equiv (v_{A,o}/v_o)^2 \equiv \alpha$, $v_o/c \equiv \beta$, $(\Omega_o z_o/v_o)^2 \equiv \varepsilon$, $P_o/\rho_o v_o^2 \equiv \eta$, $(R_o/z_o)^2 \equiv l$, $(GM/z_o)/v_o^2 \equiv \mu$, $1-\gamma \equiv b$, and $1-\gamma\alpha \equiv b_*$.

We have reduced the system of the ideal MHD equations to the following set of three first order differential equations for the functions $\tau(x) \equiv R_x'(x)$, $R_x(x)$, and $f(x)$, together with a set of algebraic equations that define λ_x, $\langle B_z \rangle$, $\langle \Omega \rangle$, and $\langle B_\varphi \rangle$ as functions of R_x, and f. The algebraic equations are:

$$\lambda_x = 1/f \ , \qquad \langle B_z \rangle = B_o/R_x^2 \ , \qquad \langle \Omega \rangle = \Omega_o \frac{fb_* - \alpha b}{fR_x^2 - \alpha} \ , \quad \text{and} \quad \langle B_\varphi \rangle = B_o \left(\frac{\Omega_o R_o}{v_o} \right) \frac{1}{R_x} \frac{b_* - bR_x^2}{fR_x^2 - \alpha} \ .$$

The three differential equations are (with initial conditions $f\big|_{x=1} = 1$, $R_x\big|_{x=1} = 1$)

$$R_x' = \tau \ , \qquad \tau' = \frac{D_1 D_6 - D_3 D_4}{D_2 D_4 - D_1 D_5} \ , \qquad f' = -f^2 \frac{D_2 D_6 - D_3 D_5}{D_2 D_4 - D_1 D_5} \ , \qquad \text{where} \qquad D_4 = \frac{\tau f^2}{R_x} \ ,$$

$$D_1 = \eta - f^2 + \alpha\varepsilon l f^2 R_x^2 \frac{(b_* - bR_x^2)^2}{(fR_x^2 - \alpha)^3} \ , \quad D_2 = \frac{\alpha l \tau}{R_x^2}(\varepsilon l \beta^2 b^2 R_x^2 - 1) \ , \quad D_5 = \frac{\left(\alpha\varepsilon l \beta^2 b^2 R_x^2 + fR_x^2 - \alpha\right)}{R_x^3} \ ,$$

$$D_3 = -\mu \frac{1}{fx^2} + \alpha\varepsilon l \tau \left[R_x \frac{b_* - bR_x^2}{(fR_x^2 - \alpha)^2} \left(\frac{b_* - bR_x^2}{R_x^2} + 2\frac{fb_* - \alpha b}{fR_x^2 - \alpha} \right) - \beta^2 \frac{b^2}{R_x} \right] + \alpha l \frac{\tau^3}{R_x^3}(2 - \varepsilon l \beta^2 b^2 R_x^2) \ ,$$

$$D_6 = 2\alpha \frac{\tau^2}{R_x^4} \left(\frac{3}{2} - \varepsilon l \beta^2 b^2 R_x^2 \right) - 2\alpha\varepsilon\beta^2 b^2 \frac{1}{R_x^2} + \frac{\mu}{fx^3} - \frac{2\eta}{l} \frac{1}{fR_x^2} + \frac{2\alpha\varepsilon}{R_x^2} \frac{(b_* - bR_x^2)^2}{(fR_x^2 - \alpha)^2} - \frac{\varepsilon}{f} \frac{(fb_* - \alpha b)^2}{(fR_x^2 - \alpha)^2} \ .$$

For a complete solution of the above system of equations however, we also need the value of $\tau\big|_{x=1}$.

II. Results

Here we present the results of our one-dimensional model for the simple case of $R_x = x$, $\eta = \mu = 0$. The system of the ideal MHD equations reduces to a first order differential equation for the velocity

$$f' \left\{ \alpha x^2 [x^2 - 1 + \gamma(\alpha - x^2)]^2 + \frac{1+l}{\varepsilon l}(\alpha - fx^2)^3 \right\} = \frac{x}{f}(\alpha - fx^2)[\alpha - f + \gamma\alpha(f-1)]^2$$
$$-2\alpha x[x^2(1-\gamma) + \alpha\gamma - 1][\alpha - f + \gamma\alpha(f-1)] - \frac{\alpha}{x}(\alpha - fx^2)[(x^2-1) + \gamma(\alpha - x^2)]^2 \ . \qquad (1)$$

We solve equation (1) for ranges of parameter values appropriate to a massive black hole model for an active galactic nucleus. We take the black hole mass to be $10^7\, M_\odot < M < 10^{10}\, M_\odot$, the jet full-opening angle $1° \leq 2\theta_o \leq 17°$, and $v_o = c/100$. We obtain Ω_o by assuming Keplerian orbits and use the Eddington luminosity and equipartition to obtain B_o. Equation (1) has two critical points, one for the "inflow" and one for the "outflow". One of the critical points is at $x_c^2 = (\alpha\gamma - 1)/(\gamma - 1)$, $f_c = \alpha(\gamma - 1)/(\alpha\gamma - 1)$ which for $\alpha = 1$ corresponds to the jet origin. Except the origin, the critical point for the outflow is an O-type singularity, and shocks are needed to connect the solutions of the two regions around the singularity. The jump-conditions at the critical point, however, require a delta-function pressure gradient, because of the assumption $R_x = x$. The range of values for α is $10^{-2} - 10^6$, and for $\delta \equiv \varepsilon l/(1+l)$ is $10^{-3} - 10^4$. Here $\delta \equiv (v_{\varphi,o}/v_{p,o})^2$, where $v_{p,o}$ is the toroidal velocity at the origin. The results are shown in Figs. 1–3.

Figure 1 shows the dimensionless jet velocity $f(x)$. For a black hole mass of $M = 10^8\, M_\odot$, $z_o \approx 5 \times 10^{13}\, cm$. Here we used $\alpha = 6000$ (initially a sub-Alfvénic flow), $\varepsilon = \alpha/2$, and $l = 8.1 \times 10^{-3}$. The five different curves labeled $\gamma_1 - \gamma_5$ correspond to $\gamma_1 = 0$, $\gamma_2 = 10^{-3}$, $\gamma_3 = 10^{-2}$, $\gamma_4 = \frac{\sqrt{\alpha(\alpha^2 - \alpha + 1)} - \alpha}{\alpha(\alpha - 1)} \approx 0.013$, and $\gamma_5 = 0.02$. Note that the value $\gamma = \gamma_4$ separates the solutions of eq. 1 (for $\alpha > 1$) into two groups.

The first one has $f'\big|_{x=1} > 0$ for $0 \le \gamma < \gamma_4$, while the second one has $f'\big|_{x=1} < 0$ for $\gamma > \gamma_4$ (i.e., a large initial φ-component of the magnetic field results in deceleration of the jet flow, as shown by the curve γ_5). Thus, for an initially sub-Alfvénic flow the MHD forces accelerate the plasma in a small region close to the source and relativistic velocities can be achieved, while far from the source the flow decelerates.

Figure 2 shows the dimensionless jet velocity $f(x)$ for a jet with $\alpha = 0.01$ (initially a super-Alfvénic flow) and $\delta = 10$ for $\gamma_1 = 0$, $\gamma_2 = 0.6$, and $\gamma_3 = 3$. The flow accelerates in a small region close to the origin, and then its velocity remains almost constant along the length of the jet. For this case, $f'\big|_{x=1} > 0$ for $0 \le \gamma < \sqrt{\frac{1+l}{\epsilon l}\frac{1-\alpha}{\alpha}} \approx 3.15$ or $\gamma > \frac{\sqrt{\alpha(\alpha^2-\alpha+1)}+\alpha}{\alpha(1-\alpha)} \approx 11$.

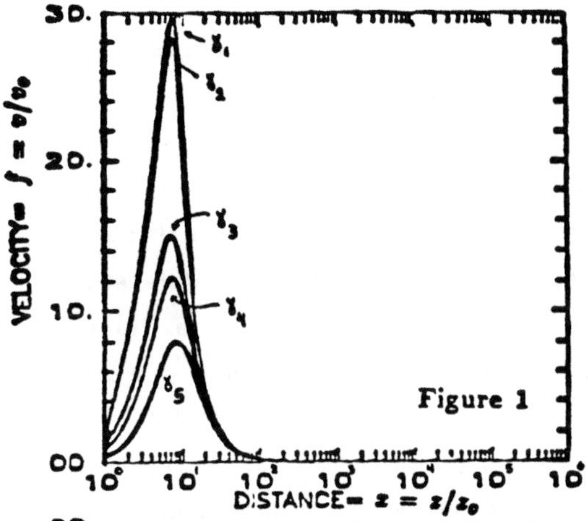

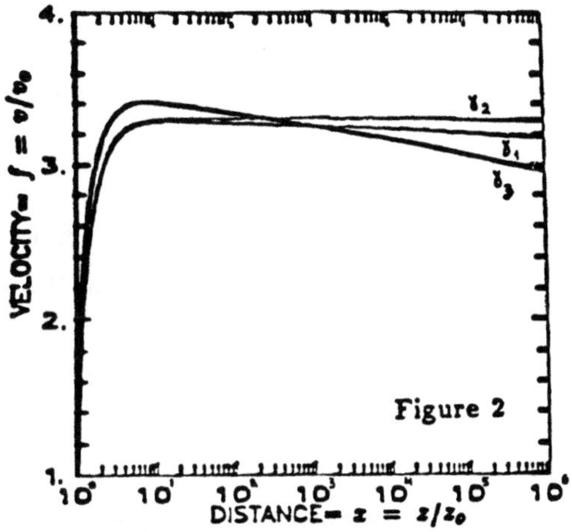

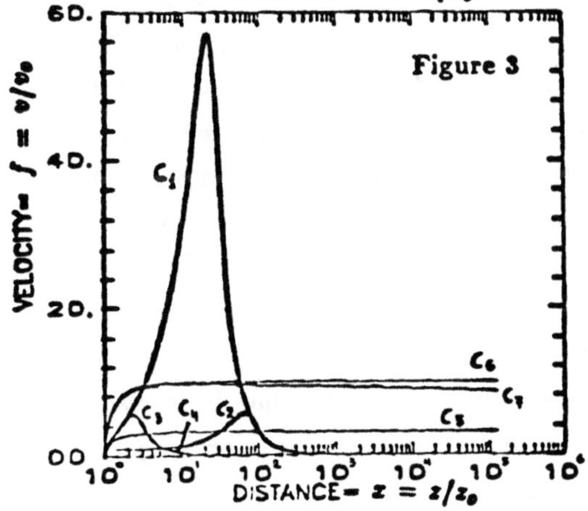

Figure 3 shows the dimensionless jet velocity $f(x)$ for $\gamma = 0$, for seven different mesh points $C:(\alpha, \delta)$, chosen in such a way as to indicate the dependence of the jet velocity on changes of the values of the two parameters α and δ. The curves labeled $C_1 - C_7$ correspond to $C_1:(10^5, 10^1)$, $C_2:(10^5, 10^{-2})$, $C_3:(10^2, 10^1)$, $C_4:(10^2, 10^{-2})$, $C_5:(10^{-2}, 10^1)$, $C_6:(10^{-2}, 10^2)$, and $C_7:(10^{-1}, 10^2)$.

Effects such as turbulence, instabilities, mass entrainment from the surroundings, and especially magnetic field reconnection may play an important role and also constrain the range of validity of the model. The simplicity of our model, however, being one-dimensional, enables us to solve it semi-analytically, to easily identify the physical consequences of different assumptions, and to reduce it to one directly comparable with existing jet models. The model could also serve as a check for the results of future 3-d MHD codes.

This research has been supported by the National Science Foundation through grant AST 87-06711, which we gratefully acknowledge.

TWO-AND-ONE-HALF DIMENSIONAL MODELS OF RADIO JETS

Josh Mitteldorf
Department of Physics, University of Pennsylvania
Philadelphia, Pennsylvania 19104
and
Paul J. Wiita
Department of Physics and Astronomy, Georgia State University
Atlanta, Georgia 30303

ABSTRACT

A numerical model is used to follow the two-dimensional shell of a three-dimensional radio jet. This model, in the spirit of previous work of Wiita and Siah, invokes a new computer code to allow internal pressure of the jet fluid to be taken as a tensor, allowing for bulk motion of the internal relativistic gas and azimuthal symmetry is no longer assumed. Three-dimensional effects of two kinds have been modeled: bending in a "wind" resulting from motion of the host galaxy through an intergalactic medium; and stability of the beam cross-sectional shape.

Results of the two-dimensional cases confirm and strengthen earlier inferences that hydrodynamics alone is incapable of accounting for observed jet collimation. Results on jet bending suggest, once again, that non-hydrodynamic effects must be invoked to explain some observations. With respect to beam cross-section, intricate oscillatory motions have been observed, some of which appear to be unstable and potentially disruptive.

I. INTRODUCTION

Although tremendous progress in understanding radio jets has been made over the past few years, many of their most fundamental parameters, such as jet fluid velocity and density, are still extremely uncertain. One of the best ways to tie the observations of jets on both small- and large-scales with theoretical models is through numerical simulations. Detailed two-dimensional hydrodynamical models have been available for a few years (*e.g.*, Norman *et al.* 1982), and more recently 2-D magnetohydrodynamical models (Clarke *et al.* 1986) and 3-D hydrodynamical models (Williams and Gull 1984, Arnold and Arnett 1986) have been attempted. Although insight into the physics of non-relativistic beams has been obtained in this way, they require much time on the largest supercomputers meaning that the necessary parameter space has not been explored in great detail.

This work is an extension of a simpler boundary following approach to the evolution of the shapes of radio jets and their surrounding cocoon first proposed by one of us (Wiita 1978) and since extended to include more complex gas distributions (*e.g.* Wiita and Siah 1986, hereafter WS) and magnetic effects (Siah 1985, Siah and Wiita 1987). The key ideas are: (1) An active nucleus pumps out an initially non-collimated relativistic plasma, with internal sound speed $c/\sqrt{3}$; (2) It propagates supersonically through a surrounding, somewhat flattened cloud of ISM; (3) The synchrotron losses are small compared to the total power. Comparisons have shown that this approach illustrated the majority of the features found later in the more detailed hydrodynamical modelling, such as the difficultly in producing hydrodynamical confinement of an initially uncollimated beam and the break-up of jets into bubbles of plasma if the powerhouse luminosity is too low.

II. APPROACH

Here we extend the previous models, which assumed axisymmetry and followed a one-dimensional boundary to simulate 2-D jet propagation. A completely new code was written which follows a 2-D boundary through 3-space, allowing us to drop the assumption of axisymmetry. Further, the fact that the passage of a fluid through a nozzle converts thermal energy directly into kinetic energy is explicitly taken into account in these new computations. Using a microcomputer allowed a very wide range of parameters to be varied, enabling us to investigate which of them have the greatest effect on jet shape and stability.

The energy source of the jet, presumably a black hole *cum* accretion disk, is modeled crudely as a point emitting a constant power output $\dot{E}$ in the form of relativistic particles. The particles are emitted as an isotropic gas, however they expand into an interstellar medium which is assumed to be non-relativistic, and isothermal, with adiabatic index $\Gamma=5/3$ and distributed with a density profile

$$\rho(r,z) = \frac{\rho_0}{1 + z^2 + (x^2 + y^2)\times(1-\epsilon^2)} \qquad (1)$$

where r and z are cylindrical coordinates centered on the source of the plasma.

The calculations described below are non-dimensional; it is envisioned that this gas distribution applies to an extended interstellar medium with a scale height of hundreds or thousands of parsecs. We assume the two gases do not mix, which is supported by two-dimensional numerical (non-magnetic) hydrodynamical calculations of the boundary between the jet and the external medium. The jet boundary expands at a speed that is much less than the internal sound speed $c/\sqrt{3}$; therefore there is just one internal pressure, independent of position in the interior fluid. This is because internal pressure gradients generated by the motion of the boundary can be smoothed out over a sound crossing time, which is small compared to the timescale of the expansion. But the expansion is typically much faster than the external sound speed; therefore, a shock front is formed and exterior fluid is "swept up" in a thin skin around the jet. This constraint, $c_{ext} \ll v_{jet} \ll c_{int}$, is generally valid.

The boundary has a surface mass density σ that is modeled with a "snowplow" algorithm: σ is increased as the piston passes through a volume of external fluid, and is diminished as the surface area expands or the local velocity diverges:

$$d\sigma/dt = (v_\perp)[\rho - \sigma(1/R_1 + 1/R_2)] - \nabla\cdot(\sigma \mathbf{v}_\parallel) \tag{2}$$

The $\rho v_\perp$ term represents swept-up mass from the exterior fluid, with $v_\perp$ the normal to the surface. The $\sigma v_\perp$ term accounts for dilution from stretching as the curved surface expands (R_1 and R_2 are the principal radii of curvature); and the divergence term accounts for fluid motion within the skin surface ($\mathbf{v}_\parallel$).

Each element of surface has its own equation of motion in this model, though *the interior points and the exterior points do not.* (It is in this respect that this model is not a full three-dimensional hydrodynamic code.) These are

$$\sigma\ d^2\mathbf{r}/dt^2 = (p - \rho c^2/\Gamma)\ \hat{\mathbf{n}} - \rho\ \mathbf{v}\ \mathbf{v}\cdot\hat{\mathbf{n}} \tag{3}$$

The LHS of this equation is acceleration times mass per unit area. The RHS contains a pressure difference, inside and out, acting in the normal direction ($\hat{\mathbf{n}}$) and a ram pressure term acting in the v direction.

We now have a closed system of equations except for the time evolution of p, which is assumed spatially uniform, and may be found by dividing the total energy in the jet interior by 3 times its volume. The total energy has a source in the $\dot{E}$ of the central point, but is diminished at each point by the work done on the piston, and is diluted as the jet volume increases. The work term is proportional to p dV/V, and the dilution term to E dV/V, where E is energy density = 3p (under the ultra-relativistic assumption) and V is the total interior volume of the jet. Thus the two terms are always in the ratio 1:3 and the evolution equations for p and V are

$$V\ dp/dt = \dot{E}/3 - (4/3) \int_{\text{surface}} dA\ p\ \mathbf{v}\cdot\hat{\mathbf{n}} \tag{4}$$

$$dV/dt = \int_{\text{surface}} dA\ \mathbf{v}\cdot\hat{\mathbf{n}} \tag{5}$$

III. RESULTS

Lack of space precludes describing the results of the dozens of computations in any detail or of presenting illustrations, many of which are in Mitteldorf (1987) and Mitteldorf and Wiita (preprint); the following types of runs were explored.

(i) Validation cases. Highly symmetric cases were run for comparison with known results, and to gain experience with the simplest of the instabilities that can arise. Some, with constant ρ, have been compared directly with analytic solutions. Others that are spherically symmetric have been compared with simple one-dimensional numerical integration. The remainder were cylindrically-symmetric cases compared to identical runs of the WS code. In all cases the accuracy was very good.

(ii) Cylindrically symmetric cases with isotropic pressure. WS have reported that in the absence of magnetic effects it requires a deeply dimpled density profile to create jets with any substantial degree of elongation. We have found that even these deeply-dimpled cases do not yield long jets, and that the WS results are due to the failure in extreme cases of their algorithm for locating their lead point.

(iii) Cylindrically symmetric cases with anisotropic pressure. Assumptions concerning internal jet pressure are relaxed to model bulk flow within the jet. These cases were never run by WS, although they could in principle have been run with minor modifications to their code. Models in groups (i)-(iii) illustrate that there is a strong tendency for elongated jets to relax toward spherical bubbles. We have found a simple analytic model that explains the effect qualitatively and can also make rough quantitative predictions of the final elongation.

(iv) Lobed quasi-cylindrical cases. Stability of the jet cross-section is explored. Non-axisymmetric perturbations are introduced by initializing jets with fluted or elliptical cross-sections. They are run with the full 2½-dimensional code.

(v) Ram pressure from intergalactic wind. The bending of jets by the ram pressure due to the motion of the central galaxy through an intracluster or intergalactic medium is frequently invoked to explain wide-angle tail and head-tail sources. We have investigated various ways of modeling this effect within the limitations of our code. For instance, in the full 2½-dimensional code, a cross-wind is applied to the upwind surface of the jet, but not to the corresponding downwind half of the surface. The result is an instability that we argue is physical (not numerical), further evidence that hydrodynamics is inadequate to explain the propagation and bending of jets. Another approach involves making the assumption that intergalactic wind pushes evenly on upwind and downwind surfaces; then the program can run stably, allowing significant bending. However, without abandoning the simplicity of cylindrically-symmetric computation, we can make a strong case that a subsonic intergalactic wind is both quantitatively inadequate to bend relativistic radio jets and qualitatively would not produce shapes comparable to those observed.

IV. CONCLUSIONS

1. This new code successfully reproduced analytical and previous numerical results whenever comparisons were possible. All disagreements with earlier work of WS were shown to be due to the previous code not following the position of the head of the beam accurately under some circumstances.

2. Unless extremely unlikely initial conditions were employed, purely hydrodynamical effects acting on relativistic fluids could not produce jets as narrow as are often observed. In general, once the jet emerged into a region where the external density became constant or spherically symmetric any earlier collimation tended to decrease significantly.

3. The stability of jet cross-sections were studied in detail. It was found that even small perturbations to the circular cross-section produced unstable oscillations near the head of the jet. These could be suppressed to study the modes along the body of the jet. In these cases some modes appeared underdamped, and oscillated through circular symmetry and back; others were overdamped, and relaxed gently into a circular cross-section. Sometimes both situations were observed at different points along the length of a single jet. In general, as jets propagate through steeper external pressure gradients these instabilities tend to grow faster.

4. In general these relativistic beams were found to be unstable to ram-pressure bending unless an arbitrary, and probably unphysical, approach to the interaction between the moving external fluid and the jet is invoked.

5. Additional forces are likely very important in producing collimation and enhancing stability. The most likely candidate is the jet's self-magnetic field.

REFERENCES

Arnold, C.N. and Arnett, W.D. 1986, *Astrophys. J. (Letters)*, **305**, L57.

Blandford, R.D. and Rees, M.J. 1974, *Mon. Not. Roy. astr. Soc.*, **169**, 395.

Clarke, D.A., Norman, M. and Burns, J.O. 1986, *Astrophys. J. (Letters)*, **311**, L41.

Mitteldorf, J. 1987, *Ph.D. Thesis, University of Pennsylvania.*

Norman, M.L., Smarr, L.L, Winkler, K.-H.A. and Smith, M.D. 1982, *Astron. Astrophys.* **113**, 285.

Siah, M.J. 1985, *Astrophys. J.*, **298**, 107.

Siah, M.J. and Wiita, P.J. 1987, *Astrophys. J.*, **313**, 623.

Wiita, P.J. 1978, *Astrophys. J.*, **221**, 436.

Wiita, P.J. and Siah, M.J. 1986, *Astrophys. J.*, **300**, 605.

Williams, A.G. and Gull, S.F. 1984, *Nature*, **310**, 33.

LINEAR SIZE VERSUS REDSHIFT AND LINEAR SIZE VERSUS POWER FOR EXTENDED EXTRAGALACTIC RADIO SOURCES

Alexander Rosen and Paul J. Wiita
Department of Physics and Astronomy, Georgia State University
Atlanta, Georgia 30303

ABSTRACT

A three-dimensional boundary-following code, altered to allow for a two-phase external medium, is used to simulate extragalactic jets as they propagate through a power-law isothermal galactic halo and on into a hotter, less dense, and uniform intergalactic medium (IGM). Good agreement is obtained with regard to the relationships between the overall linear size of such radio sources and both the total radio power (at fixed redshift) and the cosmological redshift (at fixed power). The excellent agreement of the latter with the latest observational results gives strong support to the notion that much of the X-ray background is due to a pervasive diffuse thermal component which is dense enough to be of cosmological significance.

I. INTRODUCTION

Recent analysis of X-ray observations of massive early type galaxies near the edge of the Virgo cluster and some isolated large elliptical galaxies indicates that a hot ($kT \approx 1$ keV) gaseous halo surrounds these objects (Forman, Jones and Tucker 1985; see also Trinchieri, Fabbiano and Canizares 1986). Further, an analysis of the X-ray background from 5-200 keV strongly supports the hypothesis that an even hotter (T_{IGM} = 11-25 keV at the current epoch), low-density ($n_{IGM} = 6\text{-}9{\times}10^{-7}$ cm^{-3}) pervasive intergalactic medium (IGM) exists (Guilbert and Fabian 1986; hereafter GF). We follow GF in taking its density as a constant at a given cosmological epoch. Gopal-Krishna and Wiita (1987, hereafter GW) presented a simple analytic model that considered the propagation of a beam through a pressure-matched interface at the radius of the halo (R_h).

Mitteldorf (1987) has extended the work of one of us (see, for example, Wiita and Siah 1986, hereafter WS, and Mitteldorf and Wiita, these proceedings) by developing a quasi-hydrodynamical program which follows the two-dimensional bow shock surface that envelopes a three-dimensional relativistic ($\Gamma = 4/3$) beam. We modified this program to resemble the analytic treatment of GW.

II. ASSUMPTIONS

In this picture, as the beam crosses the halo–IGM interface, two limiting cases can be solved analytically: in Model A the opening angle is assumed to remain the same, so that the abrupt decrease in the density of the confining medium causes an acceleration of the head of the beam; in Model B the velocity remains constant while the opening angle flares abruptly. Note that these two models represent extremes in the reaction of the beam, so we expect the behavior of an actual jet to lie in between those of these analytical models.

Because the halo properties are not expected to vary much out to a redshift, z, of about 1, while the IGM was much hotter and denser at earlier cosmological epochs, the radius of the halo/IGM interface was smaller in the past (GW).

Three fundamental assumptions are made in the boundary-following program: (1) a relativistic fluid ($\Gamma = 4/3$) "rises" through a heavier nonrelativistic ($\Gamma = 5/3$) confining cloud; (2) the boundary of the beam must be supersonic with respect to the external gas (halo or IGM) but subsonic with respect to the internal medium; and (3) the beam's radiation losses are negligible, at least compared to the other approximations involved.

The following parameters define our models: P, the power of the central source; An, an anisotropy parameter, which allows for an anisotropic pressure tensor for the internal medium, which is produced by the conversion of thermal energy in the beam into directed motion; R_h, the radius of the halo in units of a, the core radius of the hot galactic halo (typically measured to be about 2 kpc); T_r, the temperature ratio of the IGM/halo; e, the "eccentricity", a measure of the flattening of the gas cloud (halo); and n, the effective power law for density decrease in the halo.

The pressure–density distribution of the gaseous halo is the same as in WS

$$P(r,z) = P_o(1+e^4)^{n/2}\{(1-e^2)r^2+z^2+1+e^4[(z+1)/(r+1)]\}^{-n/2},$$

where in the numerical work the central density is scaled to 1. This corresponds to slightly dimpled and flattened isobars with limited angular momentum gradients; and, thus, δ (see Wiita and Gopal-Krishna, these proceedings) is very close to n/2.

The value of D_{max} is determined by either the timescale for nuclear activity, t_N, or by when the beam becomes subsonic. Since best estimates for t_N are quite uncertain (GW), we evaluate the results using both $t_N = 1\times10^8$ yr and 2×10^8 yr. Alternatively, if the beam becomes subsonic with respect to the external medium (either the halo or IGM), then the bow shock disappears and the hot spots are expected to dissipate quickly.

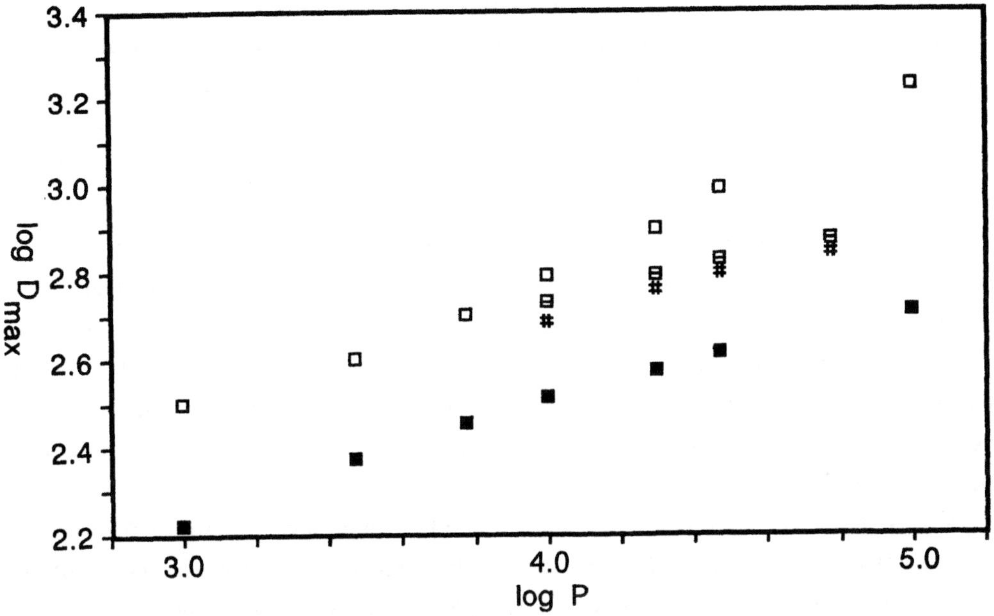

Figure 1. Plot of log D_{max} *vs* log P using the $t_N = 2{\times}10^8$ yr and subsonic criteria for two values of the IGM/ISM temperature ratio, T_r. The symbols are: ■, $T_r = 3.30$ and t_N criterion; □, $T_r = 3.30$ and c_s; #, $T_r = 31.25$ and t_N; ⊟, $T_r = 31.25$ and c_s.

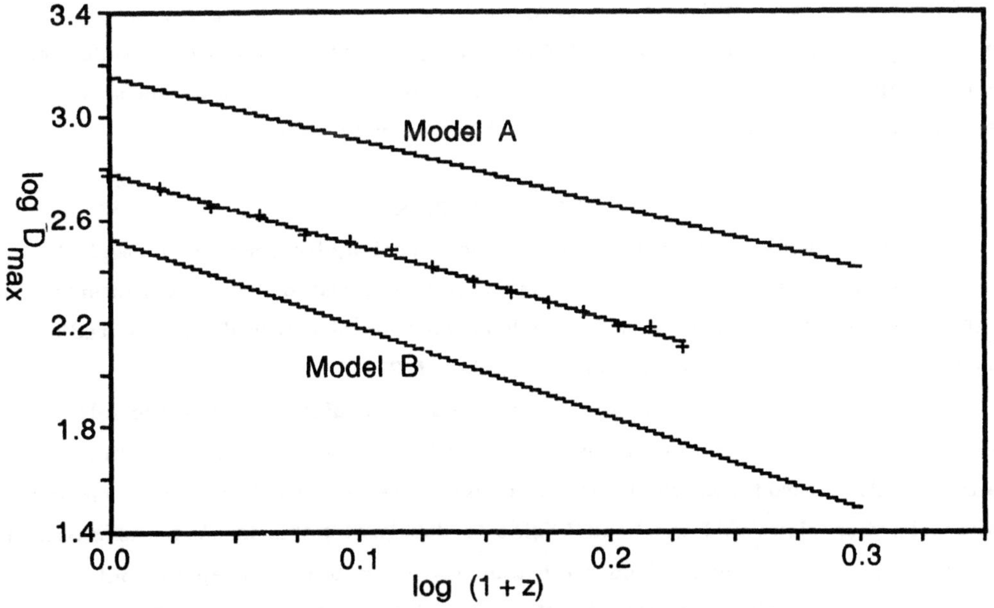

Figure 2. Plot of log D_{max} *vs* log (1+z) using only the subsonic criterion for Models A and B (see GW), and one set of runs of similar parameters.

III. RESULTS

Figure 1 plots log D_{max} *vs.* log P for the limiting values of T_r we considered (3.30 and 31.25) using the $t_N = 2\times10^8$ yr criterion as well as the subsonic criterion. While the change in the slope, q, is small for these values of T_r using the $t_N = 2\times10^8$ yr criterion, there is a definite trend toward decreasing q as the temperature ratio between the IGM and the halo is increased using the subsonic criterion. The explanation for this is a combination of two countervailing tendencies associated with the increase in T_r: (1) the density of the IGM is reduced by the same factor (to keep the pressure matched at the interface), which reduces the ram pressure of the IGM at a given velocity, and therefore the beams accelerate more quickly and to a higher maximum velocity, increasing D_{max}; and (2) the sound speed in the IGM is increased leading to a faster termination using the subsonic criterion.

Figure 2 shows the log D_{max} *vs.* log (1+z) plots of subsonic criteria for the analytic Models A and B, and a set of runs holding all other parameters constant, with a line showing the least squares linear fit to these points. The slopes of these lines are: Model A, -2.48; Model B, -3.48; line through set of data generated by the boundary following code, -2.87 ± 0.05. This clearly demonstrates that the numerical simulations yield results nearly midway in between the two extreme cases which can be treated analytically.

Both figures are similar to and compare favorably with ones previously generated (see Rosen and Wiita 1988) from a large sample of numerical simulations. In that paper we also argue why the D_{max} determined by our simulations is a very good approximation to the averaged observed value, $\langle D \rangle$.

IV. CONCLUSIONS

The main conclusions of this work are that quasi-hydrodynamical numerical simulations of a beam propagating through a pressure matched halo-IGM interface do confirm the effects of both redshift and luminosity on the overall sizes of radio sources predicted by the simple analytic model of GW.

The general behavior of the beams simulated here, after crossing the interface, can be summarized as follows: initially the head of the beam accelerates in a manner similar to Model A, which causes a higher degree of collimation. Eventually, the acceleration slows and the beam begins to decelerate; this is when the maximum aspect ratio usually occurs. Soon after this, the beam begins to spread out, analogous to the flaring predicted by Model B. Very preliminary results from a 2-D hydrocode of a nonrelativistic beam (Wiita, Rosen and Norman 1988) indicate that this overall scenario is correct. The analytic work predicts that the size of these sources should fall off as $(1+z)^{-\sigma}$, where σ is very nearly 2.5 for Model A and 3.5

for Model B. For the group of runs with similar parameters that covered the largest range of z, we obtain $\sigma = 2.87$, which agrees very well with the latest determination from observational data by Oort *et al.* (1987b), $\sigma \approx 3.3 \pm 0.5$, although earlier work tended to find somewhat lower values (*e.g.* Eales 1985). Thus the behavior appears to lie between the limiting analytical Models A and B, which is not at all surprising, although some additional computations will be necessary to further confirm and quantify this conclusion.

Although Kapahi (1985), Oort *et al.* (1987a,b), and Alexander and Leahy (1987) all found $q \approx 0.3$, we typically find $q \approx 0.25 \pm 0.05$. In light of the limited size of those data sets and the selection effects incorporated in them, we feel that the numerical simulations presented here are consistent with the luminosity-size relationship in powerful double-lobed radio sources.

If these agreements are not just coincidental, then these results are very important in both giving an insight on the physical make-up of jets, and also in indicating the existence of a cosmologically significant IGM. The properties and effects of this IGM will require additional investigation; in particular, the questions of how it was heated, its degree of uniformity, and its relationship to absorption line systems in quasars will have to be addressed. If our model continues to fit the observations well, it is possible that we will be able to constrain the values of some of the more uncertain parameters, *e.g.*, t_N. One consequence of this scenario is that we expect future observations with high resolution X-ray telescopes to find smaller galactic halos at larger redshifts because the IGM was denser and hotter in the past.

REFERENCES

Alexander, P. and Leahy, J.P. 1987, *Mon. Not. Roy. astr. Soc.*, **225**, 1.

Eales, S.A. 1985, *Mon. Not. Roy. astr. Soc.*, **217**, 179.

Forman, W., Jones, C. and Tucker, W. 1985, *Astrophys. J.*, **293**, 102.

Gopal-Krishna and Wiita, P.J. 1987, *Mon. Not. Roy. astr. Soc.*, **226**, 531.

Guilbert, P.W. and Fabian, A.C. 1986, *Mon. Not. Roy. astr. Soc.*, **220**, 439 (GF).

Kapahi, V.K. 1985, *Mon. Not. Roy. astr. Soc.*, **214**, 19P.

Mitteldorf, J. 1987, *Ph. D. thesis, University of Pennsylvania.*

Mitteldorf, J. and Wiita, P.J. 1988, these proceedings.

Oort, M.J.A., Katgert, P., Steeman, F.W.H., and Windhorst, R.A. 1987a, *Astron. Astrophys.*, **179**, 41.

Oort, M.J.A., Katgert, P. and Windhorst, R.A. 1987b, *Nature*, **328**, 500.

Rosen, A. and Wiita, P.J. 1988, *Astrophys. J.*, in press.

Trinchieri, G., Fabbiano, G., and Canizares C.R. 1986, *Astrophys. J.*, 310, 637.

Wiita, P.J. and Siah, M.J. 1986, *Astrophys. J.*, **300**, 605 (WS).

EVOLUTION OF RADIO JETS IN GALACTIC HALOS AND THE INTERGALACTIC MEDIUM

Paul J. Wiita
Department of Physics and Astronomy, Georgia State University
Atlanta, Georgia 30303
and
Gopal-Krishna
Radio Astronomy Group, Tata Institute of Fundamental Research
P.O. Box 1234, Indian Institute of Science Campus, Bangalore 560 012, India

ABSTRACT

Analytical models for the propagation of relativistic jets through both a hot interstellar medium and an even hotter intergalactic medium (IGM) are considered and shown to have surprisingly good predictive powers. We use the best observational data for the current properties of these media and make reasonable estimates of their cosmological evolutions. Limiting cases for the behavior of jets crossing the interface between them are analyzed and can quantitatively explain:

1) the current mean linear-size of double radio sources (~350 kpc);
2) the decrease in linear-size with cosmological redshift, z; and,
3) the increase in linear-size with radio power (at fixed z).

Because both very extended jets and very weak ones can allow the energy density in the lobes to become comparable to that of the microwave background, inverse Compton losses of the synchrotron emitting electrons against the background photons must also be considered. We can then argue that the most intrinsically powerful radio sources produce giant radio galaxies (those whose linear size exceeds 1.5 Mpc) and can also explain:

4) the number of observed giant sources at $z \lesssim 0.2$ (about 20 known);
5) the median observed radio power of such sources (~10^{26}W/Hz at 408 MHz); and,
6) the relative prominence of their nuclei in the radio band. We predict the existence of a significant number of giant radio galaxies in the redshift range $0.2 \lesssim z \lesssim 0.6$.

Finally, we argue that the break in the local radio luminosity function (LRLF) of isolated elliptical radio sources at about 10^{24} W/Hz at 1 GHz is mostly due to the fact that very weak jets are likely to become subsonic even within the halo. When this effect, which leads to such beams quickly fading away as radio sources, and the inverse Compton losses are convolved with an assumed universal power-law LRLF, the observed flattening at low radio powers is obtained.

I. INTRODUCTION

X-ray observations of relatively isolated elliptical galaxies indicate the presence of an extended, quasi-isothermal halo of hot gas ($kT_{halo} \approx 1$ keV; *e.g.* Forman, Jones and Tucker 1985), and the majority of the x-ray background emission is probably best explained in terms of a uniform IGM (with $kT_{IGM} \approx 18$ keV; *e.g.* Guilbert and Fabian 1986). In earlier papers (Gopal-Krishna and Wiita 1987, hereafter GW; Wiita and Gopal-Krishna 1987) we combined these observational constraints on the external media with very simple analytical models for radio jets. These jets were assumed to contain relativistic plasma, with a relativistic bulk velocity down the jet (*e.g.* Scheuer 1974) so that its power is mainly kinetic. However, the speed of advance of the heads of the jets, v_h, assumed to be associated with the hot-spots, was taken to be non-relativistic.

The key idea of these papers was that when the jet reached the pressure-matched interface between the halo and the IGM, at R_h, its propagation must be dramatically affected. Our Model A assumed that the initially constant opening angle, θ, of the jet is unchanged on crossing the interface so that the abrupt decrease in density (equal to T_{halo}/T_{IGM}) produces a major drop in the ram pressure slowing the head of the jet, which therefore accelerates. Model B had the velocity remaining continuous across the interface so that the jet would flare and actually slow down, despite the drop in ram pressure. The density of the halo is measured to decrease roughly as $n(r) \approx n_o[1+(r/a)^2]^{-3/4}$ and we argued that a conservative interpretation of the observations implies that, at least out to a redshift of ~1, the properties of the halo have not changed very much (GW). However, the IGM was both hotter and denser in the past, and we showed that $R_h \approx R_h(z=0)(1+z)^{-10/3}$ under these circumstances.

In making connections between these simple models and observations of the overall linear size of Fanaroff-Riley (1974) class II (powerful double) radio galaxies, we used average, observationally estimated, values for various parameters, *i.e.*, $\theta=0.04$ rad, $v_h \approx 10^4$km s^{-1}, $n_{IGM} \approx 7\times10^{-7}$ cm^{-3}, $n_o \approx 0.01$ cm^{-3}, $a \approx 2$ kpc. This choice of parameters leads to $R_h(z=0) \approx 171$ kpc. The key assumption we made is that the lobes become much weaker quite quickly when the nuclear engine turns off (at $t_N \approx 10^8$ yr) or the advance of the beam head becomes subsonic with respect to the external medium, whichever occurs first. If Model B is somewhat closer to the reality than Model A, the average total linear size for nearby (3CR sample, with $\langle z\rangle \approx 0.15$) radio sources is $\langle 2D\rangle \approx 350$ kpc and the dependence on redshift is predicted to be: $\langle 2D\rangle \propto (1+z)^{-3}$. Both the value of current linear size and its dependence on cosmological redshift are in excellent accord with observations (*e.g.* Eales 1985, Gopal-Krishna *et al.* 1986, Kapahi 1985, Oort *et al.* 1987). Relatively simple numerical simulations which follow the boundary between the jet plus swept up cocoon and the external media (Wiita and Siah 1986, Siah and Wiita 1987, Mitteldorf 1987, Mitteldorf and Wiita 1988, Rosen and Wiita 1988a,b) support this model. This numerical work also gives a good match to the observed relation between linear size and radio luminosity, $\langle D \rangle \propto P^{0.3}$ (Kapahi 1986, Oort *et al.* 1987). We are undertaking high resolution, two-dimensional hydrodynamical computations (Wiita, Rosen and Norman, in preparation) to continue to test this very promising picture.

II. AN IMPROVED ANALYTICAL MODEL

Our earlier work concentrated on the *average* properties of FR II radio galaxies and was based on the *average observed properties* of galactic halos and the IGM. However, there are several extensions of the model needed if both extremely powerful radio jets and very weak ones are to be considered. It is the goal of this paper to include the additional effects which allow us to treat these more extreme cases.

A. Extension to Relativistic Bulk Motion

Although it is quite clear that the vast majority of radio jets do not have relativistic velocities of advance, the evidence that internal flows remain very fast even on > 100 kpc scales is growing (Bridle 1988). Within our model, as the beam power, L_b rises, v_h eventually becomes a significant fraction of *c*, at least for early stages of the beam's life, and we must generalize our earlier work to allow for this possibility. If D is the length of a beam at time t and ρ_b, $c\beta_b$, and γ_b are its density, bulk velocity and bulk Lorentz factor, we have (*e.g.* O'Dea 1985)

$$L_b = \pi(\theta Dc/2)^2 \rho_b \gamma_b (\gamma_b - 1)\beta_b c. \tag{1}$$

Noting that the rate of advance is determined by a balance between beam thrust and ram pressure balance, we can transform to the shock frame, where the beam velocity is $c\beta_b'$. Because $\beta_b \gg 1$ Gopal-Krishna, Wiita and Saripalli (1988; GWS) find

$$\rho_b(\gamma_b' c)^2 = \rho_{ext}(\gamma_h \beta_h c)^2, \tag{2}$$

where ρ_{ext} is the density of either the halo or IGM. This can be written as

$$\beta_h/(1-\beta_h) = (2/D\theta)[L_b/\pi c^3 \rho_{ext}(D)]^{1/2}. \tag{3}$$

Within the halo, we can obtain the solution,

$$v_h(D) = cX[1+(D/a)^2]^{3/8}/\{D+X[1+(D/a)^2]^{3/8}\}, \tag{4}$$

with $X = (2/\theta)(L_b/\pi c^3 n_o m_H \mu)^{1/2}$, and, while D(t) cannot be explicitly found,

$$t = [0.8a^{3/4}D^{5/4} + XD]/(cX). \tag{5}$$

Upon crossing the interface the result depends upon whether Model A (constant θ) or Model B (constant v_h) is taken, but the general form is the same:

$$v_h(D) = cX_{A,B}/(D+X_{A,B}), \text{ with } X_A = X\{n_o/[n_{IGM}(0)(1+z)^3]\}^{1/2},\ X_B = X(R_h(z)/a)^{3/4}, \quad (6)$$

$$D_A(t) = \{X_A^2 + 2X_Act + R_h^2(z)[1-8X_A^2/(5XX_B)]\}^{1/2} - X_A, \quad (7A)$$

$$D_B(t) = [X_B^2 + 2X_Bct - 3R_h^2(z)/5]^{1/2} - X_B. \quad (7B)$$

All of the above expressions are equivalent to those given in GWS and reduce to the simpler ones found in GW when $\beta_h << 1$. At very large times we find $D \propto t^{1/2}$, $D \propto L_b^{1/4}$ and $v_h \propto 1/D$, although that limit is often not reached because the advance of the head becomes subsonic or $t_{lim} > t_N$, the period that the nucleus (and beam) remains active.

B. Inverse Compton Losses

Aside from the usually explicit invocation of equipartition, an assumption implicitly made in most discussions of extended radio sources is that the radio emission, P, is a fixed fraction, ϵ, of the total beam power ($2L_b$, in our notation). While we will retain the equipartition assumption, so that the relativistic electrons are taken to have 4/3 the energy density of the magnetic field, we will modify the latter assumption to allow for the fact that the fraction of power emitted in the radio band will decrease due to inverse Compton scattering of the relativistic electrons off the cosmic background photons. While this effect is minor for typical radio sources, both very weak and very extended sources can have the magnetic energy density in their lobes, u_B, approach the energy density of the microwave background, $u_{ph}(z) = 4.0\times10^{-13}(1+z)^4$ erg cm^{-3}. When this occurs, the efficiency at which a radio source emits radio flux decreases, while its x-ray flux rises.

This basic idea has been considered several times in the past in somewhat different contexts (*e.g.* Rees and Setti 1968) and was argued by Eales (1985) to be inadequate to explain the complete linear size *vs* redshift relation. But we stress that it is particularly applicable to very extended double radio sources since a large fraction of their emission comes from their lobes and not their hot spots (Jager 1986). In the lobes, u_B is lower than in the hot spots, so this inverse Compton "reduced radio efficiency" (RRE) is more important (GWS).

We can get a good approximation to the effect of RRE by noting that the observations indicate that the average magnetic energy density of a source, $\langle u_B\rangle \approx \eta u_{B,h}$, with $\eta \approx 0.1$ (Saripalli 1987) and $u_{B,h}$ the magnetic energy density in the hotspot. However, using the minimum energy argument, $u_{B,h} = 3u_h/7$, but ram pressure confinement of the hot spot implies $\rho_{ext}v_h^2 \approx u_h/3$, so

$$\langle u_B \rangle \approx (9/7)\,\eta \rho_{ext} v_h^2. \tag{8A}$$

But, we conservatively assume that $\langle u_B \rangle$ cannot fall below the limit set by the static confinement of the lobe due to thermal pressure of the external medium, taken to be a $\Gamma \approx 5/3$ gas, so

$$\langle u_B \rangle_{min} = (3/7)\langle u_{min} \rangle = (3/7)u_{ext} \approx (9/14)p_{ext}. \tag{8B}$$

We now generalize the total radio emission as a function of time to allow for RRE

$$L_R(t) = \epsilon'(t) L_R(t{=}0) = \epsilon'(t)\cdot\epsilon\cdot 2L_b, \text{ with} \tag{9}$$

$$\epsilon'(t) = L_R(t)/[L_R(t)+L_X(t)] = [1 + (u_{ph}(z)/\langle u_B \rangle]^{-1}. \tag{10}$$

Inserting the larger of (8A) and (8B) into (10) enables us to estimate the radiated radio power at any stage for our beam model, and thus, the observed radio flux density, S (in Jy), if the distance of the source and a cosmological model (values of H_0 and q_0) is specified. Reasonable assumptions about the source morphology also enable us to obtain typical surface brightnesses (Jy arcmin^{-2}), but as they are not usually important (GWS), we will not consider them here.

III. GIANT RADIO GALAXIES

Fewer than 20 extragalactic radio sources are known to have total linear extents, 2D, exceeding 1.5 Mpc (with H_0 taken as 50 km s^{-1}Mpc^{-1}), thereby qualifying them as giant radio galaxies (GRGs). Aside from their immense sizes, these sources have several other important properties in common (Saripalli *et al.* 1986). For instance, their radio luminosities seem to cluster in a narrow range ($P_{408} = 10^{25}$–10^{27} W Hz^{-1} at 408 MHz), which is just above the FR luminosity transition. Also, GRGs have a higher ratio of core radio luminosity to total radio luminosity than do typical double radio galaxies, although this ratio is not as large as for quasars.

Why do these sources have only moderate (observed) powers, despite their huge sizes? Within the beam model, one could argue that either: (i) GRGs are associated with modest, but very long-lived, nuclear engines, or (ii) GRGs are produced by powerful engines over relatively short times, with reduced radio outputs. We shall argue that the latter is far more likely. However, it should be noted that because most GRGs were discovered from flux limited surveys, and the total radio source luminosity function is steeply declining (see §IV) with increasing luminosity, their excess abundance just above the FR break may be a selection effect (A. Bridle, private communication).

A. Applying the Analytical Models

Using the improved analytical models of §II we have computed the time evolutions of v_h, D, L_R, and S (in Jy) for total beam powers ($2L_b$) in the range 10^{36} W through 10^{41} W and redshifts between 0 and 1. Sufficiently weak sources never become GRGs because they either become subsonic in the IGM shortly after leaving the halo or because their initial advance is so slow that they take more than t_N (usually chosen as 10^8 yr; see GW) to reach linear distances of $D \geqslant 750$ kpc.

At somewhat higher intrinsic powers the beams can make it out that far without dying, but that doesn't mean that we can *observe* that source as a GRG. The effect of inverse Compton losses on such spread-out sources can cut the actual L_R to the extent that the observed radio flux, S, falls below a detection threshold. We note that the largest sensitive survey used for discovering GRGs employed the 6C synthesis array at 151 MHz which appears to be confusion limited below a flux density of ~5 Jy (Saunders, Baldwin and Warner 1987). In this paper we refer all flux estimates to 1 GHz, so this limit corresponds to a minimum detectable flux density of $S_{lim} \approx 1$ Jy at 1 GHz, using a spectral index of $\alpha = -0.8$ ($S_\nu \propto \nu^\alpha$).

It is clear that at earlier cosmological epochs, the *fraction* of currently detectable giant sources would be smaller for several reasons. Firstly, high z implies greater distance and therefore smaller S for a given L_R. Secondly, because the microwave background energy density was higher in the past, we expect the RRE to have set in sooner, cutting down L_R. Finally, if our basic model for propagation across the halo/IGM interface is valid, R_h was smaller at earlier epochs and c_s was larger, further reducing the number of possible GRGs. Using the constraints on size, velocity, lifetime and flux density given above, the logarithm of the minimum beam power (in Watts) needed to produce an observable GRG is about 37.25 for Model A and 37.75 for Model B, for a redshift of only 0.05. But, for z = 0.45, these values rise to 38.85 and 39.55, respectively. Still, the rapid decline with redshift in the fraction of radio galaxies qualified to be giants, is counteracted by the rapid increase in the co-moving volume with z.

B. The Numbers and Powers of GRGs

These model calculations give us estimates of the minimum power required for the formation of an observable GRG at a given redshift and also provide us with their sizes and radio flux densities as functions of time. We must convolve the appropriate radio luminosity function with these estimates in order to compute the total numbers of observable GRGs in each redshift interval. We have used the recent determination of the co-moving density of steep spectrum radio sources given by Peacock (1985) for $q_o = 0$ (the final results are not very different if $q_o = 1/2$ is chosen) to obtain the radio luminosity function (RLF), $\rho(P,z)$. However, this is not immediately useful, for this RLF is based upon *observed* radio fluxes, while our models predict P(t) for initial P (the flux corresponding to $L_R(0) = \epsilon 2L_b$).

Elsewhere (GWS) we have discussed in some detail how to effectively redefine the RLF as a function of L_b instead of P. The key idea is that log P remains nearly constant and $\approx \log(2\epsilon L_b)$ until a time $t_{1/2}$, after which RRE causes a rapid drop by at least an order of magnitude. Thus, to lowest order, $P = P_0$ for $t \leqslant t_{1/2}$ and $P = 0$ for $t > t_{1/2}$, so that the vast majority of sources currently observed at a given power P turned on at essentially that power during the past $t_{1/2}$ years. Therefore, the redshift dependent rate of creation of sources per unit volume with total beam power $2L_b$ is about equal to the observed density of sources with power P divided by $t_{1/2}$ for that value of P. But a source can only be observed as a GRG if the time, t_G, it takes to grow to $2D \geqslant 1.5$ Mpc is shorter that the time t_* over which it would remain observable, bearing in mind the constraints of §III.A. Defining P_* as the minimum initially radiated power required to produce a GRG satisfying those constraints, we have the number of observable GRGs in a given *z* bin

$$N(z)\Delta z = \Delta V(z) \int_{P_*}^{\infty} [t_*(P,z) - t_G(P,z)]\ \pi(P,z)/t_{1/2}(P,z)\ dP\ \Delta z, \qquad (11)$$

with $\Delta V(z)$ the volume contained between $z - \Delta z/2$ and $z + \Delta z/2$ (*cf* Windhorst 1984).

The mean radio luminosity, ⟨P⟩, of GRGs can be roughly predicted as a function of redshift by considering the emitted radio power, P_{1Mpc}, at the value D = 1 Mpc, which is the typical half-size of known GRGs (Saripalli *et al.* 1986). We find ⟨P⟩ by inserting P_{1Mpc} into the integrand of the RHS of (11) and dividing through by $N(z)\Delta z$. We evaluated these quantities for Models A and B, but then took the geometrical mean between them ("Model AB") as our best estimate of the actual values for N(z) and ⟨P(z)⟩ in light of the results summarized in §I and the simulations of Rosen and Wiita (1988a,b). These are presented below for the range over which the RLF is well defined, and compared with observations (GWS and references therein).

z-range	$\Delta V(Gpc^3)$	N_{pred}	N_{obs}	$\log\langle P(W/Hz)\rangle_{pred}$	$\log\langle P(W/Hz)\rangle_{obs}$
0.0–0.1	0.79	13	12	26.0	25.6
0.1–0.2	4.74	14	4	26.6	26.4
0.2–0.3	11.1	12	2	26.9	26.7
0.3–0.4	18.7	10	0	27.3	----
0.4–0.5	27.0	7	0	27.5	----
0.5–0.6	35.9	5	0	27.6	----

IV. THE SHAPE OF THE LOCAL RADIO LUMINOSITY FUNCTION

It has been well known since the work of Auriemma *et al.* (1977) that the LRLF for elliptical radio galaxies is much steeper above $P \sim 10^{24}$ W Hz^{-1} at 1 GHz than it is below. Recent work (Windhorst 1984, Subrahmanya and Harnett 1987) has confirmed and extended this result. We argue that this break can be understood without invoking an *ad hoc* corresponding break in the beam power function $\Phi(L_b)$. For, within the framework developed so far in this paper, the heads of beams with $L_b \lesssim 10^{25}$ W are decelerated to sonic speeds within the halo itself, leading to a rapid decay of radio luminosity (Gopal-Krishna and Wiita 1988). Only a couple of additional effects, most important for relatively weak sources, must be added to the considerations of RRE and subsonic fading discussed above in order to understand why this is so. First, we introduce the observed dependence of the beam opening angel on the beam power: θ (radian) = $0.02 + 0.03[29 - \log(2\epsilon L_b)]$ (Bridle 1986).

We then noted that when the P(t)'s predicted for our composite Model AB are plotted for various values of P_0, the RRE decline only tended to set in just outside R_h, as long as $P_0 \gtrsim 10^{26.5}$ W Hz^{-1} at 1 GHz, at least for $z \lesssim 0.1$. For $25.0 \lesssim \log P_0 \lesssim 26.5$, the inverse Compton decline was significant inside R_h, but the beam's heads still advanced supersonically within the halo. But, for $\log P_0 \lesssim 25.0$, the weakling beams became subsonic within the halo and are likely to be subject to instabilities causing them to break apart into bubbles (*e.g.* Wiita 1978).

Our second additional point involves estimating the emitted luminosity of these bubbles, rising buoyantly with the sonic speed, c_s, and maintaining equilibrium with the thermal pressure, p, of the halo. For $(D/a)^2 \gg 1$, the $p \propto D^{-3/2}$, implying $dp/dt \propto -c_s p^{5/3}$. Using the minimum energy criterion, $P \propto U p^{3/4}$, where the pressure, p, within the bubble, which is matched by the halo pressure, is equal to its internal energy U divided by 3 times the volume *V* of the bubble; Pacholczyk 1977). Since synchrotron and expansion losses dominate we have: $dU = -p\,dV - P\,dt$. From these relations it can be shown that during this buoyant rise phase,

$$P \overset{\sim}{\propto} p^2 \propto D^{-3} \overset{\sim}{\propto} t^{-3}. \tag{12}$$

At this stage we can argue that the power-law for the beam luminosity function, $\Phi(L_b = P_0/2\epsilon) \propto P_0^{-1}$ valid at the high luminosity end of the LRLF $[\rho_L(P)]$, could actually extend to much lower beam powers and still not conflict with the observed flattening of the LRLF at $P \sim 10^{24}$ W Hz^{-1}. This is because only sources with $P_0 \geqslant P$ can contribute to $\rho_L(P)\,dP$, but they do so only for a limited time, $\Delta t(P_0)$, during which these sources decay through the observed luminosity bin centered on P. We then perform the following integration numerically to find

$$\rho_L(P) = \int_P^{\infty} \Delta t(P_0)\Phi(P_0)dP_0 \Big/ \int_P^{\infty} \Delta t(P_0)dP_0. \tag{13}$$

When evaluated at z = 0.1, this convolution yields an extremely good fit to the LRLF quoted by Windhorst (1984); see Gopal-Krishna and Wiita (1988) for details. Note that we have not yet attempted to see if our beam *cum* halo/IGM model is equally impressive in fitting the shape of the RLF at higher redshifts. This would require our being able to separate out real evolutionary effects in Φ from those introduced by RRE and subsonic motions, and current data may not allow this.

V. CONCLUSIONS

Our model for jet propagation through a galactic halo and into the IGM appears to provide an excellent way to approach a wide variety of data concerning extragalactic radio sources. In its simplest incarnation (GW), this picture yields good values for the typical current linear size of double radio galaxies as well as for the linear-size *vs.* redshift relation. *It should be stressed that in every case the value of each parameter that enters into the model is the one best supported by observations.* These parameters are unfortunately numerous, and should perhaps be gathered together here: t_N, a, T_{halo}, n_o, T_{IGM}, $n_{IGM}(0)$, and θ for the simplest version; also ϵ, ϵ' and η for the generalized version developed above. Approximate numerical models verified the basic picture, yielding results between those of the limiting analytical cases and providing a good fit to the linear-size *vs.* power relation (Rosen and Wiita 1988a,b). Although it is not strictly valid to extend our scenario back beyond $z \approx 1$, it may be an explanation of why no radio galaxies are known to extend further than about 10 kpc at very high ($z \approx 2$) redshifts (*e.g.* Lonsdale and Barthel 1985). More detailed numerical simulations using the Cray X-MP/48 at the National Center for Supercomputing Applications are being undertaken, and preliminary results do seem to support the basic model, although since the hydrodynamics code can only deal with non-relativistic motions and non-relativistic equations of state its verification capability is not ideal. One clear prediction of this scenario is that AXAF should find smaller galactic halos at larger redshifts because the IGM exerted much higher pressures in the past.

By adding relevant physics, including relativistic motions of the jet head, we now can treat both the most powerful and weaker radio galaxies. Including the inverse Compton losses of the synchrotron emitting relativistic electrons against the microwave background photons essentially turns out to be of great importance. The key conclusion of this piece of our work is that *GRGs are indeed produced by intrinsically powerful AGNs* whose radio luminosities suffer declines of about an order-of-magnitude between the time of their turning-on and their qualifying as giants. This single idea thus explains the key facts about GRGs: their moderate apparent powers; their rarity; and their relatively high core/lobe flux ratio. We predict that a significant number of unidentified GRGs exist in the redshift range from 0.25 to 0.6. But the average radiated power of these putative distant sources grows slowly with redshift, so that their flux densities would cluster just above the confusion limit; therefore, we can understand why they have yet to be found. We plan to conduct a search with the VLA for promising high z GRG candidates identified by the Ooty radio telescope. Their discovery in numbers, and with powers, close to those we predict would be powerful evidence in favor of this theory.

When we consider what should happen to weak beams that actually become subsonic within the extended galactic halos we also appear to be able to explain why there is a break in the local radio luminosity function. Thus our picture effectively does away with the need to postulate more than one underlying type of nuclear engine. Rather, a single kind of prime mover, produced with a simple power law distribution fit to the high end of the LRLF, can yield both the location and amount of the break in the slope of the LRLF.

At this stage in the development of this theory, the number of successful matches to disparate observations is roughly equal to the number of parameters in the theory, most of which are well constrained by yet other observations. If all of these agreements are not just coincidental, and we find it hard to imagine that they are, then these results are very strong evidence in favor of the three fundamental propositions *that most strong jets are predominantly relativistic even on large scales, that extended hot galactic halos are ubiquitous, and that a cosmologically significant ($\Omega \approx 0.25$) IGM exists.*

Let us conclude by noting that many, and probably most, of the differences between observed classes of radio sources (and perhaps other types of AGN too) are probably due to the interaction of their initially similar jets with different types of external media. Jets emerging through the denser, but cooler, gas of spiral galaxies, or out of elliptical galaxies embedded in intracluster media, will certainly behave differently from those emerging from a hot halo into an IGM. It would be unfortunate if the near coincidence between the luminosity which divides the more diffuse, edge darkened FR I RGs from the more symmetrical, edge brightened FR II RGs, and the break in the LRLF (which we have claimed to explain) turned out to be merely an accident. Future extensions of our model, both analytical and numerical, will be undertaken to investigate this and other exciting prospects.

We greatly appreciate the assistance of Lakshmi Saripalli and Alexander Rosen in much of the research summarized here; Rosen also suggested improvements to the presentation. We thank Alan Bridle, V. Borkar, Vijay Kapahi, Javad Siah and C.R. Subrahmanya for helpful conversations. PJW is grateful for support from a Georgia State University Research Grant and from a Smithsonian Institution Foreign Currency Research Grant.

REFERENCES

Auriemma, C., Perola, G.C., Ekers, R.D., Fanti, R., Lari, C., Jaffe, W.J. and Ulrich, M.-H. 1977, *Astron. Astrophys.*, **57**, 41.

Bridle, A.H. 1986, *Can. J. Phys.*, **64**, 353.

Bridle, A.H. 1988, *these proceedings.*

Eales, S.A. 1985, *Mon. Not. Roy. astr. Soc.*, **217**, 179.

Fanaroff, B.L. and Riley, J.M. 1974, *Mon. Not. Roy. astr. Soc.*, **167**, 31P.

Forman, W., Jones, C. and Tucker, W. 1985, *Astrophys. J.*, **293**, 102.

Gopal-Krishna, Saripalli, L., Saikia, D.J. and Sramek, R.A. 1986, in **Quasars, IAU Symposium No. 119**, eds. G. Swarup and V.K. Kapahi (Reidel: Dordrecht), p. 193.

Gopal-Krishna and Wiita, P.J. 1987, *Mon. Not. Roy. astr. Soc.*, **226**, 531 (GW).

Gopal-Krishna and Wiita, P.J. 1988, *Nature*, in press.

Gopal-Krishna, Wiita, P.J. and Saripalli, L. 1988, submitted to *Mon. Not. Roy. astr. Soc.* (GWS).

Guilbert, P.W. and Fabian, A.C. 1986, *Mon. Not. Roy. astr. Soc.*, **220**, 439.

Jager, W.J. 1986, *Ph. D. thesis, University of Leiden.*

Kapahi, V.K. 1985, *Mon. Not. Roy. astr. Soc.*, **214**, 19P.

Kapahi, V.K. 1986, in **Highlights of Astronomy, Vol 7**, ed. J.P. Swings (Reidel: Dordrecht), p. 371.

Lonsdale, C.J. and Barthel, P.D. 1985, *Astrophys. J.*, **303**, 617.

Mitteldorf, J. 1987, *Ph. D. thesis, University of Pennsylvania.*

Mitteldorf, J. and Wiita, P.J. 1988, *these proceedings.*

O'Dea, C.P. 1985, *Astrophys. J.*, **295**, 80.

Oort, M.J.A., Katgert, P., Steeman, F.W.H., and Windhorst, R.A. 1987, *Astron. Astrophys.*, **179**, 41.

Pacholzcyk, A.G. 1977, **Radio Sources** (Pergamon Press: New York), p. 197.

Peacock, J.A. 1985, *Mon. Not. Roy. astr. Soc.*, **217**, 601.

Rees, M.J. and Setti, G. 1968, *Nature*, **219**, 127.

Rosen, A. and Wiita, P.J. 1988a, *Astrophys. J.*, in press.

Rosen, A. and Wiita, P.J. 1988b, *these proceedings.*

Saripalli, L. 1987, *Ph. D. thesis, Indian Institute of Science* (in preparation).

Saripalli, L., Gopal-Krishna, Reich, W. and Kühr, H. 1986, *Astron. Astrophys.*, **170**, 20.

Saunders, R., Baldwin, J.E. and Warner, P.J. 1987, *Mon. Not. Roy. astr. Soc.*, **225**, 713.

Scheuer, P.A.G. 1974, *Mon. Not. Roy. astr. Soc.*, **166**, 513.

Siah, M.J. and Wiita, P.J. 1987, *Astrophys. J.*, **313**, 623.

Subrahmanya, C.R. and Harnett, J.I. 1987, *Mon. Not. Roy. astr. Soc.*, **225**, 297.

Wiita, P.J. 1978, *Astrophys. J.*, **221**, 436.

Wiita, P.J. and Gopal-Krishna 1987, in **13th Texas Symposium on Relativistic Astophysics**, ed. M.P. Ulmer (World Scientific, Singapore), p. 355.

Wiita, P.J. and Siah, M.J. 1986, *Astrophys. J.*, **300**, 605.

Windhorst, R.A. 1984, *Ph. D. thesis, University of Leiden.*

QUESTIONS AND COMMENTS

S. Caganoff: The analytic King model is not a good approximation to the complete King model at large radii. How does this affect your models?

P.J. Wiita: We have used the form for the gas distribution derived by Forman, Jones and Tucker from *Einstein* X-ray observations in all of the work presented here. However, numerical models using slightly different, but still essentially power-law, distributions have been computed and show no important changes.

J.O. Burns: Your pseudo-hydro simulations assume a relativistic equation of state and bulk relativistic flow, whereas the hydro simulations are non-relativistic. I'd expect this to produce differences in the way in which the jet flares near the point of termination. Can you reconcile the predictions of the two codes?

P.J. Wiita: I agree that some differences are to be expected but we do not yet have enough full hydro runs to see if they are significant. If the agreement turns out very good using very light jet fluids and high internal Mach numbers, which is as close as the hydro-code comes to relativistic flows, we naturally will be pleased. But if the differences are big we will try to modify the hydro-code to allow it to treat relativistic flows and see if the putative problems remain.

M.J. Valtonen: Where do you get your observed power-size correlation $P \propto D^{0.3}$? In the case of 3C sources this correlation is known to hold only for sources with $D \lesssim 200$ kpc, but for larger sources than 200 kpc the power declines sharply with size ($P \propto D^{-3}$).

P.J. Wiita: Much larger samples are now available and we have cited the latest results in the literature (Kapahi 1986, Oort *et al.* 1987). Some (or even most) of the decline you mention may be due to the inverse Compton losses which could be strongly affecting the largest members of the high-flux-limited 3C sample.

M. Sulkanen: Do you expect any changes in jet behavior (in the numerical modelling) if the changes in ρ and T across the halo/IGM interface aren't so rapid, and extend over a few jet radii? Is such a sharp transition realistic?

P.J. Wiita: Granted that such a sharp transition is probably not realistic, I do not believe that smoothing it out over several jet radii would have any significant effects. However, this is something we can and will check in future work.

AGNS IN CLUSTERS OF GALAXIES AND THE BOOTES VOID

Jack O. Burns
Institute for Astrophysics
The University of New Mexico
Albuquerque, NM 87131

1. Introduction

It has been suspected for quite some time that environment plays a role in the origin and evolution of extragalactic radio sources. However, the details of such a relationship remain unclear after several decades of study. Stocke (1978), Dressel (1981), and Heckman et al. (1985) all claim that the galaxy neighborhood influences the probability of radio emission and the radio power of a galaxy. On the other hand, Auriemma et al. (1977), Fanti et al. (1982), and Hummel et al. (1983) have argued that the radio luminosity function is the same for cluster and "field" galaxies. One suspects that differences in the definitions of the source samples and the galaxy environments have produced these very different conclusions. What is needed is a systematic study of radio activity in well-defined environments. Over the past dozen years, co-workers and I have been using the VLA to survey rich clusters, poor groups, and more recently, cosmic voids. With these diverse samples, we are in a position to reflect upon the influence of environment on radio emission in AGNs.

2. Observations of Rich Clusters and Poor Groups

I will begin by summarizing several of the more interesting properties of cluster radio sources. Details of the work on Abell clusters can be found in Zhao, Burns, and Owen (1988) and on poor groups in Burns et al. (1987). It is important to note that the sources in these samples are of moderate to weak radio powers, generally below the break in the radio luminosity function ($< 10^{25}$ W/Hz at 20-cm).

There are several interesting negative correlations with global, cluster-wide environment. First, radio power is not a function of cluster environment. There is no correlation between number of galaxies in an Abell cluster and the radio power at 20-cm. Extending this to poor groups with the same redshift and flux density cutoffs, we find no differences in the distributions of radio powers in rich and poor clusters. Second, in contradiction to the usual assumption, tailed radio sources are found in *both* rich and poor clusters. Because the dynamic pressures experienced by typical galaxies in poor groups is < 0.1 of that in rich clusters, it had been assumed that head-tail (HT) sources would be found almost exclusively in rich clusters. However, our survey of poor groups reveals that about 7% of the groups have HT morphologies, a fraction not very different from Abell clusters. In the case of B2 0800+42, a U-shaped HT is found in an exceptionally poor environment with only 5 galaxies within a diameter of 1.2 Mpc (Stocke and Burns, 1987).

Turning now to the direct environs of radio galaxies, I noted above that considerable controversy still surrounds the issue of environmental impact on radio emission. From analysis of a sample of relatively nearby elliptical galaxies with m < 15 and radio powers similar to our cluster sample, Heckman et al. (1985) concluded that "all measures of local density are larger (by at least a factor of 2 - 3) for radio-loud galaxies". They suspected that the larger densities enhanced the likelihood of tidal interactions, and thus stimulated nuclear activity in galaxies (e.g., fueling of central engine). Our sample of poor groups allows us to comment further on this interpretation. We find that galaxies which have both companions and are "disturbed" (according to Zwicky's visual impressions on the Sky Survey) have a factor of 4 increase in probability of being a radio source in comparison to galaxies with no signs of interaction. Clearly, such indications of recent tidal encounters must be important in activating the engines within the nuclei of ellipticals. However, this effect represents only an enhancement of the probability of activity since there are many moderate to low radio luminosity galaxies that show no signs of interaction with companions.

One type of galaxy that clearly demonstrates a strong effect of environment is a dominant galaxy at the center of a cluster. The largest of this subclass of galaxies, cDs, are radio-loud (with log P_{20} > 22 W/Hz) nearly 50% of the time. This is about a factor of 5 more than non-dominant cluster galaxies. A subset of these dominant galaxies have an excess of central x-ray emission above that expected for extrapolation of an isothermal King model from the outer portions of the clusters (e.g., Sarazin, 1986). This is usually interpreted as the result of a radiatively-driven cooling flow onto the central galaxies since the estimated cooling times in the centers are less than typical cluster ages (< 10^{10} yrs). From this x-ray excess, one can estimate the mass infall rate (e.g., Arnaud, 1987). In Fig. 1, we show a plot of mass accretion rate as a function of radio power for dominant central galaxies in nearby Abell clusters. There may be a weak correlation (Cox rank correlation test suggests that the correlation can be rejected on the 1.6% level), although the sample is still small and the error bars are large. A more significant relation is between cooling flows and radio morphologies. In Fig. 2, cooling time is plotted against radio power for central dominant galaxy clusters. We see that extended, tailed sources are associated only with dominant galaxies that do not have the classic signature of a cooling flow (central x-ray excess and cooling times < 10^{10} yrs). So-called cooling flow clusters have smaller sources that are often confined to within the optical boundaries of the galaxies and sometimes resemble small doubles. A similar relationship between radio source size and x-ray excess was noted by O'Dea and Baum (1986). One is tempted to conclude that a cooling flow quenches the propagation of extended structures such as jets, lobes, and tails.

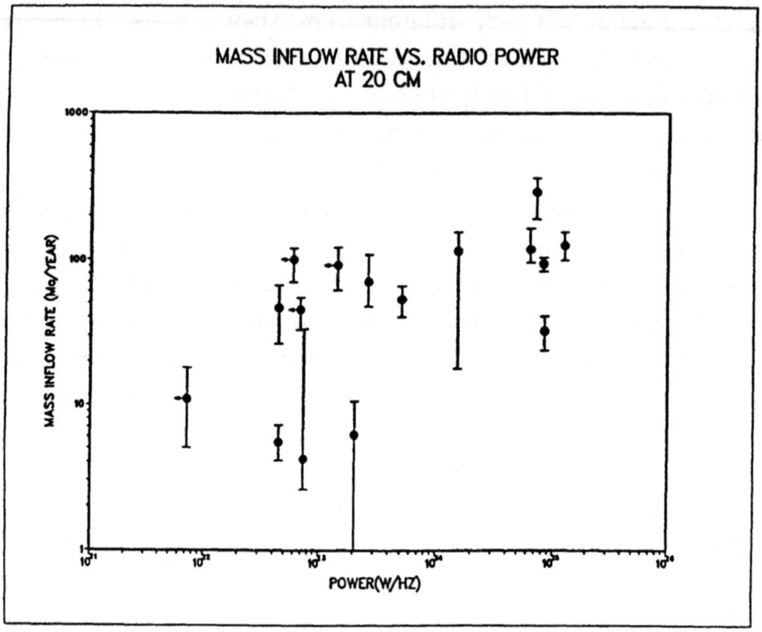

Fig. 1 - Average mass accretion rate (from Arnaud, 1987) versus 20-cm radio power for Abell clusters with x-ray cooling flows.

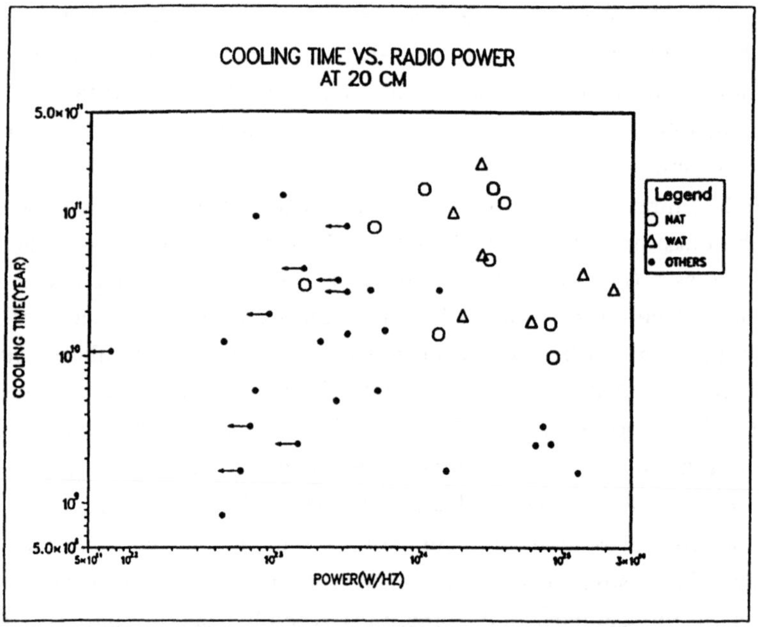

Fig. 2 - Cooling time (from Arnaud, 1987) versus 20-cm radio power for a sample of nearby Abell clusters. The points are compact or unresolved sources and the extended symbols are extended, tailed sources.

3. Numerical Simulations of Radio Sources Associated With Dominant Galaxies

In an effort to understand the dichotomy of source structures associated with dominant cluster galaxies, a series of hydrodynamical scenarios have been constructed to simulate the propagation of fluid flows through realistic central galaxy atmospheres. We have used a 2-D code called ZEUS, written by M. Norman, running on a Cray X-MP supercomputer at NCSA, to study how various environments affect radio jets. We considered both supersonic infall and galaxy wind outflow models.

First, we examined the propagation of a mildly supersonic (Mach numbers of M = 2 - 5; note that unless otherwise stated the Mach number is measured with respect to the internal sound speed of the flow) jet through a background gas that was moving supersonically in a direction opposite to that of the jet. To first order, this would mimic a jet moving through a supersonic cooling infall. Not surprisingly, the morphology that emerged was that of a small classical double with a leading termination shock (a Mach disk) at the head of the jet and a strong bow shock. The pressure behind the bow shock is greater (by roughly M^2) than that of the ambient thermal pressure and serves to dynamically reconfine the jet at a pressure higher than that of the background. We find that the head of the jet also does not progress very far due to the strong opposing ram pressure. For light, non-relativistic jets, it can be seen from momentum balance considerations that jet head velocities will be small (or even reverse directions) if the jet interacts with a hot (10^7K), supersonic background flow. This resulting morphology agrees qualitatively with that seen associated within dominant galaxies in cooling flow clusters.

Second, Burns (1986) and O'Donoghue and Owen (1986) have noticed that wide-angle tailed (WAT) sources often have jets that abruptly and symmetrically disrupt at distances of 10 to 20 kpc from the galaxy cores. The jets transform into diffuse, broad tails that extend about another 100 kpc outward. Norman, Burns, and Sulkanen (1988) have suggested that this transition in morphology is produced by a sudden change from supersonic flow in the jet to a subsonic flow in the tail. The jet becomes transonic by forming an internal planar shock that in turn is produced by passage of the jet through a shock in the ambient gas. This external shock forms when, for example, a supersonic (partial) galaxy wind encounters the intracluster medium (ICM). Numerical simulations shown in Fig. 3 confirm that this scenario can produce tail-like morphologies that look qualitatively similar to WATs.

An alternative scenario that we have examined for producing WATs involves a sharp pressure jump at the sonic radius of a supersonic cooling flow. Sulkanen, Burns, and Norman (1988) find that in clusters with sonic temperatures of 1 - 3 x 10^6 K, slightly cooler than normally assumed (e.g., Sarazin, 1986), the sonic radius lies at distances of 10 - 20 kpc from the galaxy nucleus. Near this point, the pressure jumps by orders of magnitude due to the effects of cooling. Since the cooling flow is slightly colder, most of the radiation within the sonic radius will lie outside the soft x-ray band of the *Einstein* IPC. Thus, such clusters will not have central x-ray excesses, but only smooth isothermal King model surface brightness profiles from the ICM, as is observed for WAT clusters. We are currently running numerical simulations of jets through such cooling flow atmospheres to see if they disrupt and produce tails resembling WATs.

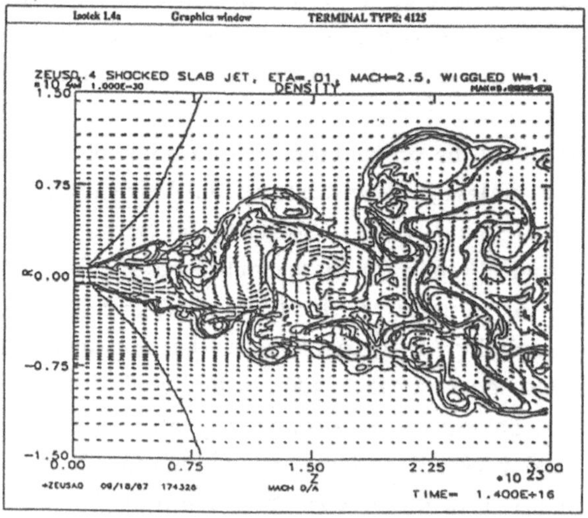

Fig. 3 - Hydro simulation of jet disruption created by passage through a shock in the ambient gas. The contours are density and the dashed lines are velocity vectors. Note the turbulent, subsonic flow past the point where the external shock intercepts the jet.

4. The Bootes Void

Kirshner et al. (1987) and Moody et al. (1987) have demonstrated that a large, 60 h^{-1} Mpc diameter region in Bootes is empty of bright galaxies and is underdense by a least a factor of 10 in comparison to the average value of surrounding regions. However, the void is not completely empty. There are 10 emission-line galaxies within the limits of the void that were found from objective prism surveys down to about m = 17. Of the 8 emission-line objects studied by Moody et al., 6 have HII region-like spectra (possibly starbursts?) and 2 have Seyfert-like spectra. These galaxies are at least 4 magnitudes brighter than the dwarfs predicted by theory to exist in voids.

Recently, Burns, Moody, and Gregory (1988) observed these 8 galaxies with the VLA at 20-cm. Five out of 8 have weak (4 - 230 x 10^{21} W/Hz) and compact radio emission. Both the optical and radio properties of these galaxies look surprisingly similar to starburst and Seyfert galaxies that lie in much denser global environments.

We also examined the local neighborhood of these emission-line objects using CCD imaging. In two cases, I Zw 81 and 1427+42, there is good evidence that these galaxies are the brightest members of poor groups. However, there are no signs of recent tidal interactions between the emission-line objects and their suspected companions. In the remaining 6 cases, there are no galaxies within a 0.5 Mpc diameter down to m = 22. I conclude that it is unlikely that galaxy interaction is the cause of the activity in these cases.

5. Conclusions

Recent evidence suggests that about 30% of high power sources may be triggered by galaxy mergers (Heckman et al., 1986). However, for sources below the break in the radio luminosity function, galaxy radio emission has a variety of causes. The abundance of gas within a galaxy and the evolutionary state of a galaxy can apparently determine if a weak radio source will "turn-on". This is often seen to be true from rich cluster sources to the Bootes void emission-line objects. External stimulation is not always required. However, there is a prominent subset of galaxies where interaction with companions has helped to generate a radio source. Such interaction appears to greatly enhance the probability that a galaxy will become active. Finally, the global environment seems to have little effect on generating radio galaxies. Weak sources are found to exist within rich clusters and cosmic voids. The exceptions are dominant galaxies that lie at the centers of clusters. These galaxies are 5 times more likely to be radio sources than typical ellipticals. The cooling flows in some clusters may help to stimulate the production of radio sources, but also suppress extended structure.

I wish to thank my many co-workers for their input into this research including F. Owen, J.-H. Zhao, M. Norman, M. Sulkanen, S. Gregory, and J. Moody. This work was supported by grants from the NSF, AST-8521115 and AST-8611511.

References

Arnaud, K. 1987 in Cooling Flows in Clusters and Galaxies, Cambridge Workshop, ed. A. Fabian, preprint.
Burns, J. 1986, Can. J. Phys., 64, 373.
Burns, J. et al. 1987, A.J., 94, 587.
Burns, J., Moody, J., and Gregory, S. 1988, A.J., submitted.
Dressel, L. 1981, Ap.J., 245, 25.
Fanti, C. et al. 1982, Astr.Ap., 105, 200.
Heckman, T., Carty, T., and Bothun, G. 1985, Ap.J., 288, 122.
Heckman et al. 1986, Ap.J., 311, 526.
Hummel, E., Kotanyi, C., and Ekers, R. 1983, Astr.Ap., 127, 205.
Kirshner, R. et al. 1987, Ap.J., 314, 493.
Moody, J. et al. 1987, Ap.J. Lett., 314, L33.
Norman, M., Burns, J. and Sulkanen, M. 1988, Nature, submitted.
O'Dea, C. and Baum, S. 1986, in Radio Continuum Processes in Clusters of Galaxies, ed. C. O'Dea & J. Uson (NRAO: Green Bank), p. 141.
O'Donoghue, A. and Owen, F. 1986, IBID, p. 155.

Sarazin, C. 1986, Rev. Mod. Phys., 58, 1.
Stocke, J. 1978, A.J., 83, 348.
Sulkanen, M., Burns, J., and Norman, M. 1988, Ap.J., submitted.
Zhao, J.-H., Burns, J., and Owen, F. 1988, A.J., submitted.

QUESTIONS AND COMMENTS

M.C. Begelman: When you try to establish a correlation between mass inflow rate and the 20-cm radio power, what inflow rate do you use? This inflow rate is often thought to vary with radius in cooling flows.

J.O. Burns: The rates that we use in Figures 1 and 2 are best characterized as global average values as determined recently by Keith Arnaud from the *Einstein* database. It is also interesting to note that Malgoli *et al.* (1987, *Ap.J.*, **319**, 632) have suggested that thermal conduction may suppress instabilities in cooling flows and thereby maintain a nearly constant mass accretion rate into the galaxy core.

A.H. Bridle: It is important, when examining correlations between radio emission and environment, to distinguish between data from above and below the break in the radio luminosity function. For example, your examples of lack of correlation of radio properties with cluster richness seem to come mainly from below the power of the break. For these models, it is likely (as in Paul Wiita's models) that the source properties are controlled by the galactic atmospheres rather than by the cluster environment.

J.O. Burns: I agree completely. It is well known that powerful classical double sources avoid nearby rich clusters. Therefore, by preselecting cluster environments, we also preselected the class of radio galaxy that we could study.

P.E. Hardee: How does flaring observed in the numerical simulation respond to the jet's Mach number? Do higher Mach number jets punch through the interface without flaring?

J.O. Burns: The flaring or disruption is a strong function of the ratio of the jet Mach number to the Mach number of the surrounding wind. When this ratio is < 1, the jet will disrupt via the formation of a Mach disk. This can be shown to be true from the Rankine-Hugoniot jump conditions and Bernoulli's theorem. For higher jet Mach numbers (as would be expected in classical doubles such as Cyg A), the jet remains collimated and is unaffected by the external shock.

M. Valtonen: The galaxy with perhaps the strongest cooling flow is Cygnus A, which is also the classical example of an extended double. Doesn't this go against the anticorrelation you mentioned between these two properties?

J.O. Burns: There are several points to be made here. First, Cyg A is not in a rich cluster, and thus would not be included in our statistics. Second, Cyg A is exceptional in several ways. It is one of the highest luminosity radio galaxies known and is in an exceptionally dense cooling flow. If such a cooling flow managed to stimulate the production of a high Mach number, high kinetic energy jet (via accretion?), then the jet would not be limited to the direct environs of the galaxy. It could easily break free as noted above. However, in general, galaxies at the centers of cooling flow clusters do not seem to be able to generate such powerful radio sources.

SPACE DISTRIBUTION AND LUMINOSITY FUNCTION OF QUASARS

Maarten Schmidt
California Institute of Technology
Pasadena, CA 91125

ABSTRACT. We discuss the evolution of the luminosity function of quasars in terms of either density evolution or luminosity evolution. The observations appear to support pure luminosity evolution in which the entire luminosity function shifts in the luminosity coordinate as a function of cosmic time. The simplest interpretation is that all quasars were formed some 10 billion years ago and that all suffer the same luminosity decrease as the time goes on. This picture encounters major difficulties in that it predicts (a) present quasar masses that are 100 times as large as those derived from the physics and dynamics of the broad-line region, and (b) a contribution by Seyfert galaxies to the 2 keV X-ray background that is considerably in excess of that observed.

The luminosity function of quasars is of fundamental importance in the interpretation of quasar counts as a function of optical magnitude and redshift. The shape of the luminosity function and its dependence on redshift potentially convey information about the energy mechanism of quasars and its dependence on environment. Depending on the life time of the quasar phase, the luminosity function can be used to estimate the space density of dead quasars or black holes, presumably located in the nuclei of galaxies. Although we will discuss only the optical luminosity function, the general luminosity function of quasars is also a function of radio and X-ray luminosity, and possibly other properties related to the environment.

1. QUASARS WITH REDSHIFT LESS THAN 2.2

In this talk we will discuss primarily the luminosity function of quasars with redshift z smaller than 2.2. For these redshifts, almost all quasars show an ultraviolet excess, $U-B < -0.4$. This allows a preselection of quasar candidates, purely on the basis of color, which eliminates more than 99 percent of the stellar foreground objects. Slit spectroscopy is required to separte the quasars from white dwarfs and hot subdwarfs, and to determine the redshift.

1.1 Density Evolution

Much of the early discussion of the evolution of the quasar luminosity function was conducted in terms of the change of space density of quasars with redshift or cosmic time. For that reason we will start with the case of density evolution, and discuss the concept of luminosity evolution in the next section.

For derivations of the space density and luminosity functions, it is essential to have surveys in hand that are complete over a given area of sky to a given magnitude limit. Such well-defined surveys were slow in coming about for many years. Braccesi *et al.* (1970) produced a list of 175 UV-excess objects, but spectroscopic follow-up work for relatively few objects was published (Braccesi, Lynds, and Sandage 1968). Around the same time, I published spectral information on 20 quasars in two Sandage-Luyten fields (Schmidt 1970). The steep slope of the counts (estimated at a factor of 6 per magnitude at the time; it is actually more like a factor of 8) required quasars to have a non-uniform space distribution. Assuming that the density varies as an exponential of cosmic time t (in units of the present age of the universe), the density was found to vary as exp(-10t).

Richard Green started his survey for stellar objects with an ultraviolet excess in the early 1970s. The final catalog of the Palomar Green survey contains 1715 objects, all with magnitudes and spectroscopic classifications (Green, Schmidt, and Liebert 1986). The 92 PG quasars with $M(B) < -23$ (for H=50) were used by Schmidt and Green (1983) in conjunction with the Hoag-Smith 4-meter grism survey (Osmer 1980) and a small sample of faint quasars collected by Kron and Chiu (1981) to derive the redshift dependence of the luminosity function.

We found that the space density of quasars increases with redshift, with strong evidence that the rate of increase depends on the absolute magnitude, quasars of higher luminosity showing a steeper increase of density with redshift. Since the Hubble diagram was still poorly covered by the few well-defined quasar samples, we still had to assume some parametric form for the density increase. We chose exp(-kt) where k was assumed to be a linear function of absolute magnitude M(B). For details of this luminosity-dependent density evolution model, see Schmidt and Green (1983).

Recently, large surveys for faint quasars such as those of Boyle *et al.* (1987) and Crampton *et al.* (1987) have improved the coverage of the Hubble diagram enormously. It is now possible to derive the luminosity function for given ranges of redshift directly, since at most redshifts survey data are available from four or five surveys to differing limiting magnitudes. For details of available surveys, I refer to a review by Green (1986) and recent articles byBoyle et al. (1987) and Crampton et al. (1987).

From these modern compilations, it is clear that the luminosity function increases with redshift at all luminosities. The density increase is steeper for quasars of higher luminosity. This confirms qualitatively the concept of luminosity-dependent density evolution introduced by Schmidt and Green (1983). Crampton et al. (1987) discussed their well-defined sample of more than 100 quasars with $z < 2.2$ in terms of one of the Schmidt-Green models and obtained a better fit by adding a smooth cutoff to the evolution at high redshifts. The inclusion of further survey material such as that from Boyle et al. (1987) will probably lead to further adjustments of the Schmidt-Green models of luminosity-dependent luminosity evolution.

1.2. Luminosity Evolution

An alternative description of the evolution of the luminosity function is based on luminosity evolution. If the individual objects have very long life times and if their luminosities all vary with time in identical fashion, then the entire luminosity function will shift in luminosity as time goes on. We will discuss below the consequences of quasars having very long life times.

In an early discussion of luminosity evolution for quasars (Schmidt 1972), I found that this was inconsistent with the steep slope of the counts. This conclusion came about as follows. Most quasars studied in the 1960s were quasi-stellar radio sources from the 3CR and Parkes catalogs, which exhibited a lower limit to their optical absolute luminosity (corresponding to redshift 0.25 at apparent magnitude 18). The luminosity evolution needed to explain the slope of the counts brighter than 18th magnitude is so steep, that 18th mag quasars at the lower limit of absolute magnitude would at redshift 2.5 still be brighter than magnitude 18.5. As a consequence, the predicted ratio of quasars of all absolute magnitudes observed at 19th mag to that of 18th mag was less than 1, in conflict with the observations (Schmidt 1972).

Several years later, Mathez (1976) noted that optically selected quasars do notexhibit the above mentioned lower limit of luminosity and proposed that quasars are subject to luminosity evolution. While the lack of a lower limit of luminosity made luminosity evolution feasible, there was at the time little solid evidence to support the concept.

Modern data for $z < 2.2$, discussed above, are remarkably well fit by luminosity evolution, as stressed by Boyle et al. (1987). The luminosity function for quasars at redshifts between 0 and 2.2 can be fitted onto each other by a shift in the luminosity coordinate. Boyle et al. (1987) find that the luminosity evolves as $(1+z)^{3.5}$. Koo and Kron (1988) recently published further quasars redshifts and a statistical discussion of their results in SA 57. They find a quasar surface density to B = 22.6 of 100 per square degree. Their compilation of luminosity functions shows no evolution at all for M(B) = -25 to -23 in the range of z = 1 to 3. It appears that their luminosity functions are, in part, not consistent with those given by Boyle et al. (1987). Since the data for individual quasars in the samples of Boyle et al. have not been published, it is difficult to evaluate the differences. For the sake of discussion, we will adopt the luminosity evolution scheme given by Boyle et al.

The concepts of density evolution and luminosity evolution lead to conclusions about the space density of quasars that sound radically different, but upon inspection they are not inconsistent. Under luminosity-dependent evolution, the co-moving space density of quasars of given luminosity rises with redshift. The directly observed range from different complete samples is as large as a factor of 300 for quasars of M(B) = -26.5 (see Green 1986). Under luminosity evolution, the entire luminosity function shifts in the luminosity coordinate only, hence the space density of quasars is the same at different redshifts. The apparent contradiction is resolved once we realize that under luminosity evolution, quasars and Seyfert nuclei are considered together. As quasars age and become dimmer, they are absorbed in the much larger pool of Seyfert nuclei.

The luminosity function has to obey the conservation equation

$$\frac{\partial n}{\partial t} + \frac{\partial}{\partial L}(\dot{L}n) = S(L,t)$$

see Cavaliere et al. (1985). The luminosity function at a given epoch is linked to that at a nearby epoch by the luminosity variation $\dot{L}$ and by the birth rate S(L,t). It is of interest to consider two extreme cases, with the life time of quasars either being short, or very long.

If the life time of quasars is short compared to the Hubble time, then the number of quasars observed reflects the number of quasars born up to a life time ago, i.e., the instantaneous birth rate. This situation is similar to that for B-type stars in our Galaxy. This case is best described by density evolution, where in its simplest form the variation of the luminosity function reflects the birth rate of quasars as a function of redshift and luminosity.

If quasars were all formed at t = 0 and have a long life time, then we have the case of luminosity evolution. The number of quasars will be constant, and if the change of absolute magnitude with time is the same for all quasars, then we have pure luminosity evolution.

We shall now discuss the consequences of a long life time and pure luminosity evolution of quasars in terms of (1) the mass of quasars, (2) the environment of quasars, and (3) the X-ray background contributed by quasars and Seyfert galaxies.

1) Mass of Quasars -- Assuming that all quasars were formed 10 billion years ago and that their luminosity evolves as $(1+z)^{3.5}$ (Boyle *et al*. 1987), we derive the present mass accreted onto the presumed black hole. In this case the zero-redshift luminosity function ranges from M(B) = -25 to -19. For 3C 273 at M(B) = -27.0, the total flux is 4.7×10^{47} erg/sec (Perry *et al*. 1987). Using this conversion, the age, and the luminosity variation, the range of luminosities in the luminosity function corresponds to a total radiated energy of 10^{63}-10^{65} ergs. For an efficiency of conversion of rest-mass energy to radiated energy of 0.1, the present mass of quasars will be at least 10^{10}-10^{12} solar masses.

Masses of quasars have been derived from the dynamics of the broad-line emission region. Wandel and Yahil (1985) found for Seyfert nuclei masses of 10^7-10^9 solar masses, and for quasars 10^8-10^{10} solar masses. Apparently, quasar masses associated with a life time of 10 billion years are about 100 times larger than those deduced from the dynamics of the broad-line region.

2) Environment of Quasars -- Hutchings *et al*. (1984) have found that one-third of nearby quasars have a disturbed host galaxy. This suggests that the quasar was formed recently, rather than at high redshift. Yee and Green (1987) find that radio quasars with redshifts larger than 0.55 are often found in a rich cluster environment. This is not seen at

lower redshifts, suggesting that radio quasars have a relatively short life time.

Environmental arguments such as those given above are still weak. It might be argued that when a host galaxy is disturbed, it produces increased accretion on the black hole central to the quasar leading to an increase in luminosity and hence visibility in surveys. Also, radio quasars constitute a small minority of quasars and therefor might not be typical for the bulk of quasars.

3) X-Ray Background Contribution -- Using a sample of 20 Seyfert galaxies in the HEAO1 A-2 surveys, Schmidt and Green (1986) find that Seyfert galaxies are responsible for 25 percent of the 2 keV X-ray background, if they have a uniform space distribution with a cutoff at redshift 3. Another 30 percent is contributed by quasars, galaxies, and clusters of galaxies. The spectra of all these contributors are softer than that of the observed X-ray background. As a consequence, none of these contributions can be increased a lot without causing conflict with the observations.

If the optical luminosity evolution of $(1+z)^{3.5}$ also applies to X-ray energies, then the Seyfert galaxies alone will produce a background out to redshift 3 of 260 percent of the observed background, or 190 percent out to redshift 2.2. As discussed, the spectrum would be a mismatch of the observed spectrum.

We conclude from the above considerations that it is unlikely that all quasars were formed 10 billion years ago and suffer luminosity evolution of order $(1+z)^{3.5}$. More specifically, in order to avoid a black hole mass larger than that derived from the broad-line region, the life time of quasars is probably not much more than 10^8 years. In this case, the finding that the evolution of the quasar luminosity function mimics pure luminosity evolution would probably be accidental.

The luminosity evolution of Seyferts in X-rays has to be less than about $(1+z)^2$, in order to avoid an excessive X-ray background contribution. In this connection, it will be of great interest to produce well-defined optical samples of Seyfert galaxies, so as to obtain a direct determination of their optical evolution.

2. QUASARS AT LARGE REDSHIFTS

Since the ultraviolet excess disappears at large redshift, searches for quasars with $z > 2$ have mostly been based on the detection of emission lines. Many hundreds of quasars have been found on photographic objective prism or grism plates, usually on the basis of visual inspection.

For statistical purposes, a major problem with these surveys is that they do not have a well-defined limit of completeness. Osmer (1980) showed that a Schmidt objective prism survey and a 4-meter grism survey were discrepant in numbers of quasars detected by a factor of 5 to 10 at overlapping magnitudes. This finding raised the specter that some surveys might not be complete at any magnitude.

Suspicions about the existence of a redshift cutoff of quasars have existed since 1972. The most impressive evidence was provided by Osmer (1982) in a search for quasars with Lyman-alpha emission in the range of redshift 3.7-4.7 down to magnitude 20 over an area of 5 square degrees. While he found a substantial number of emission-line objects, none turned out to be quasars with $z = 3.7$-4.7. Based on a smooth extrapolation from lower redshifts, around 20 quasars in this redshift range were expected. This study started the concept of a redshift cutoff at redshift 3.5. Uncertainty about the sensitivity of this survey prompted discussion about the significance of the result (Savage and Peterson 1983; Osmer 1983).

On the basis of other survey work, various estimates have been made about the redshift z(max) at which the co-moving space density of quasars reaches a maximum, as listed below:

z(max) = 2	Hazard and McMahon (1985)
3	Schmidt, Schneider and Gunn (1986b)
2.45 and 3.8	Veron (1986)
1.7	Crampton, Cowley and Hartwick (1987)

The large range of z(max) values, coupled with the uncertainty pertaining to visual searches of prism or grism exposures, makes it clear that the statistics of high-redshift quasars is not in good shape.

What is needed is an objective search algorithm for high-redshift quasars from prism or grism observations. These have been developed for photographic objective prism surveys by Clowes (1981), and for CCD grism surveys by Schmidt, Schneider, and Gunn (1986a). Two small complete samples have been published by Schmidt, Schneider, and Gunn (1986a,b). Lately, we have carried out transit observations with an array of 4 CCDs in order to cover substantial areas of sky; these yield close to 1 quasar with $z > 3$ per square degree (Schmidt, Schneider, and Gunn 1987). One complete sample has been obtained very recently, but detailed reduction and discussion is not finished yet. Three quasars with redshift larger than 4 have been detected through their emission lines: one on photographic objective prism exposures, two in CCD grism surveys.

Another method to search for high-redshift quasars is the multi-color technique. This technique was pioneered by Koo and Kron (1982,1986). Galactic stars occupy a rather well-defined tube in multi-dimensional color space. All objects outside the tube are quasar candidates, subject to confirmation by slit spectroscopy. This method is an extension of the UV excess technique which can only be employed at lower redshifts. It is not clear yet what fraction of quasars of given redshift is located in the stellar tube and hence not selected as a quasar candidate. To date, several quasars with $z > 4$ have been found with this technique. The only complete sample published is that covering S.A. 57 by Koo and Kron (1988).

In the absence of substantial complete samples of high-redshift quasars based on well-defined, objective search criteria, a discussion of the evolution of the luminosity function at high redshift would be premature. Suffice it to say that most workers in this field remain confident that the co-moving density of quasars declines with redshift for redshifts larger than 2 or 3. Establishing a solid statistical base for an evaluation of the luminosity function of quasars beyond redshift 2 is an urgent goal for the near future.

REFERENCES

Braccesi, A., Formiggini, L., and Gandolfi, E. 1970, Astron. Astrophys., **5**, 264.
Braccesi, A., Lynds, R., and Sandage, A. R. 1968, Astrophys. J. (Letters), **152**, L105.
Boyle, B. J., Fong, R., Shanks, T., and Peterson, B. A. 1987, Mon. Not. Roy. Astron. Soc., **227**, 717.
Cavaliere, A., Giallongo, E., and Vagnetti, F. 1985, Astrophys. J., **296**, 402.
Clowes, R. G. 1981, Mon. Not. Roy. Astron. Soc., **197**, 731.
Crampton, D., Cowley, A. P., and Hartwick, F. D. A. 1987, Astrophys. J., **314**, 129.

Green, R. F. 1986, in IAU Symposium 119, Quasars, ed. G. Swarup and V. K. Kapahi (Dordrecht: Reidel), p. 429.
Green, R. F., Schmidt, M., and Liebert, J. 1986, Astrophys. J. Suppl., **61**, 305.
Hazard, C., and McMahon, R. 1985, Nature, **314**, 238.
Hutchings, J. B., Crampton, D., and Campbell, B. 1984, Astrophys. J., **280**, 41.
Koo, D. C., and Kron, R. G. 1982, Astron. Astrophys., **105**, 107.
Koo, D. C., and Kron, R. G. 1986, Pub. Astron. Soc. Pac., **98**, 285.
Koo, D. C., and Kron, R. G. 1988, preprint.
Mathez, G. 1976, Astron. Astrophys., **53**, 115.
Osmer, P. S. 1980, Astrophys. J. Suppl., **42**, 523.
Osmer, P. S. 1982, Astrophys. J., **253**, 28.
Osmer, P. S. 1983a, in IAU Symposium 104, ed. G. O. Abell and G. Chincarini (Dordrecht: Reidel), p. 35.
Osmer, P. S. 1983b, in "Quasars and Gravitational Lenses", Proc. 24th Liège Symp. (Université de Liège), p. 51.
Perry, J. J., Ward, M. J., and Jones, M. 1987, Mon. Not. Roy. Astron. Soc., **228**, 623.
Savage, A., and Peterson, B. A. 1983, in IAU Symposium 104, ed. G. O. Abell and G. Chincarini (Dordrecht: Reidel), p. 57.
Schmidt, M. 1970, Astrophys. J., **162**, 371.
Schmidt, M. 1972, Astrophys. J., **176**, 273.
Schmidt, M., and Green, R. F. 1983, Astrophys. J., **269**, 352.
Schmidt, M., and Green, R. F. 1986, Astrophys. J., **305**, 68.
Schmidt, M., Schneider, D. P., and Gunn, J. E. 1986a, Astrophys. J., **306**, 411.
Schmidt, M., Schneider, D. P., and Gunn, J. E. 1986b, Astrophys. J., **310**, 518.
Schmidt, M., Schneider, D. P., and Gunn, J. E. 1987, Astrophys. J. (Letters), **316**, L1.
Veron, P. 1986, Astron. Astrophys., **170**, 37.
Wandel, A., and Yahil, A. 1985, Astrophys. J. (Letters), **295**, L1.
Yee, H. K. C., and Green, R. F. 1987, private communication.

DISCUSSION

Wiita: 1) If a long life time and a power law pure luminosity evolution law is not acceptable, what about an exponential form for luminosity evolution?
2) Although you are against life times of 10^{10} years, while presumably 10^8 years is all right, what about 10^9 years?

Schmidt: 1) The arguments against the interpretation of the apparent pure luminosity evolution in terms of a 10^{10} year life time for quasars hold equally for power law and exponential forms of the luminosity evolution.
2) While a life time of 10^8 years for quasars leads to no conflict with evidence about their masses, a life time of 10^9 years -- if mostly spent at low luminosity -- might be just possible.

Marshall: I am reluctant to abandon pure luminosity evolution models on the basis of theoretical arguments regarding mass and the poorly determined X-ray to optical relation for low luminosity AGN (required for arguments concerning the X-ray background), because the fits to the data can be annoyingly good. Can you envisage observations that could be performed to rule out long life time interpretations?

Schmidt: The contribution to the X-ray background by Seyfert galaxies can be determined quite directly from the complete X-ray sample contained in the HEAO1 A-2 data, without any recourse to X-ray to optical relations. As for direct observational evidence, if it could be shown that optically selected quasars above a particular redshift are found in rich galaxy clusters, and not below that redshift, this would be rather convincing evidence for a short life time of quasars. (The Yee-Green evidence applies to radio quasars only.)

Krolik: The argument about the validity of "pure luminosity evolution" can be split into two parts: 1) Does a function of the form $n_o f(L,t)$ in which $\int f(L,t)dL = 1$ describe the data which determine $n(L,t)$? 2) Given the evolution equation (see conservation equation in the text), the simplest physical interpretation of pure luminosity evolution would be $S(L,t) = S_o\ j(L)\ \delta(t-t_o)$ and $\dot{L} = W(L,t)$. Is this description of S and $\dot{L}$ the most likely?

To answer the second question one must introduce arguments involving physical plausibility, degree to which the evolution equation is satisfied and least number of free parameters. By studying other simple forms for S and L we may make some progress.

MacAlpine: I have several short comments. The first concerns the apparent incompleteness of Schmidt objective-prism surveys. I believe Osmer's determination that the Curtis-Schmidt survey finds a substantially smaller number of Lyman-alpha quasars than 4-m grims surveys at apparent magnitudes 18-18.5 can be simply attributed to a Schmidt survey Lyman-alpha quasar completeness limit around magnitude 17-17.5. On the other hand, the new APM survey yields surface densities which agree very well with those from UV excess surveys out to magnitude 18.5. I also wish to point out that the previously most sparsely populated central part of the Hubble diagram is now well covered by APM quasars. Finally, the APM sample contributes useful points to the brighter ends of the high-redshift luminosity functions which were poorly determined in the Koo and Kron diagram.

A PRELIMINARY EXAMINATION OF REDSHIFT AND LUMINOSITY CHARACTERISTICS FOR APM SURVEY QUASARS

G. MacAlpine and S. McGaugh
Department of Astronomy, University of Michigan, Ann Arbor, MI 48109

S. Anderson and R. Weymann
Mt. Wilson and Las Campanas Observatories, Pasadena, CA 91101

D. Turnshek
Space Telescope Science Institute, Homewood Campus, Baltimore, MD 21218

P. Hewett
Institute of Astronomy, Madingley Road, Cambridge CB3 OHA, England

F. Chaffee and C. Foltz
Multiple Mirror Telescope Observatory, University of Arizona, Tucson, AZ 85721

The APM Survey will ultimately produce roughly 1000 quasars with $m_J \leq 18.5$, detected from UK Schmidt direct and objective-prism IIIa-J plates which are scanned with the Cambridge Automated Plate Measuring (APM) facility. In the first of a planned series of papers, spectroscopic and photometric data have been reported for 192 quasars in a 102 square degree area centered on the Virgo Cluster (Foltz *et al.* 1987). Although a thorough analysis of selection effects has not yet been completed, we present a preliminary discussion of the source-count, redshift, and luminosity characteristics for this initial APM quasar sample. Our purpose here is to demonstrate the potential of the new survey, and no attempt to derive conclusive results is intended at this stage in the project.

Figure 1 shows integral quasar surface densities, comparing the APM Virgo field sample with results from various UVX surveys. The error bars represent statistical ($N^{1/2}$) uncertainties; and we have adopted $B = m_J + 0.067$. The latter comes from $m_J = B - 0.28(B - V)$ (see Harris and Smith 1981) with $B - V = 0.24$ assumed for typical APM quasars (see Barbiere *et al.* 1982). Objective prism surveys have generally been significantly less complete down to a given limiting magnitude than UVX samples; for instance, Impey and He (1986) found a factor of two deficiency of quasars to $V = 20$ toward Virgo, using eye searches of UK Schmidt plates. However, the success of the APM survey should not be surprising because the low-dispersion spectra on the discovery plates have good signal to noise characteristics considerably fainter than the chosen magnitude limit of 18.5 and because the automated search employs color as a primary selection criterion, along with emission or absorption features and spectral discontinuities.

In Figure 2, we illustrate how the APM quasars fill in a previously sparsely populated region of the Hubble diagram, complete coverage of which will facilitate the derivation of luminosity function evolution. The diagonal line represents $M_B = -23$. The "Curtis Schmidt" and "4-M Grism" data come from several surveys (e.g., MacAlpine, Smith, and Lewis 1977 or Osmer and

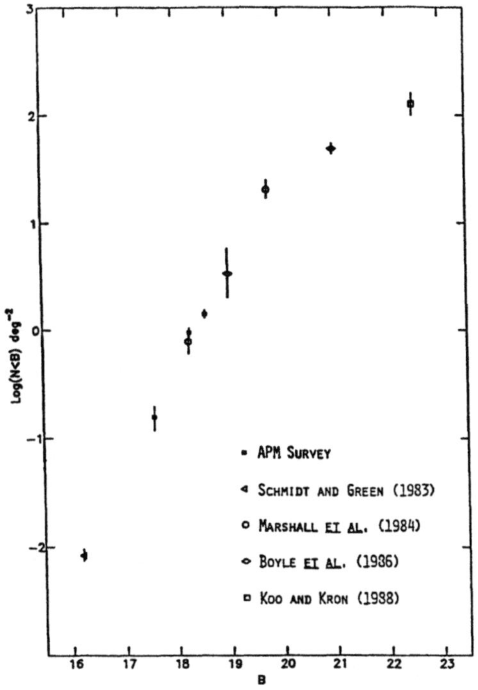

Figure 1. Quasar surface densities.

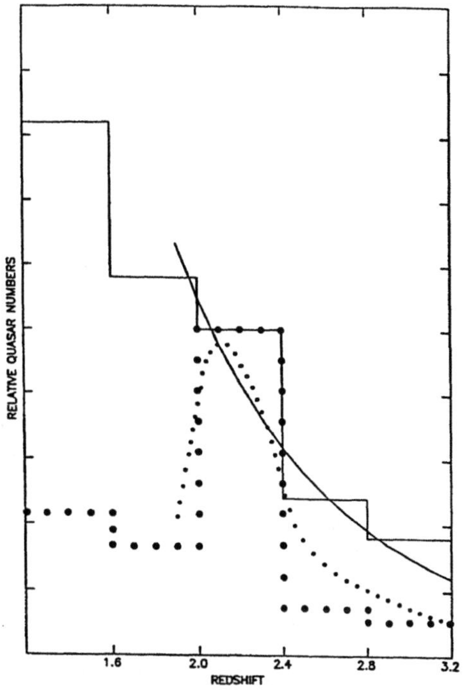

Figure 3. Partial redshift distributions for APM Survey quasars (solid-line histogram) and UM Curtis Schmidt Survey quasars (dotted histogram), normalized at $z = 2.1$. The solid-line and dotted curves represent the same evolution model, with the latter including selection effects (see text).

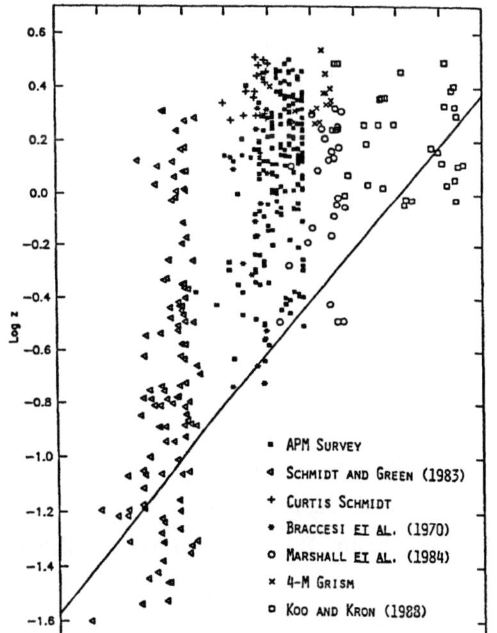

Figure 2. Quasar Hubble diagram.

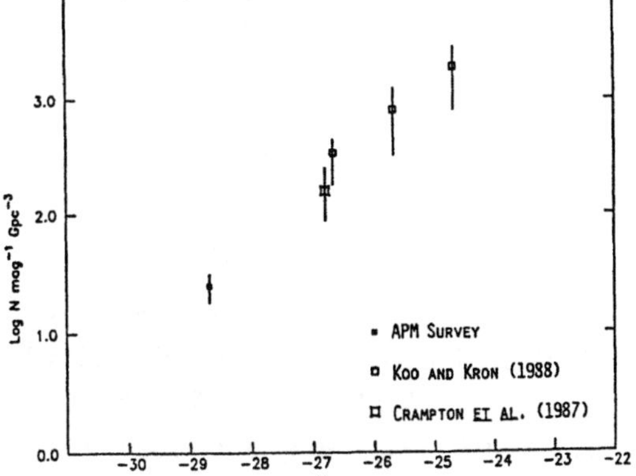

Figure 4. Composite quasar luminosity function for $q_o = 0.5$ and $2.51 < z < 3.16$.

Smith 1980 for the former and Hoag and Smith 1977 or Sramek and Weedman 1978 for the latter). The new Durham UVX Survey quasars (not shown) would be distributed primarily in the upper right part of the diagram in the magnitude range $19 < B < 21$ (see Boyle *et al.* 1987).

Figure 3 compares a broadly-binned part of the redshift distributions for APM Survey quasars (solid-line histogram) and UM Curtis Schmidt Survey quasars (dotted histogram) (see Lewis, MacAlpine, and Weedman 1979). The solid-line and dotted *curves* represent the same dN/dz evolution model, with the latter having been corrected for several potentially important selection effects which are inherent in eye searches of objective-prism plates and which (in this case) result primarily from the wavelength sensitivity of the IIIa-J emulsion. It is evident that redshift-dependent selection effects are considerably alleviated in the APM Survey. Biases that remain can be recognized and quantified by testing the quasar selection software using a wide variety of appropriate synthetic spectra. A project to accomplish this is presently underway.

Finally, Figure 4 demonstrates the importance of the APM quasar sample for determining luminosity functions. We show a luminosity function diagram, adapted from Koo and Kron 1988, for $q_o = 0.5$ and the redshift bin $2.51 < z < 3.16$. The APM survey is useful for delineating the bright end of the luminosity function at all redshifts and particularly important for $z > 2.1$ where UVX surveys are known to suffer from incompleteness effects due to shifting of strong emission lines into measured wavelength passbands.

Part of this work was supported by NSF grant AST-8615485.

REFERENCES

Barbiere, C., Capaccioli, M., Cristiani, S., Nardon, G., and Omizzolo, A. 1982, *Mem. S. A. It.*, **53**, 511.

Boyle, B. J., Fong, R., Shanks, T., and Peterson, B. A. 1986 in *Structure and Evolution of Active Galactic Nuclei*, ed. G. Giuricin (Dordrecht: Reidel), p. 491.

———. 1987, *M.N.R.A.S.*, **227**, 717.

Braccesi, A., Formiggini, L., and Gandolfi, E. 1970, *Astr. Ap.*, **5**, 264.

Crampton, D., Cowley, A. P., and Hartwick, F. D. A. 1987, *Ap. J.*, **314**, 129.

Foltz, C. B., Chaffee, F. H., Jr., Hewett, P. C., MacAlpine, G. M., Turnshek, D. A., Weymann, R. J., and Anderson, S. F. 1987, *A. J.*, in press.

Harris, W. E., and Smith, M. G. 1981, *A. J.*, **86**, 90.

Hoag, A. A., and Smith, M. G. 1977, *Ap. J.*, **217**, 362.

Impey, C. D., and He, X.-T. 1986, *M.N.R.A.S.*, **221**, 897.

Koo, D. C., and Kron, R. G. 1988, *Ap. J.*, in press.

Lewis, D. W., MacAlpine, G. M., and Weedman, D. W. 1979, *Ap. J.*, **233**, 787.

MacAlpine, G. M., Smith, S. B., and Lewis, D. W. 1977, *Ap. J. Suppl.*, **34**, 95.

Marshall, H. L., Avni, Y., Braccesi, A., Huchra, J. P., Tananbaum, H., Zamorani, G., and Zitelli, V. 1984, *Ap. J.*, **283**, 50.

Osmer, P. S., and Smith, M. G. 1980, *Ap. J. Suppl.*, **42**, 333.

Schmidt, M., and Green, R. F. 1983, *Ap. J.*, **269**, 352.

Sramek, R. A., and Weedman, D. W. 1978, *Ap. J.*, **221**, 468.

THE EVOLUTION OF ACTIVE GALACTIC NUCLEI: A MULTI-MASS MODEL

B.W. Murphy, H.N. Cohn, and R.H. Durisen
Astronomy Dept., Indiana University,
Bloomington, Indiana 47405, USA

ABSTRACT

We have developed a multi–mass energy–space Fokker–Planck code to determine the luminosity and dynamical evolution of an AGN model consisting of a dense stellar system surrounding a supermassive black hole. This model is able to produce the full range of observed AGN luminosities: $10^{44} - 10^{48}$ ergs s^{-1}.

1. METHOD

We use a multi–mass Fokker–Planck stellar dynamics code to follow the evolution of the system. This code is an extension of the multi–mass code developed by Murphy and Cohn (1988) for globular clusters and extends the work of David *et al.* (1987b) on the evolution of AGN's. The initial density distribution is a Plummer model (n=5 polytrope) with a core radius of 1 parsec. The initial stellar system contains 15 mass groups with masses ranging from 0.3 to 30 $M_{\odot}$, satisfying a power–law mass function of the form $dN \propto m^x dm$, where m is the stellar mass and **x** is the power–law slope of the mass function. Stars lose mass by three mechanisms: 1) tidal disruptions by the central black hole, 2) physical collisions between stars, and 3) stellar evolution. All of this mass is assumed to be accreted by the black hole, which is given an initial seed mass of $\sim 10^4$ $M_{\odot}$. During early times the black hole grows at an Eddington–limited rate, and the excess mass accumulates in a central reservoir. The relative importance of stellar evolution mass loss and collisional mass loss depends on the mass function and stellar density, but in all cases tidal disruptions prove unimportant.

Our treatment of stellar evolution is similar to that of Applegate (1986) in which only one mass group undergoes evolutionary mass loss at any given time, after which it becomes a degenerate stellar remnant with a smaller mass and radius. Because we are using a large range of stellar masses, we have extended the method for determining the fractional mass loss due to physical collisions developed by Spitzer and Saslaw (1966) to include stars of differing masses and radii (Murphy *et al.* 1988). We assume that stars along the main sequence have an n=3 polytropic density distribution and that only overlapping areas of both stars can be liberated. We find that colliding stars of nearly *identical* mass have the greatest fractional mass loss and that the average fractional mass loss is related to velocity by $f(v) \propto e^{-2/v}$, when a Maxwellian velocity distribution is assumed and v is the velocity dispersion in units of the escape velocity from the stellar surface. Because most stars in the stellar system are low mass stars, they provide a large fraction of the total mass which the black hole accretes from physical collisions.

2. RESULTS

Table 1 gives the parameters and results for our models. The evolution of the mass density profiles of 5 mass groups for Model 2B are shown in Figure 1. At late times, when the black hole has a mass comparable to the mass of the nucleus, collisions have depleted stars out to a radius of nearly 10^{-4} pc. Only degenerate stellar remnants can exist at smaller radii due to their small cross sections. By the time the evolution has reach 5 x 10^9 yrs the massive stars have lost a significant fraction of their mass due to stellar evolution and have become neutron stars and white dwarfs.

Table 1. Parameters and results for models.

Model	Avg. Mass ($M_\odot$)	Initial M_{core} ($M_\odot$)	ρ_{cent} ($M_\odot$ pc^{-3})	Slope of Mass Function, x	Relaxation time t_{rh} (yr)	L_{peak} (ergs s^{-1})
1	0.43	3 x 10^9	7.2 x 10^8	–2.5	1.9 x 10^{10}	6.5 x 10^{47}
2A	0.55	3 x 10^8	7.2 x 10^7	–1.5	5.4 x 10^9	3.1 x 10^{46}
2B	0.43	3 x 10^8	7.2 x 10^7	–2.5	6.8 x 10^9	2.6 x 10^{45}
2C	0.36	3 x 10^8	7.2 x 10^7	–3.5	8.1 x 10^9	3.9 x 10^{45}
3	0.43	3 x 10^7	7.2 x 10^6	–2.5	2.4 x 10^9	3.2 x 10^{44}

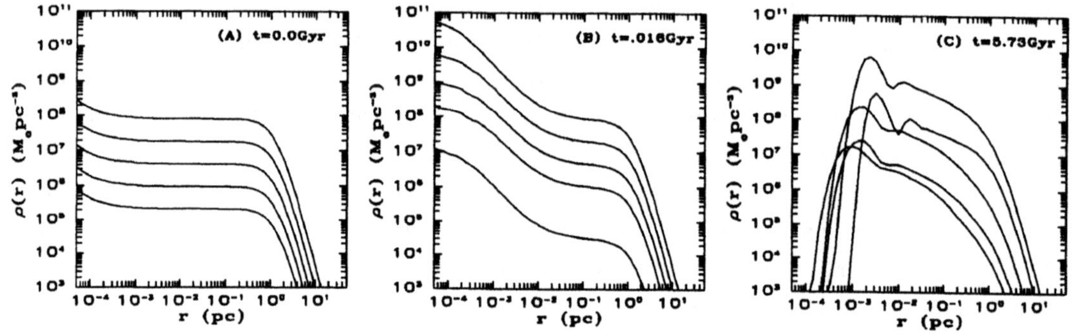

Figure 1. Stellar density evolution of Model 2B for five of the fifteen mass groups. At 1 parsec, from top to bottom, the <u>initial</u> stellar mass of each group is 0.3, 0.8, 2.16, 5.8, and 15.5 $M_\odot$.

We find that stellar evolution is the predominant mass loss mechanism for low density nuclei, while collisions dominate in high density nuclei. As can be seen in Figures 2 and 3 our evolutions show that, for identical initial central mass densities, flat mass functions will produce higher luminosities than steep mass functions in nuclei in which stellar evolution dominates all other mass injection mechanisms, while steep mass functions produce higher luminosities in nuclei which are dominated by collisions. Model 2C, which has a mass function slope of –3.5, shows that the mass injection rate from collisions increases by nearly a factor of 10 compared to Model 2B, with a slope of –2.5 (close to the Salpeter IMF value of –2.35). Model 2C has a larger percentage of low mass stars than Model 2B resulting in more collisions between nearly identical stars and thus greater fractional mass loss. Peak luminosities for our models range from 10^{44} to 10^{48} ergs/s. With a black hole photon efficiency of 0.1, as the initial central densities range from 7 x 10^6 to 7 x 10^8 $M_\odot$ pc^{-3}. The highest luminosities (~10^{48} ergs s^{-1}) are produced by nuclei that have high central densities and steep mass functions (x=–3.5), because they are dominated by physical collisions and the lowest luminosities (~10^{44} ergs s^{-1}) are produced by nuclei dominated by stellar evolution with low central densities and steep mass functions (x=–3.5).

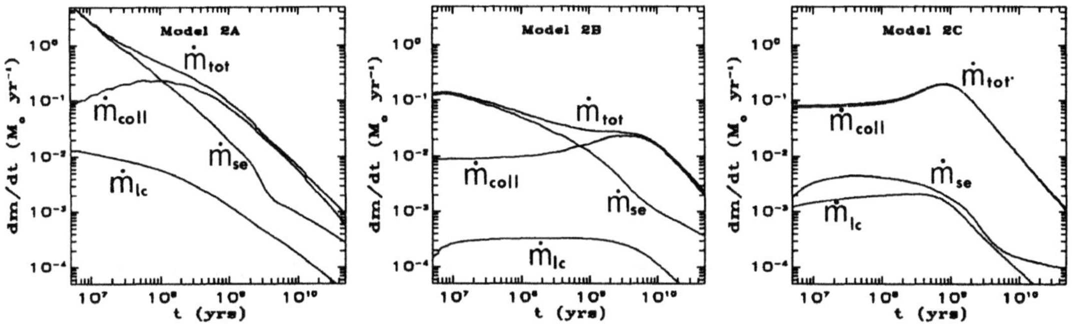

Figure 2. Mass injection rates for Models 2A, 2B, and 2C. These models have the same initial central densities and values of x=–1.5, –2.5, and –3.5, respectively.

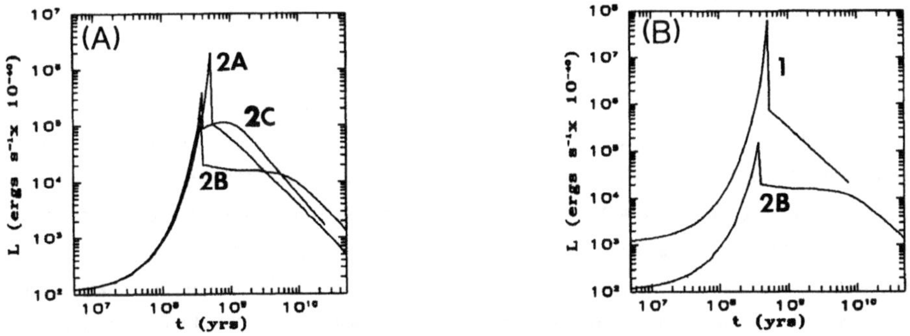

Figure 3. Luminosity evolution of models: (A) Models 2A, B, and C. (B) Models 1 and 2B.

For a wide range of parameters, dense stellar systems with seed black holes can produce the full range of AGN luminosities via mass loss intrinsic to the stellar system. Luminosities in the range of Seyfert galaxy luminosities can be maintained for many billions of years by intrinsic mass loss alone. In future models, we plan to use the accretion/wind hydrodynamics code of David *et al.* (1987a) to calculate the gas flow within the nucleus, including the effects of heating by supernovae.

REFERENCES

Applegate, J.H. 1986, *Ap. J.*, **301**, 132.

David, L.P., Durisen, R.H., and Cohn, H.N. 1987a, *Ap. J.*, **313**, 556.

David, L.P., Durisen, R.H., and Cohn, H.N. 1987b, *Ap. J.*, **316**, 505.

Murphy, B.W. and Cohn, H.N. 1988, *Mon. Not. R. Astr. Soc.* (in press).

Murphy, B.W., Cohn, H.N., Durisen, R.H., and Ostriker, J.P. 1988, to be submitted to *Mon. Not. R. Astr. Soc.*.

Spitzer, L. and Saslaw, W.C. 1966, *Ap. J.*, **143**, 400.

Evolution of Low Luminosity Quasars

Herman L. Marshall
Space Sciences Laboratory
U. of California, Berkeley

Abstract

New data samples from faint quasar surveys are added to data from brighter surveys. As shown in previous papers by other authors, there is a significant feature in the optical luminosity function. Here it is shown that this break is comparable to a feature in the luminosity function derived for Seyfert galaxies. The quasar luminosity function is determined in several redshift intervals to examine evolution questions. A pure luminosity evolution model is derived that fits all the data adequately.

1. Data

The new data are taken from Boyle (1986) and Koo and Kron (1987). The first sample is described in Boyle *et al.* (1987) and is completely identified to B = 20.9 while the second is completely identified for B < 21.1 and partially identified to B = 22.6 (assuming B = J + 0.1, as suggested by Koo and Kron). Koo and Kron claim that quasar candidates were chosen for spectroscopy on a random basis, not related to apparent magnitude for instance, so I will assume that the area of the survey is reduced to 0.156 sq. deg (32 of 59 sources are spectroscopically identified at "quality" 3 or greater). Objects identified as having narrow lines are excluded from the analyses.

2. Fit Results

A pure luminosity evolution model is assumed in what follows; luminosity functions take the form

$$\phi(L) = \begin{cases} \phi_0 \left(\frac{L}{L^*}\right)^{-\beta} & L < L^* \\ \phi_0 \left(\frac{L}{L^*}\right)^{-\alpha} & L \geq L^* \end{cases} \qquad (1a)$$

where

$$L^* \equiv L^*_Q (1+z)^k. \qquad (1b)$$

Eq. 1b incorporates all evolution effects. There are five parameters of the model: ϕ_0, α, β, L_Q^*, and k. Several of these parameters are taken or calculated from the model fits of Marshall (1985, models E, F, and G) to higher luminosity data, for which evolution is much more apparent, so that parameters ϕ_0, α, and k are "known". (See Marshall 1987 for a formula giving ϕ_0 in terms of ρ_0, α, β, and L_Q^*.) As in prior papers, luminosities and luminosity parameters are given at 2500Å in the quasar rest frame in units of 10^{30} erg^{-1} s^{-1} Hz^{-1} for H_0 = 50 km s^{-1} Mpc^{-1}. The fit method is taken from Marshall *et al.* (1983a) and is similar (except for the assumption of a cosmological model) to that of Marshall (1987). This notation differs somewhat from that of Marshall *et al.* (1983) so that the notion of a "luminosity at zero redshift" need not be introduced.

The likelihood fit for model G (where $\alpha = 3.6$, $k = 3.2$, and $\rho_0 = 25$ Gpc^{-1} for $q_0 = 0.5$) gives $L_Q^* = 0.25 \pm 0.05$ and $\beta = 1.50 \pm 0.20$. The uncertainties are somewhat correlated. The model passes the Kolmogorov-Smirnov (K-S) tests defined by Marshall *et al.* (1983a) (see Figure 1). These results are not yet directly comparable to those of Marshall (1987) for Seyfert galaxies because a smooth luminosity function was assumed there. Fitting such a model (see Eq. 1 of that paper), I find $L_Q^* = 0.18$ and $\beta = 1.13$, clearly lower values due to the different shapes of the assumed luminosity functions. These values are consistent with the results of Marshall (1987) at the 1σ level, indicating that the luminosity function of Seyfert galaxies is quite similar to that of quasars, once evolution effects are accounted using a pure luminosity evolution model. However, *this model fails the K-S test criteria*, indicating that even the model with a sharp break, defined by Eq. 1 may not be an acceptable fit since a "smooth" luminosity function is more appealing. Clearly, a more detailed examination of the luminosity function is needed.

3. The Luminosity Function in Intervals of Redshift

I now add the high luminosity quasar data from Schmidt and Green (1983), Mitchell, Warnock, and Usher (1984) and the brighter sample from Marshall *et al.* (1983b) so that the shape of the luminosity function can be readily compared in different redshift intervals. For the sake of simplicity, I take six intervals with boundaries defined by $L = 0.1$ and $z = 0.4, 0.7, 1.0, 1.3, 1.6, 1.9$, and 2.2, since I am concerned with data from UV excess surveys and most of the surveys are biased against sources with small redshifts. The luminosity cutoff accounts for the lack of objects in the Koo and Kron sample at somewhat higher redshifts (there are no broad-lined objects with $z < 0.9$).

The luminosity function is calculated by summing $1/V_a$, where V_a is defined by Avni and Bahcall (1980) as the "available" volume in which a quasar is detectable. A cumulative luminosity function is produced in each redshift interval and displayed in Figure 2. I find that the average value of V_e/V_a (cf. Avni and Bahcall) is *not* consistent with the value of 0.5 as would be expected if evolution were not present. Thus, the assumption that the intervals are small enough that evolution is not important must be false. The shape of the luminosity functions do not depend strongly on the evolution model.

Although confidence intervals are not shown for the derived distributions, it appears as though it would be difficult to construct a model that requires that the luminosity function maintain its shape at all redshifts and evolves only because of a changing scale factor. Thus models more complex than pure luminosity evolution may be required to satisfy all the data. A detailed investigation of other models and possible variations on the present one is outside the limited scope of this presentation.

References

Avni, Y. and Bahcall, J.N. 1980, *Astrophysical Journal*, **235**, 694.

Boyle, B. 1986, Ph. D. thesis, Durham University.

Boyle, B., Fong, R., Shanks, T., and Peterson, B.A. 1987, *M.N.R.A.S.*, **227**, 717.

Koo, D.C. and Kron, R.G. 1987, *Astrophysical Journal*, in press.

Marshall, H.L. 1985, *Astrophysical Journal*, **283**, 109.

Marshall, H.L. 1987, *Astronomical Journal*, **94**, 628.

Marshall, H.L., Avni, Y., Braccesi, A., Huchra, J.P., Tananbaum, H., Zamorani, G., and Zitelli, V. 1983b, *Astrophysical Journal*, **269**, 42.

Marshall, H.L., Avni, Y., Tananbaum, H., and Zamorani, G. 1983a, *Astrophysical Journal*, **269**, 35.

Marshall, H.L., Tananbaum, H., Zamorani, G., Huchra, J.P., Braccesi, A., and Zitelli, V. 1983b, *Astrophysical Journal*, **269**, 42.

Mitchell, K.J., Warnock, A., III, and Usher, P.D. 1984, *Astrophysical Journal Letters*, **287**, L3.

Schmidt, M. and Green, R.F. 1983, *Astrophysical Journal*, **269**, 357.

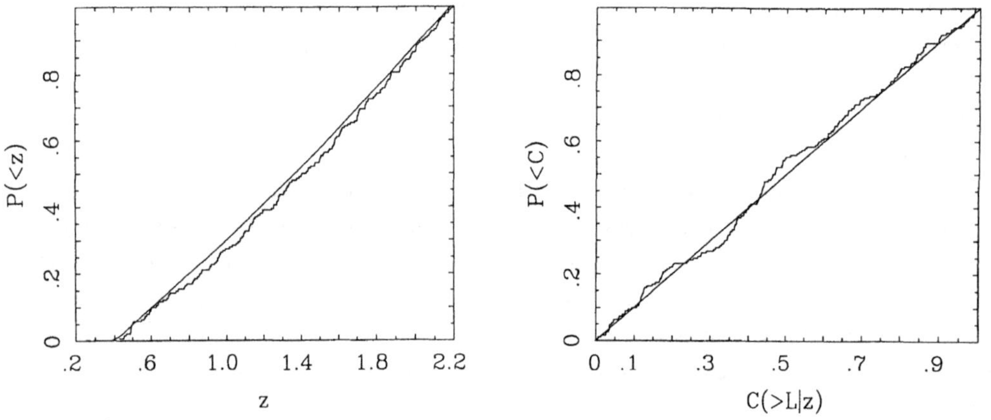

1. K-S test results for the best pure luminosity evolution model fit described in the text. These tests are described by Marshall *et al.* (1983a). The model passes both tests.

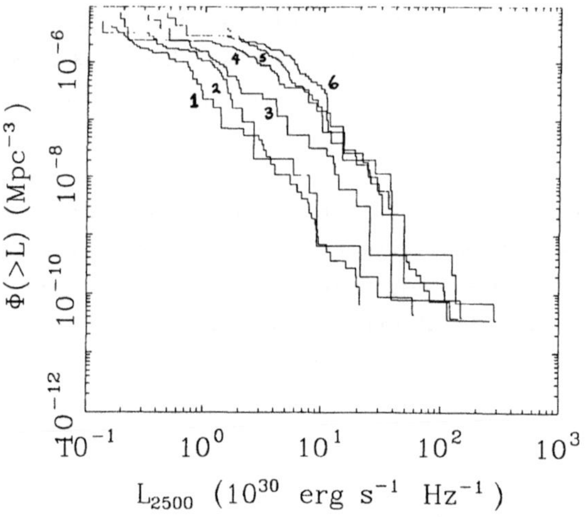

2. Integral luminosity functions for six redshift intervals. The intervals have constant widths in redshift, starting at z = 0.4 and ending at z = 2.2. The functions are marked from 1 to 6, representing the lowest through the highest redshift intervals. Each bend in a given curve represents one quasar.

EVOLUTIONARY SEQUENCE OF SEYFERT GALAXIES

David L. Blank and Noam Soker
Department of Astronomy, University of Virginia
P.O. Box 3818 University Station
Charlottesville, VA 22903, U.S.A.

Abstract

We suggest that there exists an evolutionary sequence from Seyfert 2 to Seyfert 1 galaxies. We review observations which support this suggestion and the idea that the main parameter is the dust content in and around the Broad Line Region (BLR). We propose that Seyfert 2 galaxies evolve into Seyfert 1 Galaxies by expelling the dust from the BLR. It is further supposed that the dust is due to the interaction of the host galaxy with a second galaxy. The amount of dust needed is less than that found in a typical spiral galaxy.

I. Introduction

That dust opacity is responsible for the distinction between Seyfert 1 and Seyfert 2 galaxies was first discussed by Rowan-Robinson (1977). Recently Lawrence (1987) stressed the importance of dust opacity as being one of the physical factors which produce the observed variety of AGNs. Strongly supporting the importance of obscuration in distinguishing between the Seyfert types is the presence of hidden BLRs in a few known Seyfert 2 galaxies (Antonucci and Miller 1985, J. Miller this volume). Lawrence suggests that age is the single parameter that is responsible for the fundamental difference between the Seyfert types.

In a recent paper Sanders *et al.* (1988) suggested an evolutionary connection between ultraluminous infrared galaxies and quasars. They propose that the ultraluminous infrared galaxies evolve into quasars by consumming their dust. These ultraluminous infrared galaxies, enshrouded by dust, result from the merger or interaction of gas-rich spiral galaxies. In the initial phase of the merger, starburst activity outshines the AGN, but later the AGN becomes the dominant source of luminosity. They claim that a substantial fraction, if not most, quasars form in the above scenario.

Motivated by Lawrence's idea, that Seyfert galaxies can be a one-parameter family based on age, and on the above suggestion made by Sanders *et al.*, we propose that the different Seyfert types form an evolutionary sequence from Seyfert 2 galaxies to Seyfert 1 galaxies. We suggest that at least some Seyfert 2 galaxies form as a consequence of tidal interaction between two galaxies, having enough gas to fuel the AGN and enough dust to obscure the BLR. This interaction can also cause the radio activity (Heckman 1983) , which is stronger for Seyfert 2 galaxies (Ulvestad and Wilson 1984). The supply of dust decreases with time, the obscuration of the BLR is no longer complete, and the system turns into a Seyfert 1 galaxy, or intermediate Seyfert type.

II. Supporting Observations

The importance of interaction between galaxies in fueling AGN activity has been widely recognized in recent years. Work by many observers (e.g. Keel *et al.* 1985) has established that Seyfert activity is much more likely to be found in galaxies that are interacting than in isolated field galaxies. Byrd, Sundelius and Valtonen (1987) have shown by computer modeling, that tidal triggering is physicaly sufficient to explain Seyfert activity. In addition, they give plausible reasons to suppose that the actual percentage of interacting galaxies undergoing Seyfert activity may be much higher than what is commonly believed.

Dahari (1984) has classified the optical morphology of a large sample (152) of Seyfert galaxies. In this classification, the interacting class (IAC) runs from 1 to 6. An IAC=1 is given to isolated, symetric galaxies and an IAC=6 is given to grossly disturbed spirals or overlapping systems. An interesting result is that although the nuclei of Seyfert 1 galaxies are generally brighter than the nuclei of Seyfert 2 galaxies, it is the Seyfert 2 galaxies which tend to have the higher IAC number. One can naturally fit this into our hypothesis: Seyfert 2 galaxies appear more disturbed than the Seyfert 1 galaxies since they are closer to the time of tidal triggering. As time goes on, the system relaxes, looks less disturbed and the Seyfert 2s evolve into Seyfert 1s. Considering that Seyfert activity may last for up to $\sim 10^9$ years there could be enough time for this to occur (Byrd Sundelius and Valtonen 1987).

It has been known for some time that the infrared radiation from Seyfert 1 and Seyfert 2 galaxies is dominated by nonthermal and thermal processes respectively (Neugebaur *et al.* 1976). With the results from IRAS, this basic picture has been confirmed. Edelson, Malkan and Rieke (1987) found that of the CFA Seyferts, all of the Seyfert 2 galaxies and 30% of the Seyfert 1 galaxies have a large thermal infrared excess. This may support the suggestion that the near nuclear regions of Seyfert 2s contain more dust than Seyfert 1s.

III. The Dust

We suggest that the observed difference between the Seyfert types depends on both viewing angle and dust opacity along the line of sight to the nucleus. Viewing angle by itself is not enough to distinguish the Seyfert types because the NLR properties of Seyfert 2s and Seyfert 1s are systematicaly different. These differences are the mean excitation level (Ferland and Netzer 1983), mean radio luminosity (Wilson and Willis 1980), radio size (Ulvestad and Wilson 1984) and range in line widths (Whittle 1985). The difference between the Seyfert types therefore is an intrinsic difference. We suggest that the observed difference between the Seyfert types depends on both viewing angle and dust opacity along the line of sight to the nucleus.

We propose that the dust obscurring the BLR in Seyfert 2 galaxies is blown away from a dusty torus a few parsecs from the central black hole. The existence of dusty tori has been suggested previously (e.g. Krolik and Begelman 1987). The dust that is blown away has a nonspherical distribution. So in order to see the BLR, one needs to look down a narrow cone. The opening angle of the cone increases with time as the dust content decreases. The dust distribution is patchy and so the opening angle is not well defined. We have shown (although we will not present the details here), that even in the extreme case where the dust is flowing through the BLR, and the destruction time is short, the total amount of dust needed during the Seyfert 2 phase is less than the amount of dust found in a typical spiral galaxy.

We thank Oved Dahari, Larry David, Greg Seab and Mark Whittle for helpful discussions.

REFERENCES

Antonucci, .R.J., and Miller, J.S. 1985, *Ap. J.*, **297**, 621.

Byrd, G.G., Sundelius, B., and Valtonen, M. 1987 *Astr. Ap.*, **171**, 16.

Dahari, O. 1984, *Ph.D.* thesis, University of California, Santa Cruz.

Edelson, R.A., and Malkan, M.A., and Rieke, G.H. 1987, *Ap. J.*, **321**, 233.

Ferland, G.J., and Netzer, N. 1983, *Ap. J.*, **264**, 105.

Heckman, T.M. 1983, *Ap. J.*, **268**, 628.

Keel, W.C., Kennicutt, R.C., Hummel, E., and Van Der Hulst, J.M. 1985, *Astr. J.*, **90**, 708.

Krolik, J.H., and Begelman M.C. 1987, *Ap. J.*, in press.

Lawrence, A. 1987, *Pub.A.S.P.*, **99**, 309.

Neugebauer, G., Becklin, E.E., Oke, J.B., and Searle, L. 1976, *Ap. J.*, **230**, 79.

Rowan-Robinson, M. 1977, *Ap. J.*, **213**, 635.

Sanders, D.B., Soifer, B.T., Elias, J.H., Madore, B.F., Matthews, K., Neugebauer, G., and Scoville, N.Z. 1988, *Ap. J.*, (in press).

Ulvestad, J.S., and Wilson, A.S. 1984, *Ap. J.*, **278**, 544.

Whittle, M. 1985, *M. N. R. A. S.*, **216**, 817.

Wilson, A.S., and Willis, A.G. 1980, *Ap. J.*, **240**, 429.

MULTIPLE QSO IMAGES WITH ARCMINUTE SPLITTINGS

Adeline Caulet
Department of Astronomy and Astrophysics
University of Chicago

I. INTRODUCTION

Paczynski (1986a) pointed out the puzzling cases of multiple QSO images with similar blue spectra, and with angular separations $\Delta\phi$ exceeding one arcminute. The redshifts of the QSO images, derived from broad emission lines, show discrepancies between 100 and 1000 km s^{-1} with errors of 300 up to 900 km s^{-1}. Clearly, two opposite interpretations are possible.

1. These images are examples of a multiply imaged single source. This deserves careful examination as gravitational lensing on such large scale raises difficult problems: for instance, extreme masses of matter need to be invoked to do the lensing, of order $\Delta\phi^2 \sim 4 \cdot x\ 10^{14}$ solar masses. Superclusters of galaxies, cosmic strings, black holes have been proposed (Ostriker and Vishniac 1986, Gott 1985, Paczynski 1986b).

2. These images are genuinely different quasars. If this is proved to be always the case for large $\Delta\phi$, constraints can be derived on the theoretical ideas quoted above, i.e. one can set limits on the contribution of supermassive regions and very high density condensations to Ω_0.

To discriminate between interpretations 1 and 2, it is crucial to get accurate redshift measurements and spectra of good signal to noise ratio over a broad wavelength range. This paper follows up Paczynski's suggestion (1986a) to measure the redshifts from those emission lines which are not expected to vary on a time scale of 100 to 1000 years, comparable to the time delays between the pair directions of sight. Such lines are excited in regions within a few hundred parsecs from the QSO nucleus, so the light travel time implies no detectable variability over a thousand years; relevant emission lines are the forbidden lines O II] 3727, Ne V] 2974, 3346, 3426, O III] 4363, 4959, 5007, Ne III] 3869, 3967. The redshift determinations improve as these lines are narrower (~ 500 km s^{-1}) than the broad emission lines, Ly α 1215, N V 1240, C IV 1549. When no forbidden line is detected, the emission line Mg II 2800 provides a very good tool, as it is almost always observed and has proved to give a reliable measurement of the center of mass of QSOs (Junkkarinen 1987). I present preliminary results for the QSO pair images Q0107-025 A and B (Surdej et al. 1983), and Q0307-195 A and B (MacAlpine and Feldman 1982).

II SPECTROSCOPIC OBSERVATIONS

The observations were made on the KPNO 4 meter, with the Cryogenic Camera; a grism and a blocking filter RG 610 were selected to cover the spectral range λ 6500 - 10000 Å at a resolution FWHM ~ 15 Å ~ 500 km s^{-1} at λ 8500 Å. Rms errors are less than 1/3 pixel = 55 km s^{-1}.

Spectra are presented in Figures 1 and 2. Table 1 lists emission line data; one concludes that the two QSO pairs are distinct QSOs. Q0107-025 A and B are probably physically clustered (Δv ~ 600 km s^{-1}). Shaver and Robertson (1983) had identified several absorption line systems towards Q0307-195 A and B. Table 2 is based on the identification of the Mg II absorption doublet in the new intervening metal system z = 1.4237 towards the QSO image Q0307-195 A. New identifications are made for some absorption lines listed by Shaver and Robertson. The absorption system previously identified in the spectrum of Q0307-195 A at z = 2.0919 may be bogus.

Table 1: Emission lines

QSO Images		V	Δφ arcmin	Broad lines	<z>	Δv km s^{-1}	Prog line used	<z>	Δv km s^{-1}
0107-025	A	18.2	1.3	C III],Mg II[a]	0.96 ± 0.01	1535	all	0.9577	598
	B	17.4		C III],Mg II[a]	0.95 ± 0.01		Ne III], Ne V]	0.9538	
0307-195	A	18.6	1.0	HI,NV,CIV[b]	2.1439 ± 0.0003	2130	MgII	2.1444	2097
	B	19.1		HI,NV,CIV[b]	2.1217 ± 0.0003		MgII	2.1225	

Notes: Δv velocity separation between the pair lines of sight
a) Surdej et al. 1983, λ range 3400 - 6000 Å
b) Shaver and Robertson 1983, λ range 3190 - 4970 Å
For Prog lines used, rms <z> ~ 0.0002

Table 2: Absorption line list reviewed towards Q0307-195 A

No.	λ_{obs}	Old identification	Old z	New identification system z = 1.4237	
5	3378.5	Ly α Forest		Si IV 1394	
6	3399.8	Ly α Forest		Si IV 1403	
22	3704.7	Ly α Forest		Si II 1527	blend
25	3751.3	Ly α Forest		C IV 1548	
26	3758.3	H I 1216	2.0915	C IV 1550	
34	3897.5	Si II 1260	2.0922	Fe II 1608	
39	4050.1	C II 1335	2.0349	Al II 1670	

Notes: No., λ_{obs} , and Old identification are from Shaver and Robertson (1983).

Further data analysis is in progress. A detailed discussion of these results presented elsewhere.

III. CONCLUSIONS

1. The methodology just described is adequate to rule out the gravitational lens hypothesis if narrow forbidden lines are detected with different properties, allowing accurate redshift determinations, or if the MgII emission lines show a significant shift between the pair lines of sight.

2. When no forbidden line is observed, and the low ionization emission lines have similar properties (redshifts, profiles), the issue is logically difficult, since spectral differences are likely to be observed due to the systematic trend for high ionization broad emission lines to be shifted with respect to low ionization lines (Gaskell 1982, Wilkes 1984). Q1146+111 B and C images are a good example (Turner et al. 1986, Turner 1986). For such ambiguous cases, broad and narrow band deep CCD exposures of a few arcminute square field around the multiple QSO images may supplement efficiently the spectroscopic studies of the QSO nuclei, by searching for a candidate cluster of galaxies associated with the QSOs.

3. In addition to these observations, there is growing evidence that QSO clustering may occur over a wide range of angular separations (Bahcall et al. 1986, Djorgovski et al. 1987). Further observations on this phenomenon should provide clues to the formation of large scale structures at high redshift.

I gratefully acknowledge the KPNO staff, and Richard Dreiser of Yerkes Observatory for technical assistance. I especially thank Richard Kron for his help in obtaining the observational data, and for support of this project.

REFERENCES

Bahcall, J., Bahcall, N., and Schneider, D. 1986 Nature 323, 515.
Djorgovski, S., Perley, R., Meylan, G., and McCarthy, P. 1987 Ap. J. (Letters) 321, L17.
Gaskell, C.M. 1982, Ap. J. 263, 79.
Gott, J.R. 1985 Ap. J. 288, 422.
Junkkarinen,V.T. 1987 in "*QSO absorption lines: probing the Universe*", STScIl, Ed. C. Blades, C. Norman, and D. Turnshek, p 21.
MacAlpine, G. M. , and Feldman, F. R. 1982, Ap. J. 261, 412.
Ostriker, J.P., and Vishniac, E.T. 1986 Nature 322, 804.
Paczynski, B. 1986a Nature 319, 567.
Paczynski, B. 1986b Nature 321, 419.
Shaver, P., and Robertson, J.G. 1983 Ap. J. (Letters) 268, L57.
Surdej, J. et al. 1983 in "*Quasars and gravitational lenses*", 24th Liege Colloquium, p 355.
Turner, E.L. et al. 1986 Nature 321, 142.
Turner, E.L. 1986 IAU Symposium 124.
Wilkes, B.J. 1984 M.N.R.A.S. 207, 73.

RELATIVISTIC BEAMING, LUMINOSITY FUNCTIONS, AND COSMOLOGY

Stephen P. Reynolds
North Carolina State University

INTRODUCTION

Surveys of active galactic nuclei with VLBI are becoming sufficiently sensitive that statistical analyses of the results can be performed. Furthermore, as more high-redshift superluminal sources are discovered, cosmological effects must be considered in calculating expected luminosity functions from simple source models. I present calculations of various statistical properties of an ensemble of relativistic jets, identical except for orientation, distributed with constant comoving density. These properties include: the integral luminosity function of radio sources per steradian brighter than some radio flux density S_ν, $dN/d\Omega\ (> S_\nu)$; the maximum value of redshift z at which a source of flux density S_ν can be seen; and the distribution in redshift of sources of a given flux density. The luminosity function is relatively flat at low fluxes ($dN/d\Omega\ (> S_\nu) \propto S_\nu^{-1/2}$ for flat-spectrum sources), bending to a steeper slope at a flux depending on high-frequency spectral cutoffs in the sources. The redshift distribution at a given flux is remarkably flat, predicting substantial numbers of high-redshift superluminal sources.

BASIC ASSUMPTIONS

To illustrate the effects, consider identical jet sources. Denote quantities in the rest-frame of jet emitting material with double primes, quantities in the host quasar rest-frame with single primes, and quantities in the observer's frame without primes. In the rest-frame of the jet material, its radio spectral luminosity, assumed isotropic, is $L_0(\nu''/\nu_0)^\alpha$ erg s^{-1} Hz^{-1} where α is near 0 typically, up to a cutoff $\nu_{\rm max}$ above which the spectrum is assumed to drop steeply. Let the jet have bulk Lorentz factor γ' with velocity β', and orientation θ' (in the quasar rest-frame) with respect to the line-of-sight to the observer, so that $\theta' = 0$ points toward the observer. Typical parameter values to be investigated are $\gamma' = 5$ or 10, and $L_0 = 10^{32}$ erg s^{-1} Hz^{-1}. The Hubble constant H_0 is 100 km s^{-1} Mpc^{-1}; most results are for $q_0 = 1/2$. In the rest-frame of the host quasar, frequencies are shifted by the Doppler factor

$$D' \equiv \left[\gamma'(1 - \beta'\cos\theta')\right]^{-1}.$$

Define $P_0 \equiv L_0(\nu''/\nu_0)/4\pi$ ergs s^{-1} sr^{-1}. For a source that radiates for a cosmologically long lifetime, the observer-frame flux is related to jet-frame luminosity by

$$S_\nu = (D')^{3-\alpha}(1+z)^{1-\alpha}(P_0/r^2)$$

(Ryle and Longair 1967) where $r \equiv r(z)$ is the coordinate distance,

$$r(z) = (c/H_0)\left\{q_0 z + (q_0 - 1)\left[(1 + 2q_0 z)^{1/2} - 1\right]\right\}\left\{q_0^2(1+z)\right\}^{-1}$$

and H_0 and q_0 are the local values of the Hubble constant and deceleration parameter. If the source is made up of several components with constant proper lifetime, their observed lifetimes will depend on time-dilatation and light-travel time effects, so that the total observed flux of the source will be given by

$$S_\nu = (D')^{2-\alpha}(1+z)^{1-\alpha}(P_0/r^2)$$

where P_0 is now the sum of rest-frame luminosities of jet components. If a jet source cannot be assumed to be made up of blobs emitting isotropically in their rest-frame, generalizations of the above relations become quite complicated and model-dependent (Reynolds 1982; Lind and Blandford 1985). A crude way of dealing with this model-dependence is to let the exponents to which the Doppler factors are raised be variables (Urry and Shafer 1984):

$$S_\nu = (D')^p (1+z)^\sigma (P_0/r^2).$$

The probability of observing flux greater than S_ν from such a jet is then

$$P(>S_\nu) = 1 - \cos\theta'(S_\nu) = 1 - \frac{1}{\beta'}\left[1 - \frac{1}{\gamma'}\left(\frac{P_0}{S_\nu}\right)^{1/p} r^{-2/p}(1+z)^{-\sigma/p}\right].$$

LUMINOSITY FUNCTIONS

The integrated luminosity function $dN/d\Omega\ (>S_\nu)$ can be calculated from

$$\frac{dN}{d\Omega}(>S_\nu) = \int_0^{t_0} dt\ r^2(1+z)N_0P(>S_\nu)$$

(Peebles 1971), where t_0 is the maximum proper time at which a source can be seen, corresponding to a maximum redshift $z_{\max}$, and N_0 is a normalization factor. The upper cutoff $z_{\max}$ can come about either because sources at higher redshift are too faint, or because their spectral cutoff $\nu_{\max}$ has been redshifted below the observing frequency. For this simple model, there is a minimum redshift $z_{\min}$ as well, because sources too close by may be too bright to produce the observed flux S_ν in any orientation. In the matter-dominated era, dt can be rewritten in terms of r, q_0, and H_0 (see, *e.g.*, Weinberg 1972). Choose a commonly discussed case of $p = 2-\alpha$ and $\sigma = 1-\alpha$; define $\zeta \equiv 1+z$. The integral can be done for $q_0 = 1/2$ and $\alpha = 0$, with the result

$$\frac{dN}{d\Omega}(>S_\nu) = \frac{4N_0c^2}{H_0^3}\left\{\left(1-\frac{1}{\beta'}\right)J_1 + \frac{1}{2\beta'\gamma'}\left(\frac{P_0}{S_\nu}\right)^{1/2}\frac{H_0}{c}J_2\right\}$$

with

$$J_1 = -\frac{2}{3}\left(\zeta_{\max}^{-3/2} - \zeta_{\min}^{-3/2}\right) + 2\left(\zeta_{\max}^{-1} - \zeta_{\min}^{-1}\right) - 2\left(\zeta_{\max}^{-1/2} - \zeta_{\min}^{-1/2}\right),$$

and

$$J_2 = \frac{2}{3}\left(\zeta_{\max}^{-3/2} - \zeta_{\min}^{-3/2}\right) - \zeta_{\max}^{-1} + \zeta_{\min}^{-1}.$$

Here $\zeta_{\max} \equiv 1 + z_{\max}$ is determined from the requirement that $P(>S_\nu) = 0$; at $\zeta_{\min}$, $P(>S_\nu) = 1$.

For $q_0 = \frac{1}{2}$, the requirement $P(>S_\nu) = 0$ becomes

$$\zeta_{\max}^{\sigma/2}\left(1 - \zeta_{\max}^{-1/2}\right) = \left(\frac{H_0}{2c}\right)\left[\gamma'(1-\beta')\right]^{-p/2}\left(\frac{P_0}{S_\nu}\right)^{1/2}$$

which can be inverted numerically for $\zeta_{\max}$. The value of $\zeta_{\min}$ is found by replacing $(1-\beta')$ by $(1+\beta')$ in the above expression. Results for the relative luminosity function are shown below, where $\nu_{\max}$ has been taken to be 10^{11} Hz.

The distribution in redshift of sources of a particular radio flux can be obtained by differentiating the above luminosity function with respect to S_ν and z, and will apply for redshifts between $z_{\min}(S_\nu)$ and $z_{\max}(S_\nu)$. Again for $q_0 = \frac{1}{2}$, this gives

$$\frac{d^3N}{dz\,d\Omega\,dS_\nu} = \frac{1}{\gamma'\beta'p}\left(\frac{N_0}{H_0}\right)\left(\frac{2c}{H_0}\right)^{(2p-2)/p} P_0^{1/p} S_\nu^{-(1+p)/p}\zeta^{(2-2\sigma-5p)/2p}\left(\zeta^{1/2}-1\right)^{(2p-2)/p}.$$

This function is plotted below for $p = 2-\alpha$ and $\sigma = 1-\alpha$. The S_ν dependence is a combination of the explicit factor $S_\nu^{-1/p}$ and the implicit dependence through $z_{\min}$ and $z_{\max}$.

Relative Luminosity Function of Beamed Sources

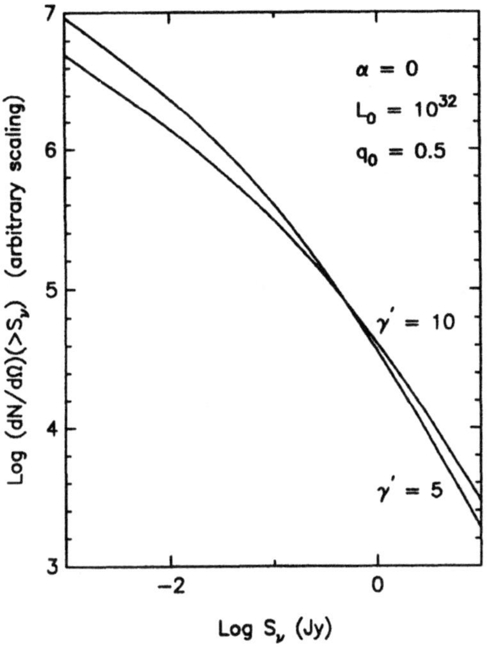

Distribution of Sources in Redshift

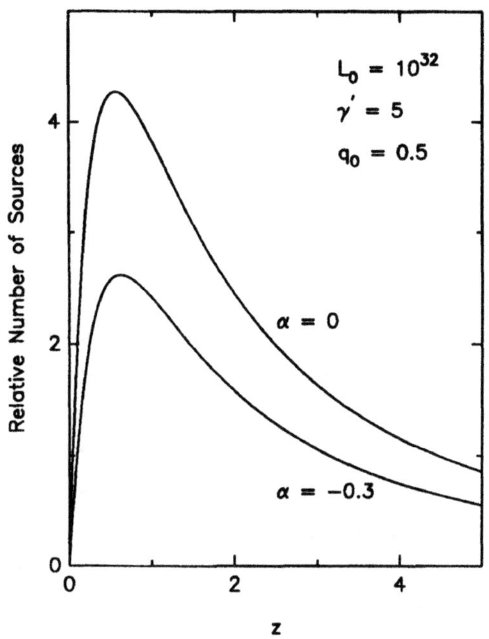

DISCUSSION

These results apply to a sample of sources all of whose flux is relativistically beamed. A more realistic calculation would involve an integral over the (unknown) intrinsic luminosity function of jet sources (that is, a distribution in P_0). However, in the limit of large and small fluxes, the observed luminosity function can be shown to vary as $P_0^{1/2}$ for faint sources, and $P_0^{3/2}$ for bright ones (again assuming $p = 2$ and $\sigma = 1$). The broad curvature evident in the figure will become even broader, but the distribution will not look significantly different. The redshift distribution will also not be strongly affected by integrating over a range of values of P_0. At very low flux levels the cutoff $\zeta_{\max}$ is due to the spectral cutoff, independent of P_0; so the distribution will just scale as $P_0^{1/2}$. At somewhat higher flux levels, $\zeta_{\max}$ is roughly proportional to P_0, making the dependence of the distribution on P_0 quite weak. This redshift distribution is remarkably flat; a good fraction of the sources contributing a given flux density will come from quite high redshifts, barring source evolution. Thus the appearance of substantial numbers of high-redshift superluminals is to be expected.

I would like to thank A. Beadles-Hay for useful discussions and Research Corporation for support.

REFERENCES

Lind, K. R., and Blandford, R. D. 1985, *Ap. J.*, **295**, 358.
Peebles, P. J. E. 1971, *Physical Cosmology* (Princeton: Princeton University Press), p. 180.
Reynolds, S. P. 1982, *Ap. J.*, **256**, 13.
Ryle, M., and Longair, M. S. 1967, *M. N. R. A. S.*, **136**, 123.
Urry, C. M., and Shafer, R. A. 1984, *Ap. J.*, **280**, 569.
Weinberg, S. 1972, *Gravitation and Cosmology* (New York: Wiley), p. 482.

LIST OF PARTICIPANTS

Margo F. Aller	University of Michigan
Robert Antonucci	Space Telescope Science Institute
William C. Bagnuolo	Georgia State University
Thomas J. Balonek	Colgate University
David L. Band	IGPP/Lawrence Livermore National Laboratory
Mitchell C. Begelman	JILA/University of Colorado
Gregory Benford	University of California, Irvine
Luc Binette	Canadian Institute of Theoretical Astrophysics
John A. Biretta	Center for Astrophysics
David Blank	University of Virginia
John M. Blondin	University of Chicago & NASA/GSFC
Alan H. Bridle	National Radio Astronomy Observatory/Charlottesville
Jack O. Burns	University of New Mexico
Gene G. Byrd	University of Alabama
Saul Caganoff	Mt. Stromlo & Siding Spring Obs. & NRAO/VLA
Michael T. Carini	Georgia State University
Timothy E. Carone	University of Arizona
Adeline Caulet	Yerkes Observatory & University of Alabama
Gerald Cecil	Institute for Advanced Study
Wayne Christiansen	University of North Carolina
Marshall H. Cohen	California Institute of Technology
William Cooke	University of Florida
D. Michael Crenshaw	Computer Science Corp. & IUE/GSFC
Oved Dahari	Space Telescope Science Institute
Darren L. DePoy	National Optical Astronomy Observatories
Charles D. Dermer	Lawrence Livermore National Laboratory
Michael M. De Robertis	York University
Richard Elston	Steward Observatory
Charles R. Evans	California Institute of Technology
Andy Fabian	University of Cambridge
C. Martin Gaskell	University of Michigan
Jean W. Goad	National Optical Astronomy Observatories/ADP
Robert Goodrich	Lick Observatory
Richard L. Hackney	Western Kentucky University
Philip E. Hardee	University of Alabama

Charlene A. Heisler	Yale University
Sethanne Howard	Georgia State University
Philip A. Hughes	University of Michigan
Judith Irwin	University of Toronto
Demosthenes Kazanas	NASA/Goddard Space Flight Center
William C. Keel	University of Alabama
Mario Klaric	University of Alabama
Anuradha Koratkar	SUNY/Stony Brook
Theodore Koupelis	University of Rochester
Julian H. Krolik	Johns Hopkins University
Robert J. Leacock	University of Florida
Gordon MacAlpine	University of Michigan
Matthew A. Malkan	University of California, Los Angeles
Herman Marshall	University of California, Berkeley
Obradovic Marzan	University of Alabama
Joseph M. Mazzarella	University of Michigan
Stacy S. McGaugh	University of Michigan
Ben Q. McGimsey	Georgia State University
Fulvio Melia	University of Chicago & Northwestern University
H. Richard Miller	Georgia State University
Joseph S. Miller	Lick Observatory
Brian W. Murphy	Indiana University
Richard F. Mushotzky	NASA/Goddard Space Flight Center
Donald E. Osterbrock	Lick Observatory
Bradley M. Peterson	Ohio State University
Richard W. Pogge	Lick Observatory
Gail A. Reichert	Computer Science Corp.
Stephen P. Reynolds	North Carolina State University
Alex Rosen	Georgia State University
Edward Rosenblatt	University of California, Santa Cruz/UCLA
Alberto C. Sadun	Agnes Scott College
John L. Safko	University of South Carolina
D.J. Saikia	Jodrell Bank/University of Manchester
Maarten Schmidt	California Institute of Technology
Richard A. Shaw	Lick Observatory
Noriaki Shibazaki	Stanford University
Isaac Sholsman	JILA/University of Colorado
Javad Siah	Villanova University
Alex G. Smith	University of Florida

Paul S. Smith	Steward Observatory
Noam Soker	University of Virginia
Wayne A. Stein	University of Minnesota
Jack W. Sulentic	University of Alabama
Martin E. Sulkanen	University of New Mexico & Cornell University
Wei-Hsin Sun	University of California, Los Angeles & NASA/GSFC
Yukio Tomozawa	University of Michigan
Kenneth C. Turner	Innovative Systems Corp.
C. Megan Urry	Space Telescope Science Institute
J. Patricia Vader	Yale University
Mauri Valtonen	University of Helsinki & University of Turku
Wil van Breugel	University of California, Berkeley
Sylvain Veilleux	Lick Observatory
Amri Wandel	Stanford University
Martin J. Ward	University of Washington
Anthony J. Wasilewski	Scientific Atlanta
Barbara Wasilewski	Georgia State University
William Washburn	Agnes Scott College
James R. Webb	University of Florida
Ann E. Wehrle	California Institute of Technology
Bjorn Westin	Georgia State University
Tor Westin	Georgia State University
Ray Weymann	Mt. Wilson & Las Campanas Observatory
Mark Whittle	University of Cambridge & University of Virginia
Paul J. Wiita	Georgia State University
Richard M. Williamon	Fernbank Science Center
John W. Wilson	Georgia State University
Diana M. Worrall	Center for Astrophysics
James P. Wright	National Science Foundation
Adam Wysota	SUNY/Stony Brook
Andrzej Zdziarski	Space Telescope Science Institute

Lecture Notes in Mathematics

Vol. 1236: Stochastic Partial Differential Equations and Applications. Proceedings, 1985. Edited by G. Da Prato and L. Tubaro. V, 257 pages. 1987.

Vol. 1237: Rational Approximation and its Applications in Mathematics and Physics. Proceedings, 1985. Edited by J. Gilewicz, M. Pindor and W. Siemaszko. XII, 350 pages. 1987.

Vol. 1250: Stochastic Processes – Mathematics and Physics II. Proceedings 1985. Edited by S. Albeverio, Ph. Blanchard and L. Streit. VI, 359 pages. 1987.

Vol. 1251: Differential Geometric Methods in Mathematical Physics. Proceedings, 1985. Edited by P.L. García and A. Pérez-Rendón. VII, 300 pages. 1987.

Vol. 1255: Differential Geometry and Differential Equations. Proceedings, 1985. Edited by C. Gu, M. Berger and R.L. Bryant. XII, 243 pages. 1987.

Vol. 1256: Pseudo-Differential Operators. Proceedings, 1986. Edited by H.O. Cordes, B. Gramsch and H. Widom. X, 479 pages. 1987.

Vol. 1258: J. Weidmann, Spectral Theory of Ordinary Differential Operators. VI, 303 pages. 1987.

Vol. 1260: N.H. Pavel, Nonlinear Evolution Operators and Semigroups. VI, 285 pages. 1987.

Vol. 1263: V.L. Hansen (Ed.), Differential Geometry. Proceedings, 1985. XI, 288 pages. 1987.

Vol. 1265: W. Van Assche, Asymptotics for Orthogonal Polynomials. VI, 201 pages. 1987.

Vol. 1267: J. Lindenstrauss, V.D. Milman (Eds.), Geometrical Aspects of Functional Analysis. Seminar. VII, 212 pages. 1987.

Vol. 1269: M. Shiota, Nash Manifolds. VI, 223 pages. 1987.

Vol. 1270: C. Carasso, P.-A. Raviart, D. Serre (Eds.), Nonlinear Hyperbolic Problems. Proceedings, 1986. XV, 341 pages. 1987.

Vol. 1272: M.S. Livšic, L.L. Waksman, Commuting Nonselfadjoint Operators in Hilbert Space. III, 115 pages. 1987.

Vol. 1273: G.-M. Greuel, G. Trautmann (Eds.), Singularities, Representation of Algebras, and Vector Bundles. Proceedings, 1985. XIV, 383 pages. 1987.

Vol. 1275: C.A. Berenstein (Ed.), Complex Analysis I. Proceedings, 1985–86. XV, 331 pages. 1987.

Vol. 1276: C.A. Berenstein (Ed.), Complex Analysis II. Proceedings, 1985–86. IX, 320 pages. 1987.

Vol. 1277: C.A. Berenstein (Ed.), Complex Analysis III. Proceedings, 1985–86. X, 350 pages. 1987.

Vol. 1283: S. Mardešić, J. Segal (Eds.), Geometric Topology and Shape Theory. Proceedings, 1986. V, 261 pages. 1987.

Vol. 1285: I.W. Knowles, Y. Saitō (Eds.), Differential Equations and Mathematical Physics. Proceedings, 1986. XVI, 499 pages. 1987.

Vol. 1287: E.B. Saff (Ed.), Approximation Theory, Tampa. Proceedings, 1985–1986. V, 228 pages. 1987.

Vol. 1288: Yu. L. Rodin, Generalized Analytic Functions on Riemann Surfaces. V, 128 pages, 1987.

Vol. 1294: M. Queffélec, Substitution Dynamical Systems – Spectral Analysis. XIII, 240 pages. 1987.

Vol. 1299: S. Watanabe, Yu.V. Prokhorov (Eds.), Probability Theory and Mathematical Statistics. Proceedings, 1986. VIII, 589 pages. 1988.

Vol. 1300: G.B. Seligman, Constructions of Lie Algebras and their Modules. VI, 190 pages. 1988.

Vol. 1302: M. Cwikel, J. Peetre, Y. Sagher, H. Wallin (Eds.), Function Spaces and Applications. Proceedings, 1986. VI, 445 pages. 1988.

Vol. 1303: L. Accardi, W. von Waldenfels (Eds.), Quantum Probability and Applications III. Proceedings, 1987. VI, 373 pages. 1988.

Lecture Notes in Physics

Vol. 286: R. Alicki, K. Lendi, Quantum Dynamical Semigroups and Applications. VIII, 196 pages. 1987.

Vol. 287: W. Hillebrandt, R. Kuhfuß, E. Müller, J.W. Truran (Eds.), Nuclear Astrophysics. Proceedings. IX, 347 pages. 1987.

Vol. 288: J. Arbocz, M. Potier-Ferry, J. Singer, V.Tvergaard, Buckling and Post-Buckling. VII, 246 pages. 1987.

Vol. 289: N. Straumann, Klassische Mechanik. XV, 403 Seiten. 1987.

Vol. 290: K.T. Hecht, The Vector Coherent State Method and Its Application to Problems of Higher Symmetries. V, 154 pages. 1987.

Vol. 291: J.L. Linsky, R.E. Stencel (Eds.), Cool Stars, Stellar Systems, and the Sun. Proceedings, 1987. XIII, 537 pages. 1987.

Vol. 292: E.-H. Schröter, M. Schüssler (Eds.), Solar and Stellar Physics. Proceedings, 1987. V, 231 pages. 1987.

Vol. 293: Th. Dorfmüller, R. Pecora (Eds.), Rotational Dynamics of Small and Macromolecules. Proceedings, 1986. V, 249 pages. 1987.

Vol. 294: D. Berényi, G. Hock (Eds.), High-Energy Ion-Atom Collisions. Proceedings, 1987. VIII, 540 pages. 1988.

Vol. 295: P. Schmüser, Feynman-Graphen und Eichtheorien für Experimentalphysiker. VI, 217 Seiten. 1988.

Vol. 296: M. Month, S. Turner (Eds.), Frontiers of Particle Beams. XII, 700 pages. 1988.

Vol. 297: A. Lawrence (Ed.), Comets to Cosmology. X, 415 pages. 1988.

Vol. 298: M. Georgiev, F' Centers in Alkali Halides. XI, 282 pages. 1988.

Vol. 299: J.D. Buckmaster, T. Takeno (Eds.), Mathematical Modeling in Combustion Science. Proceedings, 1987. VI, 168 pages. 1988.

Vol. 300: B.-G. Englert, Semiclassical Theory of Atoms. VII, 401 pages. 1988.

Vol. 301: G. Ferenczi, F. Beleznay (Eds.), New Developments in Semiconductor Physics. Proceedings, 1987. VI, 302 pages. 1988.

Vol. 302: F. Gieres, Geometry of Supersymmetric Gauge Theories. VIII, 189 pages. 1988.

Vol. 303: P. Breitenlohner, D. Maison, K. Sibold (Eds.), Renormalization of Quantum Field Theories with Non-linear Field Transformations. Proceedings, 1987. VI, 239 pages. 1988.

Vol. 304: R. Prud'homme, Fluides hétérogènes et réactifs: écoulements et transferts. VIII, 239 pages. 1988.

Vol. 305: K. Nomoto (Ed.), Atmospheric Diagnostics of Stellar Evolution: Chemical Peculiarity, Mass Loss, and Explosion. Proceedings, 1987. XIV, 468 pages. 1988.

Vol. 306: L. Blitz, F.J. Lockman (Eds.), The Outer Galaxy. Proceedings, 1987. IX, 291 pages. 1988.

Vol. 307: H.R. Miller, P.J. Wiita (Eds.), Active Galactic Nuclei. Proceedings, 1987, XI, 438 pages. 1988.

GPSR Compliance

The European Union's (EU) General Product Safety Regulation (GPSR) is a set of rules that requires consumer products to be safe and our obligations to ensure this.

If you have any concerns about our products, you can contact us on ProductSafety@springernature.com

In case Publisher is established outside the EU, the EU authorized representative is:

Springer Nature Customer Service Center GmbH
Europaplatz 3
69115 Heidelberg, Germany

Batch number: 10838062

Printed by Printforce, the Netherlands